WORKSHOP ON OBSERVING GIANT COSMIC RAY AIR SHOWERS FROM $>10^{20}$ eV PARTICLES FROM SPACE

WORKSHOP ON OBSERVING GIANT COSMIC RAY AIR SHOWERS FROM $>10^{20}$ eV PARTICLES FROM SPACE

College Park, Maryland   November 1997

EDITORS
John F. Krizmanic
USRA/NASA Goddard Space Flight Center

Jonathan F. Ormes
Robert E. Streitmatter
NASA Goddard Space Flight Center

American Institute of Physics

AIP CONFERENCE
PROCEEDINGS 433

Woodbury, New York

Editor:

SEP/AE
PHYS

John F. Krizmanic
NASA Goddard Space Flight Center
Code 661
Greenbelt, MD 20771

Email: jfk@cosmicra.gsfc.nasa.gov

L.C. Catalog Card No. 98-71564
ISBN 1-56396-788-X
ISSN 0094-243X
DOE CONF- 9711122

Printed in the United States of America

CONTENTS

Preface ix

Historical Remarks xiii
    A. Bunner

Search for the End of the Cosmic Ray Spectrum 1
    J. Linsley

Powerful Radio Galaxies as Sources of the Highest Energy Cosmic Rays 22
    P. L. Biermann

Acceleration of Cosmic Rays by Colliding Galaxies 37
    F. C. Jones

Gamma Ray Bursts and Extreme Energy Cosmic Rays 42
    L. Scarsi

Early Air Fluorescence Work: Cornell and Japan 54
    G. Tanahashi

Results from Fly's Eye and HiRes Projects 65
    P. Sokolsky

Results from AGASA Experiment in the Extremely High Energy Region 76
    M. Nagano

Feasibility Study of an Airwatch Mission 87
    C. N. DeMarzo

Orbiting Wide-angle Light-collectors (OWL): Observing Cosmic Rays
from Space 95
    R. E. Streitmatter

Russian Plans for the ISS 108
    G. K. Garipov, L. A. Gorshkov, B. A. Khrenov, M. I. Panasyuk,
    O. A. Saprykin, and V. S. Syromyatnikov

Great Science Observatories in the Space Station Era and
OWL Efforts in Japan 117
    Y. Takahashi

Bremsstrahlung and Pair Creation: Suppression Mechanisms
and How They Affect EHE Air Showers 132
    S. R. Klein

Physics Opportunities Above the Greisen–Zatsepin–Kuzmin Cutoff:
Lorentz Symmetry Violation at the Planck Scale 148
    L. Gonzalez-Mestres

From Atmospheric Electricity to the 16-Joule Cosmic Ray Air Shower 159
    G. W. Clark

Ultrahigh Energy Cosmic Rays from Topological Defects—Cosmic Strings,
Monopoles, Necklaces, and All That 168
    P. Bhattacharjee

Galactic and Extragalactic Magnetic Fields in the Local Universe:
An Overview 196
    P. P. Kronberg

Intergalactic Propagation of Heavy Nuclei—Are They
the Trans-Greisen Events? 212
    F. W. Stecker
Propagation of Cosmic Rays and Neutrinos Through Space 217
    S. Yoshida
Can Ultra High Energy Cosmic Rays Be Evidence for New
Particle Physics? 226
    G. R. Farrar
Angle-Time-Energy Images of Ultra-High Energy Cosmic Ray Sources 237
    G. Sigl
Relic Neutrinos, Monopoles, and Cosmic Rays above $\sim 10^{20}$ eV 246
    T. J. Weiler
Comments on the Physics Potential of Ultra-High-Energy Neutrino
Interactions with OWL 262
    D. B. Cline
Large Natural Cherenkov Detectors: Water and Ice 265
    F. Halzen
Ultra High Energy Cosmic Rays from Decaying Superheavy Particles 279
    V. S. Berezinsky
Air Shower Modelling 297
    T. K. Gaisser
Principles of Wide Angle, Large Aperture Optical Systems 304
    D. J. Lamb, R. A. Chipman, L. W. Hillman, Y. Takahashi,
    and J. O. Dimmock
Auger: What, Why and How? 312
    C. Pryke
Workshop Summary 321
    D. Schramm
AIRWATCH: The Fast Detector 353
    E. Alippi, A. Lenti, P. Attiná, A. Gregorio, R. Stalio, P. Trampus,
    L. Bosisio, G. Giannini, A. Vacchi, V. Gracco, A. Petrolini, G. Piana,
    O. Catalano, S. Giarrusso, and G. Bonanno
Background Measurement with UVSTAR 358
    A. Gregorio, R. Stalio, P. Trampus, and L. Scarsi
Air Fluorescence Efficiency Measurements for
AIRWATCH Based Mission: Experimental Set-up 361
    B. Biondo, O. Catalano, F. Celi, G. Fazio, S. Giarrusso, G. La Rosa,
    A. Mangano, G. Bonanno, R. Cosentino, R. Di Benedetto, S. Scuderi,
    G. Richiusa, and A. Gregorio
Balloon Borne Detector for the Cosmic Ray Energy Spectrum
Measurements and its Possibilities in Connection with
AIRWATCH Experiment 367
    R. A. Antonov, D. V. Chernov, A. N. Fedorov, E. E. Korosteleva,
    M. I. Panasyuk, E. A. Petrova
Low Frequency Radio Radiation from Clusters of Galaxies 373
    V. S. Berezinsky and P. Blasi
The OWL Detector: OWL Aperture and Resolution 382
    H. Y. Dai, E. C. Loh, and P. Sokolsky

Observation of UHE Neutrino Interactions from Outer Space.............. 390
    G. Domokos and S. Kovesi-Domokos
Focal Plane Detectors: Possible Detector Technologies
for OWL/AIRWATCH 394
    E. Flyckt
Camera for Detection of Cosmic Rays of Energy of More Than 10 EeV
On the ISS Orbit 403
    G. K. Garipov, B. A. Khrenov, V. P. Nikitsky, M. I. Panasyuk,
    O. A. Saprykin, A. V. Sholokhov, and V. S. Syromyatnikov
Observing Air Showers from Cosmic Superluminal Particles.............. 418
    L. Gonzalez-Mestres
Wide Angle Refractive Optics for Astrophysics Applications 428
    D. J. Lamb, R. A. Chipman, L. W. Hillman, Y. Takahashi,
    and J. O. Dimmock
Computer Modeling of Optical Systems Containing Fresnel Lenses 434
    D. J. Lamb, R. A. Chipman, L. W. Hillman, Y. Takahashi,
    and J. O. Dimmock
Focal Plane Reduction of Large Aperture Optical Systems................. 439
    D. J. Lamb, R. A. Chipman, L. W. Hillman, Y. Takahashi,
    and J. O. Dimmock
Connection Between the Statistical Parameters of Hadronization
and the QCD Coupling Constant...................................... 446
    L. Popova and G. Kamberov
Computer Aided Optimal Design of Space Reflectors and
Radiation Concentrators 460
    O. A. Saprykin, Y. K. Spirochkin, V. G. Kinelev, and V. D. Sulimov
A Study of the Correlation of EHE Cosmic Rays with Gamma Ray Bursts.... 469
    Y. Takahashi
On the Origin of Ultra High Energy Cosmic Ray Particles................. 483
    K. O. Thielheim
Photodetectors for OWL... 500
    J. W. Mitchell
Polymer Selection Criteria for the Orbiting Wide- angle Light
collector (OWL) Project: Lens Material 511
    T. M. Leslie, E. Burleson, J. O. Dimmock, D. J. Lamb, L. W. Hillman,
    Y. Takahashi, and M. D. Watson

Workshop Agenda 517
Listing of Poster Papers 523
Workshop Participants 525
Author Index... 535

Preface

This proceedings on the *Workshop on Observing Giant Air Showers from* $> 10^{20}$ *eV Particles from Space* held on November 13-15, 1997, documents an emerging capability to use the Earths atmosphere to study the physics of a flux of particles whose origin is unknown. It was the consensus of the experts attending this conference that the few events above 2×10^{20} eV are "real" in the sense that their energies must lie in a region where it is extremely hard to understand their origin. On the one hand they must be extragalactic as containing them in galactic magnetic fields appears to be impossible. On the other hand, if they are extragalactic in origin, then interactions with 2.7 K blackbody radiation (assuming they are charged nuclei) restrict their origin to a relatively small region (by astronomical scales) surrounding the galaxy, a volume of space in which we know of no extraordinary sites in which they might be accelerated. One obvious hypothesis for the origin of these particles is that they are accelerated in some as yet unknown manner, perhaps by objects powered by black holes. The jets in these objects are large enough to be potential sites for a "bottom-up" acceleration where they are at the extreme limits of Nature's most powerful accelerators.

Because of the difficulties associated with "bottom-up" models, there has been considerable amount of work done theoretically on a "top-down" hypothesis in which these particles are manifestations of topological defects that are remnants of the Big Bang. At that time, the universe was so hot that the forces that hold nuclear matter were combined and the same as the force that is responsible for electricity and magnetism and the weak interaction. These topological defects would have approximately 10^{24} eV of energy trapped in a pseudo-particle left over from the symmetry breaking that occurred at the GUT (Grand Unification Theories) scale as shown on the accompanying figure.

These experimental facts leave us faced with the following conundrum as enunciated by Prof. David Schramm in his closing remarks: The conservative explanation is given by some exotic mechanism of acceleration by black holes while the more physically interesting mechanism involves unknown physics from a time when the universe was only 10^{-35} seconds old!

Several groups from Italy, Japan, Russia and the United States are working on the technologies required to observe a large area of the Earth's atmosphere, use it as a giant scintillator and detect a statistically significant sample of these events, perhaps as rare as one per square kilometer per millennium, to unambiguously settle the question of which hypothesis is correct. Because the light from these showers can only be seen on moonless nights on the dark side of the Earth, this will require looking at literally millions of square kilometers of atmosphere from

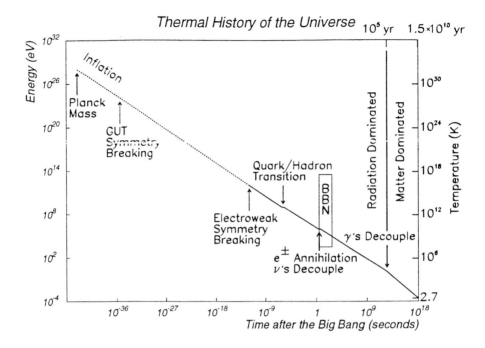

Thermal History of the Universe

low Earth orbit to see the streaks of light moving across the sky at the speed of light.

The conference allowed these various groups to come together and share their progress, and I think its fair to say that enthusiasm for the endeavor and optimism about its eventual success were greatly reinforced at this workshop. We heard a number of talks about various aspects of the theories associated with the origin of these ultra-high energy particles, talks about the existing state of the data on them, and plans for observing them from a new generation of ground based arrays to see the particles arriving at the Earth's surface and eyes to see the showers in the atmosphere. There were also a set of significant papers on the technologies emerging to study the particles from space. It was realized that while groups within the different countries are currently working independently, there will come a time in the not too distant future when these efforts should be joined together to carry out the first measurements from space. If those are successful and demonstrate that we know how to make these observations in a stable and reliable manner from space, then the potential exists to expand the effort to use the full dark side of the Earth as a detector which will be 100 times larger, perhaps large enough to detect neutrino showers in the atmosphere and do "astronomy" with 10^{20} eV neutrinos.

On behalf of the organizers, I'd like to thank all the participants in the workshop for their interesting and stimulating contributions, and thank the authors for getting their papers to us in a timely manner.

One final note: Prof. David Schramm was killed in an airplane accident during the holiday season following this meeting and was unable to prepare his remarks for publication. His concluding remarks at the conference were a memorable occasion, further etched into our memories by his untimely death. This Workshop must have been one of the final talks given by David. Since we know of no form of exhortations by the editors that can produce a paper from David, we have taken the unusual step of including herein copies of his transparencies. Unfortunately we must leave it to the reader to imagine the force of David's personality, intellectual capacity and booming voice. We dedicate this volume to his memory.

John F. Krizmanic
Jonathan F. Ormes
Robert E. Streitmatter

Historical Remarks

Alan N. Bunner[1]

NASA Headquarters, Code SA
Washington, DC 20546

It is indeed a pleasure to add my welcome to this workshop. I am delighted to recognize so many colleagues and old friends among the attendees here.

It's very exciting to me personally to be here at a workshop discussing the detection of giant air showers from space. I was lucky to be a part of the pioneering experiments at Cornell University in 1963-1967 that first attempted to observe cosmic ray extensive air showers by the fluorescence of the atmosphere.

The idea of using the atmosphere itself as a cosmic ray scintillator was first experimentally tested by a group at Cornell University, under the direction of Professor Ken Greisen, a group that included Seinosuke Ozaki, Goro Tanahashi, Ed Jenkins, Peter Landecker, Fred Ruckdeschel, Edith Cassell, Alan Zabell, Roscoe Marrs, and myself. Our inspiration and encouragement came from pioneers and experts such as John Linsley, John V. Jelley, K. Suga, A. E. Chudakov, S. Haykawa, H. Hoerlin (Los Alamos), C. B. A. McCusker, and many others.

Our first crude observatories in the area of Ithaca, New York, were characterized by huge photomultiplier tubes with Fresnel lenses, delay line analog electronics, cameras for recording pulse profiles, and overcast skies. I have the highest possible admiration for the Fly's Eye group in Utah who have made such tremendous progress with the technique of atmospheric fluorescence.

I am honored and delighted to add my welcome to these proceedings.

---

[1] Alan.Bunner@hq.nasa.gov

Search for the End
of the Cosmic Ray Energy Spectrum

John Linsley

*Istituto di Fisica Cosmica ed Applicazioni dell'Informatica
del CNR, Palermo, Italy, and Dept. of Physics and Astronomy,
Univ. of New Mexico, Albuquerque, NM 87131, USA*

Abstract. The title I was asked to speak about expresses an idea that occurred rather recently in the history of cosmic ray studies. I argue that the idea of a possible end of the cosmic ray energy spectrum came into being after a sequence of three rapid advances in knowledge which I describe, calling them 'breakthroughs'. I suggest that the present workshop be regarded as a step toward **a fourth breakthrough.** I argue that this may occur through application of **the Space Airwatch concept--the earth atmosphere as target and signal generator**--as embodied in the NASA OWL project.

INTRODUCTION

I will weave together several topics, using chronological time as an ordering principle, starting with the discovery of *höhenstrahlung* by Victor Hess in 1912. One of them is the question of composition: **what are** the particles that rain down from above? (the spectrum is a spectrum **of what?**). Another is the **origin** of the particles. I will argue that it took a long time for the role of **particle energy** to be appreciated: to make the **correct interpretation** of this radiation's distinctive penetrating power. Only after the attention of scientists had shifted from penetrating power to energy came the great discoveries in particle physics, and an approach to understanding the origin problem that permitted the steady progress which followed.

EARLY DAYS

From 1912 until 1929 there was no evidence that **energy per particle** played a very important part in explaining the Hess phenomenon. Instead, attention focussed on **penetrating power**, as shown by the name first given to the Hess radiation: '**durchdringende** höhenstrahlung'

Hess Radiation Consists of Gamma-Quanta

Of previously known ionizing rays, the ones with the most penetrating power were gamma rays from radioactivity, with energy per quantum about an MeV. So Hess

CP433, *Workshop on Observing Giant Air Showers from Space*
edited by J.F. Krizmanic et al.

particles came to be called 'ultra-gamma rays', and for the first couple of decades after their discovery they were regarded as nothing but ultra-short wavelength photons.

It was already known that the shorter the wavelength the greater the penetrating power. There was no reason at that time, however, to guess that that the energy advantage of Hess particles, over gammas from radioactivity, was very great; one or two orders of magnitude were enough to explain evidence that the new radiation could penetrate to the bottoms of deep lakes. Millikan's new name **"cosmic rays"** begged the question of **origin** by assuming that the Hess particles are uncharged. On that assumption, solar origin was ruled out by the absence of an eclipse effect, and the observed smallness of any day-night effect.

The 'Top-Down' Theory of Millikan: Energy Spectrum Cutoff 220 MeV.

The following account is from the 1964 book *Cosmic Rays* by Bruno Rossi[1]:

"Millikan noted that the actual absorption curve of cosmic rays did not correspond to a single curve [of inverse exponential form]. However it could be represented by the **sum** of the absorption curves of three groups of photons..."

"Millikan was led to the following speculation: Interstellar space is filled with very dilute hydrogen gas. Conceivably out of this gas the atoms of the heavier elements might continuously evolve by a spontaneous process of **fusion**. Once in a while, for example, four hydrogen atoms might meet and fuse to form a helium atom".

From the known mass defects of the abundant elements helium, nitrogen plus oxygen, and silicon plus neighboring elements, Millikan's fusion hypothesis explained nicely the energies of the three groups of photons.[2] This convinced Millikan that "cosmic rays" were the "birth cry" of atoms being continuously created in space. As given by Rossi, the energies of the three groups were 26, 110, and 220 MeV.

FIRST BREAKTHROUGH
Cosmic Rays ARE NOT merely Penetrating
Highest Energy goes above 1300 MeV from New Physics
then Jumps above 10,000 MeV from Geomagnetic Effect.

---

1. Bruno Rossi, *Cosmic Rays*, New York, McGraw-Hill, 1964, pp. 27,28
2. Deduced from their absorption curves using an early theory (Dirac, 1926) of gamma-ray absorption due to Compton collisions. The contribution to attenuation of pair production was unknown.

In a 1928 oral presentation (of observations he had made the preceding year, which then were published the following year) Dimitri Skobeltzyn[3] reported some evidence about cosmic rays that he had obtained quite by chance. Among magnetic cloud chamber photographs of low-energy electron tracks from radioactivity, he noticed some which showed equally thin tracks that were practically straight. He interpreted them as due to cosmic ray **electrons with energy at least 10 or 15 MeV**. In some cases there were **groups of 3 or 4 such tracks**.

In 1929 came a result by Walther Bothe and Werner Kohlhörster.[4] The experiment is well known, so I will only state the conclusion: If primary Hess rays are gamma quanta, **then their energy must be typically much higher** than Millikan proposed (so that the correspondingly more energetic Compton electrons would be able to traverse 4.1 cm of gold). **Otherwise,** the primary Hess rays must themselves be **electrons,**[5] again **with much higher energy.**[6]

These results, by Skobeltzyn and by Bothe and Kohlhörster, **suggested** that instead of being gamma rays, whose penetrating power stems from electrical neutrality, the primary Hess rays are charged particles. They didn't **prove** this, but they cast doubt on the **cutoff energy**, suggesting a value higher than 220 MeV.

Does Hess Radiation consist of Charged Particles rather than Photons?

The experiment of Bothe and Kohlhörster used the newly-invented Geiger-Müller counter tubes. Learning quickly to make them in his own laboratory near Florence, Bruno Rossi combined them with another bright idea, his own electronic logic, to go further. Even in its earliest form, the "Rossi circuit" had ten times better time resolution than the photographic method used in Bothe's laboratory, and with its help the coincidence requirement could easily be strengthened by adding a third GM tube to the 'telescope'. With these improvements Rossi soon showed that a sizable fraction of sea level cosmic rays behave like charged particles that can traverse a full meter of lead, dissipating energy by ionization in the amount **1300 MeV!**[7] Many of the primary Hess particles had to possess this much energy, regardless of their nature.

---

3. D.V. Skobeltzyn, "The Early Stage of Cosmic Ray Particle Research", in *Early History of Cosmic Ray Studies*, eds. Y. Sekido and H. Elliot, D. Reidel, Dordrecht (1985) 47-52, hereafter *Early History*.
4. W. Bothe and W. Kohlhörster, Zeits. Phys. **56**, (1929), reprinted (translated and abridged) in *Cosmic Rays* by A.M. Hillas, Pergamon Press, (1972), 148-167.
5. Muons were, of course, unknown; the other known charged particles, protons, were thought unlikely.
6. The thickness of the gold absorber was equivalent to about half a meter of water, so to penetrate it a charged particle needed to have an energy **greater than 100 MeV**.
7. B. Rossi, Naturwiss. **20**, 65 (1932). This experiment put a much stronger constraint on energy than did experiments which showed that Hess radiation penetrates to great depths in lake water, such as those of Millikan and Cameron in Southern California, and of Erich Regener in Lake Constance. Although the latter extended to a depth of 235 m, there was no requirement of **space-time coherence** to prove that the **transport of energy** was due to **charged rather than neutral** particles.

The counter experiments gave compelling evidence that the cosmic ray energy spectrum extends beyond 220 MeV by at least a decade. The specific "birth cry" origin theory of Millikan had to be given up, but it was replaced by a 'death cry' theory, that Hess particles arise from 2-photon **annihilation** of common elements in space! The notion of an energy spectrum made up of lines, or bands, held out for some years.

The question of **primary charge** became especially important at this point because the case for "cosmic" origin--the scientific basis for the new name "cosmic rays" that had been bestowed by Millikan just a few years earlier--**required that Hess particles be uncharged!** Otherwise, as could have been shown at this time although the argument wasn't advanced publically until much later, the observed constancy in time of Hess radiation (and absence of an eclipse effect) could easily have been accounted for by bending of charged particle paths in an extended chaotic magnetic field surrounding the sun. However the opportunity for presenting a solar origin model of **charged** Hess radiation seems to have been missed, so the question neutral vs charged was debated in isolation from the question of origin.

No Line Spectrum;
Geomagnetic Effect Shows Cutoffs Much Too High

A conclusion in favor of the primaries being charged was reached by using the earth's magnetic field. Work in this direction had been started by J. Clay in 1927, even before the Bothe-Kohlhörster experiment.[8] It was prompted by a qualitative analogy with the problem of the aurorae, which suggested searching for a **geomagnetic latitude effect** on Hess radiation intensity, measured the usual way with shielded high-pressure ionization chambers.

The first prediction of an **east-west effect** on Hess radiation was by Rossi, who showed, with some help from Fermi, how to apply **quantitatively** Störmer's work on charged particle orbits in the field of a magnetic dipole like the earth's.[9] Application of the Störmer theory was taken up by Georges Lemaitre and Manuel Vallarta.[10] Arthur Compton followed Clay in demonstrating that there do exist geomagnetic effects on the intensity of Hess radiation.[11]

The presence of a height-dependent latitude effect on the intensity of Hess particles proved that they are charged. It indicated that the energy spectrum of primary Hess particles **extends to at least 10-15,000 MeV!** The azimuthal or east-west effect showed moreover, and surprisingly, that **the charge is positive!**

---

8. J. Clay, Naturwiss. **15**, 356, (1927).
9. B. Rossi, Phys. Rev. **36**, 606, (930).
10. G. Lemaitre and M.S. Vallarta, Phys. Rev. **43**, 87 (1933).
11. A.H. Compton, Trans. Am. Geophys. Union, 154 (1933).

Evidence that the primary rays are charged particles opened a way to understanding their remarkably high energy as resulting from acceleration,[12] rather than direct conversion of rest-mass. The existence and role of interstellar magnetic fields wasn't yet appreciated, and difficulties with particle physics weren't yet resolved. Nevertheless, Clay speculated at this time that the primary energy spectrum has an inverse exponential shape, with a **mean energy 30 GeV** (30,000 MeV, and showed it graphically extending as far as 100 GeV.[13]

Showers and Mesons
A NEW BALLGAME
with Energy per Particle the Star Performer

In 1929 Pierre Auger had joined Skobletzyn in interpreting his cloud chamber pictures, conservatively, as showing "simultaneous production of several Compton electrons by ultra γ radiation."[14] However, coincidences observed by Rossi (with a target block and three counters in a triangular arrangement) were much too frequent to be explained in this way.[15]

The following year, by using GM counters plus electronic logic to control a cloud chamber, Blackett and Occhialini photographed events with much larger numbers of coherent tracks, and gave to them the name "showers."[16]

About this time, ion chambers were made more sensitive to sudden changes. So-called bursts were identified, due to simultaneous production of ion-pairs in numbers $\gg 10^6$.[17] A new way was opened--**calorimetry**--for evaluating energy.

Meanwhile, Rossi tried in 1931 to detect an east-west effect using inclined GM counter telescopes. When he was defeated, by the too-low altitude and too-high latitude of his laboratory near Florence, he started preparing an expedition to Asmara (2370 m.a.s.l.) in Eritrea, not far from the equator (latitude 11°N). This time he and his colleague Sergio DiBenedetti succeeded, but sadly for them they lost priority, by a couple of months, to two results obtained near Mexico City (2250 m.a.s.l., latitude 20°N), one by Thomas A. Johnson and the other by Luis Alvarez, under Compton.

---

12. W.F.G. Swann, "A Mechanism of Acquirement of Cosmic-Ray Energies by Electrons", Phys. Rev. **43**, 217-220 (1933), reprinted in *Selected Papers on Cosmic Ray Origin Theories*, ed. S. Rosen (Dover, New York 1969), hereafter *Selected Papers*.
13. J. Clay, "The Earth-Magnetic Effect and the Corpuscular Nature of Ultra-radiation, IV", Koninklijke Akademie van Wetenschappen te Amsterdam, Proceedings of the Section of Sciences, **35**, 1282-1290 (1932), reprinted in *Cosmic Rays* by A.M. Hillas, Pergamon Press, (1972), 175.
14. P. Auger and D. Skobeltzyn, *Comptes Rendus* **189**, 55 (1929); quoted from D. Skobeltzyn in *Early History*, 47-52, 51.
15. B. Rossi, Phys. Zeitsch. **33**, 304, (1932)
16. P.M.S. Blackett and G.P.S. Occhialini, Proc. Roy. Soc. A**139**, 699-727, (1933).
17. G. Hoffmann and W.S. Pforte, Physik. Zeits. **31**, 347 (1930); E.G. Steinke, Zeits. Physik **75**, 115, (1932); H. Carmichael, Proc. Phys. Soc. **46**, 169, (1934).

Rossi's full report[18] gives his conclusion about a side-effect that he observed during calibration of the counters and electronic logic that made up his system: "It would seem....that from time to time there arrive upon the equipment very extensive groups of particles (*sciami molto estesi di corpuscoli*) which produce coincidences between counters even rather distant from each other".

In English, 'groups of particles' of this nature had been given the name "showers".[19] This is the term that was used in conversations that must have taken place between Rossi and Blackett's collaborator Beppo Occhialini. Thus, translated a la Blackett, the phrase written by Rossi is "extensive showers" (of particles, omitting 'very'). This leaves no doubt about the origin of the "E" and the "S" of the modern name "EAS", for extensive air shower.

Cloud chamber photographs by Carl Anderson and Seth Neddermeyer, using a much stronger magnetic field than Skobeltzyn's, showed tracks of individual particles with much more energy, up to several GeV.[20] Anderson wrote about this In a memoir, "Millikan (who was Anderson's boss at the time) virtually hit the ceiling."[21]

Cascade Theory

One of the first theoretical explanations of showers was incorrect: that they are 'explosive' interactions by gamma-rays. Such interactions were expected, according to Heisenberg, due to high-energy breakdown of QED.[22] QED breakdown was also invoked to explain how ordinary electrons could penetrate a meter of lead in spite of energy loss due to radiation.

In the end, the 'electrons' in question were shown to be new, unstable particles, muons, which suffer much less radiation loss because they are much heavier.

In fact, the positrons observed by Blackett and Occhialini are described correctly by QED as due to pair production by gamma quanta. And very fast ordinary electrons do, in fact, lose energy rapidly by radiation, mostly by emission of a few high-energy quanta. The definitive QED predictions were worked out by Hans Bethe and Walter Heitler.[23] At once, the possibility of cascade multiplication was seen, and a race was

---

18. B. Rossi, *Supplemento a la Ricerca Scientifica*, 1, 579 (1934)

19. Was this not very brilliant choice suggested by the initial sound of the German name *stoss* (English 'burst') of a type of event whose similar nature must surely have been evident? How much better would have been "spray", the name invented by Frank Oppenheimer for a distinctive type of event that he noticed in cloud chamber pictures made much later using balloons.

20. C.D. Anderson and S. Neddermeyer, Proc. International Conf. on Physics, London, 1934, 1, 171-187).

21. C.D. Anderson in *Early History* 117-132, p. 123

22. W. Heisenberg, "Zur Teorie der 'Schauer' in den Höhenstrahlung", Zeits. Physik, **101**, 533-540, (1936).

23. H.A. Bethe and W. Heitler,"On the Stopping of Fast Particles and on the Creation of Positive Electrons", Proc. Roy. Soc. A**146**, 83-112, (1934).

on to predict the consequences. Homi Bhabha and Heitler used a step by step method, neglecting ionization loss except as a way to bring an end to multiplication.[24] John F. Carlson and J. Robert Oppenheimer did not neglect ionization loss, and they pioneered the use of diffusion equations, arguing that the step by step method "becomes probibitively laborious for thicknesses and energies of the order of those involved in large showers and bursts".[25] They also presented a "simple argument" for getting "an extremely rough indication of what to expect", which became the standard introductory exercise. But as luck would have it, the model in question is almost always attributed now to Heitler.[26]

The Carlson-Oppenheimer paper is a physics tour de force. The authors address all of the available evidence related to average cascades, in materials ranging from air to lead, including the Rossi transition effect between widely different materials, and they outline a way of dealing with fluctuations. They come down hard against Heisenberg's explosion theory, and against a conclusion by Bhabha and Heitler about ascribing sea level Hess radiation to primary electrons and photons of high initial energy. They propose, as a consequence of applying cascade theory to the totality of then-available evidence, that "the actual penetration of these ['hard'] rays has to be ascribed to the presence of a component other than electrons and photons, [made up of] particles not previously known to physics."

The method of attack by Carlson and Oppenheimer on the cascade diffusion equations was mathematically effective but far from pretty. Lev Landau heard about their work and that of Bhabha and Heitler, in Copenhagen on a visit to Niels Bohr.[27] He soon published, with Yuri Rumer, a vastly more elegant treatment.[28] The superiority of his mathematics was recognized at once, and his approach was adopted by all other

---

24. H. Bhabha and W. Heitler. "The Passage of fast electrons and the theory of cosmic showers". Proc. Roy. Soc. **159**, 432-458, (1937).
25. J.F. Carlson and J.R. Oppenheimer, "On Multiplicative Showers", Phys. Rev. **51**, 220-231, (1937).
26. W. Heitler, *The Quantum Theory of Radiation*, 2nd ed. (London, 1944), 234-235; inferior, I think, to the first edition.
27. " In 1935 [sic, but the year had to be 1936] Landau visited Bohr in Copenhagen for the last time. Pomeranchuk and I met him at the [Kharkov] railway station when he came back. In the car on the way home Landau spoke nonstop about the theoretical physics he had learned on the trip. He was particularly interested in the explanation of cascades in cosmic ray showers recently given by Robert Oppenheimer and John Carlson, and by Homi Bhabha and Walter Heitler. These investigators had explained only the essence of the phenomenon; but they didn't formulate a detailed theory of the showers. Landau himelf soon constructed the detailed theory, in collaboration with Yuri Rumer. Let me dwell a bit on this work as an illustration of Landau's mathematical talent. Being a great connoisseur of statistical physics, Landau started by formulating kinetic equations, like Boltzmann's kinetic equations, for the electrons and photons generating a shower. To solve them, he essentially reinvented the Mellin transformation. I find it curious that he didn't know about this well-established mathematical technique, and impressive that he was able to invent it when the need arose." A.I. Akhiezer, in "Recollections of Lev Davidovitch Landau", Physics Today, June 35-42 (1994),
28. L. Landau and G.[sic] Rumer, Proc. Roy. Soc. **166**, 213 (1938). One may guess that the work was carried out in Moscow, at Pyotr Kapitsa's Institute of Physical Problems, where Landau and Rumer had been given refuge after Landau was dismissed from his position (continued on footnote next page)

workers on the subject (Robert Serber and Hartland Snyder at Cal Tech), and in review works by Heisenberg[29] and by Rossi and Greisen.[30]

SECOND BREAKTHROUGH
Cosmic Rays ARE NOT merely Stellar/Solar
Highest Energy Jumps to $10^{15}eV$

A phenomenon as remarkable as the extensive air showers first noticed by Rossi is unlikely to be fully appreciated by a single investigator from a single observation. In fact, a number of steps took place before EAS were fully appreciated by Pierre Auger. Auger's work on Hess radiation seems to have started at the time of a visit by Skobeltzyn to a neighboring institute in Paris, with the multiple track interpretation referred to in an earlier section.[14] After failure of a project to set up a magnetic cloud chamber, Auger learned to make GM tubes, and used them in collaboration with Louis Leprince Ringuet as vertical coincidence telescopes to investigate the latitude effect.[31] By 1935 he had a position at the École Normale Supérieur where, alone or in collaboration with a sequence of students, he continued working with GM tubes plus electronic logic in studies of the hard and soft components of Hess radiation.

Extensive Air Showers:
the 'Discovery', and
Results by Averaging

About 1938 Auger acquired a student with exceptional gifts--with a Charpak-like golden touch--named Roland Maze. Maze invented a new way of making GM tubes that was still being used in Lodz, Poland, on my last visit there about 1994. (Maze was a longtime collaborator with the cosmic ray group in Lodz). It seems likely that he introduced Auger's group to the addition of an organic vapor to make his new-style tubes 'self-quenching'.[32] This would have simplified his next task, of designing logic

(continued) as head of the physics faculty at Kharkov University in the early days of Stalin's Terror. And that Kapitsa himself saw to it that the manuscript reached the PRS editors in London. And that its authors had barely finished their work when, in April, 1938, Landau was arrested, together with Rumer and another friend, Moissey Koretz. Although Landau was released after 12 months, on bail, Rumer and Koretz had to serve long terms of forced labor. [data are from Ref. 27 and from an article by G. Gorelik, "The Top-Secret Life of Lev Landau", in Scientific American, 72-77 (August 1997)].

29. W. Heisenberg, "The Cascade Theory" in *Cosmic Radiation*, trans. T.H. Johnson, Dover Publications, (New York, 1946), 16-25. The text, "composed by C.F. v. Weizsäcker after a symposium by W.H.",..."Closely parallels a treatment by Landau and Rumer". The list of references on pp. 181-186 suggests that, due to wartime conditions, the Physical Review issue containing the paper of Carlson and Oppenheimer was not received by Heisenberg and his colleagues in Berlin. Reference 25 of the present work does not appear on the list.

30. B. Rossi and K. Greisen, "Cosmic-ray Theory", Rev. Mod. Phys. **13**, 240-309 (1941).

31. P. Auger and L. Leprince-Ringuet, Comptes Rendus **199**, 785 (1934).

circuits with a much shorter resolving time than those of Rossi. (Rossi was obliged to hand-make 1000 megohm quenching resistors. The resolving time of his coincidence circuits was about a millisecond. It was ten times better than previously used photographic methods, but it wasn't good enough for separating EAS from background, at sea level, by a 2-fold coincidence requirement.)

Recognizing the unusual electronic ability of his 'alumnus' Maze, Auger encouraged his ambitions, forseeing the advantages of better resolving time for the kinds of studies he and his group were engaged in. Maze responded with enthusiasm. Having already produced the best GM tubes in the history of mankind, he aimed for a hundred-fold improvement in the speed of making logical choices, and in a short time he had in his hands a new brain-child. But, now, how should he **prove** that the invisible waveforms had, as he intended, an overlap in time of only a millionth of a second? Not even a Maze could imagine, at that time, the kind of technology that makes such a task easy in the laboratories of today. But of course Maze was familiar with an **indirect** method, which makes use of the rate of accidental coincidences between trains of signals that are random and time-incoherent. The customarily used source of such signals was a couple of GM tubes placed conveniently far apart on the same workbench. But what Maze observed when he tried this was totally unacceptable to him! Surely his splendid GM tubes couldn't be at fault. Was there some cross talk in his new circuit? After 'rounding up the usual suspects', and one by one eliminating them, the conclusion reached by Maze might well have been the same that Rossi had reached four years earlier. However Maze wasn't working alone; at about this stage, surely he reported what he was finding to Auger.[33]

It seems that Auger, to his credit, saw immediately the possible application of cascade theory, and set about systematically to apply it, with the goal of estimating the energies of the relevant primary particles. The first paper on his new subject, by Auger, Maze, and Térèse Grivet-Meyer, was presented in June, 1938.[34] Within a year, methods developed for finding the density of particles vs core distance had been applied in Paris and at high altitude stations of Pic du Midi and Jungfraujoch. Tracks of the showers had been photographed with cloud chambers, a barometric effect had been measured, and a variety of counter-absorber measurements had been made to determine the properties of the shower particles.[35]

---

32. Possibly, good reading habits or an efficient grapevine directed Maze to work newly published by A. Trost, Zeits f. Physik **105**, 399 (1937), "first user of polyatomic-gas counters", according to Rossi's book, *High Energy Particles*, p. 105.
33. Information that might confirm or overturn the speculations presented here may be found in a single-author paper by Maze, cited by Auger at the 1939 Chicago Symposium, that I haven't yet read: R. Maze, J. de Physique et le Radium **9**, 162 (1938). Judging from context, it seems to be an account of the new electronic apparatus.
34. P. Auger, R. Maze, T. Grivet-Meyer, "Grandes gerbes cosmiques atmosphériques contenant des corpuscules ultrapénétrant", Académie des Sciences, Paris, **206**, 1721-1723 (1938).

Auger found the size (number of particles) of a typical shower from the affected area and the particle density (deduced on the basis of statistical theory from comparative 2, 3, and 4-fold coincidence rates). In retrospect, his result, $N=10^6$, was too high by about a factor 5/2, according to later more detailed measurements, if one accepts his counting rate and his estimates of efficiency and target area. (The relevant solid angle, which isn't given, can be taken as unity without significant error). But the over-estimate of N was more than compensated by Auger's much too conservative estimate of the primary energy per sea level particle. He argued correctly that the energy belonging to the shower is about N times the critical energy, which he took as 100 MeV. But his allowance--only a factor 10--for shower energy lost above the observation level was too low by a factor 15 compared to later results (1 vs 15 ± 3 GeV/sea level particle).[36]

Thus the evidence gathered by Auger and his colleagues would have supported a claim, as to the cosmic ray energy cutoff, almost an order of magnitude higher than the claim that was made. (By the way, this would have been the result of basing the claim on mountain-altitude rather than sea-level observations, without any improvement in the correction for energy lost in the atmosphere).

In any case, the announcement by Auger was a breakthrough: his conservative value of energy was a good 4 orders of magnitude higher than the previous record high energy. It's noteworthy how **model-independent** it was, as indeed it had to be, considering that the primary particles were assumed to be electrons. (The improved later value of primary energy per shower particle quoted above, due to Greisen, is model-independent, as is a similar result by Nikolskii.[37] They are based essentially on treating the atmosphere and the earth beneath it as a **calorimeter**).

Another NEW BALLGAME
Redefinition of 'Cosmic':

The Compton-Getting prediction, and all of the theories concerning geomagnetic effects, were based on the assumption that Hess radiation consists of charged particles that impinge on the milky way galaxy from infinity. They were assumed to have an isotropic directional distribution, like starlight, and to travel **in straight lines** until they encountered the earth's magnetic field. This view seems to have been unquestioned until it was challenged by Hannes Alfvén[38] somewhat before the time of Auger's discovery. Alfvén was among the first to appreciate the problem of supply-

---

35. P. Auger, P. Ehrenfest, Jr., R. Maze, J. Daudin, C. Robley, A. Fréon, "Extensive Cosmic-Ray Showers", Rev. Mod. Phys. **11**, 288-291 (1939). The Symposium dates are June 27-30. In March, Nazi troops had occupied Czechoslovakia; in August, the Molotov-Ribbentrop agreement was announced; in September, Hitler's troops invaded Poland.
36. K. Greisen, "Cosmic Ray Showers", Ann. Rev. of Nuclear Science **10**, 63-108 (1960).
37. S.T. Nikolskii, Proc. 5th Inter-American Seminar on Cosmic Rays, La Paz, Bolivia, eds. I. Escobar et al., Lab. Physica Cosmica, La Paz, **2**, XLVIII (1962).

ing the observed cosmic ray energy, and he proposed a galactic magnetic field, with strength of order 10^{-10} Gauss, **created by the cosmic rays**, that would confine to this galaxy the ones that are observed at the earth.

This amounted to a recognition that **Hess radiation is not 'cosmic'** in the sense that had been accepted previously. The implications for theories of origin were, of course, profound.

Outer Space is not Empty!

It took some time for interstellar magnetic fields to be accepted.[39] About the time this happened, a rather extreme proposal was made by Richtmeyer and Teller, that a magnetic field belonging to the sun could confine Hess particles to its own vicinity, and at the same time stir up directions and smooth out time variations so as to agree with observations.[40] One of the experimental tests proposed by the authors is, "No particles greatly in excess of 10^{16} eV are to be expected."

Evidence from extensive air showers soon gave a reply. However, the new awareness of a distinction between cosmic rays present at some specified place in the universe, and those incident on the earth, was accompanied by a new tendency to divide the observed cosmic rays in two classes on the basis of particle energy. Clearly, arguments about supplying enough energy apply only to 'the bulk of cosmic rays', not to those with extreme energies.

In order to respect this distinction, I chose in writing this survey to adopt the name **'Hess particles'** for the former class. If this makes any sense, then it also makes sense to adopt **a new name, 'Auger particles'**, for the ones with higher energy. I know of no basis for defining a boundary value of energy, but I think that a good working definition is, **Auger particles are the ones that are observed by means of the showers they produce in the earth's atmosphere.**

Whereas the presence of Hess radiation in certain astrophysical objects can be inferred from observations of radio waves, X-rays, and gamma rays, at present there is no such capability for higher energy cosmic particles. Relevant secondary particles--neutrinos or gamma rays--will need to be detected by means of air showers, so they will also be 'Auger particles'.

---

38. H. Alfvén, Zeits. f. Physik **105**, 319 (1937); in a paper published soon afterward, "On the Motion of Cosmic Rays in Interstellar Space", Phys. Rev. **55**, 425-429 (1939), the author argued that, "most of the cosmic rays we receive on the earth are generated within less than 1000 light years from us".
39. A delightful account is given by Alfvén himself: "Recollection of Early Cosmic Ray Research", in *Early History* 427-431.
40. R.D. Richtmeyer and Edward Teller, "On the Origin of Cosmic Rays", Phys. Rev, **75**, 1729-1731 (1949).

About this time Enrico Fermi entered the game, partly it seems through influence of Alfvén, and the previous trickle of origin theories turned into a flood.[41] But the change, from a universe of objects (stars) to a universe filled with 'plasma', had started long before, with the birth of radio-astronomy.[42] By the mid-1950's, vigorous development of this new field brought in Iosif Shklovski and Vitaly Ginzburg.

But the new origin theories were aimed only at explaining Hess radiation, even those which paid attention to galaxies outside the milky way and to the metagalaxy. The ones that didn't simply disregard the Auger component[43] were few: there wasn't enough relevant evidence. From the new vantage point, the question of an end of the cosmic ray energy spectrum was devalued in favor of energy budget considerations. The role of highest-energy particles in astrophysics wasn't yet ripe for inquiry.

Some Cosmic Rays are indeed Gamma Quanta
Some are Electrons
but Most are Protons and other Nuclei.

Attention shifted to the question, what kinds of particles make up the Hess radiation? Although the essential fact was already established, that nearly all of the **field sensitive** primary Hess rays are **positively** charged particles, the habit persisted of favoring electrons over protons, even electrons with an exotic positive charge. Interpretation of ground level cosmic rays had been eased by classifying them in components, as 'hard' or 'soft'. By interpreting the hard component as secondary, made up of unstable muons, it was seen that not only the observed penetrating power but also the east-west effect could be explained. But the soft component had a much smaller east-west effect. This was hard to understand because it was seen that most of the individual particles of the ground-level soft component have low energy. Once again there was a misunderstanding about penetrating power and energy. Again there was a wrong interpretation--that primaries of the soft component are positive and negative in equal numbers[44]--because the right interpretation was paradoxical: Primaries of the soft component **are field insensitive because they have such high energy**, compared to the ones that make up the much more penetrating hard component.

The work by Marcel Schein and his collaborators that finally settled the matter in

---

41. This is indicated by the Contents of Rosen's *Selected Papers* (see Ref. 12), iii-vii. It lists 16 titles in 17 years, 1932-1948; then the number jumps to 15 in the 2 years 1949, 1950. In the following 15 years, through 1966, the rate falls back to less than 3 per year.
42. Carl Jansky was active about the time of the 'first breakthrough'; Grote Reber, about the time of the 'second breakthrough'.
43. G. Cocconi, "Intergalactic Space and Cosmic Rays", Il Nuovo Cimento 3, 1433-1442 (1956).
44. T.H. Johnson, "Evidence that Protons are the Primary Particles of the Hard Component", Rev. Mod. Phys. 11, 208-210 1929).

favor of proton primaries is well known.[45] So is the discovery, by Phyllis Freier, Ed Lofgren, Ed Ney and Frank Oppenheimer, and on the same balloon flight by Helmut Bradt and Bernard Peters,[46] that some of the primary Hess particles are heavier nuclei.

Cloud chamber photographs from similar balloon flights hinted at the presence of primary electrons or perhaps gamma rays.[47] Predictions about the gamma ray secondaries of galactic cosmic rays had been made earlier by Eugene Feenburg and Henry Primakoff,[48] G.W. Hutchinson,[49] and Satio Hayakawa.[50] Evidence for relativistic electrons in supernova remnants was pointed out in 1960 by Iosip Shklovski.[51] Electrons incident on the earth were first observed the following year, by James Earl[52] and independently by Peter Meyer and Rochus Vogt.[53] Not until the 1968 OSO-3 experiment by George Clark, Gordon Garmire and William Kraushaar was there evidence that some of the primary Hess rays are indeed gamma quanta.[54]

THIRD BREAKTHROUGH
Cosmic Rays ARE NOT merely Galactic
Highest energy jumps to $10^{20}eV$

For a decade following the discovery by Auger's group, observations of the same type were made using GM tubes in various configurations, with appropriate electronic logic, at different altitudes. Results for average EAS were derived by statistical analysis.

Individual Extensive Air Showers

A first step toward finding energies and directions of **individual showers** was made by Bob Williams, working in Rossi's group at MIT. Williams showed how to find individual core positions from samples of particle density in an array of ion chambers.[55] In this way he found 3 or 4 points on the lateral distribution curves of single showers, which enabled him to find individual values of shower 'size' (= number of particles at the observation level).

---

45. M. Schein, W.P. Jesse and E.O. Wollan, "The Nature of the Primary Cosmic Radiation and the Origin of the Mesotron", Phys. Rev. **59**, 615-618 (1941).
46. P. Freier, E.J. Lofgren, E.P. Ney and F. Oppenheimer, Phys. Rev. **74**, 213 (1948).
47. C.L. Critchfield, E.P. Ney and S. Oleksa, Phys. Rev. **79**, 402, (1950).
48. E. Feenberg and H. Primakoff, Phys. Rev. **73**, 449- (948).
49. G.W. Hutchinson, Phil. Mag. **43**, 847 (1952).
50. S. Hayakawa, Progr. Theor. Phys. **8**, 571 (1952).
51. J.S. Shklovski, in *Cosmic Radio Waves*, Cambridge, Harvard Univ. Press, 197 (1960).
52. J.A. Earl, Phys. Rev. Lett. **6**, 125 (1961).
53. P. Meyer and R. Vogt, Phys. Rev. Lett. **6**, 193 (1961).
54. G.W. Clark, G. Garmire and W.L. Kraushaar, Astrophys. J., **153**, L203 (1968).
55. R.W. Williams, "The Structure of Large Cosmic Ray Air Showers", Phys. Rev. **74**, 1689-1700 (1948).

A few years later, Pietro Bassi and George Clark joined Rossi in showing that the arrival directions of individual showers can be derived from particle arrival times at non-collinear points.[56] The time information was obtained by the first use of scintillators in an air shower experiment.

Since scintillators and ion chambers have a similar response to groups of shower particles, the obvious next step was an experiment in which the first two steps were combined. This was done successfully in the so-called MIT Agassiz experiment.[57] The advantage of measuring individual showers, rather than averaging together mixed groups of them, can be seen from the authors' conclusion that, "The existence of primary particles with energies greater than 10^{18} eV is established by the observation of one shower with more than 10^9 particles". (One may note, comparing these numbers with those used by Auger, that the energy per sea level shower particle was taken as 10 GeV, a more conservative value than Greisen's,[36] so as to allow for some fluctuation).

The core-finding step, in the MIT method, depends mainly on the particle distribution being radially symmetric, but it also makes use of a 'trial function' describing the variation of signal amplitude with core distance. This aspect of showers was investigated, using electromagnetic cascades as a model, by Jun Nishimura and Koichi Kamata.[58] Kenneth Greisen called attention to a simple analytical expression, the 'NKG formula', that can be fitted to air shower data from various types of detectors by adjusting a parameter--the 'age parameter'--taken over from cascade theory.[59] In the MIT method such an adjustment is made, rendering this step model-independent.

After a near-disaster when one of the original Agassiz toluene-terphenyl-Popop scintillators was set on fire by lightning, George Clark did some R&D that resulted in replacement of the whole lot of nasty, smelly, poisonous tanks with lovely disks of stable, efficient plastic (similarly doped polystyrene).[60] Greisen took advantage of Clark's 'factory' to make a set of disks for an array at Cornell. Greisen's layout gave an energy response that began lower and extended higher than the one used at Agassiz, but he used the same methods for finding shower axis direction and shower size.[61]

---

56. P. Bassi, G. Clark and B. Rossi, "Distribution of Arrival Times of Air Shower Particles", Phys. Rev. 92, 441-451 (1953).

57. G. Clark, J. Earl, W. Kraushaar, J. Linsley, B. Rossi and F. Scherb, "An experiment on air showers produced by high-energy cosmic rays", Nature, 180, 353-356 and 406-409, (1957).

58. J. Nishimura and K. Kamata, Progr. Theor. Phys. 7, 185 (1952). This is the most recent of the works cited by Greisen in ref. 59. A later summary is given by K. Kamata and J. Nishimura in Suppl. Progr. Theor Phys. 6, 93-185 (1958).

59. K. Greisen, "The Extensive Air Showers", in Progress in Cosmic Ray Physics, Ed. J.G. Wilson, Interscience Publishers, Inc. New York, Vol. III (1956). This article, the 'Old Testament' of 'Greisen's Bible', also contains his formula for the 'profile' of an average photon-initiated electromagnetic cascade (p. 17). The corresponding 'New Testament', Ref. 36, contains invaluable expressions for the energy and core distance dependence of air shower muons.

60. G.W. Clark, "Air Shower Experiments at MIT", in Early History, 239-246 (1985).

As data-taking at the Agassiz Station wound down, thoughts of some group members turned to making a similar but much larger array. Actions that resulted began in 1956 and reached full intensity the following year, when I was joined by Livio Scarsi, sent from Milan on loan to Rossi's group by Beppo Occhialini. Exciting years followed, during which a giant array was set up and put in operation at Volcano Ranch. In 3 months of data-taking, with the detector spacing not yet increased to the 0.9 km final value, a shower was observed with an estimated primary energy more than 10^{19} eV.[62] The highest energy observed in three years of full-scale operation was 10^{20} eV.[63]

It was the first time, I believe, that an equally well-founded claim could be made regarding **the primary energy of an individual EAS**. The credibility resulted from two features of the design: (1) The array location was high enough so that the conversion from shower size to primary energy was made ·without any large extrapolation, using the relation--essentially calorimetric--between primary energy and size **at maximum development**. (2) Thin scintillators were used for sampling the shower front.

This claim requires some explanation. A side result of the Agassiz experiment was a comparison between 'scintillator density'[64] and the particle density derived in the customary way from the multiplicity of fired GM tubes. Little by little it was understood that scintillators aren't counters; they don't count particles like GM tubes; they behave more like ion chambers: they register the amount of **deposited energy**. Plastic resembles air in its chemical makeup, so if it is used in the form of **thin slabs**, and if disturbances from roofs and the like are avoided, it yields proper samples of the **local rate of energy loss** by incident shower fronts.

Thus the reported energy did not depend on any theoretical model: it depended only on interpolation by means of two curves: one for the empirically determined lateral distribution of energy flow, and the other, a bell-shaped curve describing the longitudinal development. The energy was found **by calorimetry**.

By integration of an empirically determined 'lateral distribution of scintillator density', fitted to observed samples, one obtains the entire rate of energy loss by a shower at the level of observation.[65] Multiplying this by some factor, to obtain **the total energy deposited in the atmosphere**, is equivalent to another integration over some bell-shaped curve representing the shower profile. One can be sure that the empirical

61. J. Delvaille, F. Kendziorski and K. Greisen, "Spectrum and Isotropy of EAS", J. Phys. Soc. Japan **17**, Suppl A-III, 76-83 (1961)

62. J. Linsley, L. Scarsi and B. Rossi, "Extremely Energetic Cosmic-Ray Event", Phys. Rev. Lett. **6**, 485-487 (1961).

63. J. Linsley, "Evidence for a primary cosmic-Ray Particle with Energy 10^{20} eV", Phys. Rev. Lett. **10**, 146-148 (1963).

64. derived from the observed collector charge and the deposited energy, per unit collector charge, due to traversals of the plastic by unaccompanied relativistic muons.

profile **resembles** the profile of electromagnetic cascades. The result of the second integration will be more or less uncertain depending on the location of the observation plane with respect to an individual shower profile. Both the systematic and the random error (due to development fluctuations) are a minimum, for showers of a given size, when the level of observation corresponds to the maximum in the relevant profile curve.

The primary energy given to neutrinos is of course ignored, and the energy given to muons is under-estimated. However the amounts are almost certainly small: the under-estimation of primary energy from these defects is unimportant.

Giant Arrays Catch On

The next two giant arrays had some advantages over Volcano Ranch, but they did not have the advantage of measuring the energy calorimetrically. Instead of scintillators, the Haverah Park array used relatively thick water calorimeters, whose response to showers cannot be described in simple terms. The Sydney array SUGAR used scintillators buried in earth so that they responded to EAS by counting the muons above some threshold. It follows that in both cases the primary energy estimates depend on models of the hadronic interactions which mediate between the primary particles and those that are observed. In case of Haverah Park, the modelling was done with exceptional skill by Michael Hillas, whose rare insight was supported by carefully analyzed auxiliary experiments. In case of SUGAR, time had to pass, while knowledge accumulated from all sides about the EAS muon component, before final energy values could be assigned to the large number ($>10,000$) of giant EAS that were observed during 11 years of operation.[66]

Is There a Better Way?

Operating giant arrays of particle detectors for long periods of time proved to be difficult and expensive. New methods of observation were sought. At the same conference where the first giant array results were reported, Greisen wrote of the need

---

65. Assumptions are made, similar to those of Auger, about continuity and the statistical properties of EAS secondary particles. At first there was uncertainty in this step due to imperfect knowledge about 'scintillator density' vs core distance for small core distances. Later work at Volcano Ranch reduced this uncertainty. The initial estimates turned out to be on the low side. The estimated primary energies of the Ref. 62 and 63 events increased to 5×10^{19} eV and 1.4×10^{20} eV, as given in the *Catalogue of Highest Energy Cosmic Rays, Giant EAS*, Ed. M. Wada, Tokyo, World Data Center C2 for Cosmic Rays, hereafter *Catalogue*, **No. 1, Volcano Ranch**, pp. 3-59 (1980). At this time the probable value of the 'height factor' was still over-estimated by about 40 percent, according to evidence and analysis presented in a review a few years later: J. Linsley, Proc. 18th ICRC (Bangalore) **12**, 135-191 (1983).
66. M.M. Winn, J. Ulrich, J.S. Peak, C.B.A. McCusker and L. Horton, "The cosmic ray spectrum above 10^{17} eV", J. Phys. G: Nucl. Phys.**12**, 653-674 (1986). Events from this experiment are also described in the World Data Center *Catalogue*, **No. 2, SUGAR** (1986).

for new methods of observation, remarking that "Several groups throughout the world are now studying new methods using radio waves and atmospheric scintillation light."[67] The countries he had in mind were revealed the following year in La Paz. A talk by Koichi Suga[68] about interest in Japan prompted A.E. Chudakov to comment, "In 1955-57 I examined a possibility of using scintillation of the air in order to record high energy air showers. Some experimental investigations were made to establish the energy threshold of this method".[69] By this time Greisen, who was present in La Paz, had already started similar investigations, in preparation for an attempt to detect large air showers by the new method. First Seinosuke Ozaki and then Goro Tanahashi assisted in this work as visitors from Japan. They were joined by a series of Greisen's students, of whom the first and most noted was Alan Bunner.

GZK, the Missing Cutoff

The work on detecting EAS by fluorescence was motivated by a widely held belief that by this means the search for a cosmic ray spectrum end might be extended rather quickly by another decade, to 10^{21} eV. Although results of the first attempt at Cornell were disappointing, by 1965 a second, much more ambitious assault was under way with a new instrument called a fly's eye. With an estimated range of more than 50 km its counting rate would equal that of a 1000 km^2 array of particle detectors, in spite of the on-time limitation to moonless nights in fine weather. That would have done it, if the rate above 10^{21} eV had indeed been one per 1000 km^2 per year, as Greisen allowed himself to conclude by an extrapolation.[70]

67. J. Delvaille, F. Kendziorski and K. Greisen, Proc. 7th ICRC (Kyoto, September 1961), J. Phys. Soc. Japan **17**, Suppl. A-III, 76-83 (1962).

68. K. Suga, Proc. 5th Interamerican Symposium on Cosmic Rays, La Paz, Bolivia, eds. I. Escobar, et al., Laboratorio de Fisica Cosmica de la Universidad Mayor de San Andres, **2**, XLIX-1-5 (1962). Suga, who by that time was a leading figure in the BASJE project, described a "scintilation method" and a "radio echo method". The former was based on production of UV light by the nitrogen molecules of air. Evidently he drew on a 1957 discussion he had with Minoru Oda, Goro Tanahashi and others of the Univ. of Tokyo I.N.S. group. As a resilt of that discussion, a talk on this subject was given the following year at a symposium at Mt. Norikura Observatory. A number of years later, the story was told in a Japanese language publication by Tanahashi, a copy of which was given to me by Gene Loh. I am indebted to Yoshiyuki Takahashi for providing an English translation.

69. A.E. Chudakov, *ibid.* His comment continues, "1) The output of the ionization light for 1 MeV electrons in air for different density was obtained in the form y = (1000 light quanta)/(1 + P/P$_o$) (MeV)$^{-1}$, where P is pressure of air and P$_o$ = 10 mm Hg, (in some disagreement with presented result). 2) The experimental attempts to improve the ratio (signal/noise) by using filters which absorbed night-sky light did not give practical results. 3) The proposed arrangement was not the same as presented here, and consisted of a great number of photomultipliers looking toward the zenith, with opening angle $\pm 90°$ and separated one from another by a distance about 7 km. In this case the calculated threshold of the method came out to be about 10^{20} eV primary energy. A very complete information about the individual shower (position of the axis, angles, number of particles at different levels of penetration) can be obtained for energgy $> 10^{21}$ eV".

70. K. Greisen, "Highlights in Air Shower Studies, 1965", Proc. 9th ICRC (London) **2**, 609-615 (1965. By taking the effective solid angle to be 3 steradians rather than 1, he made a factor of 3 overestimate of the intensity at 10^{20} eV. Above that energy he took for the integral slope a beyond ankle value -1.6.

In order to arrive at earth, ultra high energy particles must first be produced, and then they must travel a great distance through space that is not empty. Greisen investigated the limits on the bottom-up production of protons by astrophysical engines, and found no cause for a cut-off lower than 10^{21} eV. But in the same year came the discovery by Arno Penzias and Robert W. Wilson of a **truly cosmic radiation**, the 2.7K microwave background.[71] Its effect on the propagation of ultra high energy Auger protons was soon pointed out by Greisen,[72] and independently by Zatsepin and Kuzmin.[73] The cutoff they predicted is strong but not perfectly sharp; values given as a 'limiting energy' are in the range 4-6x10^{19} eV. But even a single 10^{20} eV particle was unexpected.

It wasn't certain at first that the Penzias-Wilson radiation was universal, but it didn't take long for that loophole to be closed. Meanwhile, the giant arrays at Haverah Park and Pilliga Forest (SUGAR) were recording showers that seemed to be fully as energetic as the Volcano Ranch event. A skeptic could point out that the Volcano Ranch probable error was a factor of two at that time, and that anyway the particle's history may not have been typical; the equally high or higher energies claimed for some Haverah Park and SUGAR events might be contested as being hadron model dependent. An even stronger basis for questioning the apparent GZK cutoff violations may have been, from Greisen's viewpoint, results of the first test of the especially model-independent fluorescence method. On the other hand, the **presence** of a cutoff is difficult to prove. For whatever reason, in 1972 work on the Cornell Fly's Eye ended abruptly.

Giant Arrays Get Better

At Yakutsk in Siberia, work began about this time on a new giant array with energy calibration more calorimetric than ever. The workhorse detectors would be scintillators like those of Volcano Ranch, and the less favorable location at sea level would be compensated by including observations of atmospheric Cerenkov light, as well as of muons and even of low-energy nucleons. Its results are indeed hadronic model independent over an especially wide energy range.[74]

At Haverah Park, Alan Watson and his colleagues from Leed and Durham added more deep-water Cerenkov tanks to improve the EAS energy resolution,[75] and scintillators

71. A.A. Penzias and R.W. Wilson, Astrophys. J. **141**, 419 (1965).
72. K. Greisen, Phys. Rev. Lett. **16**, 148 (1966).
73. G.T. Zatsepin and V.A. Kuz'min, JETP Letters **4**. 78 (1966).
74. N,N Efimov, T.A. Egorov, D.D. Krasilnikov, M.I. Pravdin and I. Ye. Sleptsov in *Catalogue*, **No. 3, Yakutsk** (1988); N.N. Efimov, T.A. Egorov, T.A. Glushkov, A.D. Pravdin and I. Ye. Sleptsov, in *Astrophysical Aspects of the Most Energetic Cosmic Rays*, eds. M. Nagano and F. Takahara, Singapore, World Science, p.20 (1991).
75. D.M. Edge, J. Lapikens, T.J.L. McCombe, R.J.O. Reid, S. Ridgeway, K.E. Turver, A.A. Watson and A.M. Wray, Proc. 15th ICRC (Plovdiv) **9**, 137 (1977).
76. A.J. Bower, G. Cunningham, J. Linsley, R.J.O. Reid and A.A. Watson, J. Phys. G: Nucl. Phys. **9**, L53 (1983). See also *Catalogue*, **No. 1 Haverah Park** (1980).

were added whose readings, compared with Yakutsk and Volcano Ranch, showed good agreement.[76] In Japan, an array at Akeno began operation at an altitude significantly above sea level (depth 900 g/cm^2). Although not 'giant' in comparison with others, it had a large density of scintillators and a number of large muon detectors. It forms the nucleus of AGASA. the largest array that is active at the present time.

FLUORESCENCE MAKES A COMEBACK
FLY'S EYE Succeeds in Utah
SOCRAS

Soon after the Cornell fly's eye shut down, a group at the University of Utah decided to try its hand, using larger light concentrators and the advantage of Utah's clear night skies.[76] Led by Haven Bergeson after the untimely death of Jack Keuffel, in November 1976 the group made a successful test at Volcano Ranch using 3 mirror units out of 67 that were being constructed for the complete EAS observing station.[77] After full operation as a monocular device, a second somewhat smaller eye was added some distance away. By means of binocular viewing a subset of showers can be reconstructed with especially great accuracy, so that the longitudinal development profiles are correspondingly well resolved. The Utah Fly's Eye measures calorimetrically the electromagnetic energy of individual EAS, but hadronic models or data from other experiments are needed for estimating the primary mass dependent energy retained by muons.

About the time the Utah detector began operating, an astrophysics planning committee met to hear presentations, including one by the Utah group. Being present in the role of a godfather I was asked by someone at coffee, "Why not try to detect the fluorescence from space, from a satellite?" Why not, indeed? I thought on the way back to the hearing room. Back-of-the-envelope scaling calculations showed how the inverse-square signal loss might be compensated through combining the incoherently directed Utah mirrors into a single mirror with about the same area. Perhaps 100 square meters isn't out of the question, since the optical quality doesn't need to be very good. Thus a casual suggestion led to filling out a response form[78] and to invention of the acronym SOCRAS: Satellite Observatory of Cosmic Ray Air Showers. Intended to look downward toward the night sky like an owl, rather than outward in

77. H.E. Bergeson, G.L. Cassiday, T.-W. Chiu, D.A. Cooper, J.W. Elbert, E.C. Loh, D. Steck, W.J. West, J. Boone and J. Linsley, "Measurement of Light Emission from Remote Cosmic-Ray Air Showers", Phys. Rev. Letters **39**, 847-849 (1977).
78. J. Linsley, "Study of 10^{20} eV Cosmic Rays by Observing Air Showers from a Platform in Space", response to "Call for Projects and Ideas in High Energy Astrophysics for the 1980's", Astronomy Survey Committee (Field Committee) (1979).
79. J. Linsley, "Detection of 10^{10} GeV Cosmic Neutrino with a Space Station", Proc. 19th ICRC (La Jolla) **3**, 438-441, lists a 9 item bibliography. (1985).

all directions like a fly, the primary advantage was that **uniformity of response over a very large sensitive area is achieved with a single compact instrument**. The counting rate for the highest energy cosmic rays would be 50-100 times greater than the combined rate of all ground based arrays then in existence. Compared to a ground based fluorescence detector, the counting rate would again be much greater. Moreover, **the sensitivity would be inherently energy-independent**, whereas the sensitivity of a ground based fluorescence detector depends strongly on the primary energy, in a manner that changes with changing atmospheric conditions. But the idea was ahead of its time. Failing for 10 years to gain support for it I let it sleep.[79]

NEW OBSERVATIONS
Yakutsk, Fly's Eye, AGASA

In recent years extremely large events were reported by two groups whose previous observations had supported a GZK cutoff. The Yakutsk group estimated the energy of its new shower was 1.1×10^{20} eV:[80] the Utah group claimed for its shower a record-breaking 3×10^{20} eV.[81] Soon afterward the Akeno group reported 2.2×10^{20} eV as the energy of a very clean event just recorded at its new giant array AGASA.[82] The new arrival directions, like those of similarly energetic showers observed previously, did not point to any good candidate accelerating object, and they did not form statistically significant clusters. Theorists began paying attention to 'top-down' origin models.[83]

Surface Arrays are Pushed to the Limit
EAS-1000, Auger Project

Experimentalists responded by proposing ever larger arrays of particle detectors. A Soviet proposal for EAS-1000--named for its size in square km--was given formal approval, and construction began[84], but the project was hit hard by the political and economic problems that came with *glasnost* and *perestroika*, and its future is uncertain.[85] The even more ambitious Auger Project is well known.[86] Its goal is to

80. B.N. Afanasiev, M.N. Dyankonov, T.A. Egorov, V.P. Egorova, A.N. Efimov, N.N. Efremov, A.V. Glushkov, S.P. Knurenko, V.A. Kolosov, A.D. Krasilnikov, I.T. Makarov, A.A. Mikhailov, E.S. Nikiforova, V.A. Orlov, M.I. Pravdin, I.Ye Sleptsov, G.G. Struchkov, "Recent Results from Yakutsk Experiment", in Proc. Tokyo Workshop on Techniques for the Study of Extremely High Energy Cosmic Rays, ed. M. Nagano, Tokyo, Institute for Cosmic Ray Research, pp. 35-51
81. D.J. Bird, et al, Phys. Rev. Letters **71**, 3401 (1993).
82. N. Hayashida, et al, Phys. Rev. Letters **73**, 3491(1994).
83. C.T. Hill, Nuclear Physics **B224**, 459-490 (1983); P. Bhattacharjee, C.T. Hill and D.N. Schramm, Phys. Rev. Letters **69**, 567 (1992); G. Sigl et al, "Implications of a Possible Clustering of Highest Energy Cosmic Rays", FERMILAB Pub-96/121A (1996).
84. G.B. Khristiansen et al, in Proc. 14th Texas Symposium on Relativistic Astrophysics, Annals of the New York Academy of Sciences, **571**, 640 (1987).
85. G.B. Khristiansen et al, in Nuclear Physics (Proceedings Supplement) **B28**, 40 (1992).
86. J. Cronin, *ibid.* 213 (1992). See also the Pierre Auger Project Design Report, 2nd Edition (1996).

build and operate two air shower detectors, one to be placed in the Northern Hemisphere and the other in the Southern Hemisphere, each one consisting of an array of 1600 particle detectors spread over 3000 km^2.

FOURTH BREAKTHHROUGH?
MASS, OWL
Space Airwatch

On 15 May, 1995, my wife Paola telephoned me in Palermo that Yoshiyuki Takahashi was trying to get in touch with me from Marshall Space Flight Center. His message was, "I have written a paper about a Maximum-energy Air Shower Observing Satellite. The technology and neutrino detection capability relate to John's original idea of 1979. I would like to send my text to John, and talk with him." As soon as we had talked, I told Livio Scarsi what had happened. "It sounds as if it might be fun", he said. "But it will be hard for me to explain 'MASS' to the Italian Space Agency. We should call it something more general." After a few tries we came up with "Airwatch", short for "Space Airwatch".

Of the improvements over SOCRAS that were called for by Takahashi in his MASS proposal, the most important are reducing the aperture and enlarging the field of view of the optical system. His order of magnitude aperture reduction means a size reduction that moves the proposal within striking distance of systems that meet present launch requirements. His order of magnitude field of view enlargement increases by the same amount his counting rate. The resulting numbers are large enough so that, if formidable technical problems are solved and if institutional support is forthcoming, it is not unreasonable to imagine **a fourth breakthrough, in the search for an end of the cosmic ray energy spectrum.**

Takahashi moved fast. managing to organize a workshop that took place in Huntsville even before the Rome ICRC where he presented his paper on MASS.[87] By the following spring a group formed at GSFC and put together a proposal for OWL, a pair of Orbiting Wide angle Light Concentrators. The first Airwatch Symposium was held early this year in Catania. A number of contributions to the present workshop from Italy and Russia are the result.

ACKNOWLEDGEMENT

Thanks are due to Alan Watson, who over the years has supported my efforts to understand earlier events by supplying copies of hard-to-find documents, and to my daughter, Amina Quargnali-Linsley, for help in dealing with problems of so-called home publishing.

---

87. Y. Takahashi, "Maximum-energy Auger (Air) Shower Satellite (MASS) for Observing Cosmic Rays in the Energy Region 10^{20-22} eV", Proc. 24th ICRC (Rome) **3**, 595-598 (1995); see also *Huntsville Workshop Report*, Huntsville, Alabama, August 7-8, 1995, The University of Alabama in Huntsville.

Powerful Radio Galaxies as Sources of the Highest Energy Cosmic Rays

Peter L. Biermann

Max Planck Institute für Radioastronomie
Auf dem Hügel 69, D-53121 Bonn, Germany
plbiermann@mpifr-bonn.mpg.de

Abstract. We summarize the status of the search for the origin of the highest energy cosmic rays. We briefly mention several competing proposals, such as Gamma Ray Bursts also giving rise to energetic protons, high energy neutrinos and cosmological defects, and then concentrate on the possibility that powerful radio galaxies can provide the sources. We describe several tests, some of which have been performed already. First, powerful radio galaxies must be able to accelerate protons to such energies; this entails that there is sufficient space for the Larmor motion. Second, we require at least one candidate radio galaxy with sufficiently strong shock fronts to be the source, at a sufficiently close cosmological distance. Third, the distribution of arrival directions of the highest energy particles on the sky ought to reflect the source distribution as well as the propagation history. The present status can be summarized as inconclusive: Powerful radio galaxies have been tested more than any other candidate source class, but a definitive confirmation is still outstanding. If we were able to confirm this particular theory - or any other - these particles at beyond 10^{20} eV may be turned into tools of high energy physics.

INTRODUCTION

The detection of several cosmic ray events with energies well beyond 10^{20} eV [14,33] is challenging us [4,28,84,10,58]. These energies surpass by far anything reachable in particle accelerators on Earth.

Ever since the detection of the microwave background we know that high energy particles cannot come to us across the entire universe [31,98,89,1,3,68,29,70]; for a homogeneous distribution of sources, extragalactic protons come from nearby. The summed spectrum of such hypothetical sources should cut off near $5 \, 10^{19}$ eV, if the dominant particles are all protons. This is the Greisen-Zatsepin-Kuzmin or GZK cutoff.

Over the last few years it has become clear that we have to take the detection of particles clearly beyond this energy very seriously. Therefore we can ask, which distances can these particles come from, if they arrive truely with several 10^{20} eV. Using average energy loss, the answer is 50 Mpc, if we are generous, and is 100 Mpc,

CP433, *Workshop on Observing Giant Air Showers from Space*
edited by J.F. Krizmanic et al.

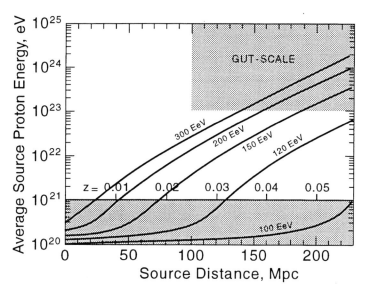

FIGURE 1. What energy does a proton need to have at a given distance to provide some specific energy at Earth (courtesy T. Stanev).

if we really push - provided that the initial energy is no more than a few times 10^{21} eV, which appears to be the maximum for a model such as a radio galaxy as a potential source.

There seems to be a weak correlation of arrival directions with the supergalactic plane, and any theory needs to either argue that the correlation is not significant, or needs to explain it. This correlation appears just at the GZK-cutoff, as expected, since near that cutoff the source distribution becomes cosmologically local, where sources are no longer isotropically and homogeneously distributed in the universe. Any known sources correlate with the supergalactic plane on scales much larger than our Galaxy, and smaller than about 100 Mpc.

On the other hand, if we find that there is no plausible source for such particles, and they need to originate further away, then the required initial energy approaches the energy scale of the Grand Unified Theories (GUT), and we begin to discuss quite different physics. For the Fly's Eye event this happens already at a rather small distance of 200 Mpc, where the required initial energy is 10^{24} eV.

The two key problems therefore are

- To generate particles at these extreme energies.

- To get these particles to us.

In this brief review I will discuss both these issues, and concentrate on the pro-

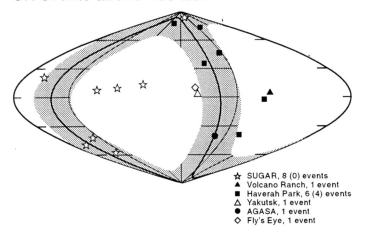

CR events above 100 EeV

☆ SUGAR, 8 (0) events
▲ Volcano Ranch, 1 event
■ Haverah Park, 6 (4) events
△ Yakutsk, 1 event
● AGASA, 1 event
◇ Fly's Eye, 1 event

FIGURE 2. The sky above 10^{20} eV in galactic coordinates, with the Galactic anticenter at the center of the figure. The supergalactic plane sheet is shaded.

posal that powerful radio galaxies may be the sources we are looking for. In order to save space, I suggest to the readers to refer to another recent review to find helpful graphics and more relevant literature [10].

I VARIOUS PROPOSALS

There are many proposals how to explain the origin of the high energy particles, many of which were also discussed at this meeting.

Among the proposals there are:

- An unknown or little known particle. At the meeting Glennys Farrar from Rutgers proposed the S_0 particle to be a candidate; this particle is neutral, and so its path would be straight to us. Therefore, if such particles were to be emitted by an AGN in a flux commensurate with its electromagnetic output, then we ought to see a quasar just "behind" the most powerful events, with a quite appreciable electromagnetic flux. Surprisingly, this is in fact true for the two most powerful events known [9]: The radioquasar 3C147 is in the positional error range of both the Fly's Eye and the Yakutsk event. In the error box of the highest energy Akeno event there is the optically selected quasar PG 0117+213. This quasar is very luminous, with an estimated accretion disk luminosity of 2×10^{47} erg/sec. To check further we require good positional error boxes for all the very high energy cosmic ray events.

- Protons from the decay of cosmological defects or other unusual particles [5,83,84,16,72,42,64,65]. This proposal is certainly the most fascinating, since it ties these particles directly to the origin of the universe. There is an issue whether the produced gamma-ray background is close to or actually exceeds the measured background fluxes.

- Monopoles, as a direct explanation of these events [43]. It is an open question whether monopoles can mimick a proton event at any reasonable energy.

- Cosmic ray neutrinos can annihilate on relic neutrinos (work by Tom Weiler, also reported at this meeting), and then locally produce very high energy protons. This proposal requires neutrinos to have mass, and the new limits from experiments allow the required range of masses.

- Particle acceleration in cosmological structure formation shock waves [59,40,41,18]. There is some evidence now that such shocks have actually been detected through radio emission [19], and so this option becomes viable. On the other hand, it is not clear whether the highest energies required can really be reached.

- Galaxy collisions [15]: This option was dicussed at the meeting at some length by Frank Jones, and dismissed. It cannot give the maximum energy required. It does have the beauty, that there are many such collisions. The ultimate collision is a merger of galaxies. Merging galaxies are quite common, and easy to detect through their far-infrared emission.

- Galactic wind shocks [39]. Galactic winds may exist, and they may have shocks just as expected for the Solar wind. But again, in analogy with the Solar wind, the energies of such particles cannot really be high enough to explain the data.

- Pulsars remain viable candidates [37,82], but then the arrival directions cannot be interpreted, because at these energies the path through the Galaxy is straight, and the events do not cluster around the plane of the Galaxy. They actually seem to avoid the Galaxy.

- Gamma Ray Bursts(GRBs) [32,26]. Since the theory of GRBs [56] is beginning to stand on a firmer footing with the recent detection of afterglows at all classical wavelengths, e.g. [27], and with the first redshift determination [57,2], GRBs are all the rage. They may explain many things, also high energy cosmic rays [95,96]. GRBs are believed to be the product of gigantic explosions, with initial Lorentz factors of several hundred. Both internal and external shocks ought to exist. Both kinds of shocks can accelerate protons: For the internal shocks it has been shown by Rachen and Meszaros (presented at the GRB-meeting at Huntsville in the fall of 1997) that the particles can escape the acceleration site only as neutrons - then the parameters may fit. For external shocks the required particle energies can also be reached, but are then strongly dependent on the environmental density. It is an open issue for some, whether

25

the cosmological local density of GRBs can be sufficiently high to explain the cosmic ray events. Others ask whether the afterglow spectrum can be understood with the kind of acceleration theory, that is used to obtain protons of 10^{21} eV.

- Active Galactic Nuclei (AGN) [7,63]: The recent detection of several AGN of the blazar class at photon energies near 1 - 10 TeV [66,67,61,62,46,36] has revived interest in hadronic models to explain this emission [50–53]. In a hadronic model protons are accelerated first, and then produce the observed emission as a result of their interaction. Two difficulties here are a) that the energetic protons are produced close to the central engine, and there the photon field is high, and so the maximum proton energy limited, and b) that the protons suffer adiabatic losses on the way out to us, and so lose even more energy.

- Last, the powerful radio galaxies, [7,71]: These sources have many advantages, among them that this proposal ought to be easily refutable. Therefore, I will discuss this particular option in the main part of this review, and then point out some critical observations, that may decide on the merits of this concept.

II RADIO GALAXIES

The proverbial radio galaxy is Cyg A; it shows the canonical structure in radio emission, large radio lobes, and hot spots. In general such radio lobes are powered by jets, that emanate from a central object at relativistic speeds. The jet dissipates in gigantic shock waves probably, as the hot spots are interpreted. There is reason to believe that the power in these jets is commensurate to the thermal emission from the central compact accretion disk around a supermassive black hole. The isotropic powers from the core region are of order 10^{46} erg/sec, with the most extreme reaching 10^{48} erg/sec, but then probably anisotropic (relativistically boosted) emission. The size of radio galaxies is typically 300 kpc, but can reach several Mpc. The overall radio structure with lobes and hot spots is only seen for powerful objects, dubbed "Fanaroff Riley class II" radio galaxies [23]. The radio (and sometimes detected optical) emission from radio lobes, jets, and hot spots is all non-thermal, *i.e.* Synchrotron emission from highly relativistic electrons.

The notion that the hot spots are weakly relativistic shock waves leads to a successful explanation of many of their properties. These shock waves are expected to accelerate particles to very high energy.

We propose to explore this concept to explain the very high energy events observed. To be successful, such a theory must account for the following:

- Can radio galaxies in general accelerate particles to 10^{21} eV?

- A very practical test: Does the Larmor radius of a particle at 10^{21} eV fit inside the source region?

- Can a specific source accelerate particles to the energy required, about 10^{21} eV? Since radio galaxies cannot be hidden, even through our Galaxy, we ought to know them, and we ought to be able to identify, which single radio galaxy or which few radio galaxies can do it.

- Are there any testable predictions? Best would be a prediction, that has been made, and that is tested subsequently.

- Are there any specific observations, that require particles at 10^{21} eV? Just as in our Galaxy, where the acceleration of electrons is "visible" in the remnants of supernova explosions, the corresponding argument for protons remains elusive and rather indirect.

I will try to convince the reader that radio galaxies pass all these tests.

A Acceleration versus losses

Let us consider a shock of velocity U_1, with an upstream magnetic field strength B, a fraction b of the magnetic energy in turbulence, a spectrum of turbulence of $I(k) \sim k^{-1}$ in wavenumber k, assuming isotropy in wavenumber space, and with $I(k)$ energy per wavenumber and per volume. We then adopt a maximum and minimum wavenumber and define $\Lambda = \ln(k_{max}/k_{min})$. This last assumption is quite different from what Biermann & Strittmatter [7] did, where a Kolmogorov spectrum was used; we use a k^{-1} spectrum here to maximize the resulting particle energy, inspired by arguments on the TeV emission from the Crab nebula [38]. A Kolmogorov spectrum may be a much better assumption to describe the data, and we will return to it below. Balancing then gains and losses in standard shock acceleration theory [17] we obtain

$$E_{p,max} = 7.4 \times 10^{19} \left(\frac{b}{\Lambda} \right)^{1/2} \frac{1}{B^{1/2}} \frac{U_1}{c} \, \text{eV} \tag{1}$$

An absolute limit corresponds here to $(b/\Lambda)^{1/2}(U_1/c) = 1$. Using also a typical value for the magnetic field strength in hot spots of $B \approx 10^{-4}$ Gauss, we obtain an absolute limit of $E_{p,max} \approx 3. \times 10^{22}$ eV.

Using the Kolmogorov spectrum, which may be appropiate, we obtain in a similar approximation the maximum energy of [7]

$$E_{p,max} \approx 1. \times 10^{21} \, \text{eV}. \tag{2}$$

B Larmor limit

In a series of papers, Falcke et al. [20–22] have developed the concept, called "jet-disk symbiosis", that we need to consider the compact accretion disk, the jet, and the black hole as a system, which as a whole obeys the conservation laws for mass, energy and angular mommentum. This then leads to a predicted behaviour of the compact jet radio emission versus the thermal disk emission; for the maximum of efficiency the relation matches observed properties of jet-disk systems over more than 12 orders of magnitude [22].

In this approach one can ask how the magnetic field scales with the power of the system, and can then determine the maximum energy of a particle that fits inside a jet with its Larmor motion. This then leads to a scaling already discovered by Lovelace [49]; putting in numbers it gives

$$E_{p,max} \approx 5. \times 10^{20} L_{disk,46}^{1/2} \, eV, \tag{3}$$

where $L_{disk,46}$ is the thermal accretion disk luminosity in units of 10^{46} erg/sec.

Using the most powerful radio galaxy studied by Falcke et al. [20,21] with a disk luminosity of 3×10^{47} erg/sec we obtain

$$E_{p,max} \approx 4. \times 10^{21} \, eV. \tag{4}$$

Therefore, radio galaxy hot spots can accelerate particles (protons, to be specific) to the required energies.

However, using average energy losses, even for this energy the maximum distance allowed is only about 50 Mpc.

If we assume, that the shock waves are intermittently relativistic, and also point in the appropiate direction, then we may push this energy up by a small factor [60]. Another factor of 3 in energy would push the distance allowed to 75 Mpc, again using average energy losses. I emphasize that this distance limit does not include any discussion about the flux that can reach us, just a) the interaction with the microwave background, and b) the argument, that most particles should reach us, not a miniscule fraction. If we allow for a very small fraction only to reach us, then the distance limit becomes larger yet, obviously.

We note that for charged energetic particles confined to a sheet such as the supergalactic plane - see below - the flux would diminish with distance d not as $1/d^2$, but as $1/d$, and so the flux may be appreciable even for a large distance.

Therefore, radio galaxy hot spots can accelerate particles to the required energy; but at distances where we ought to know each and every one of them.

C Candidates

Which radio galaxies are candidates, without any allowance for their direction in the sky [85,9,10]. I list a few candidates:

- 3C 134: 3C134 may be the closest radio galaxy with powerful hot spots, but it is obscured by clouds in our Galaxy, and so no redshift is known. From the size - power correlation one can estimate the distance to be in the range 30

28

to 300 Mpc. A similarly uncertain estimate can be made on the basis of its optical detection in the red continuum (Naval Observatory, see [10]). 3C134 is fairly close in the sky to the Fly's Eye and Yakutsk events.

- NGC 315, at $z = 0.016$; we can estimate the maximum energy reachable by this galaxy to $1.2 \, 10^{21}$ eV. So it may be a reasonable candidate. NGC 315 is a very asymmetric radio source. NGC 315 is fairly close in the sky to the Akeno event.

- M87, at $z = 0.0043$, and with an estimated maximum particle energy of $1.5 \, 10^{20}$ eV. M87 as well is a very asymmetric radio galaxy. M87 is fairly close in the sky to two Haverah Park events.

- 3C 31, at $z = 0.017$; however, this radio galaxy has no powerful hot spots.

- Cen A, at a distance of about 5 Mpc; and again, it has no powerful hot spots. Both 3C 31 and Cen A are low power radio galaxies.

Obviously, there are many interacting galaxies in the nearby universe, such as M82 and NGC 2146; we do not propose to discuss them here.

Therefore, there is only one prime candidate radio galaxy known at present, 3C 134, which has all the tell-tale signs of a powerful object, but we do not know its redshift.

III PREDICTIONS AND TESTS

A   The supergalactic plane

Is there any prediction which has been made, and which has been verified subsequently - for this or any other proposal?

For radio galaxies I pointed out at the Fermi-Lab meeting, organized by J. Cronin in March 1995, that for approximately straight paths to us, the arrival directions on the sky ought to correlate with the supergalactic plane sheet, the locus of most cosmologically local galaxies (to distances of order 100 Mpc) [91–94,80,81]. The supergalactic plane is most pronounced in the northern sky. Radio galaxies happen to define best the supergalactic plane as clearly demonstrated by P. Shaver [80,81].

There is one difficulty with such a prediction, and it may be the reason that no correlation had been found before: Such a correlation can only be true for energies for which the particles cannot come from very far, *i.e.* for energies beyond the GZK-cutoff. This means that any such effect has a pronounced threshold. Therefore, the energy of any event considered must be very well determined, in order to avoid washing out any effect due to the flux steeply falling with energy. This last point proved to be critical, and may have prevented any search in the past to be more successful.

Therefore, the prediction was, that from 5×10^{19} eV there should be a noticeable correlation with the supergalatic plane.

This prediction was checked in three projects:

- Stanev et al. [87] used the distance of the arrival directions to the galactic plane, and to the supergalactic plane as an indicator. They were forced to eliminate half of all Haverah Park events to improve the energy determination; this then avoided the wash-out effect due to the threshold of the correlation searched for. The data used were all Haverah Park data and a subset of Akeno data. They found the effect as predicted, but it was marginal. They ran various consistency checks, such as correlating with the flux at slightly lower energies, and did find agreement.

- Hayashida et al. [35] used another technique, with the entire data set from Akeno, the occurrence of pairs and triplets of events. Again they found a weak effect, with more pairs and triplets along the supergalactic plane than expected by chance.

- Finally, Uchihori et al. [90] used all existing data, from all experiments, and also used the pair correlation technique. Again, they found a weak effect. It appears as if at least 15 % of all events beyond the GZK cutoff were coming from the supergalactic plane. Interestingly, some of the pairs had very disparate energies; if one assumes that such events come from the same source, it is possible to show that they must be singly charged.

Therefore, the prediction was verified, albeit at a marginal level. Further data will clearly shed a lot more light on this question.

B   Inhomogeneity in cosmic magnetic fields

Immediately, it was pointed out [97] that the correlation, however weak, was still too good to be true: Using a realistic source distribution in the sky it was possible to demonstrate that such a correlation ought to be invisible with the available data set. In fact, it is invisible in the South [44]. However, this calculation assumed that the path of any high energy particle is actually straight from the source to us [97,88].

At this point it is important to remember how the supergalactic plane sheet actually formed: This is a very large scale accretion flow towards a sheet in the cosmological structure formation. The accretion velocity is of order 1000 km/sec, and one expects shock waves to form on both sides of the sheet of a thickness of a few Mpc, and a longitudal scale of several tens of Mpc. It is these shocks that some have proposed to accelerate the highest energy cosmic rays.

In this accretion flow there is surely also an embedded magnetic field, and so we can expect that particles cannot penetrate upstream below a certain energy and momentum vector. This is akin to the cosmic ray modulation done by the solar wind: Particles below several hundreds of MeV never make it to the location of the Earth. At the same time the solar wind shock accelerates the anomalous cosmic rays.

Simulations [79,40,48,41,77,78,11,12] of the cosmological structure formation show that the magnetic field is highly inhomogeneous across sheets and voids, the regions between sheets. Using then the Faraday rotation of the radio signal from cosmologically distant sources we can determine the upper limit to the magnetic field in the sheets. This upper limit is of order $1\,\mu$Gauss, three orders of magnitude larger than derived previously by using a homogeneous universe [47]. Interestingly, this magnetic field strength is close to equipartition in the sheets with such values. There is observational confirmation for a strength of the magnetic field close to such a range in the case of a bridge between the Coma cluster and its neighbour through the work of Kronberg and his collaborators [45,47]. Therefore the Larmor radius for a 10^{20} eV particle is of order the thickness of such a sheet.

As a corollary I note that the phenomenon of accretion shocks in cosmological structure formation has found a nice consistency check in observation recently: Accretion shocks are most pronounced around clusters of galaxies, where they can reaccelerate electrons and so produce radio emission far from any visible source. Radio relic sources around clusters appear to be nicely explained by this notion, and so provide the first observational direct evidence of the existence of accretion shocks on these large scales.

In a sheet, we expect to have two accretion shocks with a converging flow - Fermis classical situation [24,25]. Particles with a transverse momentum below some limit will be confined to the sheet. This critical transverse momentum may correspond to 10^{20} eV/c. Then any particles with an initial path sufficiently close to parallel to the sheet will never be able to escape, and so it will be confined to the sheet.

Such a confinement entails that the flux of any source decreases with distance d only as $1/d$; this in turn means that we can detect sources in cosmic rays to a much larger distance than possible with a $1/d^2$ law.

On the other hand, it also means that any particle with a transverse momentum above the threshold will be strongly scattered, and all memory of its origin is lost.

These two extremes may explain the weak correlation we find in arrival directions of high energy cosmic rays with the supergalactic plane.

C Are 10^{21} protons required?

Is there any observation out there, other than the detection of the particles directly, which requires to assume that protons at energies of 10^{21} eV really exist in a specific class of sources?

The answer is probably "YES".

Compact AGN sources, jets and hot spots often show observationally a cutoff in their nonthermal emission near 3×10^{14} Hz, seen since 1976, mostly by the Riekes and their collaborators [73–76].

31

The emission is polarized, extends from optical through to radio waves, and there can be little doubt that this emission is synchrotron emission from highly energetic electrons.

Asking then again, to what energy electrons can be accelerated in a shock wave, balancing energy gain versus losses, directly gives a required scale for the magnetic scattering of an electron on either side of a shock wave - in the shock acceleration picture. The question then arises, what determines this length scale [55] such that the resulting maximum frequency is always the same, in sources where the magnetic field must be vastly different. The length scale can be derived from a larger scale by turbulent cascading, and a larger scale can be set by energetic protons, also accelerated in the same shock, and also subject to gains and losses just as the electrons. Since the same gains and losses work on both electrons and protons, albeit on different scales for both particles, the dependence on the strength of the magnetic field drops out. This leads to a maximum synchrotron emission frequency of [7]

$$\nu_{max} \approx 10^{-2} \frac{c}{r_0} \left(\frac{m_e}{m_p} \right)^2 \text{ Hz} \tag{5}$$

where r_0 is the classical electron radius. One critical assumption made here is a Kolmogorov cascade, for which there is lots of evidence both from solar wind studies, and from theoretical work [54,8,30]. In fact, it is possible to check on this assumption, because integrating the emission over an entire downstream region we also include emission from lower energy electrons, electrons which lost most of their energy through losses in the downstream flow. This then leads to a bend in the spatially integrated spectrum at a break frequency. The ratio of observed break frequency and cutoff frequency gives a constraint on the turbulent cascade spectrum: This is consistent with a Kolmogorov spectrum [6].

This explanation requires protons to exist at 10^{21} eV, because they set the length scale for the turbulent cascade from their balance of gain and losses [7].

D   The flux

Another check on the notion, that radio galaxies can provide the extragalactic flux, beyond $3\,10^{18}$ eV, is that the known radio galaxy population can actually readily provide the flux seen below the GZK cutoff. This is the case [68,86,69]. This also provides a natural explanation for the finding, that the Fly's Eye team detects a switch in chemical composition at that energy, from heavy to light. However, there is a caveat, in that the Akeno team disputes the existence of this switch in chemical composition [13,34]. At the meeting, Tom Gaisser pointed out how difficult it can be to derive chemical composition information.

E Tests

What are the tests at this stage?

- First we need to determine what the real magnetic field strength is in the cosmological sheets. We expect from the arguments above a value near the upper limit of

$$B_{sheet} \approx 10^{-6.5 \pm 1} \text{ Gauss} \tag{6}$$

- We need to augment our database, best at present with data from AKENO, from HIRES, and some day, hopefully, with much bigger arrays, such as AUGER, OWL, or some other system.

- Last we need to identify a real source; under the hypothesis that radio galaxies do it, we just have to identify that radio galaxy responsible for the highest energy event.

IV CONCLUSION

If we can identify the origin of the events at the highest energies, beyond 5×10^{19} eV, the Greisen-Zatsepin-Kuzmin cutoff due to the microwave background, near to 10^{21} eV = 1000 EeV, say whether they are from

- A new kind of particle such as S_o,
- Topological defects,
- Gamma Ray Bursts,
- ...
- ..., or
- Powerful radio galaxies

we will obtain a **tool to do physics at EeV energies.**

ACKNOWLEDGMENTS

PLB wishes to thank all participants for the discussions at the meeting as well as before and afterwards. PLB also wishes to thank Venya Berezinsky, Jim Cronin, Heino Falcke, Glennys Farrar, Tom Gaisser, Hyesung Kang, Richard Lovelace, Abhas Mitra, Tsvi Piran, Ray Protheroe, Jörg Rachen, Dongsu Ryu, Günther Sigl, Todor Stanev, Eli Waxman and Tom Weiler for intense discussions. Finally the author expresses his gratitude to Giovanna Pugliese and Yiping Wang for a careful reading of the manuscript.

REFERENCES

1. Aharonian, F.A. & Cronin, J.C.: 1994 *Phys. Rev. D* **50**, 1892.
2. Arav, N., Hogg, D.W.: 1997 (submitted) astro-ph/9706068
3. Berezinsky V.S., GrigorUeva S.I.: 1988, *Astron. & Astroph.* **199**, 1 - 12.
4. Berezinskii, V.S., *et al.*: 1990 *Astrophysics of Cosmic Rays*, North-Holland, Amsterdam (especially chapter IV)
5. Bhattacharjee, P., Hill, C.T., Schramm, D.N.: 1992 *Phys. Rev. Lett.* **69**, 567 - 570.
6. Biermann, P.L.: 1989 in NATO-conference proceedings, NATO ASI-series C, vol. 270, Erice conference, "Cosmic Gamma Rays, Neutrinos and Related Astrophysics", Eds. M.M. Shapiro, J.P. Wefel, Kluwer, Dordrecht, p. 21 - 37
7. Biermann, P.L., Strittmatter, P.A.: 1987 *Astrophys.J.* **322**, 643 - 649.
8. Biermann, P.L.: 1995 in *Gamow Jubilee Seminar*, Eds. A.M. Bykov, R.A. Chevalier, & D.G. Yakovlev, in *Space Science Reviews* **74**, 385 - 396.
9. Biermann, P.L.: 1995 *Trends in Astroparticle Physics*, Eds. L. Bergström *et al.*, North-Holland, *Nucl. Phys. B Proc. Suppl.* **43**, 221 - 228.
10. Biermann, P.L.: 1997 *J. of Physics G* **23**, 1 - 27
11. Biermann, P.L, Kang, H., Ryu, D.: 1997 in Proc. "Extremely high energy cosmic rays", Tokyo, Ed. M. Nagano, p. 79 - 88, astro-ph/9709250
12. Biermann, P.L., *et al.*: 1997 in Proc. XXXXII. Moriond meeting (January 1997), Eds. Y. Giraud-Héraud, J. Tran Thanh Van, Editions de Frontieres, astro-ph/9709252
13. Bird, D.J., *et al.*: 1993 *Physical Review Letters* **71**, 3401 - 3404.
14. Bird, D.J., *et al.*: 1995 *Astrophys. J.* **441**, 144 - 150.
15. Cesarsky, C., Ptuskin, V.: 1993 23rd ICRC, University of Calgary, vol. 2, OG 9.3, p. 341 - 344.
16. Daum, K. *et al.*: 1995 *Zeitschrift f. Phys. C* **66**, 417.
17. Drury, L.O'C: 1983 *Rep.Prog. Phys.* **46**, 973 - 1027.
18. Enßlin, T.A., *et al.*: 1996 *Astrophys. J.* **477**, 560 - 567
19. Enßlin, T.A., *et al.*: 1997 *Astron. & Astroph.* (in press)
20. Falcke, H., Biermann, P.L.: 1995 *Astron. & Astroph.* **293**, 665 - 682.
21. Falcke, H., Malkan, M.A., Biermann, P.L.: 1995 *Astron. & Astroph.* **298**, 375 - 394.
22. Falcke, H., Biermann, P.L.: 1996 *Astron. & Astroph.* **308**, 321 - 329.
23. Fanaroff, B.L., Riley, M.: 1974
24. Fermi, E.: 1949 *Phys. Rev.* 2nd ser., **75**, no. 8, 1169 - 1174.
25. Fermi, E.: 1954 *Astrophys.J.* **119**, 1 - 6.
26. Fishman, G.J., Meegan, C.A.: 1995 *Ann. Rev. Astron. & Astroph.* **33**, 415 - 458
27. Frail, D.A., *et al.*: 1997 *Nature* **389**, 261 - 263
28. Gaisser, T.K.: 1990 *Cosmic Rays and Particle Physics*, Cambridge Univ. Press
29. Geddes, J., Quinn, T.C., Wald, R.M.: 1996 *Astrophys. J.* **459**, 384 - 389.
30. Goldstein, M.L., Roberts, D.A., Matthaeus, W.H.: 1995 *Ann. Rev. Astron. & Astroph.* **33**, 283 - 325
31. Greisen K.: 1966, *Phys. Rev. Lett.* **16**, 748
32. Hartmann, D.H.: 1994 in *High Energy Astrophysics*, Ed. J.M. Matthews, World Scientific, p. 69 - 106
33. Hayashida, N., *et al.*: 1994 *Phys. Rev. Letters* **73**, 3491 - 3494.

34. Hayashida, N., *et al.*: 1995 *J. Phys. G: Nucl. Part. Phys.*

35. Hayashida, N., *et al.*: 1996 *Phys. Rev. Letters* **77**, 1000 - 1003

36. Hermann, G. et al., XXV ICRC'97, Durban, OG 10.3.19

37. Hillas, A.M.: 1984 *Ann. Rev. Astron. Astrophys.* **22**, 425 - 444.

38. Jager, O.C. de, Harding, A.K.: 1992 *Astrophys. J.* **396**, 161 - 172

39. Jokipii J.R., Morfill G.: 1987, *Astrophys. J.* **312**, 170 - 177

40. Kang, H., Ryu, D., Jones, T.W.: 1996 *Astroph. J.* **456**, 422 - 427.

41. Kang, H., Rachen, J.P., Biermann, P.L.: 1997 *Monthly Not. Roy. Astron. Soc.* **286** 257 - 267

42. Karle, A. *et al.*: 1995 *Physics Letters B* **347**, 161 - 170

43. Kephardt, T.W., Weiler, T.J.: 1996 *Astroparticle Phys.* **4**, 271 - 279

44. Kewley, L.J., Clay, R.W., Dawson, B.R.: 1996 *Astroparticle Physics* **5**, 69 - 74

45. Kim, K.-T., *et al.*: 1989 *Nature* **341**, 720 - 723.

46. Krennrich, F. *et al.*:1997 *Astroph. J.* **481**, 758 - 763

47. Kronberg, P.P.: 1994 *Rep. Prog. Phys.* **57**, 325 - 382.

48. Kulsrud, R.M., Cen, R., Ostriker, J.P., Ryu, D.: 1996 *Astrophys. J.* **480**, 481 - 491 (astro-ph9607141)

49. Lovelace, R.V.E.: 1976 *Nature* **262**,

50. Mannheim, K., Krülls, W. M., Biermann, P.L.: 1991 *Astron. & Astroph.* **251**, 723 - 731.

51. Mannheim, K., Biermann, P.L.: 1992 *Astron. & Astroph.* **253**, L21 - L24.

52. Mannheim, K., Stanev, T., Biermann, P.L.: 1992 *Astron. & Astroph.* **260**, L1 - L3.

53. Mannheim, K.: 1993 *Astron. & Astroph.* **269**, 67 - 76.

54. Matthaeus, W.H., Zhou, Y.: 1989 *Phys.Fluids* **B 1**, 1929 - 1931.

55. Meisenheimer, K. *et al.*: 1993 *Astron. & Astrophys.* **219**, 63 - 86.

56. Mészáros, P., Rees, M.J.: 1997 *Astrophys. J.* **476**, 232 - 237

57. Metzger, M.R. *et al.*: 1997 *Nature* **387**, 878 - 880

58. Nagano, M.: 1997 Editor, "Extremely High Energy Cosmic Rays" Tokyo, University of Tokyo

59. Norman, C.A., Melrose, D.B., Achterberg, A.: 1995 *Astrophys. J.* **454**, 60 - 68

60. Peacock, J.A.: 1981 *Monthly Not. Roy. Astr. Soc.* **196**, 135 - 162

61. Petry, D. *et al.*: 1996 *Astron. & Astroph.* **311**, L13.

62. Petry, D. *et al.*: 1997, XXV ICRC'97, Durban, OG 4.3.1

63. Protheroe, R.J., Szabo, A.P.: 1992 *Phys. Rev. Letters* **69**, 2885 - 2888.

64. Protheroe, R.J., Johnson, P.A.: 1996b in "Proc. 4th International Workshop on Theoretical and Phenomenological Aspects of Underground Physics (TAUP95)", ed. M. Fratas, *Nucl. Phys. B., Proc. Suppl.* , (in press)

65. Protheroe, R.J., Stanev, T.: 1996 *Phys. Rev. Letters* (submitted, astro-ph9605036)

66. M. Punch, M. *et al.*: 1992 *Nature* **358**, 477.

67. Quinn, J. *et al.*: 1996 *Astroph. J.* **456**, L83.

68. Rachen, J.P., & Biermann, P.L.: 1993 *Astron. & Astroph.* **272**, 161 - 175.

69. Rachen, J.P., Stanev, T. & Biermann, P.L.: 1993 *Astron. & Astrophys.* **273**, 377 - 382.

70. Rachen, J.P.: 1996 *Ph.D. thesis*, University of Bonn

71. Rawlings, S., & Saunders, R. 1991, *Nature* **349**, 138 - 140.

72. Rhode,W. *et al.*: 1996 *Astroparticle Phys.* **4**, 217 - 225
73. Rieke, G.H., *et al.*: 1976 *Nature* **260**, 754 - 759.
74. Rieke, G.H., Lebofsky, M.J., Kinman, T.D.: 1979 *Astroph. J. Letters* **232**, L151 - L154.
75. Rieke, G.H., Lebofsky, M.J.: 1980 in *IAU Symposium No. 92, Objects of high redshift,* Eds. G.O. Abell, P.J.E. Peebles (Dordrecht: Reidel) , p. 263 - 268 ·
76. Rieke, G.H., Lebofsky, M.J., Wisniewski, W.Z.: 1982 *Astroph. J.* **263**, 73 - 78.
77. Ryu, D. & Kang, H., 1997a, MNRAS, 284, 416.
78. Ryu, D. & Kang, H., 1997b, In Proc. of The 18th Texas Symposium on Relativistic Astrophysics, ed. A. Olinto, J. Frieman & D. Schramm, in press (astro-ph/9702055).
79. Ryu, D., Ostriker, J. P., Kang, H. & Cen R., 1993, ApJ, 414, 1.
80. Shaver, P.A., Pierre, M.: 1989 *Astron. & Astroph.* **220**, 35 - 41.
81. Shaver, P.A.: 1991 *Austral. J. Phys.* **44**, 759 - 769
82. Shemi, A.: 1995 *Monthly Not. Roy. Astr. Soc.* **275**, 115 - 120.
83. Sigl, G., Schramm, D.N., Bhattacharjee, P.: 1994 *Astroparticle Physics* **2**, 401 - 414.
84. Sigl, G., *et al.*: 1995 *Science* **270**, 1977 - 1980.
85. Spinrad, H., *et al.*: 1985 *Publ. Astron. Soc. Pacific* **97**, 932 - 961.
86. Stanev, T., Biermann, P.L., & Gaisser, T.K.: 1993, *Astron. & Astrophys.* **274**, 902 - 908.
87. Stanev, T., *et al.*: 1995 *Phys. Rev. Letters* **75**, 3056 - 3059.
88. Stanev, T.: 1997 *Astrophys. J.* **479**, 290 - 295,
89. Stecker, F.: 1968 *Phys. Rev. Letters* **21**, 1016 - 1018.
90. Uchihori, Y. *et al.*: 1997, In Proc. "Extremely High Energy Cosmic Rays", Ed. M. Nagano, University of Tokyo meeting, p. 50 - 60
91. de Vaucouleurs, G.: 1956 *Vistas in Astronomy,* **2**, p. 1584 - 1606.
92. de Vaucouleurs, G.: 1975a *Astrophys. J.* **202**, 319 - 326.
93. de Vaucouleurs, G.: 1975b *Astrophys. J.* **202**, 610 - 615.
94. de Vaucouleurs, G., *et al.*: 1976 *Second Reference Catalogue of bright Galaxies*, Univ. of Texas Press
95. Waxman, E.: 1995a *Phys. Rev. Letters* **75**, 386 - 389.
96. Waxman, E.: 1995b *Astrophys. J. Letters* **452**, L1 - L4
97. Waxman, E., Fisher, K.B., Piran, T.: 1996 *Astrophys. J.* **483**, 1 - 7 (astro-ph/9604005)
98. Zatsepin G.T., Kuz'min V.A.: 1966, *JETPh Lett.* **4**, 78.

Acceleration of Cosmic Rays by Colliding Galaxies

Frank C. Jones

Laboratory For High Energy Astrophysics
Goddard Space Flight Center

Abstract. It has been suggested that colliding galaxy pairs could produce cosmic rays with energies $\geq 10^{19}$ electron volts. We investigate such a system to see if such energies are likely. We find that there is a typical scale energy associated with a large moving magnetic field structure that is the same whether it is in the form of a diffusive plasma shock or a more regular field structure with little or no irregularity. This scaling leads to the conclusion that such energies as suggested are unlikely to be produced by colliding galaxies.

The possibility that colliding galaxies could be the source of ultra high energy cosmic rays ($> 10^{20}$ eV) is of interest because some investigators in the field believe that they see correlations between the arrival directions of these cosmic rays and known positions of some colliding galaxy pairs. Most recently Al-Dargazelli *et al.* [1] have shown (Fig. 1) correlations between arrival directions for particles with $E > 10^{19}$ eV and their proximity to the galaxy pair VV338. This galaxy pair is only about 6 Mpc from us... close enough for such particles to reach us without serious energy loss from the GZK effect. It is, therefore, of interest to examine our notions of particle acceleration in the context of colliding galaxies to determine whether or not such processes are likely candidates for the sources of ultra high energy cosmic rays.

If the velocity of the galaxies moving through the intergalactic medium is greater than the local Alfvén speed, a bow shock will form in front and particles colliding with one or both galaxies will be accelerated by the process of diffusive shock acceleration. If the medium is sufficiently turbulent that the diffusion scale is smaller than the galaxy separation distance, single shock acceleration will be the proper picture. In Fig. 2 we see the geometry of a shock of finite extent. The question for us to ask is "will the particles have enough time to achieve an energy of the desired amount before it leaks away from the shock system". Since any real shock has a finite size, diffusion itself (which is necessary for diffusive shock acceleration to occur) will bring the particles to the edge of the system in a finite time. Consider

CP433, *Workshop on Observing Giant Air Showers from Space*
edited by J.F. Krizmanic et al.

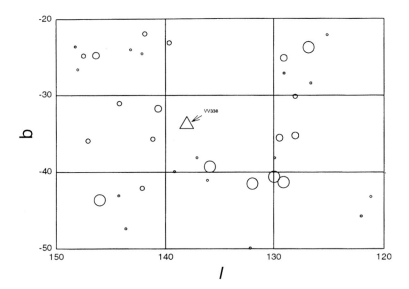

FIGURE 1. The region of the sky containing the highest concentration of particle arrival directions (for $E > 10^{19}$ eV). The world's data for individual events are shown and an indication (bubble size) is given of the particles energy (in four bands: 1-2, 2-3, 3-4, and $> 4 \times 10^{19}$ eV). The position of the colliding galaxy pair VV338 is indicated by a triangle. Data from Al-Dargazelli *et al.*. [1]

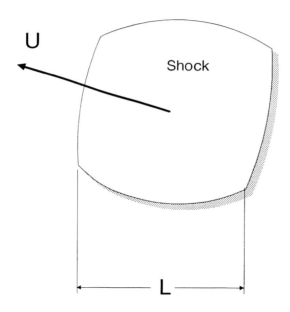

FIGURE 2. The geometry of a shock of finite lateral size

that the E-folding time for particle acceleration is given by κ_n/U^2 and the escape time to the edge is L^2/κ_p where κ_n and κ_p are the diffusion coefficients normal and parallel to the shock front, U is the shock speed, and L is the distance across the shock front as shown in Fig. 2.

These times will be equal and acceleration will end when

$$\kappa_n/U = L^2/\kappa_p$$
$$\kappa_n\kappa_p \approx \kappa_\perp\kappa_\parallel = L^2U^2 \tag{1}$$

But as was shown by Eichler [2] $\kappa_\perp\kappa_\parallel = \kappa_\parallel^2 r_g^2/\lambda^2 \approx v^2 r_g^2$ and the gyroradius, $r_g = \gamma m c v/ZeB$, where B is the magnetic field, so $\gamma mc^2 = E_{MAX} = ZeU/cBL$

Putting in the numbers gives

$$E_{MAX} = L_{100} \cdot U_{300} \cdot Z \cdot B_3 \times 10^{17} \text{ eV} \tag{2}$$

where

$$L = L_{100} \times 100 \text{ kpc}$$
$$U = U_{300} \times 300 \text{ km/sec}$$
$$B = B_3 \times 3 \text{ } \mu\text{g} \tag{3}$$

Since the values chosen in Equation 3 are nominal we see that the most we would expect from this process is about 10^{17} eV unless one pushes the constants in Equation 2 to the limits. One should note that one possibility is to consider relativistic

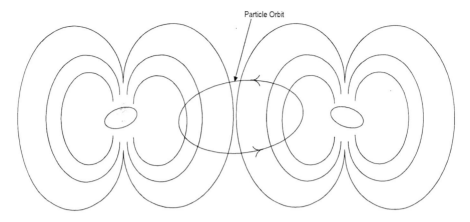

FIGURE 3. Particles interacting directly with the fields of colliding galaxies.

velocities ($U \approx c$) such as are expected in jets, but here the size is smaller so it is not clear that one gains.

We can, on the other hand, consider the case where turbulence is assumed low and diffusion scales are much larger than the separation distance of the galaxies. Here the shocks would play a much smaller role and direct interaction between the particles and the galactic magnetic fields would be important (Fig. 3). In the Calgary ICRC Cesarsky and Ptuskin [3] proposed just such a "low entropy model".

In this picture particles go back and forth between the magnetic fields of the individual galaxies and are accelerated by the electric fields induced by the moving magnetic fields, $E = BU/c$.

We have, therefore, the rate of momentum gain given by $dp/dt = ZeE = ZeBU/c$. According to Cesarsky and Ptuskin the particle will stay in the system until it drifts across the size of the galactic field so the acceleration time is limited to $T = L/v_d$ where the drift velocity due to the curvature and gradients of the magnetic field is approximately given by $v_d \approx cr_g/L$. Thus $p_{MAX} = ZeBL^2U/c^2r_g$ Inserting the expression given above for r_g this may be solved to obtain

$$E_{MAX} = ZeBL \left(\frac{U}{c}\right)^{1/2}$$
$$= Z \cdot B_3 \cdot L_{100} \cdot U_{300}^{1/2} \times (3 \times 10^{18}) \tag{4}$$

It would appear that, due to the fact that the term U/c appears under the square root, we have made a considerable gain in efficiency. It is true that the coherent type of acceleration described here is faster than the shock acceleration process which must rely on scattering of the particles to bring them back to the shock for further acceleration. This would seem to indicate that smooth, low turbulence, structures are better accelerators; they are if they are carefully designed. However, in nature, magnetic fields are rarely such good trappers of charged particles especially if they are dynamic. And, if the configuration *is* a trapping configuration, the question arises as to how the particles go there in the first place. In fact such configurations will usually squirt out particles at about the particle speed v.

If the fields of the two approaching galaxies are aligned the geometry is a mirror and particles will reflect out of the system with $v_d \approx c$. If, on the other hand, the fields are opposed there is a neutral line or sheet and the scale of variation of the field is r_g rather than L and once again $v_d \approx c r_g / r_g = c$.

Employing this value yields:

$$p_{MAX} = \left(ZeB \frac{U}{c} \right) \left(\frac{L}{c} \right)$$

(5)

and

$$E_{MAX} = ZeBL \frac{U}{c}$$

(6)

Which is just the expression we obtained earlier for a diffusive shock.

We are thus back to the shock acceleration picture (or equivalent). It has been pointed out many times (*e.g.* Jokipii [4]) that for a perpendicular shock of extent L the quantity BLU/c is just the electric potential drop across the shock and is therefore the available energy for a charged particle interacting with the shock. However, we have seen that this parameter sets the energy scale for a parallel shock as well when there is no actual electric field. We see that this parameter is, very generally, the available energy of a large , moving magnetic field, and it does not matter whether the particle interacts with it in a diffusive or non diffusive setting. Thus the scale of the energetics of a typical moving galaxy is of the order of 10^{17} eV and it is difficult to beat this by three orders of magnitude without excessive tweaking of all of the parameters involved. It, therefore, appears unlikely that colliding galaxies are the source of $\geq 10^{20}$ eV particles in the cosmic rays.

REFERENCES

1. Al-Dargazelli, S. S.,Lipski, M., Smialkowski, A., Wdowdczyk, J., and Wolfendale, A. W., 1997, 25th *ICRC* (Durban), **4**, 465
2. Eichler, D.,, 1981, *Astrophys. J.*,**244**, 711
3. Cesarsky, C. and Ptuskin, V., 1993, 23rd *ICRC* (Calgary), **2**, 341
4. Jokipii, J. R., 1987, *Astrophys. J.*, **313**, 842

Gamma Ray Bursts and Extreme Energy Cosmic Rays

Livio Scarsi

Istituto di Fisica Cosmica e Informatica - CNR - Palermo
and
Dipartimento di Energetica e Fisica Applicata.
University of Palermo

Via Ugo La Malfa, 153 - 90146 Palermo - Italy

Abstract. Extreme Energy Cosmic Ray particles (EECR) with $E > 10^{20}$ eV arriving on Earth with very low flux (~ 1 particle/Km^2 - 1000yr) require for their investigation very large detecting areas, exceeding values of 1000 km^2 sr. Projects with these dimensions are now being proposed: Ground Arrays ("Auger" with 2x3500 km^{2}sr) or exploiting the Earth Atmosphere as seen from space ("AIR WATCH" and OWL", with effective area reaching 1 million km^{2}sr). In this last case, by using as a target the 10^{13} tons of air viewed, also the high energy neutrino flux can be investigated conveniently. Gamma Rays Bursts are suggested as a possible source for EECR and the associated High Energy neutrino flux.

INTRODUCTION

Cosmic Ray particles, arriving on Earth from Outer Space with energy $> 10^{20}$ eV have been detected by ground arrays; they are referred to as Extreme Energy Cosmic Rays (EECR).

Open problems arise:

- A new component is appearing at $E \geq 5 \times 10^{18}$ eV ("ankle" feature in the energy spectrum)?
- There exists an E_{max} for Cosmic Rays?
- There exists a violation of the GZK cut-off for sources at Cosmological distance?
- What is the origin for the EECR?

For what the EECR origin is concenerd, two main classes of models have generally been considered:

- **Botton Up Processes,** based on acceleration mechanisms from lower energy injection sources. Several panoramas are proposed and some of them are illustrated at this Workshop: acceleration by Rotating AGN, in Halo of AGN Jets, by colliding Galaxies,...
- **Top Down Processes** : cascading and energy loss of " new physics" particles such as decay and annihilation of topological defects like Cosmic Strings, Monopoles,...

CP433, *Workshop on Observing Giant Air Showers from Space*
edited by J.F. Krizmanic et al.

- We want here to bring attention to the acceleration of particles (protons) in the relativistic shocks generating the GRB: **Gamma Ray Burst EECR Origin**.
Although, in a sense, this could be considered as a chapter of the Botton-Up class, its peculiarity suggests a class by itself. Its growing interest is related to the recent observational evidence which suggests an extragalactic origin of the GRBs:

GAMMA RAY BURSTS

Since the discovery GRBs have represented a challenge for the identification of their origin; after the initial surprise in the "Vela" project that they were not man made representing the gamma-ray glow of nuclear explosions occurring above the atmosphere, the intriguing mistery covering their nature has remained undisclosed for several years, up to our days. Systematic observations, mainly due to BATSE on board of CGRO, have outlined a picture characterized by a rate of occurence of about 1/day, an apparent high degree of isotropy for the arrival direction from the sky, but with a marked deviation from uniformity in the log N/log P distribution (P identifies the "fluence" which defines the energy content in the Burst).

The fluence distribution , as given by BATSE, starts from a lower thereshold at $\sim 10^{-6}$ erg going to upper values around 10^{-3} erg at the other extreme; GRBs show very structured light curves, with duration ranging from milliseconds to minutes; the energy range observed spans generally from 20-30 keV to the MeVs, occasionally stretching well inside the GeV region.

These features could be explained equally well, or, let's say with equal number of difficulties, by **Galactic** (halo) or **Extragalactic** (cosmic) models, the energy release being in this second case 10^{51-52} erg, and possibly more.

The main difficulty limiting the "model" choice was deriving from the lack of identification of the object releasing the GRB with a counterpart known at other wavelengths: the unsurmontable obtacle derived from the indetermination in the angular direction given by the Burst monitors (e.g. by BATSE with an error of the order of the degree), too large to pilot the pointing of x-ray, optical or radio telescopes.

With the year 1997, a break through has arrived with the observations of Beppo-SAX, which has produced initial source localization at 3'-5' precision within few hours from the GRB occurence, allowing the follow-on in the X-ray, optical, IR and Radio ranges (1), and the subsequent discovery of the afterglow in the various energy ranges, as predicted by theoretical models (2).

Beppo-SAX Observational Strategy

The Beppo-Sax scientific package is constituted by a set of Narrow Field Instruments (NFI) covering the energy range (0.1-300) KeV, with angular resolution of 1' in the interval (0.1-10) KeV, coaligned along the Z axis of the Satellite and two Wide Field Cameras (WFC) with FOV 40°x40°, energy range (5-30) KeV and angular resolution 3') with axis at 90° to Z and oriented in opposite directions (-Y and +Y).
The WFCs cover at any time $\sim 2.5\%$ of the sky.

A Gamma Ray Burst Monitor (GRBM) with sensitivity (30-600) KeV and 4π FOV is obtained by using the side anticoincidence of the Phoswich Detector System (PDS) of Beppo-SAX (Fig.1).

Beppo-SAX Observational Strategy for GRB

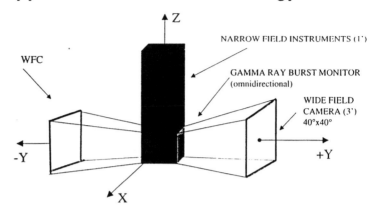

Fig.1

WFCs cover at any time 2.5% of the sky.

Detect ~ 1 GRB/(1-2) months.

Given the GRB rate of (350-400)/year registered by BATSE for the omnidirectional flux, the Beppo-SAX WFCs are expected to detect (8-10) events/year.
Following the alert by the onboard Beppo-SAX GRBM (or by an external trigger as e.g. from Batse) for an event in the FOV of one of the WFCs, the WFC interested is scanned for the Burst image and the direction determined with the precision of ~3 (Fig.2).

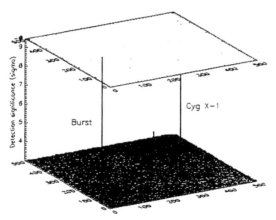

Fig.2 - GRB 960720 as seen in the WFC

As soon as available, the information is trasmitted to Observatories world wide and telecomands relay to Beppo-SAX for pointing of the Narrow Field Instrument package to the GRB position to catch the GRB after glow.

Fig.3 shows the typical Time-sequence to catch a GRB After Glow.

Beppo-SAX Time Sequence to catch a GRB After-glow

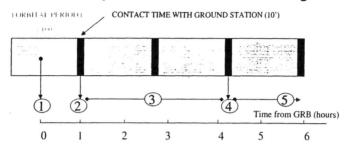

- **❶** GRBM detects a GRB in the FOV of WFCs
- **❷** Data dumped on ground station and relayed to OCC in Rome
- **❸** Data analysis. GRB position determined in WFC within 3'. Info transmitted to Observatories world wide
- **❹** TC relayed to B-SAX for NFI pointing to GRB
- **❺** Starts GRB After-glow observation by B-SAX NFI (1') (time elapsed from GRB onset ~ 5 hours)

Of the GRB events for which the source has been localized and the after-glow followed in the different energy bands, two have been of particular interest for the information provided: that occurred on February 28, 1997 and that occurred on May 9, 1997 identified respectively as 970228 and 970509

Fig.4,5,6 refer to 970228; Fig 7,8 to 970509.

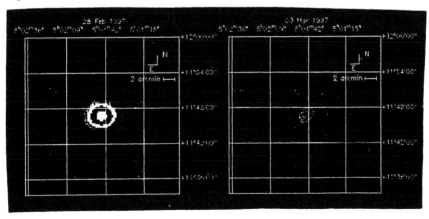

Fig. 4 - GRB 970228. NFI (1-10keV) at $(t_0 + 10h)$ and $(t_0 + 3$ days$)$

45

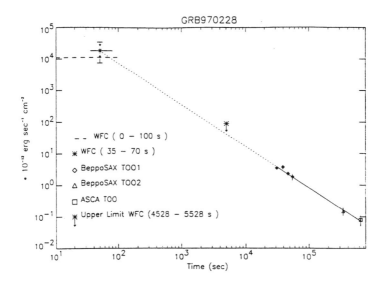

Fig. 5 X-RAY AFTER GLOW (1-10)KeV LUMINOSITY
vs time after t_0.

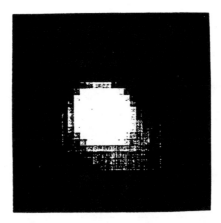

Fig. 6 HST image of the GRB 970228 TRANSIENT REGION. A close-up of the
Optical Transient shows both a point-like source (the bright emission) plus the
extended emission (below and to the right) from what may be the distant host-
galaxy.

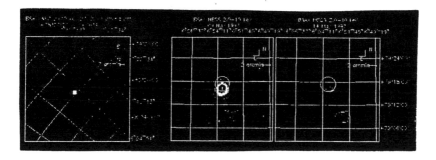

Fig. 7 GRB 970509. X-RAY AFTER GLOW (2-10)KeV LUMINOSITY
at (to + (27-200))s; (t$_0$ +6h); (t$_0$ +5d).

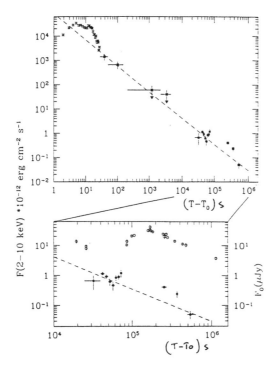

Fig. 8 GRB 970509. X-RAY AFTER GLOW LUMINOSITY as a function of time after
t$_0$. The open circles represent optical fluxes.

The main observational results:

- The After Glow luminosity in X-rays (2-10) KeV shows an increase in the first seconds after t_0 and then a decrease with an average law $\alpha\, t \sim^{-1}$, with occasional flarings observed during the few days (up to 5-6) for which the source remains above the threshold of visibility for the Beppo-SAX detectors.

- The optical light curve shows a somewhat longer lasting initial increase and then a decay with a slope similar to X-rays; flarings occur in coincidence with the x-ray curve.

- Correspondingly the radio light curve shows the same characteristics; flarings initially evident, disappearing with the evolution of the after-glow.

- Altogether, the energy release during the first 5-6 days of the afterglow compares at the same level with that emitted during the Gamma ray burst itself.

- HST observation of persistent diffuse emission around the GRB point-like counterpart in 970228 is interpreted as evidence of a possible host galaxy. For the same GRB, HST has found no evidence of proper motion on the optical counterpart for observations made 5 month apart.

- For 970509, absorption lines corresponding to $Z \sim 0.8$ have been observed in the optical counterpart, putting the source at $Z > 0.8$.

From what above, the following main conclusions can be inferred:

- GRB sources are located at Cosmological distance (more than a Gigaparsec), with a luminosity of 10^{50-52} ergs.

- An indication can be drawn supporting the Fireball model for GRBs

- The hyperrelativistics shocks responsible for the GRBs can accelerate protons to the highest energies. Support is given to the indication of GRBs as sources of Ultra High Energy Cosmic Rays and UHE neutrinos ($E_V \geq 10^{19}$ eV).

GRB Fireball Model (2)

GRBs are generated when two or more hyperrelativistic shells, issued by a source (e.g. collision of two degenerated objects like Neutron-Stars or Black-Holes) collide with each other. A relativistic shock forms, where non-thermal electrons are accelerated and then dissipate through Syncrotron (and possibly Inverse Compton) processes.
After the collision, the resulting shell will collide with the Interstellar medium forming a second relativistic shock.
Both shocks can contribute to the GRBs. An "After-glow" in Radio, Optical and X-Ray ranges is quantitatively predicted by the theoretical model.

GRB as source of Extreme Energy Cosmic Rays (3)

The hyperrelativistics shocks responsible for the GRBs are ideal for accelerating protons to the highest energies (> 10^{20} eV).
Usual diffusion approximation breacks down in relativistic shocks: the Cosmic Ray momentum distribution function is strongly peaked toward the direction perpendicular to the shock and the scattering is nearly forward/backward, so that particles to be accelerated do not have to diffuse slowly across the shock.
At every cycle the particle is not incremented infinitesimally like in non-relativistic shocks, but is multiplied by a factor γ^2 (γ= shock Lorentz factor).
For GRBs $\gamma = 10^2-10^3$ and the acceleration is very fast despite the short timescale of the Burst.

GRB origin for the Ultra High Energy Cosmic Rays could explain.

- The total energy striking Earth observed as UHECR (E>3×10^{18} eV) accounting to ~5×10^{-13} erg s^{-1} cm^{-2} sr^{-1}, to be compared with ~4×10^{-13} erg s^{-1} cm^{-2} sr^{-1} obtained by considering GRBs within 100 MPc (GZK limit) occurring at a rate of 30 yr^{-1} GPc^{-3}, each releasing 10^{51} erg in equipartition between UHECR and Gammas at low energy;

- Incidence from directions roughly isotropic from the sky.

The direction of arrival of UHECRs is expected to differ $\leq 1°$ from that of the progenitor GRB, but the delay in arrival time could be as high as 10^2-10^3 yr, washing away any possible "time" correlation.

The maximum energy to which protons can be accelerated at shocks with the ISM ic:

$$\varepsilon_{max} = 10^{20}eV \times \Theta^{-5/3} \times \eta^{1/3} \times E_{52}^{1/3} \times n_1^{1/6}$$

where : explosion energy E=$E_{52} \times 10^{52}$ erg

η= expansion Lorentz factor

Θ= beaming angle

n= ISM density= n_1 cm^{-3}

Popular values inferred from After-glow are Θ=1/3; E_{52}=1 (4), implyying

ε_{max} ~5×10^{20} eV

ULTRA HIGH ENERGY NEUTRINOS FROM GRB

- (M.Vietri - Phys.Rev. Letters- submitted November 1997)

The EE protons acceleratd to high energies in the relativistics shocks that generate GRBs can interact with the syncrotron photons surrounding the Fireball and photoproduce pions. The decay of charg ed pions then produces electrons and muon neutrinos.
The UHENs (E>10^{19} eV) can be detectable by AIR-WATCH experiments in the correlation with GRBs.

The p-p collision mechanism yield much lower energy neutrinos (~30GeV) not detectable in the same experiments.

Expected Fluxes of Ultra High Energy Neutrinos

Converting the total probability for energy loss through photopion production in the internal shock model for GRB given in (5), to the external shock scenario:
(proton of energy Ep immersed in a radiation field with turnover frequency ε_γ ~300 KeV)

$$f_\pi = 0.02 \left(\frac{L_\gamma}{4 \times 10^{51} \text{erg s}^{-1}} \right) (300 \text{KeV}/\varepsilon_\gamma) \times (100/\eta)^{7/3} \times \begin{cases} 1 \text{ if } \varepsilon_p > \varepsilon_b \\ \varepsilon_p/\varepsilon_b \text{ otherwise} \end{cases}$$

This applies for average energy of GRBs in the γ band of 4×10^{51} erg and median Burst duration ~1 s ($\varepsilon_b = 4 \times 10^{15}$ eV)

Experiments of the AIR WATCH class (target mass ~10^{13} tons of air; acceptance area ~10^6 km^2 sr) have appreciable detection efficiency for Neutrinos exceeding a threshold energy ε_ν ~10^{19} eV.
To estimate the flux of ν with $\varepsilon=10^{19}$ eV emitted by a GRB, we have to compute the energy release in protons with energies exceeding $\varepsilon_1=2 \times 10^{20}$ eV.

Since νs emitted through photopion processes typically carry away a fraction q~0.05 of the proton energy, defining the total energy released by a GRB in UHECR ($\varepsilon > 10^{19}$ eV) as E_{UHECR}, considering as ε^{-2} the proton energy spectrum, the energy E_ν released in ν becomes:

$$E_\nu = f_\pi^{(o)} \frac{E_{UHECR}}{\ln \varepsilon_{max}/\varepsilon_1} \ln \varepsilon_{max}/\varepsilon_1$$

where: $f_\pi^{(0)}$ is a numerical coefficient assumed $=0.02$ as a first approximation; ε_1 assumed$= 10^{19}$ eV.

The total, UHENs neutrino flux $\dot{n}_V$ is obtained integrating over the distances:

$$\dot{n}_V = \dot{n}_{GRB} \times (E_V/\varepsilon_V) \times (c/H_0) = f_\pi^{(0)} \times (1/\ln \varepsilon_{max}/\varepsilon_1) \times (c/H_0) \times \dot{n}_{GRB} \times E_{UHECR}$$

where: $\varepsilon_V = \varepsilon_{V,1} \times \ln \varepsilon_{max}/\varepsilon_1$ is the average neutrino energy.

The key factor is given by $E = n_{GRB} \times E_{UHECR}$ representing the injection rate per unit volume of non-thermal proton energy.
Assuming that GRBs emit about as much energy in the gamma ray band than in UHECR, the flux of UHECR at Earth is roughly explained.

Waxman (6) finds: $E = 4.5 \times 10^{44}$ erg yr^{-1} MPc^{-3}
                              for $10^{19} < \varepsilon < 10^{21}$ eV

Taking the minimum neutrino "observable energy" $\varepsilon_1 = 10^{19}$ eV

$$\dot{n}_V = 1.5 \times 10^{-10} \times (f_\pi^{(0)}/0.02) \times (50 \text{km s}^{-1} \text{ MPc}^{-1}/H_0) \text{ yr}^{-1} \text{ cm}^{-2} \sim 1.5 \times 10^{-10} \text{ yr}^{-1} \text{ cm}^{-2}$$

(a) This flux account for that produced during the GRB itself

(b) Additional flux is produced during the "after glow" (at least 6 days or so)

We can call (a)+(b): the neutrino flux produced "in situ".
Because of the "after glow" intervention, only an upper limit can be obtained for the v mass.

UHECR deriving by GRB at distances $>d$ compatible with the GZK effect, will produce v in flight rather than "in situ" with typical mean free path of order ~ 10MPc, given rise to a background flux UHENs ($\dot{n}_V^{(bg)}$) uncorrelated with observable GRBs.

$$n_V^{(bg)} = 7.7 \times 10^{-9} \text{ yr}^{-1} \text{ cm}^{-2}$$

Visibility of GRBs by an AIR WATCH experiment - ($\varepsilon A\Omega$(total) $\geq 10^6$ km^2)

Bursts with fluence $\geq$ few $\times 10^{-4}$ erg/cm^2

AIR WATCH detectability of UHENs from GRBs

Assumptions:

- energy of GRBs in equipartition between γ and UHECR.

- ~0.05 of proton energy in neutrinos from photopion processes.

- N_b of ν with $E\sim 10^{19}$ Ev => N_b of protons with $E\sim 2\times 10^{20}$ eV

in situ:

Assuming for $E_n \geq 10^{19}$ eV:

$$\sigma = (10^{-32}\text{-}10^{-31})\ cm^2; \quad X = 10^3\ gr/cm^2; \quad \text{interaction probability} \sim 10^{-5}$$

$$A\varepsilon\Omega = 10^6\ km^2\text{-sr}$$

$$\dot{N}_\nu \sim 15\ yr^{-1}\ \text{(delay of }\nu\text{s from GRB: 0 to 6 days)}$$

Bkd flux

$$\dot{N}_\nu{}^{(bkd)} \simeq 7.50\ yr^{-1}$$

If the combined directional accuracy of ν and GRB is of the order of few degrees, the rate of casual association is negligible.

ACKNOWLEDGEMENTS

I wish to acknowledge suggestions and contributions from several AIR WATCH and Beppo-SAX people and in particular: C. Butler;, O. Catalano, J.Linsley, L. Piro, B. Sacco, Y. Takahashi, M. Vietri.

REFERENCES

1 Costa E. et al., Nature, 387,783, (1997)
   Van Paradijs et al., Nature, 386, 686, (1997)
   Frail D.A. et al., Nature, 389, 261, (1997)

   Several Authors. IAU Circulars, (1997).

   Proceedings of the "Symposium on the Active X-Ray Sky. Results from
   Beppo-SAX and RXTE Satellites". - Rome 21-24 oct. 1997.To be published.

2- Meszaros P. and Rees M.J.,   Ap. J. 405, 278, (1993)
                                 MNRAS, 269, 41p, (1994)

   Rees M.J. Proceedings of the "Symposium on the Active X-Ray Sky.
   Results from Beppo-SAX and RXTE Satellites" - Rome 21-24 oct. 1997.To be
   published.

3 Vietri M. Ap.J., 453, 883 (1995)

4 Waxman E. et al. Astroph. n. 9709, 199 (1997)

5 Waxman E. and Bahcall J., Phys. Rev. Letters, 78, 2292, (1997)

6 Waxman E., Astroph J. Letters, 452, 1, (1995).

Early Air Fluorescence Work
Cornell and Japan

Goro Tanahashi

Miyako Prefectural College of Iwate, 027 Japan

Abstract. In the early days before the Fly's Eye project start in 1974, K.Greisen began the pioneering work to detect air fluorescence light from the very hight energy cosmic ray showers. 5 years late after this, G.Tanahashi started the same work in Tokyo. Both works had 2~3 versions. These works are summarized.

INTRODUCTION

Extensive air shower study by means of air fluorescence is producing many interesting results and is opening a new research window complementally with the surface array experiments. The experience of Fly's Eye showed the importance of the fluorescence observation, and in this workshop the related ambitious projects are to be discussed. On the occasion of this workshop, the summary of early works before the Fly's Eye will be given, and this summary starts with the chronology of these works.

1958	Norikura	air shower symposium (1)
1962	Bolivia	cosmic ray seminar in La Paz (2)
	K.Greisen	begins air fluorescence project with wide angle system(Cornell-W)
	S.Ozaki	joins in Cornell
1963	A.Bunner	experiment of air fluorescence efficiency (3)
1964	Cornell-W	#1 station starts the run
	G.Tanahashi	joins in Cornell instead of Ozaki(till 1966)
1965	Cornell-W	#2 station starts the run
	Greisen	proposes new imaging system(4)(Cornell-I)
1966	Cornell-I	begins
	Cornell-W	#3 station starts the run

CP433, *Workshop on Observing Giant Air Showers from Space*
edited by J.F. Krizmanic et al.

1967	Cornell-I	Further progress reported (5), starts the partial run
1968	Tokyo	begins new system (Tokyo-1)
	Tokyo-1	starts the run on Dodaira
1969	Tokyo-1	detects the fluorescence light from EAS (10)
1970	Tokyo-2	wide view system begins
1971	Tokyo-2	starts the run on Dodaira
1972	Tokyo-3	large collection area system begins
1973	Tokyo-3	starts the run on Izu, setback
1974	Fly's Eye	proposal presented

BEGINNING IN CORNELL

In the two meeting held before Cornell work, the possibility to detect the fluorescence was discussed by K.Suga, M.Oda and the others in Norikura (1), and by K.Suga and A.Chudakov in La Paz (2). K.Greisen is the first physicist who understand the potentiality of the air fluorescence light in the cosmic ray study and put this into practice. In 1962 he started his work in Cornell. A.Bunner of Cornell University measured the fluorescence efficiency and spectrum of nitrogen molecule in air produced by deuteron beam (3). This data has been the basis in designing the apparatus until recently. This result encouraged Greisen to go his way and he arranged the first detector together with S.Ozaki, Bunner and E.Jenkins. The detector is shown in Figure 1 (3). The first #1 station was set on the hill of Mt.Pleasant near Ithaca in 1964, and started to watch the night sky for the first time. G.Tanahashi joined in the group in succession of Ozaki after this start. In 1965 they installed #2 station on Brooktondale.

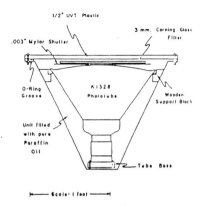

FIGURE 1. Cross-section of a Phototube Unit, from A.Bunner(3)

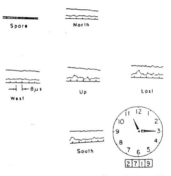

FIGURE 2. Sketch of one event as recorded on film, from A.Bunner(3)

In 1966 they installed #3 station on Richford, and the 3 stations made triangle with the distances of 11,16 and 12km. Each station consists of 5 15inch PMT(PhotoMultiplier Tube)s directed to vertical, and to east, west, north and south with 30degrees elevation. The observed pulse shapes of each 15PMT of 3 stations must be consistent with the isotropic radiative light source moving with the light velocity along the straight line. Almost all pictures on CRT are the Cerenkov light signals from local or small showers or the variuos kinds of night sky noise. After the hard task to select the fluorescence-like event from the flood of noise, 10 events of the final selection were examined carefully, but none of them could succeed to be consistent with the above requirement and all of them were discarded (11). Figure 2 represents one of such events. This result showed us the the limits of the performance of this type of detectors.

NEW IMAGING SYSTEM FOR EAS IN CORNELL

In 1965 Greisen emphasized the importance of observing the air fluorescence of EAS in his talk titled "Highlight in air shower studies, 1965" in London conference, and proposed a new system to devide the celestial sphere into about 500 mosaic segments and to record the light intensity and its arrival time of each segment (4). This design principle has been succeeded in almost all similar apparatus after this time. Cornell started to design and to arrange the work at the end of this year. In 1966 the devices were prepared and the construction of 25 sided observatory was completed (Figure 3). In 1967 Greisen presented the progress of Cornell work(5)(12). Cornell started the partial run this year and continued the fluorescence search since, but the severe climate of Ithaca in winter time and the low statistical chance to have very high energy showers prevented them to catch good events.

56

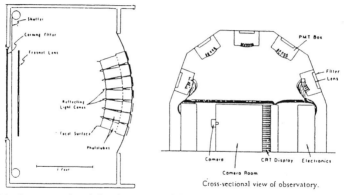

Cross-sectional view of observatory.

FIGURE 3. Detector unit and the observatory of Cornell imaging system, from A.Bunner et al (5)

ANOTHER APPROACH IN JAPAN

At the time around 1965, the study of EAS was limited within the data obtained from the ground array plane. At the same time it was also pointed out that the shower size fluctuation gives an enhanced modification on the relation between shower size and the other observables on the array plane (6)(7)(8)(9). It was thought that only the direct measurement of the developing process of EAS could overcome this problem, and the shower curve measurement by means of the fluorescence was desired. The calculation shows the light pulse shape (Cerenkov and fluorescence) observed from the distance more than 3km is Cerenkov light free, ans so, the development of EAS of around 3km has the highest probability to be observed as the fluorescence profile. Another new observing system was examined from this point of view. The signal to noise ratio (s/n) of the fluorescence is proportional to the root of integration time T.(4) For smaller value of T, s/n becomes worse, however, the differential recording of the light intensity within each PMT view has the following merit. The essential part of the shower development is contained in its early part where the energy partition does not yet progress. In order to observe this part, for example, to observe the starting depth of the virtical shower of 3km distance with the accuracy of $(10\pm5)gcm^{-2}$, we need the angular resolution of $(0.096\pm0.013)rad$ which is almost impossible to attain. If we improve the time resolution, we can attain the same thing with $(0.48\pm0.15)\mu s$ resolution. This is the reason to set $T=0.1\mu s$ at the sacrifice of s/n in Tokyo-1.

In the Table we show the parameter of 5 systems which appears in the text and the some corresponding EAS data, the maximum distance which the fluorescence of 10^{18}-$10^{19}ev$ EAS can reach the station with s/n>5, and the expected observing rate etc.

TABLE Paremeters of 5 systems and the corresponding EAS data

	Cornell-W	Cornell-I	Tokyo-1	Tokyo-2	Tokyo-3
Lens diameter(m):D	0.37	0.45	1.6	0.7	4.0
Lens area(m^2):A	0.11	0.16	2.0	0.38	12.5
Focal iength(m):F	--	0.45	2.0	0.8	4.5
Angular resolution(rad):W	1	0.1	0.08	0.21	0.03
Integration time(us):T	1.0	1.0	0.1	0.05	0.1
Number of PMT [full]	5	500	55	114	216
Number of PMT [run]	5		24	42	42
Field of view (sr) [full]	6	6	0.31	4	0.22
Field of view (sr) [run]	6		0.13	1.5	0.07
Rmin(km) (s/n>5)10^18ev	0.91	3.16	3.3	1.1	8.5
Rmin(km) (s/n>5)10^19ev	2.9	10	10	3.3	27
Rate/100hrs (s/n>5)10^18ev	6		1.6	2	6
Rate/100hrs (s/n>5)10^19ev	0.19		0.05	0.06	0.18
s /n ratio(EAS ;10^18ev, 3km)	0.46	5.6	6.0	0.71	40

TOKYO-1(PRELIMINARY SYSTEM)

The work of Tokyo-1 started in 1968 on the above consideration. The design parameters are shown in Table. The Fresnel lens was made in the machine shop of Institute for Nuclear Study(INS), University of Tokyo. The stuff is PMMA organic glass plate (MITSUBISHI#001) and the spectral transmission is shown in Figure 4. The diameter

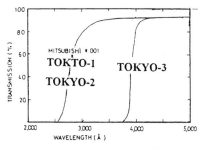

FIGURE 4. Spectral transmission of plastic lens for Tokyo-1,Tokyo-2 and Tokyo-3

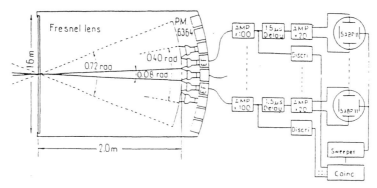

FIGURE 5. Sckematic sketch of telescope and the circuit diagram of Tokyo-1

and the pitch of the Fresnel lens are 1.6m and 16mm respectively. The block diagram of the system is shown in Figure 5. The telescope was set with the elevation of 30 degrees in the area of Dodaira Observatory(876m), University of Tokyo(10). The observation started in December 1968 until April 1969 with the trigger condition of 4 fold coincidence among 7 times of n.s.noise level of 27 PMTs output. CRTs recorded about 6000 events during 90 hours, and almost all of them were due to the Cerenkov light from nearby showers. 15 events were selected with the condition that the angular velocity of signals on each PMT is less than 0.08rad/0.15μs which nearly corresponds to the distant(>0.6km) shower. The angular velocities of these showers are shown in Figure 6.

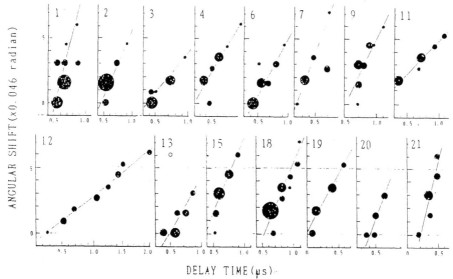

FIGURE 6. Angular velocities of selected 15 events in Tokyo-1 experiment

59

FIRST DETECTION OF
AIR FLUORESCENCE LIGHT FROM EAS

Event #12 in Figure 6 has two distinct features from the others. The first is that the angular velocity of the light signal is slow, and the second is that the light intensities from each PMT are almost the same. The CRT picture of #12 is shown in Figure 7. As for the first, the angular velocity is about 0.08rad/0.45µs which corresponds to the shower distance of 1.7km if the shower fell at right angle with the line of sight, and the atmospheric depth of this cross point is deeper than 800gcm^-2. As for the second, it is very probable that we observed the shower at around the shower maximum because we don't see any remarkable attenuation of light intensity over the field of view of about 100gcm^-2. In Figure 8 several posibilities of different injection angles Q are illustrated where Q is the angle between line of sight and the shower axis. In 1964 the depth of the shower maximum of 10^18~10^19ev was thought to be about 600~700gcm^-2, so we estimated the features of this shower as around Q=40, R=3km, E=10^19ev(10). At present time the shower maximum depth obtained by Fly's Eye is about 680~760gcm^-2 for 10^18ev~10^19ev, and so this event can also be allowd to be Q=70, R=2.2km, E=5x10^18ev EAS. These new values can increase the probability a little bit to catch such a high energy EAS like #12.

Anyway, it is natural to interpret this light as the air fluorescence, and this is the first observation of the air fluorescence from EAS.

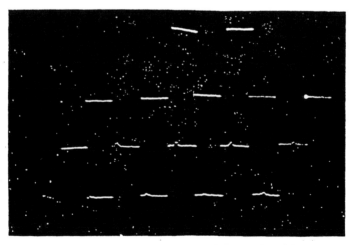

FIGURE 7. CRT picture of event #12. The full sweep length is 3.6µs

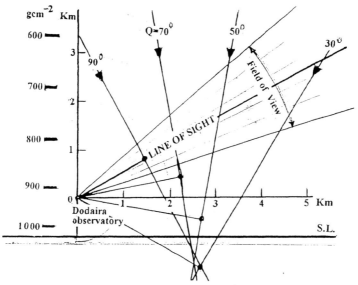

FIGURE 8. Reconstruction of shower axis of the event #12.
The injection angle between 40~70 degree is the most probable

AFTER TOKYO-1

After the detection of the air fluorescence of Tokyo-1, two kinds of new system were planned with the idea that it is better to begin more efficient experiment for the proper objective than to continue the previous observation. One is the enlargement of the field of view(Tokyo-2) to see the very high energy shower, and the other is the enlargement of the area of light collection (Tokyo-3) to see the detailed shower structure.

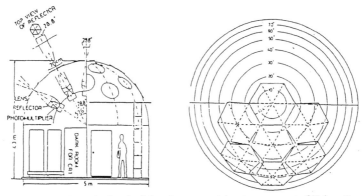

FIGURE 9. Schematic sktech of the telescope and the view segment of Tokyo-2

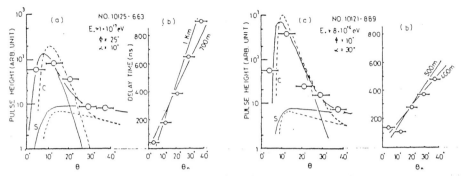

FIGURE 10. Two examples of Cerenkov light profile and its angular velocity of Tokyo-2 experiment from T,Hara (13)

The design of Tokyo-2 becomes necessarily similar to Cornell-I. However, as the number of available PMTs is limited from the economical reason, Tokyo-2 adopted the fast time resolution together with the convenient angular resolution as mentioned before. That is to say , the field of view of each PMT is not the hexagonal but the triangle shape, and this improves the accuracy of the light trajectory determination by recording the passage time duration of light(pulse width) in each PMT. Tokyo-2 parameters are in the Table and the structure is shown in Figure 9. The station was set on Dodaira in 1970 and had 82hrs run. Among 13000 events 7 showers whose light signals spread on the straight line over more than 25 degree view were selected. As shown in Figure 10 these

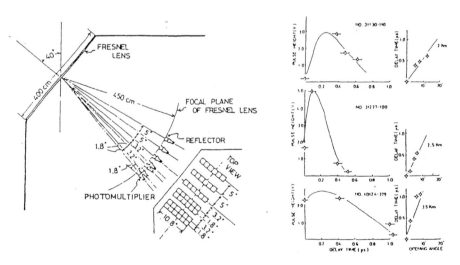

FIGURE 11. Apparatus and the view field of PMTs of Tokyo-3 experiment are shown on the left.

3 Cerenkov light profiles are also shown on the right, from T.Hara (13)

62

events have the Cerenkov peak pulse and have the distances near 1km. The observed light intensities at the tail of the Cerenkov image are more than calculated Cerenkov intensity. The excess is consistent with the calculated fluorescence intensities (13), and to get full fluorescence profile we needed a long run observation. As for the effect of the triangle view, we needed more refinement of the shorter integration time in the whole circuit.

The structure of Tokyo-3 is shown in Figure 11. The final objective of Tokyo-3 is to follow the very early part of each EAS profile. We thought two series experiments for this objective. The first one is for the distant showers of more than 2.5km and the fluorescence profiles are looked for. The second one is for the near showers of less than 2.5km and the contribution of the Cerenkov data is expected positively together with the fluorescence just like the data in Figure 10. However, unexpected trouble of the Fresnel lens was found in early time of the test run, and the Tokyo-3 experiment was collapsed.

This lens trouble was a serious management miss to give a loose specification of the plastic(PMMA) lens stuff at some preparing process. The used one was of general purpose and was mixed with UV stopper(Figure 4). Late after the lens trouble was found, we had a short period Cerenkov experiment because the lens was still sensitive for some of Cerenkov light. The data is also shown in Figure 11, and if the lens had our normal function, we could expect much larger superposed fluorescence profile over these Cerenkov one.

SUMMARY

The works driven by the passion to want to use the air fluorescence as a tool of cosmic ray research were summarized. Even though one has such passion, nothing happens unless he actually carry out. In this sense the work in Cornell directed by Greisen was a pioneer. They did not get nice fruit, but many valuable experiences of Cornell were transfered to many people. In Tokyo they succeeded to detect the air fluorescence from EAS, but not to develop further. It is delightful that those old efforts now bloom as Fly's Eye work and as the support of the cosmic gamma ray research, and that they are further developing to telescope array or to the space experiment------OWL.

ACKNOWLEDGMENTS

The author is grateful to Professor Y.Takahashi for his interest and encouragement to this work. The author is grateful to Professor K.Greisen for his invitation to Cornell in 1964~1966 and for the valuable collaboration there, and also for his interest of Tokyo works. The author wish to thank Dr. A.Bunner for his old collaboration in Ithaca and the present arrangement of this workshop. Thanks are due to Professor T.Hara and Mr. F.Ishikawa for their valuable collaboration in series of Tokyo experiments. The author also thank Mr. S.Hanazono and the other old members of the machine-shop of INS for their support of Tokyo experiments.

REFERENCES

1. Norikura Symposium Report, Uchusen Kenkyu, **3**, 5, 449pp, 1958

2. Suga.K., 5th. InterAmerican Seminar on Cosmic Rays, La Paz, XLIX, 1962

3. Bunner.A.N., Thesis, Cornell University, 1967

4. Greisen.K., Proc. 9th Int.Conf. on Cosmic Rays, Invited Paper 609pp. 1965

5. Bunner.A.N., Greisen.K., Landecker.P., Can.J.of. Phys., **46**, 10, s266pp. 1967

6. Miyake.S., Prog.Theor.Phys. **20**, 844pp. 1958

7. Grigorov.N., Shestoperov.V., ZETF. **34**, 1539pp. 1958

8. Kraushaar.K., Supp. Nuovo Cimento, **8**, 649pp. 1958

9. Tanahashi.G., Miura.Y., Proc. of the 14th Int.Conf.Cosmic Rays, 4358pp. 1975

10. Hara.T.,Ishikawa.F.,Tanahashi.G. et al., ActaPhys.Acad.Sci.Hung,**29**,s3369pp. 1970

11. Jenkins.E., Thesis, Cornell University, 1966

12. Landecker.p., Thesis, Cornell University, 1968

13. Hara.T., Thesis, Osaka City University, 1977

Results from Fly's Eye and HiRes Projects

Pierre Sokolsky
High Energy Astrophysics Institute and Department of Physics
University of Utah
Salt Lake City
Utah

Abstract

We summarize the history of the Flys Eye detector project and discuss results on the UHE cosmic ray spectrum and composition produced by this experiment. The design and present status of the High Resolution Fly's Eye is discussed. Results on the UHE composition from the HiRes prototype/CASA/MIA detector are presented. Plans for a new collaboration with the Telescope Array group to build the Snake detector are discussed.

1. Early History of Fluorescence Detection

Detection of fluorescence light produced by extensive air showers (EAS) in the atmosphere was proposed independently by Chudakov, Suga and others [1]. A group led by K. Greisen at Cornell attempted a measurement using a multi-faceted lens and photomultiplier tube system [1]. This attempt was unsuccessful but it stimulated a Utah group (included G. Cassiday, H. Bergenson and E. Loh) to build an improved version. The first detection of fluorescence light from an EAS was done in a coincidence measurement with John Linsley's Volcano Ranch array in 1978 [1]. This success lead to a proposal for a full-scale detector, to be built at Dugway Proving Grounds near Salt Lake City, Utah.

The Utah Fly's Eye experiment took data from 1981 thru 1992[2]. This experiment consisted of two detectors, Fly's Eye I and Fly's Eye II, spaced 3.4 km appart. Near 100 EeV, the Fly's eye I detector had a monocular aperture of 1000 km^2str and a duty factor of 10 percent. Fly's Eye II began taking data in 1986 and had an aperture of 40 percent of Fly's Eye I. Events viewed in stereo by both detectors had the most reliable reconstruction and best energy and shower profile resolution. In addition, the two independent measurements of each showers energy and shower maximum were used to experimentally determine the detectors energy and shower maximum resolution.

A number of significant results were published based on this data[3]. The composition of cosmic rays from .1 to 10 Eev was determined using the shower maximum method. The distribution of shower maxima (X_{max}) in the atmosphere is sensitive to the primary particle composition. Based on lower energy data one expects the CR

CP433, *Workshop on Observing Giant Air Showers from Space*
edited by J.F. Krizmanic et al.
© 1998 The American Institute of Physics 1-56396-766-9/98/$15.00

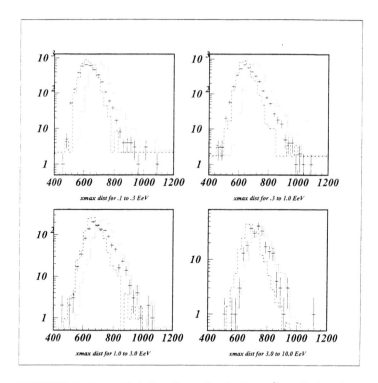

FIGURE 1. Stereo Fly's Eye X_{max} distribution and predictions for a pure Fe and a pure p flux.

composition to range from protons to FE nuclei. Experiments near the knee of the spectrum show some evidence of an increasingly heavy composition[4]. Because of limited detector resolution we compare our data to two extreme cases. Fig. 1 shows the shower maxima predicted for pure protons or Iron nuclei for four different energy bins. This figure also shows the data distribution. Predictions for Fe and p primaries show that the mean values of X_{max} are about 70 gm/cm^2 apart in any model of hadronic interactions. The stereo resolution of this detector was 45 gm/cm^2 which was sufficient to resolve the two components clearly. Fig. 2 shows the elongation rate (mean shower maximum as function of energy) compared with expectations for protons and Fe showers. The conclusion is that there is a gradual transition from a mainly heavy composition near .1 EeV to a mainly light composition near 10 EeV.

2. Results from the Fly's Eye Experiment

The stereo cosmic ray spectrum is shown in Fig 3. Other experiments observe a differential spectrum with a power law index of -3 below .1 Eev[6]. The observed

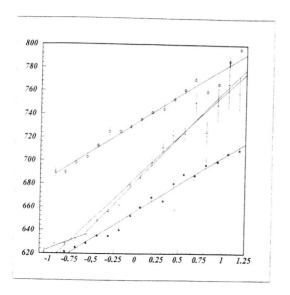

FIGURE 2. Stereo Fly's Eye data elongation rate and prediction for a pure Fe and pure p flux.

spectrum exhibits significant structure (at the 5 sigma level) between 1 and 10 Eev. The Fly's Eye stereo data shows a steepening from .3 EeV to near 3 EeV, followed by a flattening with a power index of - 2.7. This structure may be correlated with the change in the cosmic ray composition as determined by the X_{max} method. A consistent interpretation of this data is the appearance of a new, light, possibly extragalactic, component which dominates over the heavier galactic flux near 10 EeV.

The monocular F.E. I data has a significantly greater aperture, though worse energy resolution than the stereo data. Fig. 4. shows the monocular spectrum. Although a flattening of the spectral index is evident above 10 EeV, the pronounced dip structure near 3 EeV seen in the stereo data is absent.If one folds in the monocular energy resolution with the stereo spectrum, one gets a result very similar to the observed monocular data. This confirms the importance of good energy resolution if one is to detect structure in the spectrum. There is some evidence for the Greisen-Zatsepin cutoff (at the two sigma level) but the simple picture is complicated by the existence of an outstanding event at 320 EeV, well above the 60-100 EeV region predicted for the cutoff.

3. The High Resolution Fly's Eye Detector

Given these results, the next step for the Fly's Eye group was to form a new collaboration with the University of Adelaide, Columbia University and the University of Illinois to build a detector with ten times greater sensitivity to the > 10 EeV region

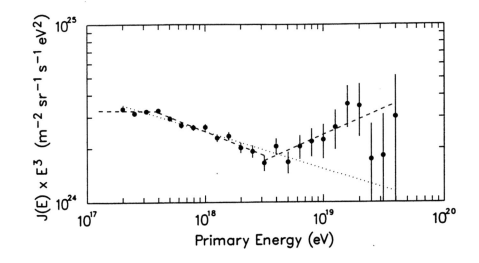

FIGURE 3. Stereo Fly's Eye spectrum.

of the spectrum. This collaboration has recently been joined by a group from the University of New Mexico. All showers in the new detector will be observed in stereo and the X_{max} resolution will be improved to better than 30 gm/cm^2. This makes the detector resolution better than the expected natural fluctuations in X_{max} of pure Fe and proton cosmic ray induced showers. The energy and X_{max} resolution will be established by comparing two independent measurements - a technique whose power was demonstrated in the previous work.

The Stage I High Resolution Fly's Eye Detector (HiRes) incorporates these goals. It was approved as a construction project in 1994 and will be completed in the summer of 1999. It consists of two detectors (HiRes I and HiRes II) located at two sites, Five Mile Hill (the original F.E. I site) and a new site located on Camel'sback mountain. The two sites are separated by 12.5 km. The HiRes detector gets its increased aperture and improved resolution through decreasing the photoube aperture from 5 degrees by 5 degrees (the tube aperture of the Fly's Eye detector) to one degree by one degree and by incresing the mirror diameter from 1.5 to 2.0m. The resultant increased signal to noise allows showers at 20 km to 30 km impact parameter distance to be clearly detected. The improved sampling along the track and full stereo coverage makes it possible for even short tracks to be well reconstructed.

A mirror is composed of four spherical glass segments, making a single spherical

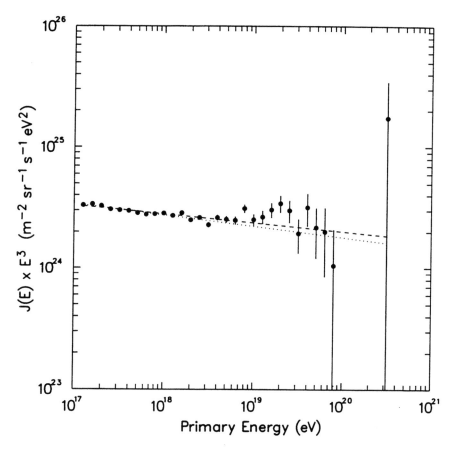

FIGURE 4. Fly's Eye monocular spectrum.

mirror. The phototube cluster consisits of 256 close packed 40mm hexagonal photo-tubes. These tubes are manufactured by Phillips. The tubes are held in a cluster box which provides the high voltages and routes out the tube signals. The front of the cluster box consists of a piece of UV filter glass which cuts out light below 300 and above 420 nm. At Camelsback, 42 mirrors are distributed in two rings, covering all azimuthal angles and elevation angles from 3 to 33 degrees. The 22 Five Mile Hill site mirrors are distributed in one ring, covering all azimuthal angles and elevation angles from 3 to 18 degrees.

The mirrors and tubes are housed in steel buildings with garage doors. The buildings come in two sizes, corresponding to the two rings of mirrors. Each building contains two mirrors and the associated electronics.

The electronics at HiRes II consists of an FADC system which digitizes the signals coming from the individual phototubes in 100 ns samples as well as digitizing the sums of the vertical and horizontal columns of tubes. The sum signals are used for trigger purposes. Once a minimal trigger condition is met, a higher level trigger algorithm decides on an area of tubes around the triggered tubes for which information should be saved. Digitization of the signal pulse allows us to measure the width as well as the area of the pulses and to sample the sky noise fluctuations both before and after the signal arrives. A modern version of the original Fly's Eye S/H electronic system is used at HiRes I.

The HiRes detector will have an aperture of near 10,000 km^2str at 100 EeV and a resolution in X_{max} of better than 30 gm/cm^2.

4. The High Resolution Prototype Detector.

A 14 mirror prototypre HiRes detector was built to test the adequacy of the HiRes design. The prototype mirrors at Five Mile Hill were aimed in an inverted pyramid over the CASA/MIA detectors[7] and coincident measurements of near .1 EeV showers were performed. Fluorescence photons, surface electron density and muon density were measured simultaneously for showers that trigger both detectors. Such an intercalibration of different techniques has never been done before.

An additional 4 mirror HiRes II prototype was also built at Camelsback. These mirrors point in the general direction of CASA/MIA, so that they have an overlapping field of view with the Five Mile Hill (HiRes I) prototype. The HiRes I and HiRes II prototype, operating together, provide data for testing methods of stereo reconstruction of events as well as testing our knowledge of the atmospheric propagation of light over long distances. A number of devices, including steerable YAG lasers and vertical Xenon flashers have been installed in the field of view of these detectors to help understand resolution and atmospheric attenuation.

Experience with running the prototype for over three years has confirmed that

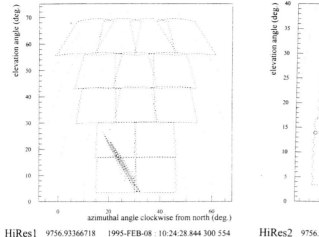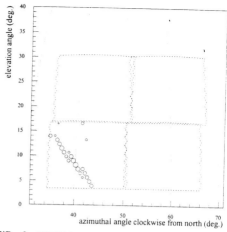

HiRes1   9756.93366718   1995-FEB-08 : 10:24:28.844 300 554   HiRes2   9756.93366718   1995-FEB-08 : 10:24:28.844 337 169

FIGURE 5. Event seen by Stereo HiResI and HiResII prototype.

the present HiRes design is sound. The phototubes gains have been stable at the 5 percent level over this period and very few tubes or channels of electronics have had to be replaced. We have tested the mechanical and electrical elements of the design over several winters and summers with no significant problems.

Fig. 5 shows a cosmic ray event seen by both HiRes I and HiRes II prototypes. This is typical of what a full HiRes track will look like - it is contained in two rings of one eye and is seen as a track in the lowest horizontal ring of the other eye.

5.HiRes Prototype/CASA/MIA results

The HiRes prototype fluorescence detector was oriented to overlook the CASA–MIA array at a distance of 3.4 km. This provided an opportunity to observe coincident EAS with an accurate determination of the muon content from the 2500 m^2 MIA array and the superior energy reconstruction from the fluorescence technique. Trajectory reconstruction, based on a hybrid method using muon shower front timing and HiRes prototype plane and timing information, was also improved. The accuracy of this method affects both energy and muon number reconstruction. Because of the flatter muon lateral distributin function (LDF) the effective aperture of MIA extended outside the 0.25 km^2 CASA–MIA array to distances of up to 2km. CASA data was not used in this study.

Coincident data was collected from September 1993 through May 1996. The muon lateral distribution function was found by binning the normalized muon density in logarithmically equispaced annular bins centered on the shower core. The muon density was determined by the number of counters hit in each bin relative to the number which could have been hit after accounting for poisson statistics and expected backgound.

Fig. 6 shows the resulting LDF. It is in good agreement with the modified Greisen form proposed by the Akeno group [8] based on their observations. The results are clearly inconsistent with the Greisen form at large distances.

Fig. 7 shows the $N_{mu} - E$ correlation. The total N_{mu} determined by MIA relies on a maximum log likelihood fit to the Akeno muon LDF form. A strong correlation is present even before the quality and antibias cuts. After these cuts, the best fitted line has a slope 0.80 ± 0.03 (stat) with a $\log_{10}$ intercept of 5.91 ± 0.05 (stat) ± 0.09 (sys) at 10^{17} eV. This compares with the Akeno result (Hayashida, 1995) over the range $10^{16.5}$–$10^{17.8}$ eV:

$$\log_{10} = (0.84 \pm 0.02) \log_{10} E - (8.38 \pm 0.35) \tag{1}$$

The Akeno result is equivalent to a $\log_{10}$ intercept of 5.90 ± 0.35 at 10^{17} eV. This data is in agreement with their intercept and consistent with their slope.

Final results on this data sample will be forthcoming shortly.

6. Present Status of the HiRes Project.

Construction of buildings and installation of mirrors and power is essentially complete at Five Mile Hill and Camel's Back. All 64 mirrors are in place and all 32 buildings have power and are operational. The HiRes I detector is now taking monocular data. Trigger event rates are as expected from monte carlo predictions.

FADC electronics has been undergoing extensive tests over the last year. A separate mirror whose tubes are read out by the FADC system and which looks in the same direction as one of the mirrors in the HiRes I prototype is used to compare stability, uniformity, linearity, noise suppression and trigger efficiency between the FADC and S/H systems. Installation of FADC hardware at HiRes II will begin later this year.

7. The HiRes/Telescope Array Collaboration - The Snake Array

If the cosmic ray flux continues well beyond the Greisen cutoff, even larger detector apertures will be required. The HiRes group has entered into a collaboration with the Telescope Array group, based at the University of Tokyo, to build a chain of fluorescence detectors with a total aperture of 80,000 $km^2 str$. In this scheme, ten or more detectors will stretch from Dugway Proving Grounds south and then east to Millard County, where two sites will overlook the proposed Auger ground array. This so-called Snake array will have an aperture similar to the Auger detector, even after

Lateral Distribution Function

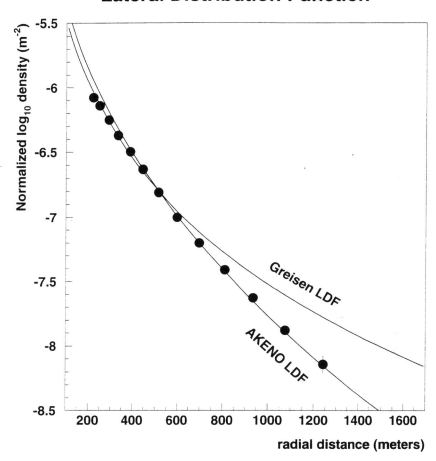

FIGURE 6. *Muon Lateral Distribution Function derived from the normalized muon density. The Greisen and Akeno functions are shown for comparison.*

N$_\mu$ -- Energy correlation

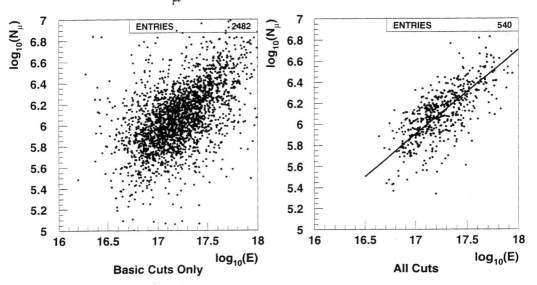

FIGURE 7. *Nmu vs. Energy correlation*

taking into account the 10 percent duty cycle inherent in fluorescence detectors and will give complementary high statistics measurements of the cosmic ray flux near 100 EeV.

5. Bibliography

1. B.Dawson, P.Sokolsky and P.Sommers, Phys. Rep. 217 (1992) and references therein.

2. R.M. Baltrusaits et al. Nuclear Methods, A240 (1985) p 410.

3. D.J. Bird et al.,Phys. Rev. Lett. 71 (1993) p 3401; D.J. Bird et al., Ap. J 424 (1994) p 491.; T. K. Gaisser et al., Phys. Rev D, 47 (1993) p 1919.

4. D. Muller et al. Ap. J. 374 (1991) 356; K. Asakimori et al., in Proc. 22 ICRC, Vol. 2 (1991) 57.

5. M.A.Lawrence et al., J. Phys. G 17 (1991) 733. N. Hayashida et al., Phys. Rev. Lett. 73 (1994) p 3491; S. Yoshida et al., Astroparticle Physics, 3 (1995) p 105.;N.N. Efimov et al., Proc. 22nd ICRC, Dublin 4 (1991) p 339.

6. A. Borione et al., Nucl. Inst. and Meth. A346 (1994) p 329.

Results from AGASA Experiment in the Extremely High Energy Region

Motohiko Nagano

Institute for Cosmic Ray Research, University of Tokyo, Tokyo 188

Abstract. Recent results on primary cosmic rays of energies above 1×10^{19}eV observed by the Akeno Giant Air Shower Array (AGASA) are summarized. It is likely that extremely high energy cosmic rays are from diffuse sources distributed isotropically in the universe. However, there are cosmic rays of energies in excess of the the predicted cutoff energy and some fraction of cosmic rays beyond 4×10^{19}eV seem to come nearby sources composing double events within a limited space angle.

INTRODUCTION

AGASA is the Akeno Giant Air Shower Array covering over 100km^2 area in operation at Akeno village about 130km west of Tokyo, in order to study extremely high energy cosmic rays (EHECR) above 10^{19}eV [1]. In this energy region, distinctive features in the energy spectrum and arrival direction distribution are expected. If the cosmic rays are of extragalactic origin, photopion production between cosmic rays and primordial microwave background photons becomes important at energies above 6 $\times 10^{19}$eV with a mean free path of about 6Mpc. Therefore a cutoff in the spectrum may be observed around several times 10^{19}eV even if the primary cosmic ray energy spectrum extends beyond 10^{20}eV. This is called as the Greisen-Zatsepin-Kuzmin (GZK) cutoff [2]. In this senario, the expected arrival direction distribution of EHECR may be quite isotropic. On the other hand, if they are galactic origin, their expected arrival direction distribution is no more isotropic, since the gyroradius of protons of energy above 10^{19}eV exceeds the thickness of the galactic disc. Therefore a study of correlations of EHECR with the galactic structure and with the large scale structure of galaxies is very important.

So far the experiments in this energy region have been made at Volcano Ranch [3], Haverah Park [4], Narabrai [5], Yakutsk [6], Dugway [7] and Akeno [8], and the significance of evidence for the *GZK cutoff* and their isotropic arrival direction distributions have increased. Therefore detection of a few $\times 10^{20}$eV cosmic rays, by the Fly's Eye [9] and AGASA [10] well beyond the predicted cutoff energy, has posed a puzzle concerning its origin. Recent observation of AGASA events with energies above 4×10^{19}eV, coming within a space angle of 2.5° from the direction

CP433, *Workshop on Observing Giant Air Showers from Space*
edited by J.F. Krizmanic et al.

of supergalactic plane [11] and a possible correlation of Haverah Park data with supergalactic plane [12] should be also remarked.

As accerleration mechanism of these EHECR, the diffusive shock acceleration is most widely accepted and shocks at radio lobes of relativistic jets from Active Galactic Neuclei, termination shocks in local group of galaxies, etc. are discussed to be possible sites [13]. The acceleration above 10^{20}eV in dissipative wind models of cosmological gamma-ray burst is also proposed [14]. Since various difficulties are still anticipated in accelerating cosmic rays up to the highest observed energies, it is also discussed that these cosmic rays are possiblly the decay products of some massive particles produced at the collapse and/or annihilation of cosmic topological defects which could have been formed in a symmetry-breaking phase transiton in the early universe [15].

In this report results on AGASA experiment published before are summarized. New results are now in preparation and will be published in the near future.

EXPERIMENT

The Akeno Air Shower Experiment started in 1979 with an array covering 1km^2 area, and the array area was expanded to 20km^2 in 1985. From 1990, the array has been expanded gradually to 100km^2 and is named as AGASA.

The AGASA consists of 111 scintillation detectors of $2.2m^2$ area each, which are arranged with inter-detector spacing of about 1km over 7km×15km area. The whole area is divided into four branches, Akeno(AB), Sudama(SB), Takane(TB) and Nagasaka(NB) for the data acquisition and trigger purposes. The two-way communication between the detectors and the central station of each branch for data transmission is carried out through two optical fiber cables. The triggering requirement is more than 5 fold coincidence of neighbouring detectors.

The data acquisition has started independently in each branch from 1990 and the whole branches are unified into one in December 1995. The details of the AGASA is described in [1] and [20].

ENERGY SPECTRUM

In order to estimate the primary energy of giant air showers observed by the AGASA, the particle density at a distance of 600m from the shower axis (S(600)) is used as an energy estimator, which is known to be a good parameter [16]. The conversion factor from S(600) [per m^2] to primary energy (E_0 [eV]) at Akeno level is derived by simulation [17] as

$$E_0 = 2.0 \times 10^{17} \times S(600)^{1.0}. \tag{1}$$

The lateral distribution of electrons and the shower front structure far from the core are quite important to determine S(600) and it is determined with many

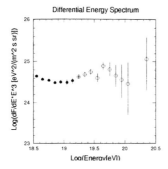
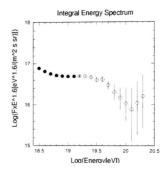

FIGURE 1. The differential and integral energy spectra determined by AGASA. New data ($\bigcirc$) are normalized to our previous spectrum ($\bullet$) at $10^{19.25}$ eV.

detectors of 1km² array (A1) by using showers hitting inside the Akeno Branch. The density at core distance r is expressed by the function as

$$\rho = N_e C_e R^{-\alpha} (1 + R)^{-(\eta - \alpha)} (1.0 + (\frac{r}{1000})^2)^{-0.6}, \tag{2}$$

where $R = r/R_M$, C_e is a normalization factor and R_M is a Moliere length (91.6m at Akeno) [18]. A fixed value of $\alpha = 1.2$ is used and η is expressed as

$$\eta = 3.97 - 1.79(sec\theta - 1) \tag{3}$$

The effective collection area has been estimated through analysis of artificial showers which are simulated by using equation (2) with the observed fluctuation of density at each core distance [18].

The differential and integral primary energy spectrum around the *ankle* is shown in Figure 1 [19]. The bars represent statistical errors only and the error in the energy determination is about 30% above 10^{19} eV, which is estimated from the analyzing the artificial showers simulated considering shower development fluctuation and experimental errors in each detector [8].

The cosmic ray of the highest energy observed up to November 1997 is still the one reported in Hayashida et al. [10]. There are several candidates above 10^{20} eV so far, two of them are displayed in Figure 2 [19]. The updated spectrum including these events is now in preparation and will be published soon.

In order to estimate the systematic error of primary energy determined from (1), the energy spectrum determined at Akeno in the lower energy is shown by open circles in Figure 3, along with the energy spectrum determined by the direct observations [21–23] and the results determined at 4300m.a.s.l. [24].

The Akeno energy spectrum between $10^{14.5}$ and 10^{18} eV is based on the number spectrum of total charged particles (shower size, N_e) in EAS, determined by the '1km² Array'(A1) [25] which is in the southeast corner of the AGASA.

The Akeno energy spectrum thus determined coincides very well with that determined by the Tibet group (closed circles in figure 3 [24]) which may be the best

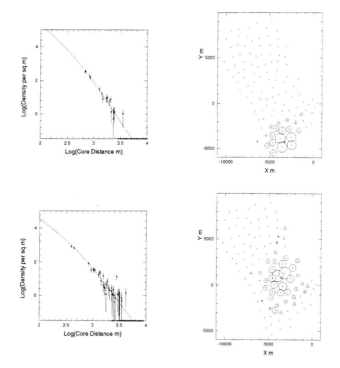

FIGURE 2. Two candidates of $> 10^{20}$ eV events. The zenith angles are 14.2° and 44.5°, respectively. Left : Lateral distributions of charged particles. Right : Density map. Radius of each circle corresponds to logarithm of particle density at each site. Dots are detectors of no particle incidence.

one around the *knee* region, since the shower sizes at Tibet altitude don't depend much on the shower development fluctuation and different primary composition. The small difference around 10^{15} eV between Akeno and Tibet spectra may be due to the fact that some showers from heavy composition are attenuated considerably and can't be reached at Akeno level.

The energy spectrum in the highest energy region from four different experiments are compiled in Figure 4 with spectra from Haverah Park [4], Yakutsk [6] and Fly's Eye [7]. The spectra are normalized to the present AGASA result at around 10^{18} eV, since the Akeno spectrum coincides with the results at the lower energy region. The normalization factors for energies in the stereo Fly's Eye, the Yakutsk and the Haverah Park are 1.1, 0.9 and 1.0, respectively. It should be emphasized here that the agreement of the calorimetric method used by Fly's Eye and Yakutsk with the Monte Carlo based method used by Haverah Park and AGASA implies that the energy determination is rather good in these experiments.

From the above discussion, the systematic error in energy determination in Akeno

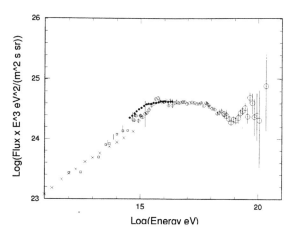

FIGURE 3. Differential energy spectrum of primary cosmic rays. Large open circles:AGASA, small open circles: Akeno 1km^2 array, closed circles:Tibet, crosses:Aoyama-Hirosaki, squares:Proton satellite, Bars:JACEE.

experiment may not be large and the energy spectrum shown in Figure 3 may be used as a standard one for discussion of the origin of primary cosmic rays.

The observed spectrum of Figure 3 is compared with the simulated results under an assumption of diffuse sources distributed isotropically in the universe and sources at various distances (z=0.004, 0.016 and 0.1) [8] in Figure 5. The uncertainty in energy estimation is convoluted to the expected curves. For comparison, the best fitted line of a power law spectrum is also drawn.

Though the fitting with power law spectrum with a slope parameter of -2.59 is

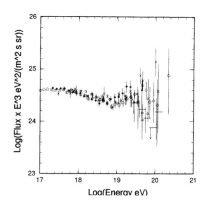

FIGURE 4. Folded energy spectra from the Haverah Park, Yakutsk, stereo Fly's Eye and AGASA experiments. All spectra are normalized to the AGASA result at around 10^{18}eV.

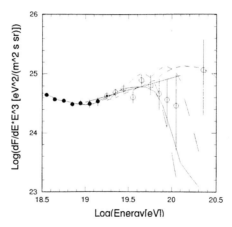

FIGURE 5. The observed energy spectrum by AGASA is compared with expected ones from sources distributed uniformly in the universe (solid curve), from sources at constant distances (dashed curves from right to left z=0.004, 0.016 and 0.1). Flux is normalized to the observed one around 10^{19}eV and the experimental error is convoluted in the expected curve. Best fitted power law spectrum is also drawn by a solid line.

also accepted up to 10^{20}eV, number of expected events above 10^{20}eV is too large. If the sources are very far or uniform in universe, we may not observe any anisotropy in the arrival direction distribution of these events.

ARRIVAL DIRECTION

The arrival direction distribution around 10^{19}eV is almost uniform as is discussed in our report in the Proceedings of the 25th ICRC in Durban [26].

The AGASA data are plotted in Figure 6 for events with energy $> 4 \times 10^{19}$eV [11]. It is seen that arrival direction of significant fraction of EHECR are uniformly distributed over the observable sky, supporting the interpretation of energy spectrum at the highest energy end. However, it should be also remarked that two pairs of showers, each clustered within 2.5°, are observed among the 20 events above 5×10^{19}eV, corresponding to a chance probability of 1.7%. If we refer to 36 showers with energies above 4×10^{19}eV, another pair is observed at high supergalactic latitude with 2.9% chance probability. It should be noted that two pairs of them are within 2.0° of the supergalactic plane.

There is no experimental evidence for the primaries of pair events to be gamma-rays. Since decay length of a neutron is about 1Mpc at 10^{20}eV, neutrons after being produced and escaping from the large magnetic field environment around a source must travel most of their way through intergalactic space as protons.

If the primary particles are protons, significant constraints are imposed on the

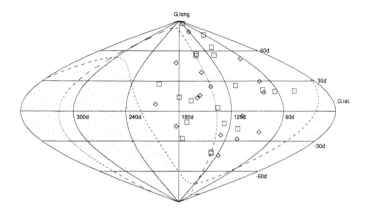

FIGURE 6. Arrival direction distribution of 36 cosmic rays above 4×10^{19}eV in galactic coordinates. Open squares represent events above 5×10^{19}eV (20 events) and open diamonds between 4×10^{19} and 5×10^{19}eV(16 events). The dashed curve shows the supergalactic plane and sky not observed by AGASA due to a zenith angle cut at $45°$ is shown as cross-hatched area.

scale, the strength and the direction of the magnetic field configuration and its turbulence in the interstellar and intergalactic space to explain pair events with angular separation of only a few degrees. Recently the method of extracting the information on the source and the intergalactic magnetic field from such a clustering of events with much higher statistics anticipated in the next generation experiments is discussed by Sigl et al. [27].

COMPOSITION

The University of Utah group performed the experiment by observing nitrogen fluorescence excited by electrons of EAS [28]. A major advantage of the optical method is its capability of measuring the depth for the maximum shower development (X_{max}). Since X_{max} correlates well with the depth of first interaction (mean free path) of primary cosmic rays, the distribution of X_{max} depends on the chemical composition of the primary flux.

From the comparison of the experimental results from Fly's Eye with the expected energy dependence of X_{max} for proton and iron primaries obtained from simulations [29] [30],they interpreted their result that the proportion of protons increases with increasing energy and is about 90% at 10^{19}eV. However, the average X_{max} differs largely among the interaction models used in the simulation as shown in Figure 7 [31].

If we use superposition model on nucleus-nucleus interaction, X_{max} of a primary particle of energy E and mass A is expressed as

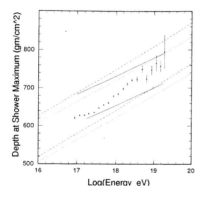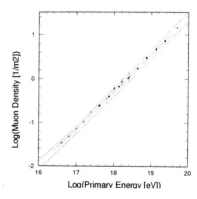

FIGURE 7. The left figure shows the energy dependence of X_{max} from the Fly's Eye experiment. Solid lines are expected X_{max} from proton (upper line) and iron (lower line) primaries cited from their original paper. Other lines are expected one from QGSJET model (dotted lines) and SYBILL (dashed lines) cited from the rapporteur paper by Watson [31]. The right one is the energy dependence of $\rho_\mu(600)$'s. The expected lines are relations of proton (lower line) and iron (upper line) from simulation using COSMOS.

$$X_{max} = a \cdot ln\left(\frac{E}{A}\right) + b, \tag{4}$$

where a and b are constants. Therefore,

$$\frac{\delta X_{max}}{\delta ln E} = a\left(1 - \frac{\delta ln A}{\delta ln E}\right) \tag{5}$$

$\frac{\delta X_{max}}{\delta ln E}$ is called an elongation rate of X_{max} [32] and is dependent only on the rate of change of mass composition with energy.

In a similar way, an elongation rate of total number of low energy muons (N_μ) is derived from

$$N_\mu = kA\left(\frac{E}{A}\right)^\alpha \tag{6}$$

as

$$\frac{\delta ln N_\mu}{\delta ln E} = \alpha + (1 - \alpha)\frac{\delta ln A}{\delta ln E} \tag{7}$$

By putting $\frac{\delta ln A}{\delta ln E}$=-0.58 from estimating the result of Fly's Eye experiment [29] and using α=0.82 from AGASA below $10^{17.5}$eV, $\frac{\delta ln N_\mu}{\delta ln E}$ =0.72 above $10^{17.5}$eV.

That is, the slope of N_μ vs E relation may change from 0.82 to 0.72 around $10^{17.5}$eV, if the above Utah group's result is due to a change in the primary composition. In Figure 7, the AGASA result [33] on the energy dependence of muon density at 600m from the core ($\rho_\mu(600)$) is plotted, along with the expected relation from a Monte Carlo simulation based on the COSMOS program [34]. The change of the composition above $10^{17.5}$eV has not been detected beyond the experimental uncertainties and assumptions used in the simulation.

DISCUSSION

It is likely that EHECR's are from diffuse sources distributed isotropically in the universe. It is also likely that there are cosmic rays of energies in excess of the GZK cut-off energy and some fraction of cosmic rays beyond 40EeV seem to come nearby sources composing double events within a limited space angle. However, observational data is still not enough to judge whether the spectrum extends beyond 10^{20}eV without the GZK cutoff or there are EHECR's of different origin beyond GZK cutoff energy.

In any case the sources of events in excess of 10^{20}eV can't be very far and must be within a few tens of Mpc, if they are protons. There are no candidate astronomical objects within a few tens of Mpc in the direction of the highest energy events which may be able to accelerate particles to more than 10^{20}eV.

These observations have led to suggestions that particles may be directly produced by decay from objects produced in phenomenon occurring on a higher energy scale. For example, topological defects (TD's) left over from the phase transitions in the early universe, caused by the spontaneous breaking of symmetries, have been proposed as a candidate for production of extremely high energy cosmic rays [35]. Such TD's are magnetic monopoles, cosmic strings, domain walls, superconducting strings etc. The remarkable result from this scenario is that bulk of cosmic rays above 10^{20}eV may be gamma-rays rather than protons. Also, there is expected to be a gap around 10^{20}eV and the energy spectrum is expected to extend up to the GUT scale $\sim 10^{25}$eV with a hard exponent such as 1.35 [36], which is based on the exponent of hadronization in QCD.

The cosmological gamma-ray burst model for the production of EHECR's is also proposed [14] and the predictions which will be tested in the future experiments are discussed [37] [27]. In order to clarify the origin of EHECR's, the HiRes Detector(phase II) [38] is under construction and the cosmic ray Telescope Array Project [39] and the Auger Project [40] are under preparation.

In order to design the surface array, it is important to examine lateral distributions and arrival time distributions of muons, electrons and photons, separately. Two detectors with two scintillators sandwiching a lead plate of 1 cm thickness (leadburger) have been built and placed at two sites of the AGASA and the experiment has been continued by K.Honda et. al. [41]. Two prototype water Čerenkov detectors are also installed near the leadburger sites [42] to understand the performance of the detector far from the core of giant air showers by comparing with data by the leadburger and muon detectors. These detectors will be helpful to understand the composition of primary particles in the highest energy region in the AGASA. The AGASA experiment will be continued for several years expecting that the next generation experiments will be in their steady operation at that time.

ACKNOWLEDGMENT

The author is grateful to the AGASA collaborators. He wishes to thank Professors J.F.Ormes and Y.Takahashi for their warm hospitality at this meeting. This work is supported in part by the Grant-in-Aid for Scientific Research No.08454061 from the Japanese Ministry of Education, Science and Culture.

REFERENCES

1. Chiba N. et al., *Nucl. Instr. and Meth.* **A311**, 338 (1992).
2. Greisen K., *Phys. ReV. Letters* **16**, 748 (1966), Zatsepin Z.T. and Kuzmin V.A., *Pisma Zh. Eksp. Teor. Fiz.* **4**, 144 (1966).
3. Linsley J., *Proc. 13th ICRC, Denver* **5** (1973).
4. Lawrence M.A. et al., *J. Phys. G: Nucl. Phys.* **17** 733 (1991).
5. Winn M.M. et al., *J. Phys. G: Nucl. Phys.* **12** 653 (1986).
6. Afanasiev B.N. et al., *Proc. Tokyo Workshop on Techniques for the Study of the Extremely High Energy Cosmic Rays, ed. by Nagano (ICRR)* **35** (1993).
7. Bird D.J. et al., *Ap. J.* **424** 491 (1994).
8. Yoshida S. et al., *Astroparticle Phys.* **3** 105 (1995).
9. Bird D. et al., *Ap. J.* **424** 491 (1995).
10. Hayashida N. et al., *Phys. Rev. Letters* **73** 3491 (1994).
11. Hayashida N. et al., *Phys. Rev. Letters* **77** 1000 (1996).
12. Stanev T. et al., *Phys. Rev. Letters* **75** 3056 (1995).
13. Various acceleration models are discussed in *Astrophysical Aspects of the Most Energetic Cosmic Rays, ed. by M.Nagano and F.Takahara* (World Scientific, Singapore) 252-334 (1990).
14. Waxman E., *Phys. Rev. Letters* **75** 386 (1995), Vietri M., *Ap. J.* **453** 883 (1995), Milgrom M. and Usov V., *Ap. J* **449** L37 (1995).
15. Recent summary and references are described in P.Bhattacharjee, *Proc. of ICRR Symp. on Extremely High Energy Cosmic Rays : Astrophysics and Future Observatories, ed. by M.Nagano,* (Inst. of Cosmic Ray Research, University of Tokyo) 125 (1997).
16. Ohoka H. et al., *Nucl. Instr. and Meth.* **A385**, 268 (1996).
17. Hillas A.M. et al, *Proc. 12th ICRC, Hobart* **3** 1001 (1971).
18. Dai H.Y. et al., *J. Phys. G: Nucl. Phys.* **14** 793 (1988).
19. Yoshida S. et al., *J. Phys. G: Nucl. Part. Phys.* **20** 651 (1994).
20. Hayashida N. et al., *Proc. 25th ICRC, Durban* **4** 145 (1997).
21. Grigorov N.L.et al, *Proc. 12th ICRC, Hobart* **5** 1760 (1971).
22. Asakimori K. et al, *Proc. 23rd ICRC, Calgary* **2** 25 (1993).
23. Ichimura et al, *Phys. Rev.* **D48** 1949 (1993).

24. Amenomori M. et.al., *Ap. J.* **461** 461 (1996).
25. Nagano M. et al., *J. Phys. Soc. Japan* **53** 1667 (1984).
26. Hayashida N. et al., *Proc. 25th ICRC, Durban* **4** 177 (1997).
27. Sigl G. el al., *Phys. Rev.* **D56** 4470 (1997), Sigl G., This proceedings (1997).
28. Baltrusaitis R.M. et al., *Nucl. Instr. Meth.* **A240** 410 (1985).
29. Bird D. et al., *Phys. Rev. Lett.* **71** 4301 (1993).
30. Gaisser T.K. et al., *Phys. Rev.* **D47** 1919 (1993).
31. Watson A., *Rapporteur paper of 25th ICRR, Durban* (1997).
32. Linsley J., *Proc. 15th ICRC, Plovdiv* **12** 89 (1977).
33. Hayashida N. et al., *Proc. 25th ICRC, Durban* **5** 217 (1997).
34. Kasahara M., private communication.
35. Bhattacharjee P., *Phys. Rev.* **D40** 3968 (1989).
36. Bhattacharjee P., Hill C.T. and Schramm D.N., *Phys. Rev. Lett.* **69** 567 (1992).
37. Waxman E. and Miralda-Escude J., *Ap. J.* **472** L89 (1996), Waxman E., *Proc. of ICRR Symp. on Extremely High Energy Cosmic Rays : Astrophysics and Future Observatories ed. by M.Nagano*, (Inst. of Cosmic Ray Research, University of Tokyo) 114 (1997).
38. Loh E.C., *Proc. Tokyo Workshop on Techniques for the Study of Extremely High Energy Cosmic Rays, ed. by M.Nagano*,(Inst. of Cosmic Ray Research, University of Tokyo), 105 (1993).
39. Teshima M. et al., *Nucl. Phys. B (Proc.Suppl.)* **28B** 169 (1992).
40. Pierre Auger Project Design Report, The Auger Collaboration (1997).
41. Honda K. et al., *Phys. Rev.* **D56** 3833 (1997).
42. Sakaki N. and Nagano M., *Proc. of ICRR Symp. on Extremely High Energy Cosmic Rays : Astrophysics and Future Observatories, ed. by M.Nagano*, (Inst. of Cosmic Ray Research, University of Tokyo) 402 (1997), Pryke C., op cit., 407.

Feasibility study of an Airwatch mission

C. N. De Marzo* for the Airwatch Collaboration[†]

*Università di Bari, Dipartimento di Fisica and INFN, Bari, Italy

Abstract. The design strategy of an Airwatch mission intended to study cosmic rays at energies above 10^{19} eV is presented. Cosmic Rays having so Extreme Energies (EECR) can be studied through the fluorescence light their showers excite in the Earth's atmosphere, as seen by a detector mounted on a space platform. According to the Airwatch concept a single detector can be used for measuring both intensity and time development of the streak of fluorescence light produced by the atmospheric shower of an EECR. Both an explorative mission on a small free flyer and a high statistics mission are considered.

In order to optimize the design of this space mission, preliminary measurements are forseen with the UVSTAR apparatus flying on the Shuttle and looking at the fluorescence light excited in the atmosphere by a laser. These measurements will provide data for signal to background evaluation.

A laboratory measurement of the fraction of ionization energy going in fluorescence light in the air, as a function of pressure, temperature and chemical composition, is also planned. A high intensity X-ray beam will be used for the purpose. This measurement is important in order to evaluate the possibility of detecting GRB impacting on the entire atmosphere.

INTRODUCTION

Measurement with good statistics of Cosmic Rays at the extreme end of the energy spectrum, i.e. at energies above 10^{19} eV, is a severe challenging issue in experimental High Energy Astrophysics. These particles are detected through the giant showers they produce in the Earth atmosphere and the observable quantities are longitudinal shower profile, size and arrival direction, depth of the shower maximum, from which energy of the primary particle and an indication on its nature can be derived. The main experimental difficulty in this study is given by the very low flux, which is at the level of 1 event km^{-2} sr^{-1} y^{-1} at 10^{19} eV and quickly decreases with energy. Nonetheless, the study of the Extreme Energy Cosmic Rays (EECR) has the greatest interest and their very existence raises important scientific questions in connection with their origin and propagation. Because of the very high energy, EECR should point back to their sources where "bottom-up" acceleration mechanisms are expected to be at work. Viceversa, in case an isotropy of their incoming directions should be established, a "top-down" model considering

CP433, *Workshop on Observing Giant Air Showers from Space*
edited by J.F. Krizmanic et al.

the EECR as a decay product of topological defects, or grand unification particles, should be took into account [1]. In this last case neutrino and photon initiated showers should be present also [2].

It has been proven by detectors like Fly's Eye [3] that the longitudinal development of the cosmic ray showers can be measured through the fluorescence light induced in the atmospheric nitrogen. But, because in more than thirty years of data taking less than 10 EECR events with energy $\geq 10^{20}$ eV have been collected by all the detectors active all over the world, the application of the fluorescence technique to study EECR needs to watch atmospheric masses at the level of $10^{12} \rightarrow 10^{14}$ tons for some years. This requirement can be fulfilled by detecting from a space platform the fluorescence light induced in the atmosphere by the EECR's, as it was first proposed by J. Linsley in 1979 [4]. More recently the idea has be rejuvenated by Y. Takahashi et al. [5] and has been developed in the OWL Proposal [6]. In fact it appears that the observation of the air from space, from orbit altitudes above 300 km, gives the possibility to monitor an Earth surface of the order of $10^{5} \rightarrow 10^{6}$ km^2, corresponding to atmospheric masses of the order of $10^{12} \rightarrow 10^{14}$ tons as large, depending on the detector field of view (FOV) and on the orbit height.

In Italy the same idea has been proposed to the Italian Space Agency (ASI) and it is now actively investigated by this Collaboration in order to work out a complete feasibility study of an Airwatch space mission, along with preliminary, explorative measurements. As a matter of fact we are convinced that a high statistics Airwatch space experiment needs to be prepared by some ancillary measurements and possibly by an explorative space mission as well. For what concerns us, the forseen ancillary experiments are:

i) to measure UV fluorescence light background in the real atmosphere and its production under laser excitation. Two Shuttle missions will be devoted to this task using the UVSTAR apparatus;

ii) to measure in laboratory N_2 fluorescence light yield as a function of pressure, temperature and chemical composition of the air. The LAX Laboratory in Palermo is suitable for the purpose.

Moreover a complete Monte Carlo simulation code for Airwatch–like missions is under development in order to test and optimize design solutions concerning counting rates and energy threshold.

So far the activity of the Airwatch Collaboration has been organized in a certain number of working groups considering: optics, photon detection and conversion, trigger readout and data handling, signal and background evaluation, fluorescence light production measurements, simulation, spacecraft configuration. From the beginning, industrial support to the designing effort has been provided by Alenia and Laben.

FEASIBILITY STUDY OF THE AIRWATCH EXPERIMENT

The feasibility study under way aims at two objectives:

1) to design an Airwatch explorative mission on a 'small satellite' with a capability of the order of ~ 80 events/year above 10^{20} eV. The ASI's definition of 'small satellite' is given in Table 1;

2) to design a high statistics mission with a capability of the order of ~ 300 events/year above 10^{20} eV, considering all possible platforms – i.e., single free flyer, cluster of satellites, space station.

The fundamental point in the designing strategy of these missions is the detection of the space and time development of the shower using the same and unique detector. This will avoid a second satellite for stereoscopic view but is more demanding on electronics. A detector with single photon sensitivity and photon counting technique is required, both for better background rejection and for registering the faintest signals. These requirements ask for a technology based on photon–electron conversion at single photon level and subsequent electron multiplication.

The technologies of the Hybrid Photo Diode Tubes (HPDT) or that of the Silicon Intensified Micro–strips are looked for the solution of the detection problem, as it is shown in more details in a poster presented at this Workshop [7]. The requirements for the optical collector are: i) UV transparency; ii) large collecting area; iii) minimum light absorption and iv) large field of view (FOV). These requirements have a different influence on the detector capabilities. The counting rate depends on iv), whereas the minimum energy threshold depends on the combination of i) ii) and iii). A Fresnel lens collector is looked for as capable of providing the most fruitful solution to this problem. In particular Fresnel lens systems can provide larger FOV than reflecting mirrors. A status report of the Airwatch optics design is presented in a poster at this Workshop [8].

The orbit height determination has to satisfy contrasting requirements. From one side, at fixed FOV, the mass of the watched atmosphere increases with the

TABLE 1.

Mass	total = 420 kg
	payload = 220 kg
Power	total = 400 W
	payload = 200 W
Orbit	400 km, circular
	$0°$ equatorial
3 axis stabilized	APE < 260 arcsec (1/2 pixel)
Terminat. point.	AMA < 50 arcsec (1/10 pixel)
Position	from GPS; ~ 50 m
Operative life	2 years

height of the detector and so increases the counting rate. On the other hand the possibility of detecting fainter showers increases as the height decreases so that a lower energy threshold is achievable. We think that in case of an explorative mission an orbit as low as possible must be chosen in order to have the lowest possible energy threshold. This requirement is important for instrument calibration because it allows comparison with measurements from the existing fluorescence detectors operating on the ground. Moreover, a lower orbit will result in the possibility of a more complete measurement of the shower development as far as its initial, less luminous parts are detected.

The interest for an Airwatch detector implemented on a "small satellite" comes also from the possibility to solve the counting rate problem by segmenting the mission into a certain number of small satellites. This 'constellation of small satellite' scenario would in general be cheaper and its schedule faster than the traditional single or double observatory, presenting the possibility to be implemented with methodology used in manufacturing large number of satellites for telecommunication purposes. In this case a further cost reduction due to a single launch of the whole system is to be considered.

The 'constellation of small satellite' solution presents some sensible advantages. By distributing the spacecrafts along the orbit the atmosphere at night can be continuously monitored. Moreover both higher mission reliability and reduced problems with design of the focal plane detector are achieved. This solution can be approached through an exploratory mission on a single small satellite that will be very useful both from the engineering and the physical point of view. Moreover a spacecraft in the range of 200 kg mass and 400 watts power can have a collection capability still interesting for the EECR study.

Finally the possibility to locate the Airwatch from Space Mission on the Space Station cannot be neglected as one having several technical advantages; not least the opportunity to use the large collecting mirror technology yet available by the Russian colleagues of this collaboration and presented at this Workshop [9].

SIGNAL TO BACKGROUND MEASUREMENT

The Ultraviolet Spectrograph Telescope for Astronomical Research (UVSTAR) consists of an optical system having two channels, each made of a telescope and a Rowland concave–grating spectrograph with intensified CCD detector. It is born by a steerable platform controlled by a tracking system capable of fine pointing.

The construction of the UVSTAR instrument has been supported by NASA and ASI and its operations span over five Shuttle flights. Two of this have been performed yet, the third is in preparation for the 1998 flight. After that, the instrument will be modified in order to measure the Earth night side background in the UV range around 330 nm.

To further improve the knowledge of the atmospheric N_2 fluorescence phenomenon, the possibility of exciting the fluorescence by a laser is also considered.

This could be done either by a laser mounted on the Shuttle and aligned with the instrument or by a laser operating from the Earth. The measurement of the laser excited fluorescence light will give the possibility to evaluate the signal to background conditions in an Airwatch mission. More details on these measurements are presented in a poster at this Workshop [10].

X–RAY INDUCED FLUORESCENCE

The yield of fluorescence light induced by X–rays is another ancillary measurement planned in order to obtain data useful for the Airwatch experiment. The point is to measure the fraction of X–ray energy absorbed by the air and going in fluorescence photons as a function of pressure, temperature and chemical composition. To this purpose a high intensity X–ray beam with energy in the range $5 \rightarrow 22$ keV will be used. The beam will be provided by the LAX facility in Palermo.

The interest of this measurement depends on the possibility of detecting Gamma Ray Burst (GRB) [11]. While EECR showering in the atmosphere will appear as UV tracks spatially well defined and short in time, a GRB will arrive as a plane wave investing the entire atmosphere and exciting a diffuse fluorescence via photon interactions. More details on this point are presented in a poster at this Workshop [12].

SIMULATION

The fundamental importance of a simulation code for design optimization can hardly be overstressed. We started to work on this task assuming the following objectives:

- to write down an algorithm with easily changeable parameters;

- to obtain evaluation of shower patterns and background on the detector;

- to estimate counting rates;

- to determine geometrical corrections on the raw images in order to derive real shower profile;

- to evaluate angular resolutions;

- to evaluate minimum energy thresholds.

As a first approximation a minimum acceptable uncertainty on the simulation results of the order of 30% will be considered. For what concerns the Monte Carlo flow chart, the following points are notable:

- atmosphere description with realistic geometry (3–D, Earth curvature, ...);

- detector description;

- choice of the primary spectrum shape and composition;

- usage of various interaction models (SIBYLL, MOCCA,...);

- photon transmission onto the detector;

- background generation;

- output production (shower images, plots,...)

The codes used for the interaction model are those reported in [13].

SUMMARY

At November 1997, the organization of the feasibility study of an Airwatch mission is the following:

- Working Groups on the various aspect of the mission started activity since the beginning of the year.

- Detector design and R&D started.

- Background and signal/background measurements by UVSTAR on the Shuttle have been planned.

- N_2 fluorescence yield, as induced by soft X-rays, started in LAX Laboratory.

- A code for full simulation of Airwatch type missions will be available soon for design assistance.

† **The Airwatch Collaboration:**
M. Ambriola, R.Bellotti, F. Cafagna, F. Ciacio, M. Circella, C.N. De Marzo, N. Mirizzi, T. Montaruli – *Dipartimento di Fisica, Università e Sezione INFN, Bari, Italy.*
G. Giovannelli, I. Kostandinov – *FISBAT–CNR, Istituto Fisica dell'Alta e Bassa Atmosfera, Bologna, Italy.*
G. Bonanno – *Osservatorio Astrofisico, Catania, Italy.*
R. Fonte – *Sezione INFN Catania, Italy and University of New Mexico, Albuqueque,USA.*
O. Adriani, G. Becattini, P. Spillantini – *Dipartimento di Fisica, Università di Firenze, Italy.*
P. Mazzinghi, G. Toci – *IEQ–CNR, Istituto di Elettronica Quantistica, Firenze, Italy.*
F. Fontanelli, V. Gracco, A Petrolini, G. Piana, M. Sannino – *Dipartimento di Fisica, Università e Sezione INFN, Genova, Italy.*
G. D'Alì, L. Scarsi – *Dipartimento di Energetica ed Applic. della Fisica Università di Palermo, Italy.*

G. Agnetta, O.Catalano, S. Giarrusso, M.C. Maccarrone, B. Sacco – *IFCAI–CNR, Istit. di Fisica Cosmica e Applic. dell'Informatica, Palermo, Italy.*
P. Lipari – *Sezione INFN, Roma Italy.*
M. Stefani – *Dipartimento di Fisica, Università di Roma 3, Italy.*
G. Giannini – *Dipartimento di Fisica, Università di Trieste, Italy.*
V. Bratina, A. Gregorio, R. Stalio, P. Trampus, B. Visintini – *CARSO, Trieste, Italy.*
C. Cepek, A. Laine, E. Mangano, M. Sancrotti – *Laboratorio TASC–INFM, Trieste Italy.*
J.N. Capedevielle – *L.P.C. Collége de France, Paris, France.*
B. Khrenov, M. Panasyuk – *Skobelisyn Institute of Nuclear Physics, Moscow State University, Russia.*
J. Linsley – *Physics Department, University of New Mexico, Albuquerque, USA.*
Y. Takahashi – *Physics Department, University of Alabama, Huntsville, USA.* L.A. Broadfoot – *Lunar and Planetary Laboratory, University of Arizona, Tucson USA.*

REFERENCES

1. Sigl G. *et al.*, *Science* **270**, 1977 (1995).
2. Vietri M., *Ultra high energy neutrinos from gamma ray burst*, subm. to *Phys. Rev. Lett.*, Nov. 1997.
3. Baltrusaitis R. M. *et al*, *Nucl. Instr. & Meth.* **A240**. 410 (1985).
4. Linsley J., *USA Astronomy Survey Committee Documents* (1979); *MASS/AIRWATCH Huntsville Workshop Report*, 34-74 (1995).
5. Takahashi Y., *Proc. 24th ICRC*, Rome, vol. **3**, 595 (1995); *MASS/AIRWATCH Huntsville Workshop Report*, 1-16 (1995); Takahashi Y. *et al.*, *SPIE Conference*, Denver (1996).
6. Barbier L. M. *et al.*, *Proposal of an Orbiting array of Wide-angle Light collectors (OWL)*, subm. to NASA, private communication (1996); Ormes J. *et al.*, *Orbiting Wide-angle Light collectors (OWL): a pair of Earth orbiting "Eyes" to study Air Showers Initiated by $> 10^{20}$ eV particles*, Proc. 25th ICRC, Durban, vol. **5**, 273 (1997) [OG 10.4.19].
7. The Airwatch Detector Group, *The Fast Detector*, these Proceedings.
8. The Airwatch Detector Group, *The Optical System*, these Proceedings.
9. Khrenov B., *Russian Plan for the Detection of CR $> 10^{19}$ eV on ISS*, these Proceedings.
10. The Airwatch Collaboration, *Background Measurements with UVSTAR*, these Proceedings.
11. Catalano O. *et al.*, *Gamma Ray Burst detection by Air Scintillation observations from Space (GRASS)*, Proc. 25th ICRC, Durban, [OG 10.6.21].
12. The Airwatch Collaboration, *Air Fluorescence Efficiency Measurements*, these Proceedings.
13. Gaisser T. K., Lipari P., Stanev T., *Longitudinal Development of Large Air Showers*, Proc. 25th ICRC, Durban, vol. **6**, 281 (1997); Lipari P., *A 'Flexible' Monte Carlo*

Code for the Simulation of Extremely High Energy Cosmic Ray Showers, Nucl. Phys. B (Proc. Suppl.) **52B**, 161 (1997), Proc. of the 9th Int. Symp. on Very High Energy Cosmic Ray Interactions, Karlsruhe, Germany, 19–23 Aug. 1996.

Orbiting Wide-angle Light-collectors (OWL): Observing Cosmic Rays From Space

Robert E. Streitmatter
for the OWL Study Collaboration[1]

Laboratory for High Energy Astrophysics
NASA Goddard Space Flight Center
Greenbelt, Maryland 20771

Abstract. High statistics observation of the highest energy cosmic rays is needed to extend the present data and resolve the astrophysical question of the origin of these particles. Use of the nitrogen fluorescence technique to make observations from space of the giant air showers induced by these cosmic rays is a promising approach. We consider the technical requirements upon an instrument capable of such measurements.

INTRODUCTION

The observation of the highest energy cosmic rays [1] beyond 10^{20} eV presents us with the challenge of understanding their origin. They are unlikely to be galactic and, in light of the serious energy loss processes of the Greisen-Zatsepin-Kuzmin [2] (GZK) effect, protons and nuclei with such energies can not reach us from distant extra-galactic sources. That is, if the highest energy cosmic rays were produced in the galaxy, they could be expected to show a marked anisotropy, as their gyroradii in the few microgauss galactic magnetic fields are of galactic or larger scale. On the other hand, if they are extragalactic they are not expected to reach us with their large energies from distances much in excess of 50 - 100 megaparsecs [3].

If the particles observed are protons of extragalactic origin and if their sources are correlated with luminous matter, then the inhomogeneity of the large scale galaxy distributions on scales less than 50 Mpc - 100 Mpc should be imprinted on the arrival directions of these particles. Stanev *et al.* [4] have noted that the arrival directions of events with energy > 4 x 10^{19} eV show a concentration in the direction of the supergalactic plane which is marginally inconsistent with the hypothesis that the sources are distributed isotropically. However, Sigl et al. [5] have argued that the arrival directions of the very highest energy events preclude an explanation in which protons are accelerated by active galactic nuclei (AGN's) or galaxy clusters; there are no candidate

1) Louis M. Barbier, Kevin Boyce, Eric R. Christian, John F. Krizmanic, John W. Mitchell, Jonathan F. Ormes (PI), Floyd W. Stecker, Donald F. Stilwell, and Robert E. Streitmatter: *NASA / Goddard Space Flight Center;* Hun-yu Dai, Eugene C. Loh, Pierre Sokolsky, and Paul Sommers: *University of Utah;* Russell A. Chipman, John Dimmock, Lloyd W. Hillman, David J. Lamb and Yoshiyuki Takahashi: *University of Alabama*; Mark J. Cristl and Thomas A. Parnell: *NASA / Marshall Space Flight Center.*

CP433, *Workshop on Observing Giant Air Showers from Space*
edited by J.F. Krizmanic et al.

sources. The mystery of the origin and nature of the highest energy cosmic rays has prompted numerous suggested explanations. The reader is referred to other papers in this volume which represent the range of these possibilities.

It is in any case clear that the additional experimental data required to resolve the matter includes measurements with 1) high event statistics, 2) energies extending well above 10^{20} eV, 3) determination of arrival directions to better than one degree, 4) energy resolution of 25% or better. We consider below these experimental desirata and their technical implications for an instrument capable of making such measurements.

REQUIRED EXPOSURE

We begin by first estimating the exposure (in km^2 sr year) required to obtain 1000 events above 10^{20} eV. Figure 1 shows three versions of the extrapolated integral all-particle cosmic ray spectrum. The "tie-point" spectrum is obtained by using the mnemonic that the integral flux above 10^{16} eV is 1 per m^2 sr year and assuming a simple E^{-2} extrapolated integral spectrum. From Hayashida et al. [1], the differential all particle spectrum from $10^{16} - 10^{18}$ eV is well represented by $J(E) = 10^{24.6} \times E^{-3} / m^2$ sr s eV. Converting to an integral spectrum and extrapolating yields the "Below Ankle" spectrum shown here in Figure 1. Finally, we take the Fly's Eye stereo spectrum, from Bird et al. [1] Table 2, between $10^{18.5}$ eV and $10^{19.6}$ eV, convert to an integral spectrum and extrapolate with an index of -1.71.

Figure 1 indicates that should the cosmic ray spectrum continue unabated, an exposure of 10^5 km^2 sr year would yield 700 to 1000 events above 10^{20} eV. Such event statistics would allow any significant deviation, structure or features between 10^{20} and 10^{21} eV to be well observed.

OBSERVATION FROM ORBIT

John Linsley was the first to suggest observation of cosmic-ray-induced air showers from space [6], noting that the nitrogen fluorescence technique then being introduced for the Fly's Eye array could be used from space to observe air showers over areas of atmosphere far larger than is possible for any plausible ground array. The possibility of observing air showers from space lay fallow for nearly a decade until Yoshiyuki Takahashi revived interest in the idea, holding the Huntsville workshop and advocating the idea at the Rome ICRC [7]. In addition to the present study, groups in Italy, Russia and Japan are pursuing the possibility of observing the highest energy cosmic rays from space [8].

GENERAL DESIGN CRITERIA

The general requirements for observing the highest energy cosmic rays from space derive from straightforward considerations. In this extended section, we make a semi-quantitative examination of the factors driving the instrument design. Following that, these requirements are incorporated into a strawman OWL whose performance is evaluated with a Monte Carlo.

Area of Atmosphere Observed

We have already noted above that the anticipated flux of cosmic rays above 10^{20} eV indicates an exposure of the order of 10^5 km^2 sr year. It is of course desirable to obtain much larger exposures if possible. In order to translate this exposure into an area (effectively a volume) of the atmosphere observed from above, the observing efficiency must be known. The overall observing efficiency has two components: (1) that due to observing conditions which preclude effective operation, such as sunlight, moonlight, lightning, light from cities, ships, oceanic biofluorescence, and high altitude clouds; (2) that due to instrument characteristics, such as light collecting ability, quantum efficiency in the 300 - 400 nm range, optical resolution and time resolution.

Note that the factors in (1) may be expected to be largely independent of the incident cosmic ray energy (i.e. observe or not-observe), while the instrument characteristics (2) when convolved with the atmospheric nitrogen fluorescence of individual events from the cosmic ray spectrum will be strongly energy dependent. At sufficiently high event energies, one expects the latter efficiency to approach 100%, while at lower energies threshold effects will occur. This is dealt with in some detail in sections below.

As anyone who has seen a night photograph taken from orbit of the American, European or Asian continents will appreciate, the lights of civilization make a polar orbit undesirable for OWL. As a baseline we have adopted an orbit lying within 28 degrees or less of the equator. After a survey of global equatorial thunderstorm activity and the frequency of clouds above 3 km altitude, as well as accounting for sunlight and moonlight, we have arrived at 8% as a minimum observing efficiency due to the factors in (1) above. It may in fact be possible to eliminate lightning by time signature and to observe under some conditions of moonlight. Here, however, in arriving at the 8% figure we have made the conservative assumption that these factors when existing completely preclude observation.

Assuming the requirement of an *effective* (i.e. equivalent to continuous 24 hour/day operation) observing capability of $\geq 10^5$ km^2 sr, an 8% actual observing efficiency leads to an *instantaneous* observing capability (instrument geometry factor) requirement of $\geq 1.2 \times 10^6$ km^2 sr. It follows, presuming for the moment a solid angle of π, that the typical minimum dimension of the area of atmosphere to be observed is 700 km. Again, even larger values are desirable.

Optical Considerations:
Resolution, FOV and Orbit Altitude

The fluorescence light from a cosmic-ray-induced air shower appears as a luminous disk, a few meters in depth with a radius less than a kilometer, moving through the atmosphere at the speed of light. The baseline OWL is a pair of orbiting electronic cameras, sensitive in the 300 - 400 nm range, which simultaneously observe a volume of the atmosphere so as to obtain a stereoscopic image with correlated position, time and photon flux information.

Mass in the atmosphere is distributed exponentially with a scale height of about 8.6 km. As a result, cascade development and nitrogen fluorescence from most cosmic-ray-induced air showers occurs largely within a few tens of kilometers of the ground. The need to trace the cascade longitudinal profile of individual events by obtaining multiple measurements along the incident trajectory leads to a natural scale for the requirement of resolving position along the trajectory of ≈ 1 kilometer. The concomitant instrumental time resolution scale is that required to travel one kilometer at the speed of light, $\approx 3 \mu s$.

The corresponding optical angular resolution required depends upon the orbit altitude; e.g. observation from 500 km implies an angular resolution of $\approx 1/500$ radian $= 0.1$ degree, while observation from 5000 km would require an angular resolution of ≈ 0.01 degree. Optically, both of these would be considered low resolution imagers; the 0.1 degree requirement is about 4 orders of magnitude higher than the diffraction limit.

The choice of orbit altitude also affects the required angular field of view (FOV). Figure 2 shows schematically the two baseline OWL satellites, each with a field of view of $\pm 30^o$, orbiting at an altitude of 0.1 Earth radius (640 km). The tactic of inclining the satellites' camera axes produces a common viewed area which is roughly elliptical. In the example shown, the major axis of the ellipse is about 1200 km, and the corresponding geometry factor is $\approx 2 \times 10^6$ km^2 sr. Larger inclinations will produce larger ellipses, geometry factors and satellite separations at the cost of required improvement of the optical angular resolution.

Higher altitudes will allow smaller fields of view. Since optical systems which are both lightweight and have a large field of view present difficult design problems, it is attractive to consider altitudes much larger 600 km which would allow optical systems which have a smaller field of view.

However, aside from optical resolution and FOV, another requirement enters into the choice of altitude: the necessity of having an optical entrance aperture (diameter) of a size sufficient to collect a number of photons which allow the shower to be well characterized. Since the photon flux from a shower will vary as $1/L^2$, where L is the distance from the shower, the entrance aperture diameter required to collect a fixed number of photons will vary as L. While increasing the orbit altitude by a factor of 10 will allow an FOV approximately a factor of 10 smaller, it will also require an entrance aperture diameter a factor of 10 larger.

Optical and Signal Considerations

An estimate of the required optical entrance aperture, and hence photon collecting power, may be obtained by making an envelope numerical calculation of the photoelectron signal expected from the maximum of a 10^{20} eV air shower.

The calculation follows [9] and is similar in approach to that in [7]. We make a number of assumptions. The UV fluorescence photons from the cascade are assumed to exit the atmosphere having suffered Rayleigh scattering resulting in a transmission factor $T_R = 0.42$. The photons are collected by an OWL "eye" located a distance L meters from the cascade and having an entrance optical aperture of A m^2. They then pass through a UV filter, required to eliminate extraneous background photons outside the 300 - 400 nm range, with an efficiency $\varepsilon_f = 0.56$. The photons pass through the optical system with an efficiency $\varepsilon_o = 0.7$ and are collected by a focal plane detector sensitive to single photons with a quantum efficiency $\varepsilon_q = 0.25$. The focal plane detector is assumed to be segmented into pixels, each of which images an object area of diameter D_p kilometers. Assume N_p shower particles following a trajectory at angle θ with respect to the camera axis, where $\theta = 90^o$ indicates a path transverse to the axis.

The expected photoelectron signal from N_p particles traveling a distance corresponding to one pixel at the camera focal plane is then

$$S_{pe} = N_p \, (D_p / \sin \theta) \, Y_F \, T_R \, \varepsilon_f \, \varepsilon_0 \, \varepsilon_q \, A / (4 \pi L^2)$$

where Y_F is the fluorescence yield from a single particle [10], taken as 4700 photons / kilometer. For present purposes, we use the Greisen shower parameterization [11] to obtain the number of particles at the maximum of a 10^{20} eV shower, $N_p = 6.9 \times 10^{10}$. While the Greisen parameterization is for electromagnetic showers, it is convenient and adequate as a first order approximation for hadron-electromagnetic showers in the present context. The hadron-electromagnetic shower parameterization described by Gaisser [12] would give an N_p about 14% larger for a 10^{20} eV shower.

As an example, take $D_p = 1$ km, $\theta = 45^o$, $L = 7.5 \times 10^5$ m, and the entrance aperture area to be 1 m^2. In this case, the expected average number of photoelectrons collected in one pixel at the shower maximum is $S_{pe} = 2.7$. Considering that this is the *maximum* of the cascade, a 1 m^2 entrance aperture is inadequate.

As the baseline OWL, we have taken a 4.9 m^2 entrance aperture (2.5 m diameter). The baseline optical system utilizes Fresnel lens to obtain the large aperture and wide field of view ($\pm 30^o$) required. The University of Alabama in Huntsville and NASA Marshall Space Flight Center are pursuing development of this technology. The reader is referred to the several papers by D. Lamb et al. [13] in this volume. An alternate optical system using a large (≈ 25 m diameter) inflatable mirror orbiting at large altitudes (≈ 5000 km) with a much smaller FOV ($\approx \pm 5^o$) is still under consideration. The present technology of space-inflatable devices [14] is marginally inadequate for the purposes of OWL. Future developments may make this option viable.

Focal Plane Detector

The photon detector at the OWL focal plane must meet a number of requirements. First, it is evident from the analysis above that it must be capable of registering single photons. Second, it must be segmented into a large number of pixels, each giving an independent response. The number of pixels depends upon the specified angular resolution. For a nominal 0.1^o resolution, each OWL eye would require $\approx 430,000$ pixels; an 0.05 degree resolution would require 1.7×10^6 pixels. Third, the area of the focal plane will be large. It follows from fundamental optical principles (conservation of etendue [13]) that the focal plane of an OWL eye, with an FOV of $\pm 30^o$ and an entrance aperture of 4.9 m^2, will have an area between 6 and 20 m^2, depending upon what effective f/# can be achieved. Finally, the focal plane detector must be capable of time response $\leq$ the minimum time for the cascade image to cross one pixel, ≈ 3 μs. While the photon detection and time response requirements might indicate use of conventional photomultipliers, considerations of the large number of pixels and consequent mass preclude this solution. A possible solution is development of custom multi-anode microchannel arrays (MAMAs) which can be combined in a mosaic arrangement. The reader is referred to the paper by J. Mitchell [15] for further discussion of possibilities and technical developments required for the OWL focal plane detectors.

OWL PERFORMANCE MONTE CARLO

While the semi-quantitative analytic arguments above aid understanding baseline requirements and suffice to make a rough OWL design, refinement depends upon use of a Monte Carlo program to evaluate performance and parameter tweaking. Here, we present the results of a Monte Carlo of OWL performance, Dai et al. [16], largely based upon the criteria arrived at in the previous sections above. The Monte Carlo includes the following: shower longitudinal profiles [17], Rayleigh and aerosol scattering, ozone absorption, Earth curvature, standard atmosphere, photoelectron statistical fluctuations, wavelength dependence of assumed bi-alkali photocathode response with peak quantum efficiency 0.28, entrance aperture 4.9 m^2, UV filter of efficiency 80%, FOV = $\pm$ 30°, pixel size corresponding to 0.05^o and the orbit altitude = 0.1 Earth radius (640 km).

Two OWL satellites, required to obtain the requisite information for stereoscopic event trajectory reconstruction, are assumed. The satellites are separated by a variable distance ranging from 500 to 2000 km. At each distance, the viewing directions of the satellites are adjusted to maximize the geometry factor for stereoscopic observation. In all the results presented below an event is registered as a trigger in one satellite when five or more contiguous pixels exceed a specified threshold number of photoelectrons, N_{tpe}. When *both* satellites trigger, an event is recorded and "data" generated. Using analysis procedures similar to those employed by the Fly's Eye, the event energy, direction and depth of shower maximum are then reconstructed from the "data".

Geometry Factor and Resolution

In Figure 3, Monte Carlo results on the OWL stereo geometry factor (in km^2 sr) are shown as a function of energy for a range of satellite separations, with the threshold set at N_{tpe} = 3 photoelectrons/pixel. It is seen that the high energy geometry factor increases with the detector separation up to a separation of 2000 km. Beyond 2000 km separation (not shown), the geometry factor decreases. When the detector separation is 500 km, the geometry factor is saturated, i.e. fully effective, at an energy of 3 x 10^{19} eV with a geometry factor of 1.3 x 10^6 km^2 sr. For larger separations, the geometry factor varies with energy, saturating at higher energies and larger geometry factors.

The effect of threshold photoelectron requirement is seen in Figure 4, where Monte Carlo results on the OWL stereo geometry factor (in km^2 sr) are shown as a function of energy for a range of threshold N_{tpe}, with a fixed satellite separation of 2000 km. Even requiring 10 photoelectrons/pixel, the geometry factor exceeds 10^6 km^2 sr at an energy just above 10^{20} eV.

The effect of energy dependence of the geometry factor upon the measured spectrum is shown in Figure 5, where the geometry factor for 2000 km separation and N_{tpe} = 3 shown in Figures 3 and 4 is convolved with the extrapolated Fly's Eye stereo spectrum.

Figures 6, 7 and 8 show the Monte Carlo results on energy and shower maximum (X_{max}) resolution for event energies of 3 x 10^{19} eV, 10^{20} eV, and 10^{21} eV respectively. A separation of 2000 km and a threshold N_{tpe} = 3 were used. The systematic shifts in energy are an artifact of a non-optimal event reconstruction algorithm and not fundamental. The energy resolution contains components due to both signal fluctuations and geometrical resolution. Both the energy and X_{max} resolution improve with energy, with the energy resolution typically being 15% and the X_{max} resolution improving from $\approx$ 65 g/cm^2 to 43 g/cm^2. The "pointing" angular resolution, i.e. the error in determining the incident trajectory direction, was typically 0.15^o.

An important omission in the Monte Carlo work to date is noninclusion of ultraviolet atmospheric background light arising from scattered starlight and atmospheric chemistry. Using an atmospheric model of Loh and Sommers [18], data from the Fly's Eye and published data [19], we have estimated [9] that the the background UV signal is approximately 10% of the shower maximum signal at 10^{20} eV. We stress that the "dark sky" background UV light level observed from space is not well established and, further, is known to vary with time and geographical location. In operation of the OWL detectors, it will possible to carry out real-time monitoring and correction for the mean UV background. In the meantime, further measurements to establish the range of variation are desirable. Ultraviolet atmospheric background light is the factor which will most likely determine the useful energy threshold of OWL.

SUMMARY

While there is a great deal of work to be done before a functional OWL instrument can be designed, study of the technical requirements and performance of a baseline device is very encouraging. Technology development work on wide field of view optical systems, large area photon detectors and data handling for a system which may contain a million channels is required.

REFERENCES

1. See, e.g., results in M. A. Lawrence, R.J. Reid and A.A. Watson, *J. Phys. G* **17** 773 (1991); D. J. Bird *et al.*, *Ap.J.* **424** 491 (1994); N. Hayashida, *et al.*, "Recent Results of AGASA Experiments," in *Proceedings of International Symposium on Extremely High Energy Cosmic Rays*, Tanashi, Tokyo, Japan, Ed. M. Nagano (1996).
2. K. Greisen, *Phys. Rev. Lett.* **16** 748 (1966); G.T. Zatsepin and V.A. Kuz'min, *JETP Lett.* **4** 78 (1966).
3. F. W. Stecker, *Phys. Rev.* **180** 1264 (1969); J.L. Puget, F.W. Stecker and J.H. Bredekamp, *Ap.J.* **205** 638 (1976); F.W. Stecker, *Phys. Rev. Lett.* **80** 1816 (1998)
4. T. Stanev *et al.*, *Phys. Rev. Lett.* **75** 3056 (1995)
5. G. Sigl, D.N. Schramm and P. Bhattacharjee, *Astropart. Phys.* **2** 401 (1994)
6. J. Linsley, Astronomy Survey Committee (Field Committee) documents; J. Linsley, *Proc. 19th International Cosmic Ray Conference (La Jolla)* **3** 438 (1985); Also see J. Linsley, this volume.
7. Y. Takahashi, ed., *Huntsville Workshop Report, MASS/AIRWATCH,* The University of Alabama in Huntsville, (August 7-8, 1995); Y. Takahashi, *24th Int. Cosmic Ray Conf.* **3** 595 (1995)
8. See, e.g., papers by C. DeMarzo, B. Khrenov, R. Antonov, and Y. Takahashi, this volume.
9. J.F. Krizmanic and R. Streitmatter, "Signal, Background and Energy Threshold Estimates for the OWL Experiment", Goddard Space Flight Center, OWL Note GSFC 6, (1997)
10. T. Kakimoto, E.C.. Loh, M. Nagano, H. Okono, M. Teshima, S. Ueno, *Nucl. Instr. Methods A* **372** 527 (1996)
11. Greisen, K., *Prog. Cosmic Ray Physics* **3** 1 (1956)
12. Gaisser, T.K., *Cosmic Rays and Particle Physics,* Cambridge: Cambridge University Press, 1990, ch. 16, pp. 238-240
13. D.J. Lamb, R.A. Chipman, L.W. Hillman, Y. Takahashi, and J.O Dimmock, this volume.
14. See, e.g. R. Freeland *et al.*, "Inflatable Antenna Technology With Preliminary Shuttle Experiment Results and Potential Applications", 18th Annual Meeting & Symposium Antenna Measurement Techniques Association, Seattle (1996)
15. J.W. Mitchell, this volume.
16. H.Y. Dai, E.C. Loh and P Sokolsky, "The OWL Detector: Aperture and Resolution", University of Utah, UUHEP 97-7 (1997)
17. T.K. Gaisser and A.M. Hillas, *15th Int. Cosmic Ray Conf.* Plovdiv **8** 353 (1977)
18. E. Loh and P. Sommers, "Signal and Noise Analysis for the Crystal Eye", University of Utah, UUHEP 96-1 (1996)
19. R.R. Meier, *Space Science Reviews* **58** 1 (1991)

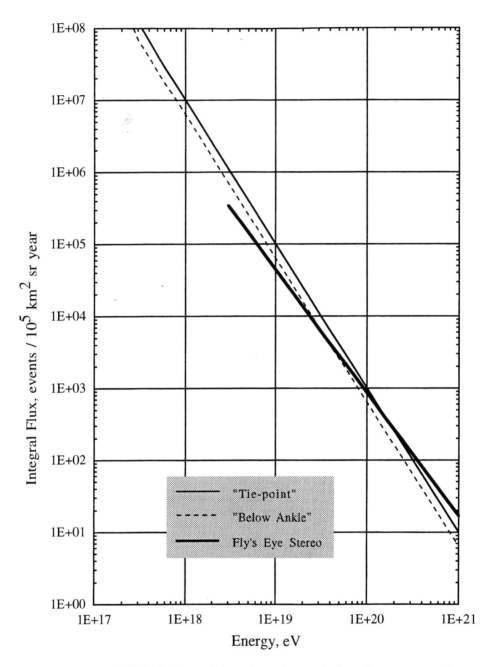

FIGURE 1. Extrapolations of several spectra indicate that an exposure of 10^5 km^2 sr year would yield 700 to 1000 events above 10^{20} eV if the spectra were to continue unabated.

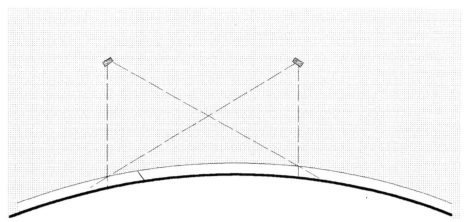

FIGURE 2. OWL concept and viewing tactic. Two satellites stereoscopically view a large volume of the Earth's dark-side atmosphere, searching for the UV signatures of cosmic-ray-induced air showers. Inclination of the cameras' axes with respect to local vertical and varying the separation of the satellites allows the commonly viewed volume to be varied and optimized for different energy ranges.

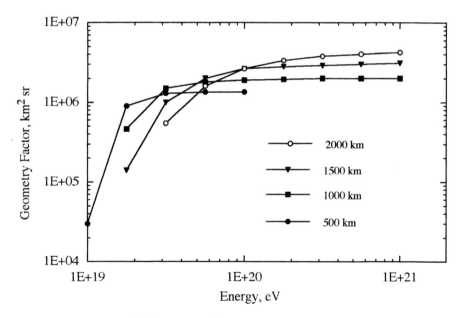

FIGURE 3. OWL observing Geometry Factor vs Energy for 4 different satellite separations ranging from 500 to 2000 km.

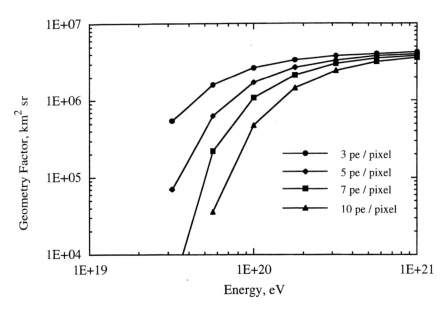

FIGURE 4. OWL observing Geometry Factor vs Energy for a range of threshold N_{tpe} photoelectrons / pixel.

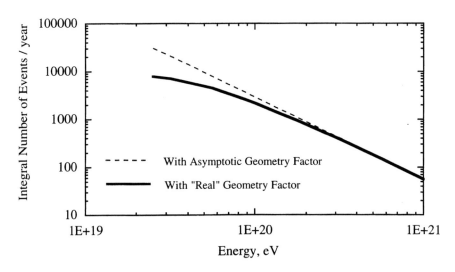

FIGURE 5. The Integral Number of Events/year vs Energy. OWL satellite separation is assumed to be 2000 km, with $N_{tpe} = 3$. The Flys Eye stereo spectrum extrapolated is also assumed. The "asymptotic" spectrum assumes the high energy geometry factor 4.2×10^6 km^2 sr and 8% observing-time efficiency. The "real" spectrum assumes the geometry factor energy dependence shown for the 2000 km curve shown in Figs. 3 and 4.

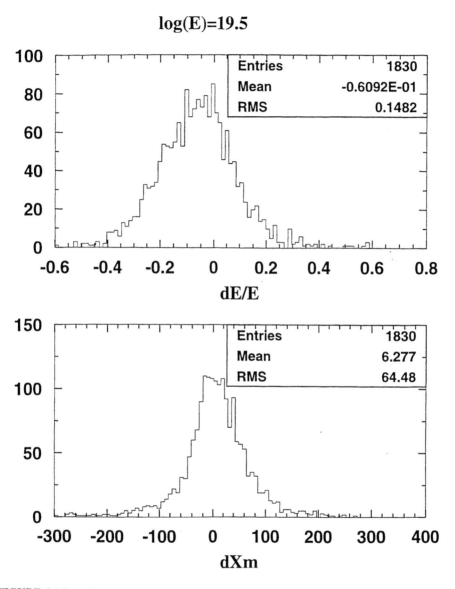

FIGURE 6. Monte Carlo simulation of OWL resolution in energy (relative) and X_{max} (in g/cm^2), for incident events of energy of 3 x 10^{19} eV. The energy resolution is ≈ 15%. The X_{max} resolution is 64 g/cm^2.

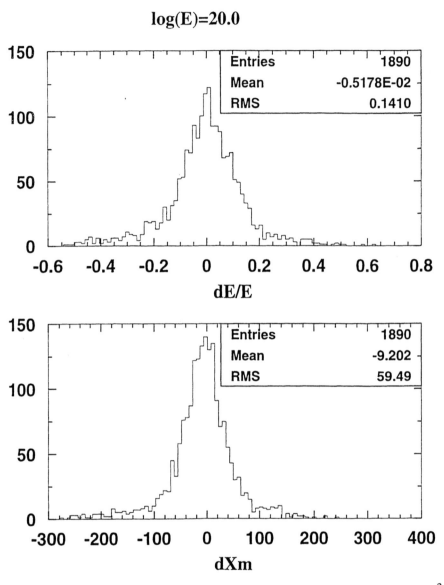

FIGURE 7. Monte Carlo simulation of OWL resolution in energy (relative) and X_{max} (in g/cm^2), for incident events of energy of 10^{20} eV.

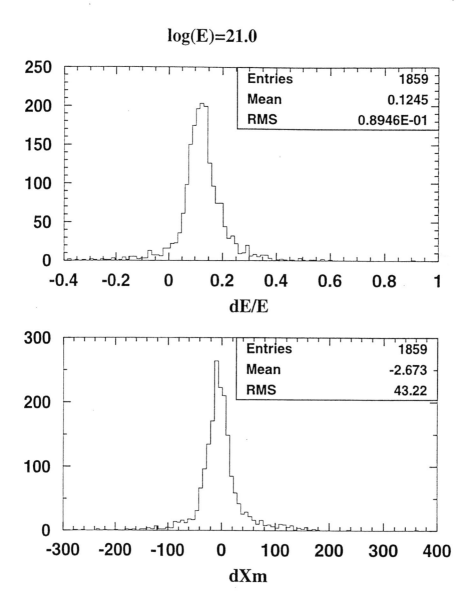

FIGURE 8. Monte Carlo simulation of OWL resolution in energy (relative) and X_{max} (in g/cm^2), for incident events of energy of 10^{21} eV.

Russian Plans For The ISS

Garipov G.K.[i], Gorshkov L.A.[ii], Khrenov B.A.[i], Panasyuk M.I.[i], Saprykin O.A.[ii] and Syromyatnikov V.S.[ii]

[i] D.V. Skobeltsyn Institute of Nuclear Physics of the Moscow State University, Vorobjevy Gory, Moscow, 119899, Russia
[ii] RSC "Energia", Korolev, Moscow region, 141070, Russia

Abstract. In this paper scientific goals of the project to operate a large mirror camera in the Russian segment of ISS are discussed.

Introduction

In opinion of many astrophysicists the Russian segment of the ISS being covered from the sky by the large area solar panels (Fig. 1) is not in a position for the major astrophysical projects. The Airwatch concept suggested by Linsley (1) gives a new way for performing an astrophysical research by looking "downward" to the Earth and observing the atmospheric glow that "reflects" information on many interesting astrophysical phenomena. The basic physical process here is a fluorescence of the air bombarded by the charged particles or γ-rays. The information on the particles and γ-rays could be revealed both in observation of time and space variation of the average glow, and in observation of the light "track" correlated in time to the passage of the primary particle or γ-ray.

With this paper we open the discussion on the scientific goals of the project suggesting to operate a large mirror camera (60 m^2 mirror area, 15°x15° field of view, 5 mrad angular resolution) in the Russian segment of the ISS.

The camera design is described in a separate paper (2).

Extremely High Energy Cosmic Ray (EHE CR) Spectrum And A Search For EHE CR Sources

At present one of the most important and interesting astrophysical problems is the origin of the EHE CR. Scientific goals of the research in this direction were formulated in two major projects for the study of EHE CR on the Earth surface: 1. "The EAS-1000" suggested by Khristiansen et al in 1987 (3) and (2). "The Pierre Auger Observatory" suggested by Cronin (4) and being performed by the Auger collaboration from 1995 (5).

The present day knowledge of the EHE CR and the interest in the problem is based on the experimental results of the large surface arrays (area of about 15-100 km^2) built from 1959 up to 1991 and operated in many countries:

CP433, *Workshop on Observing Giant Air Showers from Space*
edited by J.F. Krizmanic et al.

FIGURE 1. Russian segment of the ISS.

Volcano Ranch (USA, 1959), Haverah Park (United Kingdom, 1964), SUGAR (Australia, 1965), Yakutsk (Russia, 1965), Fly's Eye (USA, 1975) and Akeno-AGASA (Japan, 1979,1991).

With those arrays it was shown that the energy spectrum of Cosmic Rays is extended to energies as high as 10^{20} eV (the first event of that energy was recorded by the Volcano Ranch array (6) in 1963). The intensity of primary particles of such high energies is very low: it is of about one particle per century per 1 km^2 sr - and collecting data with the existing arrays is too slow. The overall statistics of recorded particles is of about 10 and there are only three events of energies well over 10^{20} eV (recorded in Fly's Eye, Yakutsk and AGASA arrays). But even with these low statistics the existence of the recorded EHE particles and their angular coordinates on the sky are making a puzzle for astrophysics.

Immediately after discovery of the black body radiation in the Universe in 1966 Zatsepin and Kuzmin (7) and independently, Greisen (8) found that particles with energies more than 5 10^{19} eV will be absorbed in collisions with photons of the black body radiation if the distance between the EHE CR source and the observer is more than 50 Mps. So the observed EHE CR particles should be accelerated in the "local" sources: in Galaxy or in the local Cluster of galaxies. The known values (or better to say their estimates) of the magnetic fields in Galaxy and in the local Cluster are weak enough not to deflect primary protons from the direction to the source (Fig. 2) and the observed directions of EHE particles should indicate the source. At the same time the experimental events do not coincide with any objects capable of accelerating particles to such energies. It is also difficult to suppose that among known local sources there are objects capable of confining particles of energy higher than 10^{20} eV (Hillas (9).

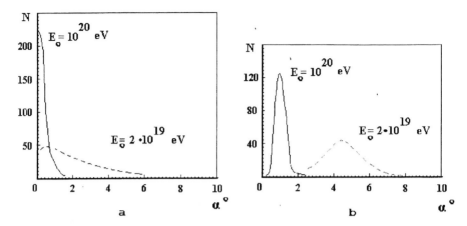

FIGURE 2. Effect of the Galactic magnetic field on the protons accelerated in a hypothetical source: a - Cyg X-1. and b - Crab. Data from paper (3).

This puzzle could be solved if the "exotic" sources of the fundamental super heavy (10^{24} eV) GUT particles- the topology defects of the Universe- are introduced for explanation of the observed events (10). The GUT particles emitted in those sources decay to hadrons, leptons and photons of much less energies- down to 10^{20} eV. Theoretically photons are the most probable particles that should be observed. In this scenario the directions on the sky indicated by the observed events with energies well beyond 10^{20} eV are directions to the positions of the topological defects in the Universe.

So the arisen EHE CR puzzle is not only the astrophysical problem but also a problem of cosmology and elementary particles physics.

For the solution of that fundamental problem the arrays of the gigantic geometrical factor S (km^2) Ω (sr) multiplied on the duty cycle T (years) are needed- SΩT $>10^5$ km^2 sr years. The closest to this figure installation on the Earth surface is suggested to be built by the Auger collaboration: two arrays in both Earth hemispheres with SΩ =6000 km^2 sr each with the field of view covering the whole sky. In 10-20 years of operation this array can solve the problem.

There is another approach to the problem - to use as much atmosphere surface as possible in comparatively short observations with the help of the instruments observing EHE CR signals from the Earth satellites. The original idea was suggested by Linsley (11) in 1981 and was developed into several projects, see (12). The approach is to observe the fluorescence track of the shower initiated by the EHE primary particle in atmosphere as it is done on the Earth surface by the Fly's Eye array but making the observations from the satellite orbit of height R so that the camera with field of view Ψ will observe the atmosphere area S=(RΨ)^2 . Estimates of the energy threshold for observations with the camera (2) with the UV sensitive (wavelengths 320-400 nm) pixel mosaic at the ISS orbit (R of about 400 km) show that the threshold energy is less than 10^{19} eV. The rate of events with the energy threshold 10^{19} eV in camera (2) is of

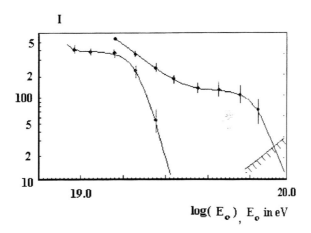

FIGURE 3. Expected result on the energy spectrum of the EHE CR in half of a year of the ISS optical camera operation. Number of events in 0.1 of logE$_o$ is presented for two possible extra Galactic spectra with different cutoffs: 1- distance from the sources 1200 Mps and 2- distance from the sources 130 Mps (both spectra are normalized to the experiment at energies 10^{18} eV). The dashed range is the hypothetical CR from exotic sources.

about 3 per hour and in half of a year of operation the shape of the energy spectrum in the range 10^{19}-10^{20} eV would be certain (Fig. 3).

With orbit height R increasing, the threshold energy is increasing in proportion to R and the observational atmosphere area is increasing as R^2 so the rate of EHE CR of a risen energy threshold stays high (for the exponent of the integral energy spectrum of EHE CR $\gamma < 2$).

So the general trend of the space observations of the EHE CR (the Space Airwatch program) is to start measurements at orbits near 400 km, to check the results at energies 10^{19}-10^{20} eV with the results of the Earth surface arrays (which measure not only fluorescent tracks but other shower components as well) and then to put the camera (or cameras) on the satellite (or satellites) with higher orbits so that the geometry factor of the instruments will be of the order of the whole atmosphere area.

Observations from the orbit are additive to the surface observations in many other aspects:

1. Observing the cascade of secondary particles (EAS) at the "entrance" side of the atmosphere it is easy to distinguish the extreme cases of primary dust grains and primary protons by the difference in the early stage of the cascade (13). The existence of the relativistic dust grains (energy per grain 10^{20} eV) is not excluded (14).

2. Primaries entering the atmosphere in near horizontal directions are registered with higher efficiency (additional information on the EAS development in the low density atmosphere will be added). The accuracy in directional angles for near horizontal directions are higher than in the surface detectors.

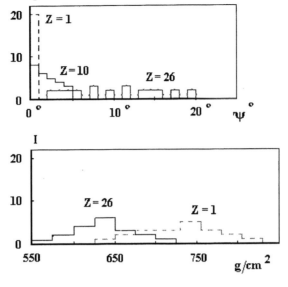

FIGURE 4. Illustration to the statistical approach to the measurement of the particle mass composition in the source. a- expected angular distributions of registered EAS to the direction of the source for primary photons, protons (charge Z=1), nuclei with Z=10 and Z=26, b- expected distribution in position of the EAS maximum for protons and iron nuclei. Statistics of 20 events for one source is assumed. $E_\wedge = 10^{20}$ eV.

3. Observations would be possible in various directions of the primary to the Earth magnetic field - it could be useful in distinguishing primary photons: production of electron-positron pairs by the EHE photons strongly depends on the transverse component of the magnetic field.

4. The joined event statistics of the surface and space detectors will allow to apply the statistical methods of evaluation of the primary particle mass composition in the source. In Fig. 4 two statistical approaches are illustrated: a. the particle direction distribution around the source (sensitive to the mass composition) and b. the distribution of the cascade maximum depths in events directed to one source. Both methods are effective only when statistics of particles directed to the source is high enough. The instrumental errors in Fig. 4 are as estimated for the ISS camera (2).

Other Scientific Goals

The most important advantage of the ISS camera is in the wide spectrum of other problems to be solved with the same instrument having an extraordinary sensitivity to the "upward" atmospheric light . The short list of them is:

FIGURE 5. Artistic view of the mirror camera in the Russian segment of the ISS.

- A search for the EHE neutrino with an extremely large target (of about 10^{11} t of air) and detection of their sources (15). It is possible to distinguish between electron and muon neutrinos. Neutrino events will be separated from EHE photons and barions by the depth of shower maximum in atmosphere. The neutrino events should be near horizontal with a shower maximum deep in atmosphere. The electron neutrino-induced cascade will be similar in shape to the "ordinary" EAS. The muon neutrino-induced event is very different: EHE muon from the first neutrino interaction will produce muon cascade (in process of muon pair production) and these cascading muons will produce many "low energy" electromagnetic showers in process of electron-positron pair production by muons. The resulting "track" will be much longer and possibly with a number of maximums as the low energy cascades are quickly absorbed in atmosphere. To compare with the surface optical detectors the space camera is more effective in observation of the long tracks due to uniform detection efficiency in the camera field of view.
- Observations of the Gamma Ray Bursts (GRB) producing fluorescent flashes in atmosphere. Camera in space can detect the direction of GRB by time measurements and in observation of the GRB flash pattern in atmosphere (16).
- Observations of the solar CR flashes as a fluctuation of the atmospheric glow. The glow from the average CR intensity is of the order of 1-10% of the overall night glow which is mostly the reflected star light.
- Observations of light flashes caused by the particles escaping the Earth magnetic traps due to plasma turbulence. There are experimental data on the light flashes covering the Earth regions of 100-1000 km long both in polar and in equatorial regions (in the equatorial region the correlation with the Brazilian magnetic anomaly was observed) (17). It was suggested (as early as 1962) by Tverskoy

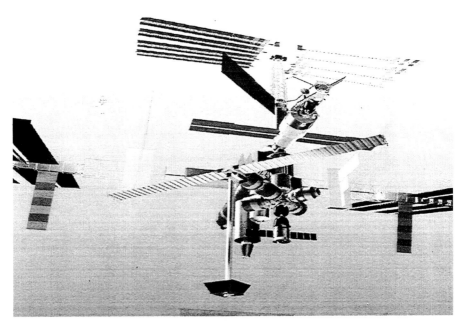

FIGURE 6.

(18,19) that plasma turbulence could be caused by the Earth seismic activity. Later the experimental indications of this effect were collected in observation from the satellites (17,20). Prediction of earthquakes in measurements of the atmosphere glow is in question (21). The ISS camera in this line can make a significant progress giving maps of the flash light in the active seismic areas. Observations in the wide spectrum of light would be important.

- Mapping of the atmospheric glow at the night side of the Earth (in visual spectrum as well as in UV) with the aim of detection of plankton migration or solving other biological and human problems related to the light fluctuations at the night side.
The light sensitivity of the camera is very high: the camera pixel photo cathode (quantum efficiency 20%) has internal dark rate of single photo electrons (p.e.) of about 5-50 kHz (depending on the type of the cathode) so that the light intensity of 0.2 photons per nsec m^2 sr could be detected. The real glow of atmosphere is thousand times higher and fluctuation of this glow could be detected in very short time intervals (for example the pixel average p.e. number in 2 μs interval is of about 100 p.e. and 10% fluctuations of light intensity could be followed in those time intervals). In time intervals of about 0.1 s the pixel field of view is fixed on the area of 2x2 km on the Earth surface. In this time interval the pixel average signal in p.e. is high ($5 \; 10^6$ p.e.) and the light intensity could be measured with statistical accuracy better than 0.1% in every pixel of the retina covering the surface area of 100x100 km (the measurement accuracy will be limited by the instrumental error).

Construction Of The Camera In The Russian Segment Of The ISS

One of the possible technical solutions of the large mirror construction is shown in Fig. 5.

The 9.5 m parabolic mirror designed for the solar energy generator is placed between modules EM of the ISS Russian segment. The mirror has to be launched to the orbit in a compact cylinder and there a special mechanism will open the mirror to its full size. The mirror comprises 36 hard "petals" packed for transportation into the cylinder with the diameter 2.2 m and the length 4.2 m. As it is shown in the figure, the pixel retina is a part of the mirror construction, but a better solution is a separate platform for the retina: then in the focal plane of the mirror various sensors could be used. Other technical approaches to the construction of the mirror are also possible. Figure 6 shows a variant of the plane Fresnel type mirror (2).

The international character of the project is evident. The petal parabolic mirror (as it is designed) could be transported to the ISS only on the USA Space Shuttle. The process of moving the mirror to the place in the Russian segment could be accomplished with the help of several robots starting with the Shuttle SRMS and ending with the Holland manipulator ERA.

All suggestions for international cooperation are welcome.

ACKNOWLEDGMENTS

Authors are grateful to J. Linsley, L. Scarsi and O. Catalano for many discussions in line of the Space Airwatch concept. One of the authors (B.A.K.) is grateful to support from RFFI, grant number 96-15-96783.

REFERENCES

1. J. Linsley et al, Proc. ICRC 25-th, Durban, 5, 385 (1997).
2. G.K. Garipov et al, this workshop, (1997)
3. G.B. Khristiansen et al, (14-th Texas Symp. on Rel. Astrophys.), Ann.. N-Y Ac. Sci., 571, 640 (1987).
4. G.B. Khristiansen et al, Nucl. Phys. (Proc. Suppl), B 28, 40 (1992).
5. J. Cronin, Nucl. Phys. (Proc. Suppl.), B 28, 213 (1992).
6. The Pierre Auger Project, Design Report, 1995 (1-st edition), 1996 (2-nd edition).
7. J. Linsley, Phys. Rev. Letters 10, 146 (1963).
8. G.T. Zatsepin and V.A. Kuzmin, JETP Letters, 4, 78 (1966).
9. K. Greisen, Phys. Rev. Letters, 16, 748 (1966).
10. A.M. Hillas, Ann. Rev. Astron. Astrophys., 22, 425 (1984).
11. C.T. Hill, Nucl. Phys. B224, 469 (1983),
12. G. Sigl, Space Sc. Rev. 75, 375 (1996).
13. J. Linsley and R. Benson, Proc, ICRC 17-th, Paris, 8, 145 (1981).
14. Mass-Airwatch, Proceedings of the workshop in Huntsville, Alabama,(1995).
15. J. Linsley, Proc. ICRC 17-th, Paris, 2, 141, (1981).
16. V.N. Tsytovich, private communication, (1997).
17. J. Linsley, Proc. ICRC 19-th, La Jolla, 3, 438, (1985).

18. J. Linsley, in (12)
19. L.V. Tverskaya and V.I. Tulupov, Geomagnetism and Aeronomia, Acad. Nauk USSR, №4, 695,(1984).
20. B.A. Tverskoy, Acad Nauk USSR Doklady, **144**,№2, 338, (1962).
21. B.A. Tverskoy, Geomagnetism and Aeronomia, Acad Nauk USSR, №1,61, (1962).
22. Yu. I. Galperin et al, Space Research (Russian), **30**, №1, 89, (1992).
23. A.S. Biryukov et al, Engineering Ecology, (1996), **6**, 92-115.

Great Science Observatories in the Space Station Era and OWL Efforts in Japan

Yoshiyuki Takahashi
(for OWL-Japan team)

Department of Physics, University of Alabama in Huntsville, Huntsville, AL35899, USA;
and LINAC, RIKEN, 2-1 Hirosawa, Wako-shi, Saitama 351-01, Japan

Abstract. A concept of "Space Factory" on the International Space Station Alpha (ISSA) is described. By following the four great observatories that purposefully took advantage of the Space Transportation System (STS), the next generation of great observatories is considered. These new astronomical projects require a very large optical telescope whose diameter is of the order of 10 m. Space telescope of this size will require careful assembly and tuning by astronauts on orbit before deployment. Once built, it could visualize the universe to the earliest galaxies, and could explore the earth-like planet in other star-system. The "Space Factory" would permit other large-scale observatories for construction in space. A step-by-step advancement of the "Space Factory" is conceived by including four or five frontier astrophysics programs. Less demanding experiments would precede the construction project of the most demanding optical telescope.

A study in Japan to observe the highest energy cosmic rays from space is synchronized with those being carried out by the OWL team in the USA and the AIRWATCH team in Italy. The Japanese efforts are coordinated in a larger program study of the Space SUBARU, which envisages a plan of orbital construction, fine-tuning and eventual deployment of large scale astrophysical instruments into the desired free-flying orbit. A space observatory of the highest energy cosmic rays can be maximized by a cluster of 6 or 7 units of the wide-angle OWL telescopes, each having a field-of-view (FOV) of ~ 60°. The ultimate viewing area could be up to 6,000 km x 6,000 km, the entire horizon for a 1000 km orbit. Within this large detection area about 10^5 cosmic hadronic events at above 10^{20} eV would be observed in a year. It also makes possible to observe the highest energy neutrinos from the known source mechanisms; including cosmic photo-production. Neutrino events from Topological Defects, Gamma Ray Burst fireballs and Blazers are observable as frequently as ~100 events per year above 10^{20} eV, if any of these hypothetical sources is real.

The OWL-JAPAN team: T. Ebisuzaki, T. Kohno, T. Otani, H. Shimizu,** K. Sunouchi, Y. Takahashi,* I. Tanihata,* A. Yoshida,** Computational Science Laboratory, LINAC Laboratory,* and Cosmic Radiation Laboratory,** RIKEN, Wako-shi, 351-01, Japan; Y. Mizumoto, National Astronomical Observatory, Mitaka, Tokyo 181, Japan; M. Nagano and M. Teshima, Institute for Cosmic Ray Research, University of Tokyo, Tanashi, Tokyo 188, Japan

CP433, *Workshop on Observing Giant Air Showers from Space*
edited by J.F. Krizmanic et al.

INTRODUCTION

The upcoming era of the International Space Station can potentially provide superb opportunities to humankind of making unprecedented, very great observatories for various space and astrophysical sciences and for other activities into deep space. Space Station has advantageous environment for constructing large structures and instruments. Weightlessness and complete freedom from any aerodynamical interference provide a powerful platform for constructing large and heavy observatories and spaceships. These are impossible to assemble and launch in a single rocket. Intelligent and powerful aids exist on ISSA. Examples are Extra-Vehicular-Activities (EVA) of experienced astronauts, various automatic sensors and monitors, as well as robotics arms. A space telescope of a 10 m size can reach 30th magnitude and 10 billion light years, and could be assembled, tuned, and tested on Station. They can be deployed to ideal orbits from the ISSA by using a booster rocket. Other scientific great observatories would be for radio, UV, X-ray, gamma-ray, and cosmic rays. They can equally be built and deployed to their best orbits. The new era starting after 2003 could see a fleet of greatest observatories that humankind has ever dreamed to have.

A total of 83 sessions of EVA's will be performed by 2003 for constructing habitat modules and truss of the ISSA. Once the ISSA is constructed, these valuable assets of astronauts and the ISSA platform can turn into a powerful space factory. The ISSA could become a true station to go to other orbits or planets.

When the Space Shuttle orbiter was envisaged in mid 70's, a powerful concept of "Great Observatories" emerged. Following the first great observatory concept of HEAO in 1960's, NASA planned four great observatories, taking advantage of the size of the Shuttle cargo-bay that could carry a large satellite. They were (Hubble) Space Telescope (HST), (Compton) Gamma Ray Observatory (CGRO), Advanced X-ray Astrophysical Facility (AXAF) and SIRTF. HST and AXAF were projects of Marshall Space Flight Center, while others were those of Goddard Space Flight Center or other Center.

This vision of STS Great Observatories was well advocated and realized by Charles Pellerin, who was the head of the Division of Astrophysics at the NASA Headquarters, and many others. Very successful performance of HST and CGRO yielded unprecedented wealth of knowledge of the Universe, as well as wonderful visual images to the general public. Indeed, both provided really revolutionary new horizon of observational universe. With the scheduled launch of AXAF in late 1998 or early 1999, the science achievements of great observatories will become greater.

Considering the tremendous new knowledge provided by these great observatories, the science community can now begin planning far greater observatories with virtually unlimited size and capacity, if we build these on the ISSA. The concept was originally envisioned by Werner von Braun right after the success of the Apollo lunar mission. After three decades since that vision, humanity is going to have a real platform. It can convert the current Space Station into the original and extremely powerful Station. It is a natural way to go beyond the limited orbit of the ISSA and to realize a large scientific observatory in space. Science community have kept complaining of Space Station's incapability for astrophysical sciences. A new era will surely revolutionize a vision.

This paper describes an opportunity of the next generation great observatories that use a "Space Factory" on ISSA. One of such astrophysical observatories, the Multiple Orbiting-array of Wide-angle Light-collector (Multi-OWL), shows that the extreme energies of the universe can be explored with decisively large statistics.

SPACE-SUBARU: SPACE TELESCOPES AND MULTI-OWL

The current NASA program does not include "Space Factory." However, the second phase projects of ISSA seem to incorporate soon an in-space assembly plan for lunar or inter-planetary spacecraft. These spacecraft's necessarily require a large and complex structure. The first phase ISSA structure will soon be completed by the year 2003 or 2004. It is not too premature to begin now a planning of a spacecraft assembly and experiments, making use of the natural infrastructure of the ISSA. Space-SUBARU [15] is a feasibility study in Japan. It studies, as its ultimate goal, how to construct a large optical and IR telescope of 10 - 20 m diameter.

NASA and ESA have been designing a 4 - 8 m telescope (NGST) for a post-HST mission. The main mirror is segmented and folded for launch and the telescope will be automatically deployed by unfolding the mirror segments. The NGST assumes an autonomous fine optical tuning to the diffraction limit. It requires an adaptive optics and many actuators to correct the deformation of the mirror. Its full-automatic deployment scheme is an ideal. Active autonomous control system stimulates the industries for the most modern technologies. Nevertheless, very many sequential steps of the automatic deployment procedure are not free from a high risk of catastrophic failures. A very large optical telescope demands very challenging optical accuracy of 0.01 arcsecond or better. Actuators are indeed successful on ground, but the technology must prove in space an efficient and rapid damping of the inertia of proper resonances. Unlike on ground, no high mass absorber of micro-vibrations exists during actuation in free space, excepting ISSA. Most of the well-planned ground telescopes, built recently by high technology, actually needed repeated access and repairs before its achievement of the designed spec.

The Space-SUBARU planning in Japan seeks a different approach to realize a very large space telescope and other scientific spacecraft. This Space-SUBARU study includes assembly of several other spacecraft experiments for different astrophysical purposes, considering a step-by-step advancement of the in-space assembly technology. The assembly and testing of less demanding spacecraft will help guiding the way to the Space-SUBARU or NGST that requires the diffraction limited optical accuracy.

The Orbiting-array of Wide-angle Light-collector (OWL) is one of the candidates for the next generation great observatories. It is dedicated to explore the highest energy universe. The OWL was approved as a NASA's mission concept for start in 2005 - 2010. The 1997 Breckenridge Workshop, which is NASA's Strategic Mission Evaluation Committee, has given this blessing to the OWL within the "Structure and Evolution of Universe (SEU)" program. The OWL is not conceived to use the ISSA, because none of the NASA Astrophysics and Space Science Divisions envisaged a use of ISSA, and an option that we can use ISSA was somewhat not assumable at the time.

Nevertheless, an assembly of OWL units into a single spacecraft is recognized as a natural first candidate for the early step experiment of Space-SUBARU. The OWL demands only 0.1 degree optical resolution, which is 10^4 times the diffraction limit and relatively easily achievable. Thus, it will be a good step for Space-SUBARU to pave the way to a very demanding Space-SUBARU 10 m optical telescope.

Other candidates, being discussed so far for the intermediate steps, are a 20 m Submilimeter Radio and IR telescope, a small-size prototype optical telescope, and a large $(10 \text{ m})^3$ Cherenkov telescope for gamma ray astronomy:

1. MULTI-OWL (120° FOV, 2.5 m D x 6)
2. Gamma-Cherenkov Observatory ~ 100 m^2 (GeV - 10 TeV)
3. LMSA (20 m Sub-Milimeter-radio & IR Telescope)
4. Prototype of Optical/IR Telescope (1 - 2 m D), which later serves for education
5. SPACE-SUBARU/NGST (10 m Optical/IR Telescope)

The above five programs are in the order of the degree of required optical resolution.

A SPACECRAFT-ASSEMBLING FACTORY ON ISSA

Significant advantages of new logistics (EVA + Robotics) will exist on the ISSA. The current Space Station work are limited to the habitat experiments and micro-gravity materials study. The von Braun's original concept of the "Space Station" was far more insightful and outreaching for the eventual quest of the space. The activities on the "Space Station" included factory-like EVA's to assemble and tune-up a large and advanced spacecraft for deployment from the low earth orbit to other orbits or planets. Two memorable Space Shuttle missions and their in-space logistics already proved the indisputable value of the von Braun's vision. Robotics arms and EVA's were critical for the success of the very demanding optical telescope, Hubble Space Telescope. Purposeful advancement and extended version of "Space Factory" on the ISSA can be very naturally conceived for more sophisticated optical telescopes. Considering the next generation of large spacecraft and astrophysical satellites, we find the following characteristics of the "Space Factory" very useful.

- INCREMENTAL TRANSPORTATION OF THE PARTS TO FORM A VERY LARGE SPACECRAFT.

- SURER ASSEMBLY AND TESTING BY EVA AND ROBOTICS.

    1. *Weightlessness* and *no aerodynamical interference* for orbital assembly would reduce unwanted mechanical stress and distortion for large optics.
    2. *High intelligence and trained human assistance* are available for perfection.
    3. *Active communication* is available for getting cooperative, ground feed-back and backup resources to solve unexpected troubles.

4. _Repair and replacement_ of bad parts are possible by re-supply logistics.
5. Some fundamental _infra-structure should already exist_ in the first phase of the ISSA construction.
6. A total of _83 EVA sessions_ will be performed by many astronauts before 2003. A large number of highly experienced astronauts will emerge during this period. This enormous and indispensable human resources would be best dedicated for more challenging, monumental achievements in space. On-orbit assembly of great scientific observatories and advanced spaceships can be perceived straightforwardly as a powerful means to realize a high value mission.

- NEAR-PERFECT DEPLOYMENT FROM ISSA.

    1. _Fine-tune-up and_ _perfection before deployment_ will promise a greater success.
    2. Attachment and installation of redundant _high power solar-panels_ and a _booster rocket_ will be feasible. Repair and rescue missions of satellites and spacecraft will become more feasible by implementing a large booster rocket during the assembly at the "Space Factory."

- THE SPACE FACTORY IS A NATURAL STEP TO DEPLOY A LARGE SPACESHIP TO THE MOON AND PLANETS.

    * Next-generation of the sophisticated manned-missions to the moon and planets is obviously more conceivable by the "Space Factory" scheme than by the past scheme of a single rocket launch. If not an already performed miniature-size probe, planetary mission is really demanding.

Assembling Platform - Possibility with JEM

Guided by the above prospect of an assembling factory on the ISSA, we consider a gradual development of the "Space Factory" for great scientific observatories. Tentative candidates (1 - 5) listed in the previous page are in the order of less stringent optical demands. **FIGURE 1** shows the Multi-OWL assembly on the JEM platform of the ISSA. **FIGURE 2** shows a 3-dimensional rendition of an assembly of the Space-SUBARU optical telescope (10 m) at the same JEM site. Attachments such as detectors, housing and solar panels are not illustrated. Crispy star images in space help fine tuning of the system.

The present ISSA was not purposely planned for in-space construction of a large mechanical structure suitable for an assembly of a10 m class telescope. Nevertheless, there are several attachment points that would allow an extension of the Truss for an adequate assembly platform. The Japanese Experiment Module (JEM) is directly attached to the Japanese habitat module and has a long robotics arm as shown in **FIGURES 1**

FIGURE 1 Artist view of an assembly of the Multi-OWL at the JEM site on the ISSA.

FIGURE 2 A concept of a construction of the Space-SUBARU (10 m) on the JEM platform [15].

and 2. This site accommodates 12 exposure modules of 1.8 m x 90 cm x 70 cm. It is possible to produce an assembly platform of about 10 m square, where a mechanically flexible mother base could be attached within a partial reach of a robotics arm. Other attachment points on the ISSA currently do not have planned robotics arms. The OWL-Japan team consider an engineering study sharing the same mechanical needs with the other projects in the Space-SUBARU efforts. An international common design for a dedicated assembly site might eventually emerge in future plans by NASA, ESA and other space agencies for the second phase of the ISSA.

Great Observatory for Highest Energy Astrophysics - OWL

One of the most significant astrophysical mysteries is the existence of extremely high energy cosmic rays, and an inquiry into the ultimate highest-end of the particle energies in the universe awaits an answer. The eventual quest towards this ultimate energy begun in early 1960's when Volcano Ranch detected the first air shower event with energy exceeding 10^{20} eV [1]. The discovery of the cosmic microwave background followed in a few years and led to the Greisen-Zatsepin-Kuzmin (GZK) cut-off of the nucleon spectrum at 5×10^{19} eV [2]. Large-area ground Air Shower arrays were built and operated since then at several stations, including SUGAR [3], Haverah Park [4], Yakutsuk [5], Fly's Eye [6] and AGASA [7]. Their detector area ranged from a few to 10 km square. Several events around 10^{20} eV were observed by them. Higher energy events at 3×10^{20} eV and 2×10^{20} eV, recorded in early 90's by Fly's Eye and AGASA, respectively, further boosted the research interest. Enlargement of the ground-based air shower arrays to about $(50 - 70 \text{ km})^2$ is being considered by the groups of Hi-Res [8], Auger Laboratory plan [9], and Telescope Array plan [10]. They can increase the statistics by 5 - 70 times over the past experiments.

The MASS [11] was the first to study a large-area observation from space. Its study turned into the subsequent OWL study [12]. The MASS/OWL satellites would observe from space the atmospheric fluorescence of the air shower created by Extremely High-Energy Cosmic Rays (EHECR). The Orbiting Wide-Angle Light-collector (OWL) [12, 13] and the AIRWATCH [14] will challenge this physics of the extreme condition in the universe, planning to have decisively large statistics. The orbiting air shower observatories extend the detector size to 1,000 km x 1,000 km or beyond, while the ground-based observatories cannot observe an area larger than 100 km x 100 km. The flux of the highest energy cosmic rays above 10^{20} eV is so low *(1 particle per km^2 per century)* that a very large detector size (area coverage) is necessary. The OWL team envisages the detector size as large as $(1000 \text{ km})^2$ that would allow to observe more than 1,000 hadronic events per year at energies above 10^{20} eV. This idea can enhance the observable scale further, if we can use the ISSA platform for "Assembling Factory." This paper describes below an extended OWL observatory that would use the "Assembling Factory" on ISSA.

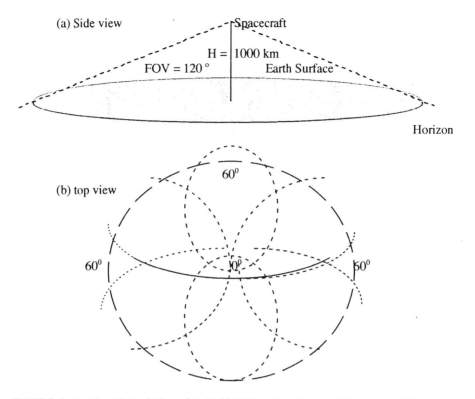

FIGURE 3 Combined Field-of-View of the Multi-OWL units: (a) horizon (side view), and (b) a projection.

TABLE 1 Target mass of atmosphere and area size [18] for observation with an orbiting Multi-OWL.

Altitude	Angle of horizon (from Nadir)	Radius of the earth's viewing horizon	View of Earth surface area & mass	Fraction of the total Earth surface
1000 km	59.8°	3242 km	3.46 x 10⁷ km² 3.5 x 10¹⁴ tons	6.78%

EXTENDED FIELD OF VIEW BY USING EVA'S ON ISSA

The Multi-OWL can consist of 6 (or 7) units of the MASS/OWL strawman optics (double Fresnel lens system) as designed by Lamb et al. [16]. The main optics would be either double Fresnel lenses of ~ 2.5 m entrance pupil, or Maksutov-type catadioptric system. Each OWL unit can observe FOV of 60 degrees [16] with 1000 x 1000 pixels UV sensors like Multi-Anode Micro-channel Array (MAMA) [17]. **FIGURE 1** illustrates an artist view of a Multi-OWL configuration at the JEM platform using a catadioptric unit for optics. Our current preference of optics is a Double-Fresnel-Lens system and the individual unit differs from this illustration.

The FOV of the Multi-OWL can cover the entire horizon of the earth viewed from a 1000 km orbit. This possibility to extend the target area by using multiple OWL optical units was pointed out by Spillantini et al., as discussed in reference [18]. **FIGURE 3** shows a multiple OWL that covers the earth's horizon. **TABLE 1** summarizes the Multi-OWL area size and the target volume of the atmosphere as a detector for the EHECR and neutrinos.

The extended area size of $(6000 \text{ km})^2$ would provide extremely high statistics of cosmic ray events at above 10^{20} eV. It would be 10^5 events per year. About 10 events with energy greater than $\mathbf{10^{22}}$ **eV** can be imagined if the GZK cut-off somehow fails.

Large FOV for Neutrino Astrophysics

The baseline objective of the OWL is to obtain a statistically decisive number of events of cosmic ray protons, nuclei and gamma rays at above 10^{20} eV, as reported by the MASS [11], OWL group [11 -13] and the Airwatch group [14]. Gamma ray bursts (GRB's) are also observable with OWL/AIRWATCH by triggering a global surge of atmospheric fluorescence, as reported already by Sacco [19] in Airwatch Symposium and by Scarsi [20] in the OWL Workshop. Scarsi also called for an attention on neutrinos at around $10^{19 \cdot 20}$ eV, considering their possible origin in GRB fireballs [22, 22, 23].

EXTREMELY-HIGH ENERGY NEUTRINO ASTROPHYSICS

QCD Parton Densities at the Highest Energies

The neutrino flux is extremely so low at EHE that no events can be observed even by using the whole atmosphere of the earth as targets. Unless the neutrino cross section rises as the modern QCD predicts [24], this remains solidly negative. The old scheme, universal 4-Fermi interaction or weak boson theory, was to predict a constancy of νN cross section ($< 10^{-35} \text{ cm}^2$) at above $E_{CMS} \sim 100$ GeV. The advanced parton structure functions of nucleons established at DESY predicts $\ln \sigma \propto 2\sqrt{\ln E_\nu}$, which can be empirically approximated by $\sigma_{\nu N} \propto E_{LAB}^{0.45 \pm 0.1}$. The likely discovery of *any* neutrino events at EHE with the OWL or Multi-OWL would clearly mean increased $\sigma_{\nu N}$ ($> 10^{-31}$ cm^2) at EHE and would become a proof of the QCD theory of neutrino-nucleon interactions that incorporated the weak bosons and the energy evolution of parton densities.

The Multi-OWL would have 3.6×10^{14} tons of atmospheric target within $S = (6000 \text{ km})^2$. With $\sigma_{\nu\text{-Air}}^{CC} \approx 14 \times 10^{-31} \text{ cm}^2$, the charged-current interaction probability we are interested in is within the atmospheric height of 500 g/cm^2, namely, $p_{int} (\theta) \approx 500 \sec\theta /$ $\lambda_{\nu\text{-Air}}^{CC} = 3 \times 10^{-5} \sec\theta \ g/cm^2$. The effective observational solid angle for detecting neutrinos with zenith angle up to $\theta = 90°$ is

$$\Omega_{eff} = 2\pi \int p_{int}(\theta) \cdot \cos\theta \cdot \sin\theta \; d\theta \approx 2 \cdot 10^{-4} \text{ sr.} \qquad (1)$$

Assuming one year flight operation with 10% duty cycle, we estimate the effective exposure time $T_{eff} = 3.2 \cdot 10^6$ sec. The event rate N (per year) is,

$$N \equiv I_\nu (E > 10^{20} \text{ eV}) \cdot (S\Omega T)_{eff} = I_\nu (/cm^2 \text{ sec sr}) \cdot 2.3 \times 10^{20} \text{ cm}^2 \text{ sec sr,} \qquad (2)$$

where we ignored, for simplicity, the events coming from below the horizon ($\theta > 90$).

Greisen Neutrinos and Conventional Sources

The conventional estimate of the highest-energy neutrino flux ($I_\nu \sim 10^{-20}/cm^2$ sec sr) is uncertain to a factor of about 10. It is not sufficiently high even for the Multi-OWL. The number of observable events would be only several events per year at 10^{20} eV for Greisen neutrinos. Because they are produced in the Universe by protons in photo-production process with 2.7° K microwave background [25, 26] (**FIGURE 4**), their population would be isotropic and very uniform. Other known plausible sources are those from AGN and Galactic Center, which are much less than the Greisen neutrinos at energies above 10^{20} eV, but they should be highly localized. Multi-OWL will have statistically meaningful Greisen-neutrino events at 10^{19} eV (~ 100 events per year). The optical aperture of the Multi-OWL currently considered is several meters for reducing the threshold to 10^{19} eV.

Clear Test of the Topological Defects

Topological defects model predicts $I_\nu (E > 10^{20}$ eV$) \sim 10^{18}/cm^2$ sec sr. About (200 - 2000) events per year at above 10^{20} eV could be observed by Multi-OWL. If such a high rate of neutrino events is indeed found, it could become a critical evidence for the topological defects [27] (**FIGURE 5**).

The flatter air shower spectrum already being observed above 5×10^{18} has been considered as an evidence for extragalactic origin. The increasing observability of neutrinos under the Standard Theory scheme would make the observable spectrum of this component flatter by $E^{0.45 \pm 0.1}$ relative to the already flatter spectrum of (presumably) hadrons at above 10^{19} eV. Neutrinos do not suffer from the GZK cut-off and may dominate observations at the highest energies above 10^{21} eV. Although the predicted neutrino intensity from Topological Defects is high, no current and future experiments are capable to detect them and this cosmological conspicuous hypothesis must await a powerful observatory like Multi-OWL.

A satellite view is most effective for horizontal or large angle air showers, unlike the Fly's Eye or ground air shower arrays. A long streak of light would be seen across the image-plane of the OWL. We recognized that high-resolution imaging of large-angle showers is quite feasible. Even a single "snapshot" alone (projected shower profile without temporal differentiation of pixels) would allow determination of energy, zenith angle and altitude of the shower maximum to a certain degree of resolution. Separated successive showers can be seen for ν_τ events. Two units of OWL or Multi-OWL, stereo-satellites, with fast timing imaging devices at each pixel are useful for maximal information on the shower development. Optimization of the stereo configuration will require a careful examination, considering the desirable spatial and temporal resolution. Preliminary Monte Carlo simulation by Dai et al. indicates the angular resolution of 0.17° (rms) and the shower maximum height resolution of ~ 50 g/cm^2 for a stereo OWL [28].

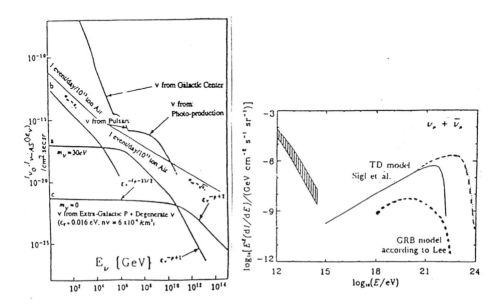

FIGURE 4 Neutrinos expected from standard sources. **FIGURE 5** Neutrinos from TD's and GRB's.

Probing Blazers and GRB Sources

Other potential high intensity astrophysical sources are Blazers and GRB's. They are very powerful high energy gamma ray sources and naturally suspected for high energy acceleration. Ultra-relativistic shock ("0-th order" Fermi) by Quenby and Lieu [29] shed light on a possibility of efficient EHE acceleration mechanism in fireballs and AGN jets where the Lorentz factor of the shock exceeds $10^{3 \cdot 4}$. Though still very speculative, some predictions of neutrino emission from GRB's have been made, at the level of 3 x 10^{19} /cm^2 sec sr at E = 10^{20} eV. These point sources could yield a number of Air Shower events for the Multi-OWL. Rough estimates suggest that the Multi-OWL should be able to see about 30 events per year from either Blazers or GRB's.

Neutrino Event Rate with Multi-OWL

Assuming only conventional Greisen neutrinos from $p + \gamma_{2.7k} \rightarrow X + \pi (\rightarrow \nu)$ process, we expect that the Multi-OWL would obtain a marginal number of events at above 10^{20} eV. However, several hundred events per year are expected at 10^{19} eV for Greisen neutrino events.

The neutrino events at $E > 10^{19}$ eV from a GRB model [22 - 23] and the reduced Topological Defects model (p = 1 model) [27] would be ~ 100 and $\sim (500 - 5000)$ events per year, respectively. They are ~ 30 and $\sim (200 - 2000)$ events per year at 10^{20} eV.

TABLE 2. Comparison of the neutrino event rates with Multi-OWL for three major origins. The energy dependence of the neutrino cross section ($\sigma \propto E^{0.5}$) is applied .

MULTI-OWL	N_ν /yr $E > 10^{19}$ eV	N_ν /yr $E > 10^{20}$ eV	Source Location	Spectral index β (E^β dE)
Greisen ν's	~ 100	~ 2	isotropic	> 3
TD ν's	$\sim (500 - 5000)$	$\sim (200 - 2000)$	discrete	~ 1.5
GRB ν's	~ 100	~ 30	point sources	~ 2
Blazer ν's	500	~5	point sources	~3.5

Due to significant difference of the spectral indices of these different neutrino predictions, the source discrimination would be possible in the energy regime 10^{19} eV - 10^{21} eV (**TABLE 2**).

TESTING FUNDAMENTALS OF RELATIVITY

Observable highest energy or Lorentz factor is likely that of the EHE neutrinos. These neutrinos in the inverse system are interesting for testing the Equivalence Principle and the fundamental absolute system.

The $\nu + p$ interaction at $E_\nu = 10^{20}$ eV is equivalent to the process of $p + \nu$ at $E_p = 10^{29}$ eV, if we accept the finite neutrino mass, 0 eV $\leq M_\nu \leq 1$ eV, and the Equivalence Principle. By a neutrino event at 10^{20} eV, we would experience an observation of the highest Lorentz factor, γ_ν (or γ_p) $\geq 10^{20}$. The argument to test the Equivalence Principle by EHE protons is valid for Lorentz factor of the order of $\gamma_\nu \geq 5 \times 10^{10}$ by examining if the GZK cut-off fails for protons. The EHE neutrinos provide far greater Lorentz factors in this sense.

Extreme Lorentz factors are interesting, because it can test the absolute system denied long time ago by the Michelson-Morley's experiment. Any asymmetry in the directional distribution (θ, ϕ) of v-induced Air Showers could lead to a test of the cosmological rest frame to the highest sensitivity of the velocity: $v \leq 1.5 \times 10^{-30}$ cm/s.

SPECIFIC MULTI-OWL TASKS FOR ASSEMBLY ON ISSA

The initial study is considered for 2 years from 1998, regarding.the following tasks:

TASK 1. Concept development of MULTI-OWL assembly on ISSA and computers
         - RIKEN with NASDA and industries (Mitsubishi, IHI and Toshiba).
TASK 2. Detector R&D for OWL - RIKEN with HAMAMATSU
TASK 3. Upgrading of the Monte Carlo code - NAO and others.
TASK 4. Science developments with the ground arrays (AGASA/TA) - ICRR

SUMMARY

A new series of great observatories is envisaged for the Space Station era by using powerful infrastructure of robotics arm and astronauts' EVA. It was shown that an orbiting "Factory" to assemble advanced spacecraft for eventual deployment to appropriate orbits and/or moon and planets would be most naturally conceived for the space program, once Space Station became operational. Several science projects that use large optical systems appear to be useful for the advanced spacecraft to be manufactured on ISSA. They are promising for success by a new scheme with the "Space Factory," while a single-rocket launch plan of them would remain highly risky or nearly impossible.

Maximum OWL configuration is considered by a Multi-OWL assembly scheme that uses the EVA's and robotics arms on the Space Station. It can provide the target area and volume of up to (6000 Km)^2 and 3.6×10^{14} tons. It would enable the observation of extremely high energy cosmic rays to 10^5 events per year. Neutrinos would become observable with Multi-OWL. The statistics of Greisen neutrino events remains poor, but those from hypothetical sources such as the Cosmic Topological Defects, Blazers and GRB fireballs, would be as frequent as 100 events per year at energies above 10^{20} eV.

The Multi-OWL configuration and designing proposal was submitted to NASDA for engineering study from April 1998 to March 2000.

ACKNOWLEDGMENTS

The author is grateful to Dr. Mamoru Mohri of NASDA for cooperative work and helpful discussions to plan a "Space Factory." He wishes to thank Dr. T. Ebisuzaki of RIKEN and Dr. T. Kajino of National Astronomical Observatory for cooperative work in planning the "Space Factory" and the Space-SUBARU. He is also indebted for valuable advice and suggestions to Dr. N. Kaifu of National Astronomical Observatory, Dr. I. Tanihata of LINAC Laboratory, RIKEN, and Drs. J. Arafune and T. Kifune of the Institute for Cosmic Ray Research, University of Tokyo. Thanks are due to the members of the Space-SUBARU, OWL and AIRWATCH teams for various discussions. Encouragement and supports from the Computational Science Laboratory and the LINAC Laboratory of RIKEN are gratefully acknowledged. Discussion in 1995 with Dr. Y. Totsuka of the Institute for Cosmic Ray Research, University of Tokyo, was helpful in the initial stage of the OWL efforts in Japan.

REFERENCES

1. J. Linsley, Phys. Rev. Lett. **10**, 146 (1963); Scientific American, **239**, 60 (1978).
2. K. Greisen, Phys. Rev. Lett., **21**, 1016 (1966); G.T. Zatsepin and V.A. Kuzmin, JETP Lett. **4**, 78 (1966).
3. M.M. Winn, et al., J. Phys. G. Nucl. Phys. **12**, 653 (1986) and **12**, 675 (1986); C.B.A. McCusker and M.M. Winn, Nuovo Cimento, X**28**, 175 (1963); C.J. Bell et al., J. Phys G, 7, 990 (1974).
4. M.A. Lawrence, R.J.O. Reid and A.A. Watson, J. Phys. G **17**, 733 (1991).
5. B. N. Afanasiev et al., Proc. Int. Symp. on EHECR, Univ. of Tokyo, ed. N. Nagano, 32 (1996); A. Glushkov et al., Astropart. Phys., **4**, 15 (1995).
6. P. Sokolsky, this volume; R.M. Baltrusaitis et al., Nucl. Instr. Meth, **A240**, 410 (1995); D.J. Bird et al., ApJ **424**, 491 (1994).
7. N. Hayashida et al., Phys. Rev. Lett., **73**, 3491 (1994); Phys. Rev. Lett., **77**, 1000 (1996); S. Yo-shida et al., Astropart Phys., **3**, 105 (1995); N. Chiba et al., Nucl. Instr. Meth, A311, 338 (1992).
8. P. Sokolsky, in press, "OWL", AIP Proc., June (1998).
9. J. Cronin, this volume, "OWL", AIP Proc., June (1998); Nucl. Phys. B, **28B**, 213 (1992); Fermilab Workshop (1995).
10. M. Teshima, this volume, "OWL", AIP Proc., June (1998).
11. Y. Takahashi, Proc. 24th International Cosmic Ray Conference, Rome, **3**, 595 (1995); *MASS/AIRWATCH Huntsville Workshop Report*, pp. 1 - 16 , Univ. of Alabama in Huntsville (1995); Proc. IInd Rencontres du Vietnam, "Sun and Beyond", Oct., 1995, Ho-Chi Minh City, ed. Tranh Thanh Van, World Scientific, pp. 445 - 454 (1996); Y. Takahashi et al., SPIE 2806, pp. 102 - 112, (1996).
12. J. Ormes et al., Proc. 25th ICRC, Durban, (1997).
13. R.E. Streitmatter, this volume, "OWL", AIP Proc., June (1998).
14. L. Scarsi (ed.), Proc. First Airwatch Symposium, Catania, 1997; Di Marzo, this volume (1998).
15. T. Ebisuzaki et al., proposal submitted to Japan Space Forum of NASDA , "Space-SUBARU", RIKEN, 1997 and 1998.
16. D. J. Lamb et al., 4 papers, this volume, "OWL", AIP proc. June (1998); R. A. Chipman, *MASS/AIRWATCH Huntsville Workshop Report*, pp. 276-282 (1995).
17. J. Mitchell, this volume, "OWL", AIP Proc., June (1998).
18. L. Scarsi, J. Linsley, P. Spillantini, and Y. Takahashi, Proc. 25th ICRC, Durban,
19. B. Sacco, in Proc. First Airwatch Symposium, Catania, 1997.
20. L. Scarsi, this volume, "OWL", AIP Proc., June (1998).

21. E. Waxman, ApJ **452**, 1 (1995); E. Waxman and J. Bahcall, Phys. Rev. Lett, **78**, 2293 (1997).

22. M Milgrom and . Usov, ApJ **449**, L37 (1995); Lee, in S. Colafrancesco, "High-E Phenomena & Large Scale Structure," Proc. First Int. Symposium on Airwatch, Catania, 1997.

23. M. Vietri, Phys. Rev. Lett., submitted, 1997; ApJ, **453**, 883 (1995).

24. C. Quigg et al, Phys. Rev. Lett. 57 (1986); R. Gandhi et al., Astropart. Phys. **5**, 81 (1996), and Nucl. Phys. **B78**, 475 (1996).

25. T. Hara and H. Sato, Prog. Theor. Phys. **65**, 477 (1981); F. W. Stecker, ApJ 228, (1979).

26. S. Yoshida et al., ApJ. **479**, 547 (1997).

27. C. T. Hill, Nucl. Phys. **B224**, 469 (1983); C.T. Hill, D. Schramm and T.P. Walker, Phys. Rev. **D36**, 1007 (1987); P. Bhattacharjee and N.C. Rana, Phys. Lett., **B246**, 365 (1990); G. Sigl, S. Lee, D. Schramm and P. Bhatttacharjee, Science, **270**, 1977 (1995); G. Sigl., Space Sci. Rev. **75**, 375 (1996).

28. H.Y. Dai et al.; and R.E. Streitmatter, this volume, "OWL" AIP Proc. June (1998).

29. J.J. Quenby and R. Lieu, Nature **342**, 654 (1989); R. Lieu and J.J. Quenby, ApJ **350**, 692 (1990).

Bremsstrahlung and Pair Creation: Suppression Mechanisms and How They Affect EHE Air Showers

Spencer R. Klein

Lawrence Berkeley National Laboratory, Berkeley, CA, 94720

Abstract. Most calculations of air shower development have been based on the Bethe-Heitler cross sections for bremsstrahlung and pair production. However, for energetic enough particles, a number of different external factors can reduce these cross sections drastically, slowing shower development and lengthening the showers. Four mechanisms that can suppress bremsstrahlung and pair production cross sections are discussed, and their effect on extremely high energy air showers considered. Besides lengthening the showers, these mechanisms greatly increase the importance of fluctuations in shower development, and can increase the angular spreading of showers.

INTRODUCTION

The electromagnetic portion of high energy air showers is governed by bremsstrahlung and pair production. Although the formulae for these process have been around for over 60 years, it is not well known that, in many situations, these formulae can be very wrong. The medium in which the bremsstrahlung or pair production occurs can drastically affect the cross sections. This contribution will discuss four different ways in which the medium can reduce the bremsstrahlung and pair creation cross sections.

These suppression mechanisms can affect air showers by increasing the effective radiation length, lengthening the showers, and moving the position of the shower maximum deeper into the atmosphere. For ground based arrays like the proposed Auger Observatory, even a small change in the depth of shower max can affect the energy hitting the ground, especially for non-vertical showers where shower maximum is considerably above ground level. Air fluorescence detectors like Flys Eye and the proposed OWL can measure the shower profile, and so their energy measurement would be less affected by unforeseen changes in the shower development. However, a change in the position of shower maximum can affect measurements of the composition of the highest energy cosmic rays. Moreover, these mechanisms will drastically reduce the number of particles in the early stages of the shower, changing the shower development profile.

CP433, *Workshop on Observing Giant Air Showers from Space*
edited by J.F. Krizmanic et al.

I   BREMSSTRAHLUNG AND PAIR PRODUCTION SUPPRESSION MECHANISMS

Suppression mechanisms for bremsstrahlung and pair production are possible because of the unusual kinematics in these processes. For ultrarelativistic particles, the momentum transfer between the radiating electron or converting photon and the target nucleus is very small, especially in the longitudinal direction [1]. For bremsstrahlung where $E \gg m$,

$$q_{||} = p_e - p'_e - k/c \tag{1}$$

$$= \sqrt{(E/c)^2 - (mc)^2} - \sqrt{((E-k)/c)^2 - (mc)^2} - k/c = \frac{m^2 c^3 k}{2E(E-k)} \tag{2}$$

where p_e and p'_e are the electron momenta before and after the interaction respectively, k is the photon energy, m is the electron mass and $\gamma = E/m$. For ultrarelativistic electrons, $q_{||}$ can be very small. For example, for a 1 EeV electron emitting a 100 PeV photon, $q_{||} = 10^{-9}$eV/c. Because $q_{||}$ is so small, by the uncertainty principle, it must take place over a long distance, known as the formation length:

$$l_{f0} = \frac{\hbar}{q_{||}} = \frac{2\hbar E(E-k)}{m^2 c^3 k}. \tag{3}$$

For the above example, l_{f0} is 200 meters; for a 1 PeV photon from the same electron, l_{f0} rises to 20 km. This distance is the distance required for the electron and photon to separate to become distinct particles. It is also the path length over which the emission amplitude adds coherently to produce the emission probability. If something happens to the electron or nascent photon while it is traversing the formation zone, then the coherence can be disrupted, reducing the effective formation length and hence the emission probability. Even weak forces, acting over a long formation length, can be strong enough to destroy the coherence required for emission. The mechanisms discussed here work by disrupting the electron or photon, reducing the effective formation length.

A   Multiple Scattering (The LPM Effect)

Multiple scattering can cause disruption by changing the electron trajectory. If, taken over l_{f0}, the electron multiple scatters by an angle larger than the typical bremsstrahlung emission angle $1/\gamma$, then emission can be suppressed [1].

The reduction can be calculated by considering the effect multiple scattering has on $q_{||}$; as the electron changes direction, it's forward velocity is reduced, and, with it, producing a change in $q_{||}$. This can be modelled by dividing the multiple scattering evenly between p_e and p'_e. Then,

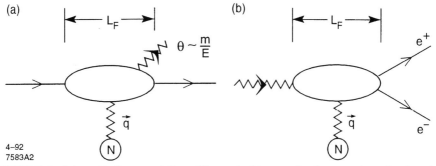

(a)   (b)

$\theta \sim \dfrac{m}{E}$

4-92
7583A2

FIGURE 1. Schematic Representation of bremsstrahlung and pair conversion, showing the formation zone.

$$q_{\parallel} = \sqrt{(E \cos \theta_{MS/2}/c)^2 - (mc)^2} - \sqrt{((E-k) \cos \theta_{MS/2}/c)^2 - (mc)^2} - k/c \quad (4)$$

where $\theta_{MS/2}$ is the multiple scattering in half the formation length, $E_s/E\sqrt{l_f/2X_0}$, where $E_s = m\sqrt{4\pi\alpha} = 21$ MeV, and X_0 is the radiation length. Scattering after the interaction is for electron energy $E - k$. This leads to a quadratic in l_f:

$$l_f = \frac{2\hbar E(E-k)}{km^2c^3(1 + E_s^2 l_f/m^2c^4X_0)} = l_{f0}\left[1 + \frac{E_s^2 l_f}{m^2c^4X_0}\right]^{-1}. \quad (5)$$

If multiple scattering is small, this reduces to Eq. (3). When multiple scattering dominates

$$l_f = \sqrt{\frac{2\hbar cE(E-k)X_0}{E_s^2 k}} = l_{f0}\sqrt{\frac{kE_{LPM}}{E(E-k)}}. \quad (6)$$

where E_{LPM} is a material dependent constant, given by $E_{LPM} = m^4c^7X_0/2\hbar E_s^2 \approx 3.85$ TeV/cm X_0. For lead, $E_{LPM} = 2.2$ TeV, while for water $E_{LPM} = 139$ TeV and for sea level air $E_{LPM} = 117$ PeV.

Since the formation length is the maximum distance over which the bremsstrahlung amplitude add coherently, the bremsstrahlung amplitude is proportional to the formation length, so the suppression factor is

$$S = \frac{d\sigma/dk}{d\sigma_{BH}/dk} = \frac{l_f}{l_{f0}} = \sqrt{\frac{kE_{LPM}}{E(E-k)}} \quad (7)$$

and the $dN/dk \sim 1/k$ found by Bethe and Heitler changes to $dN/dk \sim 1/\sqrt{k}$.

A similar effect occurs for pair production, where the produced electron and positron can multiple scatter. The two effects are closely related, as is shown in Fig. 1, and this relationship can be used to relate the bremsstrahlung and pair creation formation lengths and cross sections. For pair production

134

$$l_{f0} = \frac{2\hbar E(k - E)}{m^2 c^3 k}; \tag{8}$$

the corresponding suppression is

$$S = \sqrt{\frac{k E_{LPM}}{E(k - E)}}. \tag{9}$$

This calculation has several limitations. The semi-classical approach may fail for $k \sim E$. The calculation neglects the statistical nature of multiple scattering, instead treating it deterministically. And, it neglects many niceties like the large non-Gaussian tails on Coulomb scattering and electron-electron inelastic scattering.

Migdal developed a more sophisticated approach which avoided many of these problems [2]. He treated the multiple scattering as diffusion, calculating the average radiation for each trajectory, and allowing for interference between nearby collisions. He found the cross section for bremsstrahlung is

$$\frac{d\sigma_{LPM}}{dk} = \frac{4\alpha r_e^2 \xi(s)}{3k} \{y^2 G(s) + 2[1 + (1 - y)^2]\phi(s)\} Z^2 \ln\left(\frac{184}{Z^{1/3}}\right). \tag{10}$$

where $G(s)$ and $\phi(s)$ are the solutions to differential equations, given by [3]

$$\phi(s) = 1 - \exp\left[-6s[1 + (3 - \pi)s] + s^3/(0.623 + 0.796s + 0.658s^2)\right] \tag{11}$$

$$\psi(s) = 1 - \exp\left[-4s - 8s^2/(1 + 3.96s + 4.97s^2 - 0.05s^3 + 7.5s^4)\right] \tag{12}$$

$$G(s) = 3\psi(s) - 2\phi(s). \tag{13}$$

where

$$s = \frac{1}{2}\sqrt{\frac{E_{LPM}k}{E(E - k)\xi(s)}}. \tag{14}$$

For $k \ll E$, $s \sim 1/ < \gamma\theta_{MS} >$. For $s \gg 1$, there is no suppression, while for $s \ll 1$, the suppression is large. $\xi(s)$ accounts for the increase in radiation length as photon emission drops. Migdals solution for s and $\xi(s)$ is recursive, because s depends on $\xi(s)$. The recursion can be avoided by defining [3]

$$s' = \frac{1}{2}\sqrt{\frac{E_{LPM}k}{E(E - k)}}. \tag{15}$$

This is possible because ξ depends only logarithmically on s; a modified formulae for ξ depends only on s':

$$\xi(s') = 2 \qquad\qquad (s' < \sqrt{2}s_1) \tag{16}$$

$$\xi(s') = 1 + h - \frac{0.08(1 - h)[1 - (1 - h)^2]}{\ln\sqrt{2}s_1} \qquad (\sqrt{2}s_1 < s' \ll 1) \tag{17}$$

$$\xi(s') = 1 \qquad\qquad (s' \geq 1) \tag{18}$$

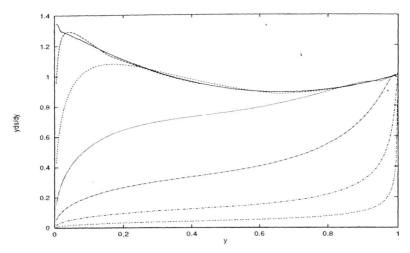

FIGURE 2. $yd\sigma_{\text{LPM}}/dy$ for bremsstrahlung as a function of y for various electron energies in lead (E_{LPM}=2.2 TeV), showing how the spectral shape changes. Electrons of energies 10 GeV, 100 GeV, 1 TeV, 10 TeV, 100 TeV, 1 PeV and 10 PeV are shown. The upper curve (solid line) is for $E = 10$ GeV, and the emission drops as energy rises.

where $h = \ln s'/ln(\sqrt{2}s_1)$. In the strong suppression limit, $s \to 0$, $G(s) = 12\pi s^2$ and $\phi(s) = 6s$. With these approximations, the semi-classical $d\sigma/dk \sim 1/\sqrt{k}$ scaling is recovered, albeit with a different coefficient.

Fig. 2 shows how the LPM effect reduces $yd\sigma/dy$ ($y = k/E$) for electrons in a lead target. The 10 GeV electron curve is very close to the Bethe-Heitler prediction; in the absence of suppression, this curve would hold for all electron energies. As the electron energy rises, emission drops. At the highest electron energies, photons with $k \ll E$ are almost completely suppressed. For $E^2/E_{LPM} < k < 1.3E^2/E_{LPM}$, the Migdal curve rises slightly above the unsuppressed; this is a consequence of either the approach or the approximations Migdal used.

Fig. 3 shows how the pair production cross section is reduced. Compared with bremsstrahlung, pair production suppression sets in at higher energies. Symmetric pairs are suppressed the most; in the extremely high energy limit, one of the produced electrons takes almost all of the photon energy. So, where the LPM effect is extremely strong, an electromagnetic shower becomes a succession of interactions where an electron emits a bremsstrahlung photon that takes almost all of the electrons energy, followed by a very asymmetric pair conversion, producing an electron or positron with almost all of the energy of the initial lepton.

Two metrics for suppression effects in showers are the electron energy loss (dE/dx) and photon conversion cross section. Fig. 4 shows how these two markers change with energy in a lead target. As the particle energy rises above E_{LPM} (2.2 TeV for lead), the energy loss and conversion cross section fall, with bremsstrahlung

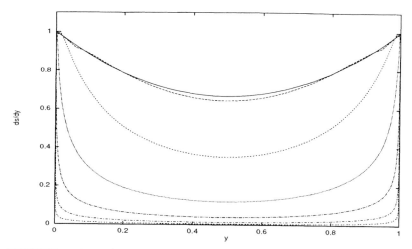

FIGURE 3. $d\sigma_{\text{LPM}}/dy$ for pair production in lead, as a function of y for various photon energies, showing how the spectral shape changes. The solid line is for $k = 1$ TeV, and the cross sections drop with energy; the other curves are for photons of energies 1 TeV, 10 TeV, 100 TeV, 1 PeV, 10 PeV, 100 PeV and 1 EeV.

affected at lower energies than pair conversion.

The LPM effect was studied experimentally by the SLAC E-146 collaboration, who sent 8 and 25 GeV electrons through thin (0.07 % to 6% of X_0) targets of materials ranging from carbon to gold. Fig. 5 shows their data for carbon [4], the material that is closest to air. Both LPM and dielectric suppression are needed to explain the data.

Since E-146, there have been a number of additional calculations of bremsstrahlung with multiple scattering, using a variety of different approaches [5–8]. Several calculations have showed that, the non-Gaussian tail of large angle Coulomb scatters introduces additional term into the cross section; in the limit of large suppression [6–8]

$$S = \sqrt{\frac{kE_{LPM}}{E(E-k)} \log\left(\frac{E(E-k)}{kE_{LPM}}\right)} \tag{19}$$

This will reduce the magnitude of suppression in extreme conditions. Inelastic interactions, involving the atomic electrons should be treated separatedly from the elastic nuclear interactions, and a different potential should be used [6,8]. This may explain the poor agreement observed by E-146 between the data and the Migdal calculations for their light carbon and aluminum targets, especially for the 25 GeV data, around $k = E^2/E_{LPM}$. In light of these results, calculations based on Migdal's formulae should be treated with caution, especially for light targets like air, where the inelastic form factor is important.

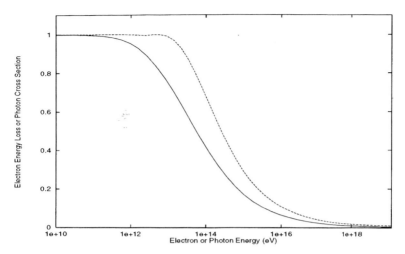

FIGURE 4. The reduction in electron energy loss ($\int_0^E (k dN/dk) dk$) due to the LPM effect (solid line) and in the photon conversion cross section (dashed line) due to the LPM effect in a lead target. The LPM effect turns on at higher energies for photons.

B   Dielectric Suppression

Dielectric suppression occurs because photons produced in bremsstrahlung can interact with the atomic electrons in the medium by forward Compton scattering [9]. The photon wave function acquires a phase shift depending on the dielectric constant of the medium $\epsilon(k) = 1 - (\hbar\omega_p)^2/k^2$ where ω_p is the plasma frequency of the medium. With this substitution, the photon momentum (k/c) in Eq. (1) becomes $\sqrt{\epsilon}k/c$, and $q_{\parallel}$ acquires an additional term $(\hbar\omega_p)^2/2ck$. This leads to a reduced l_f, and a suppression factor

$$S = \frac{k^2}{k^2 + (\gamma\hbar\omega_p)^2}. \tag{20}$$

This effect only applies for small y, $y < y_{die} = \omega_p/m$. For typical solids, $\omega_p \sim 60$ eV, so $y_{die} \sim 10^{-4}$. For $y < y_{die}$, the photon spectrum becomes $d\sigma/dk \sim k$, suppressing the emission by $(\gamma\hbar\omega_p)^2/k^2$. The effect is also clealy apparent in the E-146 data in Fig. 5.

C   Suppression of Bremsstrahlung by Pair Creation, and vice-versa

As Landau and Pomeranchuk pointed out, when l_{f0} becomes as long as X_0, then partially created photons can be pair create, destroying the coherence [1]. A simple

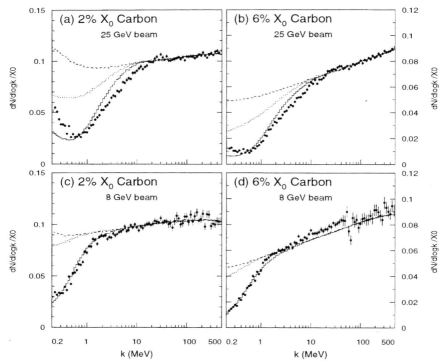

FIGURE 5. Comparison of data from SLAC-E-146 with Monte Carlo predictions for 200 keV to 500 MeV photons from 8 and 25 GeV electrons passing through 2% and 6% X_0 thick carbon targets. The cross sections are $dN/d(\log k)/X_0$ where N is the number of photons per energy bin per incident electron, for (a) 2% X_0carbon and (b) 6% X_0carbon targets in 25 GeV electron beams, while (c) shows the 2% X_0carbon and (d) the 6% X_0carbon target in the 8 GeV beam. Monte Carlo predictions are shown for LPM plus dielectric suppression (solid line), LPM suppression only (dotted line) and Bethe-Heitler (dashed line); all include transition radiation.

calculation of this effect can be done by limiting l_f to a maximum of $1X_0$ [10]. This suppression is visible (stronger than LPM and dielectric suppression) for

$$E > E_p = \frac{2X_0\omega_p E_s}{\hbar c}. \qquad (21)$$

For $E > E_p$, there is a 'window' in k where this mechanism applies: $k_{p-} = X_0\omega_p^2/2\hbar c < k < k_{p+} = 2\hbar cE(E - k)/(X_0 E_s^2)$. In this region, the photon spectrum is suppressed by k, and $d\sigma/dk$ is constant. For $k < k_{p-}$, dielectric suppression dominates, while for $k > k_{p+}$, the LPM effect is dominant. E_p is 1.6 PeV in water or ice, and 42 PeV for air at sea level. The E_ps are so similar because the variation in X_0 and ω_p partially offset each other. In sea level air, the 'window' is 331 MeV$< k < 3.0 \times 10^{-24} E^2$ (with E in eV).

There should be a similar effect where a partially produced electron or positron emits a bremsstrahlung photon. This possibility limits the coherence over l_{f0}. The strength of this suppression has not yet been calculated, but it should be comparable to that for pair production suppressing bremsstrahlung.

This approach is overly simplistic. The 'hard cutoff' in l_f should be replaced by a probabilistic approach, where the photon interaction probability depends on the distance travelled. And, other suppression mechanisms will also be in effect, the radiation length will be longer than the naive $1X_0$. An accurate calculation should include the interplay between the two reactions to arrive at an overall effective shower distance. Still, the above equation gives a reasonable estimate of when this effect needs to be considered.

Fig. 6 summarizes these results, and shows that a 'simple' bremsstrahlung photon spectrum can have several different slopes for different photon energies.

D   Magnetic Suppression

An external magnetic field can also suppress bremsstrahlung and pair creation. The bending caused by an external field acts just as does multiple scattering. The difference is that the magnetic field bending is quite deterministic, while the scattering angles are randomly distributed. The magnetic bending angle is $\theta_B = \Delta p/p = eBl_f \sin(\theta_B)/E$ where B is the magnetic field and θ_B is the angle between the field and the electron trajectory. Emission is reduced if $\theta_B > 1/\gamma$; this happens for $y < 2\gamma B \sin(\theta_B)/B_c$, where $B_c = m^2 c^3/\hbar$ is the critical field strength [11].

The magnitude of the suppression can be found using an approach similar to that used by Landau and Pomeranchuk for multiple scattering. Because $\theta_B \sim l_f$, while $\theta_{MS} \sim \sqrt{l_f}$ the energy dependence is different. With the definition $E_B = mB_c/B \sin(\theta_B)$, the suppression factor is [12]

$$S = \left[\frac{kE_b}{E(E - k)}\right]^{2/3}. \qquad (22)$$

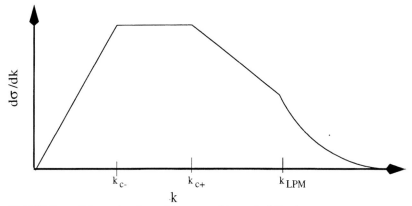

FIGURE 6. Schematic view of bremsstrahlung $d\sigma/dk$ when several suppression mechanisms are present. Below k_{c-}, dielectric suppression dominates. Between k_{c-} and k_{c+}, suppression due to pair creation is dominant. Between k_{c+} and $k_{LPM} = E^2/E_{LPM}$, the LPM effect is most important, while above k_{LPM} the Bethe-Heitler spectrum is present. For $E < E_p$, pair creation suppression disappears and LPM suppression connects with dielectric suppression.

A magnetic field will also bend pair produced electrons and positrons. The same equation holds, except that $k - E$ replaces $E - k$. For a particle trajectory perpendicular to Earth's ~ 0.5 Gauss field, $E_B \sim 45$ EeV. So, for cosmic rays with energies above 10^{20} eV, magnetic suppression must be considered.

E   Emission Angles

Besides reducing the interactions rates, these mechanisms also increase the angular spread of electromagnetic showers. When bremsstrahlung and pair creation occur with a large opening angle, the formation length is shortened. For bremsstrahlung

$$l_{f0} = \frac{2\hbar E(E - k)}{m^2 c^3 k(1 + \gamma^2 \theta_\gamma^2)} \tag{23}$$

where θ_γ is the angle between the photon and the electron trajectory. When θ_γ is included in the calculations, the average emission angle rises from $\theta_\gamma \sim 1/\gamma$ to $\theta_\gamma \sim 1/S\gamma$. If S is small enough, this spreading can dominate over multiple scattering in determining the angular spreading of the shower. If the scattering is taken over $1/2X_0$ (assuming 1 interaction per X_0, and half the particles are charged and hence subject to multiple scattering), this occurs for roughly $S < 0.05$. Most calculations of air and water showers do not include this angular spreading. It is likely to be most important for calculations for radio emission from showers in air and ice [13].

141

II SUPPRESSION IN SHOWERS

Suppression mechanisms can affect showers by slowing their development, effectively increasing the radiation length. Because they all work by reducing l_f, the effects do not add; instead, $q_{\parallel}$ must be summed, and the total suppression calculated. Usually, multiple scattering is the most significant cause of suppression, so it will be the focus of this section.

The effective increase in radiation length due to multiple scattering can be seen in Figs. 4; it shows how the area under the curve in Figs. 2 and 3 drops as the incident particle energy rises. For bremsstrahlung, energy loss is halved for electrons with $E = 22E_{LPM}$, while the pair production cross section is halved for $k = 100E_{LPM}$. Then, the radiation length in the first generation of the shower doubles; succeeding generations will show smaller effects.

However, the effects go beyond simply lengthening showers. Because soft photon bremsstrahlung is the first reaction to be affected, the number of interactions will decrease more rapidly then the electron energy loss. So, the initial part of a shower will consist almost entirely of a few high energy particles, without an accompanying 'fuzz' of lower energy particles. The initial shower development depends on a much smaller number of interactions. For example, a 25 GeV electron in sea level air will emit about 14 bremsstrahlung photons per X_0, while a 10^{17} eV electron will emit only 3. It is worth noting that, these numbers are naturally finite because dielectric suppression eliminates the infrared divergence. Pair production is similar; the pairs become increasingly asymmetric. The higher energy lepton from a 25 GeV photon pair conversion takes an average of 75% of k; for a 10^{19} eV photon, the average is more than 90%.

Because of this, shower to shower fluctuations become much larger. Misaki studied shower development in lead and standard rock. For $E \gg E_{LPM}$ he found that the position of shower maximum was shifted to larger depths, and that the position of shower maximum varied greatly from shower to shower, and that the shower to shower variations overshadowed the average shower development [14].

In the limit $E \gg E_{LPM}$, the initial part of an air shower becomes a succession of asymmetric pair production, where the higher energy of the pair loses most of it's energy to a single bremsstrahlung photon, re-starting the process. In short, the paradigm that successive generations of air showers have twice as many particles with half as much energy as the current generation fails completely.

III AIR SHOWERS

The composition of the highest energy cosmic rays is still a mystery. This article will consider two possibilities: protons (the most popular) and photons. In both cases, we will take 3×10^{20} eV as a standard energy. For incoming heavy ions, the effects are greatly reduced because of the lower per particle energy, and neutrons can be treated as protons. Because these are toy models, the likely possibility of the photon pair converting in the earth's magnetic field will be neglected.

Most current works have considered proton initiated showers. There has been disagreement as to whether the LPM effect is important in air showers. Capdevielle and Atallah found that it had a large effect on 10^{19} eV and 10^{20} eV showers [15]. However, Kalmykov, Ostapchenko and Pavlov found a much smaller effect; for a 10^{20} eV incident proton, the number of electrons at shower maximum decreased by 5%, while the position of shower maximum shifted downward by 15 ± 2 g/cm^2 [16]. This shift is less than the error on a typical measurement of shower maximum in a single shower, but it can have a significant effect on composition studies. Of course, as experiments probe higher energies, suppression becomes more important. None of these authors considered the effects of fluctuations or other suppression mechanisms.

Air showers studies are complicated by the fact that density, and hence E_{LPM} and E_p depend on altitude and temperature. Ignoring temperature changes, pressure decreases exponentially with altitude, with scale height 8.7 km. For showers, it is convenient to work with column depth measured in g/cm^2. In an isothermal model, then $E_{LPM} = 117$ PeV(A_0/A), where A is the column depth and A_0 is ground level, 1030 g/cm^2. Temperature will modify the relationship; with a temperature correction this E_{LPM} is 2.25 EeV at 36 g/cm^2 (1 X_0) depth, and 1 EeV at 90 g/cm^2 (1 hadronic interaction length, Λ). Neglecting temperature, $y_{die} = 1.3 \times 10^{-6} A/A_0$. For pair creation suppression, $E_p = 42$ PeV$\sqrt{(A_0/A)}$. The corresponding photon window is 331 MeV $< k < 3.0 \times 10^{-24} E^2 (A_0/A)$. The two window 'edges' have different A dependencies because all three mechanisms have a different dependence on ω_p and X_0.

Incoming photons react by pair production, while protons interact hadronically. A central hadronic collision will produce a shower of several hundred pions; the neutral pions will decay to photons. The highest energy π^0 will have a rapidity near to the incoming proton, and their decay photons will have energies around 2×10^{19}eV. Many diffractive processes, such as Δ production can produce photons with similar energies. Overall, photons from central interaction will have an average energy of about 2×10^{17} eV.

Although a complete Monte Carlo simulation is required to understand the effects of suppression in air showers, simple calculations can provide some indications where it matters, and should differentiate between the results of Capdevielle and Atallah and Kalmykov and collaborators. Because E_{LPM} decreases with depth, in concert with the average particle energy, suppression mechanisms can actually become stronger as one moves deeper in the atmosphere. The solid curve in Fig. 7 shows E_{LPM} as a function of altitude. The dashed curve shows the average particle energy, for an idealized Bethe-Heitler shower from a 3×10^{20} eV photon. In each successive radiation length, there are twice as many particles with half the energy. The curve with the short dashes shows a similar cascade, from a 2×10^{19} eV photon starting at 1Λ. The electromagnetic interactions at a given depth are determined

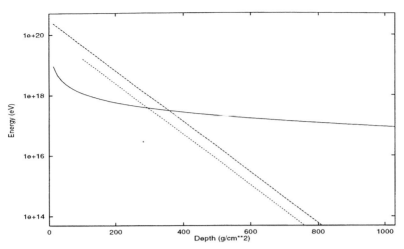

FIGURE 7. E_{LPM} (solid line), average particle energy for a 3×10^{20} eV photon shower (dashed line) and average particle energy for a 2×10^{19} eV photon created at 1Λ (short dashes). At altitudes where the average particle energy is larger than E_{LPM}, suppression is important; with the ratio of the two energies determining the degree of suppression.

by the ratio of the two curves, which gives E/E_{LPM}. For the photon case, the maximum suppression occurs around 75 g/cm^2, where $E \sim 80 E_{LPM}$. Electron dE/dx is reduced about 80%, and pair production cross section is down by 60%; the radiation length has more than doubled. For the hadronic case, the effect is smaller, and, of course, these high energy photons are only a small portion of the total shower. On the other hand, hadronic interactions are only partially inelastic, and the proton may carry a significant fraction of it's momentum deeper in the atmosphere, where suppression is larger. Because of the large variations in energy deposition depths, it is difficult to give more quantitative estimates.

This model underestimates the effect of suppression, because, with suppression shower development is slower than the idealized model, further increasing the amount of suppression in the next stage. However, it does show that suppression is important in photon shower, and in at least parts of proton initiate showers. The effect is clearly smaller than that predicted by Capdevielle and Atallah, but is consistent with Kalmykov and collaborators.

Beyond the affect on average showers, fluctuations must be considered. because the cosmic ray energy spectrum falls as $dN/dE \approx 1/E^3$, it is important to understand the tails of the energy resolution distribution; without accurate simulations, showers whose energy is overestimated can easily skew the measured spectrum.

Fluctuations can affect both ground based arrays as well as air fluorescence detectors that optically measure the shower development. For ground based arrays, suppression can change the relative position of shower maximum and the detector array. Although a sea level detector is near shower maximum for 10^{20} eV vertically

144

incident proton showers, for non-vertical showers, where the detector is deeper, it may be significantly behind shower maximum. The angular spreading discussed in Subsection I E must also be considered.

Air fluorescence detectors will observe far fewer particles in the early stages of the shower than current simulations indicate. LPM, dielectric and pair creation suppression must all be considered. Although LPM suppression affects the widest range of energies, the other mechanisms produce a larger reduction in cross section where they operate. For $E > E_p$, for example, dielectric suppression will reduce the number of bremsstrahlung photons with $k < 331$ MeV by two orders of magnitude. Pair creation suppression is more energy dependent, but, for example, at 200 g/cm^2 depth, emission of 10^{10} eV photons from 10^{16} eV electrons will be reduced by a factor of 10. Monte Carlo simulations are required to find the actual shower profile, but the initial stages of the shower will be much less visible than current simulations indicate; neither the profiles produced by Capdevielle and Atallah nor by Kalmykov will be accurate.

Even relatively late in the shower, there will be a reduction in the number of low energy particles. For example, where the average particle energy is $\sim 10^{13}$ eV, LPM suppression will reduce the number of photons below ~ 500 MeV. So, there can be measurable effects even relatively close to shower maximum. if a particle counting detector is too far above shower maximum, then it may underestimate the size of the shower.

Another way to visualize how suppression works is to consider the probability of photons penetrating deep into the atmosphere. Because, for a given photon energy, suppression increases with depth, high energy photons have a non-negligible probability of penetrating deep into the atmosphere. This can drastically change the development of photons showers. For proton showers, it can create small, dense subshowers within the main shower. Fig. 8 shows the interaction probability as a function of depth for a 3×10^{20} eV vertically incident photon. The solid and dashed curves show the LPM and Bethe-Heitler cases respectively. With Bethe-Heitler interaction probabilities, essentially all of the photons have interacted by $3X_0$ in depth, while with LPM interaction probabilities, the photon has a 7% chance of surviving to $10X_0$.

For protons, the effect is smaller, The other curves in Fig. 8 are for a 2×10^{19} eV photon, produced by a hadronic interaction at 1Λ in depth. With Bethe-Heitler (short dashes), almost all have interacted by 6 X_0, while with LPM (dots) cross sections, it has a 1.2% chance of surviving to $10X_0$. The effect on the average shower is small, but, because of the steeply falling spectrum, it is necessary to consider even relatively atypical showers.

Not considered here is magnetic suppression, which can dominate over other mechanisms for interactions in the upper 3 g/cm^2 of the atmosphere.

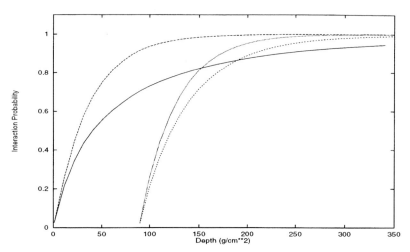

FIGURE 8. Interaction probability for photons from cosmic rays, as a function of depth in the atmosphere. The solid and dashed line show the interaction probability for a 3×10^{20}eV photon incident on the top of the atmosphere for LPM and Bethe-Heitler interaction probabilities respectively. The short dashed and dotted lines show the interaction probability for a 2×10^{19} eV photon produced in an interaction at a depth of 1Λ, also for LPM and Bethe-Heitler interactions respectively.

IV CONCLUSIONS AND ACKNOWLEDGEMENTS

Multiple scattering, the dielectric of the medium, bremsstrahlung and pair production themselves and external magnetic fields can suppress bremsstrahlung and pair production, with bremsstrahlung of low energy photons the most subject to suppression. Suppression also broadens the angular spread of emitted bremsstrahlung photons and produced pairs.

For energetic enough particles, these mechanisms reduce electron dE/dx and photon pair production cross sections. When this happens, electromagnetic showers are lengthened, and shower to shower fluctuations become much larger. When suppression gets very large, then the shower angular development can be dominated by interactions, rather than multiple scattering.

The importance of suppression in air showers depends on the incident particle type and energy. If the incoming particles are photons, then at 3×10^{20} eV, the effects are large. For 3×10^{20} protons, the effects on the average shower is smaller, but the effects of fluctuations may be considerable, particularly when particular detector are included. As experiments probe to higher energies, of course, the effects will grow larger.

The calculations presented here show the need for a complete Monte Carlo, including all relevant suppression mechanisms, in order to correctly determine the shower profile for showers with energies much above 10^{20} eV.

I would like to thank my E-146 collaborators for many useful discussions. This work was supported by the USDOE under contract number DE-AC-03-76SF00098.

REFERENCES

1. Landau, L.D and Pomeranchuk, I.J., Dokl. Akad. Nauk. SSSR **92** 535 (1953); Dokl. Akad. Nauk. SSSR **92**, 735. These papers are available in English in *The Collected Papers of L. D. Landau*, Pergamon Press, 1965.

2. Migdal, A. B., Phys. Rev. **103**, 1811 (1956).

3. Stanev, T. *et al.*, Phys. Rev. D **25**, 1291 (1982).

4. Anthony, P. *et al.*, Phys. Rev. Lett. **75**, 1949 (1995); Phys. Rev. Lett. **76**, 3550 (1996); Phys. Rev. **D56**, 1373 (1997).

5. Blankenbecler, R. and Drell, S., Phys. Rev. **D53**, 6285 (1996).

6. Zakharov, B. G., JETP Lett. **64**, 781 (1996).

7. Baier, R., Dokshitzer, Yu. L, Mueller A. H. and Schiff, D, Nucl. Phys. **B478**, 57 (1996).

8. Baier, V.N., and Katkov, V.M., hep-ph/9709214, Sept., 1997.

9. Ter-Mikaelian, M. L., Dokl. Akad. Nauk. SSR **94**, 1033 (1953); 0*High Energy Electromagnetic Processes in Condensed Media*, John Wiley & Sons, 1973.

10. Galitsky, V.M. and Gurevich, I.I., Il Nuovo Cimento **32**, 396 (1964).

11. Klein, S. R. *et al.*, in *Proc. XVI Int. Symp. Lepton and Photon Interactions at High Energies*, Eds. P. Drell and D. Rubin, AIP Press, 1993.

12. Klein, S. R., to be published.

13. Alvarez-Muniz, J. and Zas, E., preprint astro-ph/9706064.

14. Misaki, A., Nucl. Phys. B (Proc. Suppl.) **33A,B**, 192. (1993).

15. Capdevielle, J. N. and Atallah, R., Nucl. Phys. B (Proc. Suppl.), **28B**, 90 (1992).

16. Kalmykov, N. N., Ostapchenko, S. S. and Pavlov, A. I., Phys. Atomic Nuclei **58**, 1728 (1995).

Physics Opportunities Above the Greisen-Zatsepin-Kuzmin Cutoff: Lorentz Symmetry Violation at the Planck Scale

Luis Gonzalez-Mestres[*,†]

*Laboratoire de Physique Corpusculaire, Collège de France, 75231 Paris Cedex 05, France
†L.A.P.P., CNRS-IN2P3, B.P. 110, 74941 Annecy-le-Vieux Cedex, France

Abstract. Special relativity has been tested at low energy with great accuracy, but these results cannot be extrapolated to very high-energy phenomena: this new domain of physics may actually provide the key to the, yet unsettled, question of the ether and the absolute rest frame. Introducing a critical distance scale, a , below 10^{-25} cm (the wavelength scale of the highest-energy observed cosmic rays) allows us to consider models, compatible with standard tests of special relativity, where a small violation of Lorentz symmetry (a can, for instance, be the Planck length) leads to a deformed relativistic kinematics (DRK) producing dramatic effects on the properties of very high-energy cosmic rays. For instance, the Greisen-Zatsepin-Kuzmin (GZK) cutoff does no longer apply and particles which are unstable at low energy (neutron, some hadronic resonances like the Δ^{++}, possibly several nuclei...) become stable at very high energy. In these models, an absolute local rest frame exists (the **vacuum rest frame**, VRF) and special relativity is a low-momentum limit. We discuss the possible effects of Lorentz symmetry violation (LSV) on kinematics and dynamics, as well as the cosmic-ray energy range (well below the energy scale associated to the fundamental length) and experiments (on earth and from space) where they could be detected.

STATUS OF THE RELATIVITY PRINCIPLE

H. Poincaré was the first author to consistently formulate the relativity principle on the grounds of experiment, stating in 1895 [1]:

"Experiment has provided numerous facts justifying the following generalization: absolute motion of matter, or, to be more precise, the relative motion of weighable matter and ether, cannot be disclosed. All that can be done is to reveal the motion of weighable matter with respect to weighable matter"

The deep meaning of this law of Nature was further emphasized by the same author when he wrote in 1901 [2], in connexion with Lorentz contraction:

"Such a strange property seems to be a real coup de pouce presented by Nature itself, for avoiding the disclosure of absolute motion... I consider quite probable

CP433, *Workshop on Observing Giant Air Showers from Space*
edited by J.F. Krizmanic et al.
© 1998 The American Institute of Physics 1-56396-766-9/98/$15.00

that optical phenomena depend only on the relative motion of the material bodies present, of the sources of light and optical instruments, and this dependence is not accurate... but rigorous. This principle will be confirmed with increasing precision, as measurements become more and more accurate"

The role of H. Poincaré in building relativity, and the relevance of his thought, have often been emphasized [3,4]. In his June 1905 paper [5], published before Einsteins's article [6] arrived (on June 30) to the editor, he explicitly wrote the relativistic transformation law for the charge density and velocity of motion and applied to gravity the "Lorentz group" (that he introduced), assumed to hold for "forces of whatever origin". From this, he inferred that "gravitational waves" propagate at the speed of light. However, his priority is sometimes denied [7,8] on the grounds that *"Einstein essentially announced the failure of all ether-drift experiments past and future as a foregone conclusion, contrary to Poincaré's empirical bias"* [7], that Poincaré did never *"disavow the ether"* [7] or that *"Poincaré never challenges... the absolute time of newtonian mechanics... the ether is not only the absolute space of mechanics... but a dynamical entity"* [8]. It is implicitly assumed that A. Einstein was right in 1905 when *"reducing ether to the absolute space of mechanics"* [8] and that H. Poincaré was wrong because *"the ether fits quite nicely into Poincaré's view of physical reality: the ether is real..."* [7]. A basic physics issue (whether ether and an absolute rest frame exist or not), perhaps not definitely settled, underlies the debate on priority. Actually, modern particle physics has brought back the concept of a non-empty vacuum where free particles propagate: without such an "ether" where fields can condense, the standard model of electroweak interactions could not be written and quark confinement could not be understood. Modern cosmology is not incompatible with an "absolute local frame" (the **vacuum rest frame**, VRF) close to that suggested by the study of cosmic microwave background radiation. If "ether" and the VRF actually exist, the relativity principle (the impossibility to disclose absolute motion) will become a symmetry, a concept whose paternity was attributed to H. Poincaré by R.P. Feynman [9]:

"Precisely Poincaré proposed investigating what could be done with the equations without altering their form. It was precisely his idea to pay attention to the symmetry properties of the laws of physics"

As symmetries in particle physics are in general violated at some scale, Lorentz symmetry may be broken and an absolute local rest frame may be detectable through experiments performed beyond the relevant scale. It may even happen that Lorentz symmetry be just an infrared attractor. Poincaré's special relativity (a symmetry applying to physical processes) could live with this situation, in which case the relativity principle would refer to the impossibility to disclose absolute motion *through low-energy experiments*. But Einstein's approach, such as it was formulated in 1905 (an absolute geometry of space-time that matter cannot escape), could not survive. We discuss here two issues: a) the scale where we may expect Lorentz symmetry to be violated; b) the physical phenomena and experiments potentially able to uncover Lorentz symmetry violation (LSV). Previous papers on the subject are references [10] to [20] and references therein.

RELATIVITY AS A LOW-ENERGY LIMIT

Low-energy tests of special relativity have confirmed its validity to an extremely good accuracy [21,22], in impressive agreement with Poincaré's 1901 conjecture. But the situation at very high energy remains more uncertain (see [10] to [20]): high-energy physics corresponds to a domain never covered by the experiments that motivated special relativity a century ago. Figures can change by more than 20 orders of magnitude between the highest oberved cosmic-ray energies and the scale explored by the above mentioned tests of Lorentz symmetry. If LSV follows a E^2 law (E = energy), similar to the effective gravitational coupling, it can be of order 1 at $E \approx 10^{21}$ eV and $\approx 10^{-26}$ at $E \approx 100$ MeV (corresponding to the highest momentum scale involved in nuclear magnetic resonance experiments), in which case it will escape all existing low-energy bounds. If LSV is ≈ 1 at Planck scale ($E \approx 10^{28}$ eV), and following a similar law, it will be $\approx 10^{-40}$ at $E \approx 100$ MeV . Our suggestion is not in contradiction with Einstein's thought after he had developed general relativity. In 1921 , A. Einstein wrote [23]:

"The interpretation of geometry advocated here cannot be directly applied to sub-molecular spaces... it might turn out that such an extrapolation is just as incorrect as an extension of the concept of temperature to particles of a solid of molecular dimensions".

It is in itself remarkable that special relativity holds at the attained accelerator energies, thus confirming Poincaré's conjecture far beyond expectations. But there is no fundamental reason for this dazzling success to persist above Planck scale. A typical (and natural) example of models violating Lorentz symmetry at very short distance is provided by models where an absolute local rest frame exists and non-locality in space is introduced through a fundamental length scale a [11]. Such models lead in the VRF to a deformed relativistic kinematics of the form [11,12]:

$$E = (2\pi)^{-1} h c a^{-1} e (k a) \tag{1}$$

A where h is the Planck constant, c the speed of light, k the wave vector and $[e (k a)]^2$ is a convex function of $(k a)^2$ obtained from vacuum dynamics. Expanding equation (1) for $k a \ll 1$, we can write:

$$e (k a) \simeq [(k a)^2 - \alpha (k a)^4 + (2\pi a)^2 h^{-2} m^2 c^2]^{1/2} \tag{2}$$

m being the mass, α a model-dependent positive constant $\approx 0.1 - 0.01$ for full-strength LSV at momentum scale $p \approx a^{-1} h$, and in terms of momentum p :

$$E \simeq p c + m^2 c^3 (2 p)^{-1} - p c \alpha (k a)^2/2 \tag{3}$$

The "deformation" $\Delta E = - p c \alpha (k a)^2/2$ in the right-hand side of (3) implies a Lorentz symmetry violation in the ratio $E p^{-1}$ varying like $\Gamma (k) \simeq \Gamma_0 k^2$ where $\Gamma_0 = - \alpha a^2/2$. If c and α are universal parameters for all particles, LSV does not lead to the spontaneous decays predicted in [24]: the existence of very high-energy cosmic rays cannot be regarded as an evidence against LSV. With the deformed relativistic kinematics (DRK) defined by (1)-(3), Lorentz symmetry remains valid in the limit $k \rightarrow 0$, contrary to the standard $TH\epsilon\mu$ model [25]. The above non-locality may actually be an approximation to an underlying dynamics involving superluminal particles [10,12–14,17], just as electromagnetism looks nonlocal in the potential approximation to lattice dynamics in solid-state physics: it would then correspond to the limit $c \, c_i^{-1} \rightarrow 0$ where c_i is the superluminal critical speed.

As recently pointed out [16], equation (1) is a fundamental property of old scenarios (f.i. [26]) breaking local Lorentz invariance (LLI). An ansatz based on an isotropic, continuous modification of the Bravais lattice dynamics is [11]:

$$e \, (k \, a) \;=\; [4 \, sin^2 \, (ka/2) \,+\, (2\pi \, a)^2 \, h^{-2} \, m^2 \, c^2]^{1/2} \tag{4}$$

and simple extensions of the ansatz by Rédei [26] lead [16] to expressions like:

$$e \, (k \, a) \;=\; [10 \,+\, 30 \, (k \, a)^{-2} \, cos \, (k \, a) \,-\, \\ 30 \, (k \, a)^{-3} \, sin \, (k \, a) \,+\, (2\pi \, a)^2 \, h^{-2} \, m^2 \, c^2]^{1/2} \tag{5}$$

In any case, we expect observable kinematical effects when the term $\alpha(ka)^3/2$ becomes as large as the term $2 \, \pi^2 \, h^{-2} \, k^{-1} \, a \, m^2 \, c^2$. This happens at:

$$E \;\simeq\; (2\pi)^{-1} \, h \, c \, k \;\approx\; E_{trans} \;\approx\; \alpha^{-1/4} \, (h \, c \, a^{-1}/2\pi)^{1/2} \, (m \, c^2)^{1/2} \tag{6}$$

Thus, contrary to conventional estimates of LLI breaking predictions [27] where the modification of relativistic kinematics is ignored, observable effects will be produced at wavelength scales well above the fundamental length. For a nucleon, taking $a \approx 10^{-33}$ cm and $\alpha \approx 0.1$, this corresponds to $E \approx 10^{19}$ eV , below the highest energies at which cosmic rays have been observed. With full-strength LSV, for a proton at $E \approx 10^{20}$ eV and with the above value of a , we get:

$$\alpha \, (k \, a)^2/2 \;\approx\; 10^{-18} \;\gg\; 2 \, \pi^2 \, h^{-2} \, k^{-2} \, m^2 \, c^2 \;\approx\; 10^{-22} \tag{7}$$

and, although $\alpha(ka)^3/2$ is very small as compared to the value of $e \, (k \, a)$, the term $2 \, \pi^2 \, h^{-2} \, k^{-1} \, a \, m^2 \, c^2$ represents an even smaller fraction. Although relativity reflects to a very good approximation the reality of physics at large distance scales and can be considered as its low-energy limit, no existing experimental result proves that it applies with the same accuracy to high-energy cosmic rays.

Are c and α universal? This may be the case for all "elementary" particles, i.e. quarks, leptons, gauge bosons..., but the situation is less obvious for hadrons, nuclei and heavier objects. From a naive soliton model [11], we inferred that: a) c is expected to be universal up to very small corrections ($\sim 10^{-40}$) escaping all existing bounds; b) a possible approximate rule can be to take α universal for leptons, gauge bosons and light hadrons (pions, nucleons...) and assume a $\alpha \propto m^{-2}$ law for nuclei and heavier objects, the nucleon mass setting the scale.

RELEVANCE OF COSMIC-RAY EXPERIMENTS

If Lorentz symmetry is broken at Planck scale or at some other fundamental scale, and assuming that the earth moves slowly with respect to the VRF, the effects of LSV may be observable well below this energy and produce detectable phenomena at the highest observed cosmic-ray energies. This is due to DRK [11,12,15,18]: at energies above E_{trans}, the deformation ΔE dominates over the mass term $m^2 c^3 (2 p)^{-1}$ in (3) and modifies all kinematical balances. Because of the negative value of ΔE, it costs more and more energy, as energy increases above E_{trans}, to split the incoming logitudinal momentum. The parton model (in any version), as well as standard formulae for Lorentz contraction and time dilation, are also expected to fail above this energy [12,15] which corresponds to $E \approx 10^{20}$ eV for m = proton mass and $\alpha a^2 \approx 10^{-72}$ cm^2 (f.i. $\alpha \approx 10^{-6}$ and a = Planck length), and to $E \approx 10^{18}$ eV for m = pion mass and $\alpha a^2 \approx 10^{-67}$ cm^2 (f.i. $\alpha \approx 0.1$ and a = Planck length). In particular, the following effects are predicted:

a) For $\alpha a^2 > 10^{-72}$ cm^2, and assuming a universal value of α, the Greisen-Zatespin-Kuzmin (GZK) cutoff [28,29] is suppressed [11,15,16,18] for the particles under consideration and ultra-high energy cosmic rays (e.g. protons) produced anywhere in the presently observable Universe can reach the earth without losing their energy in collisions with the cosmic microwave background radiation.

b) With the same hypothesis, unstable particles with at least two stable particles in the final states of all their decay channels become stable at very high energy [11,16]. Above E_{trans}, the lifetimes of all unstable particles (e.g. the π^0 in cascades) become much longer than predicted by relativistic kinematics [11,16,18].

c) In astrophysical processes at very high energy, similar mechanisms can inhibit radiation under external forces, GZK-like cutoffs, decays, photodisintegration of nuclei, momentum loss through collisions, production of lower-energy secondaries... potentially contributing to solve all basic problems raised by the highest-energy cosmic rays. Therefore, calculations of astrophysical processes at very high energy cannot ignore the possibility that Lorentz symmetry be violated [18].

d) With the same hypothesis, the allowed final-state phase space of two-body collisions is modified and can lead to a sharp fall of cross-sections for incoming cosmic ray energies above $E_{lim} \approx (2\pi)^{-2/3} (E_T a^{-2} \alpha^{-1} h^2 c^2)^{1/3}$, where E_T is the energy of the target [19]. As a consequence, and with the previous figures for the parameters of LSV, above some energy E_{lim} between 10^{22} and 10^{24} eV a cosmic ray will not deposit most of its energy in the atmosphere and can possibly fake an exotic event with much less energy.

e) Effects a) to d) are obtained using only DRK. If dynamical anomalies are added (failure, at very small distance scales, of the parton model and of the standard Lorentz formulae for length and time [12,15]...), we can expect much stronger effects in the cascade development profiles of cosmic-ray events.

f) Cosmic superluminal particles would produce atypical events with very small total momentum, isotropic or involving several jets [10,12,14,17].

In what follows, we discuss in more detail the implications of these effects.

THE GZK CUTOFF DOES NO LONGER APPLY

For $\alpha\, a^2 > 10^{-72}\; cm^2$, with a universal value of α, a $E \approx 10^{20}\; eV$ proton interacting with a cosmic microwave background photon would be sensitive to DRK effects [11]. After having absorbed a $10^{-3}\; eV$ photon moving in the opposite direction, the proton gets an extra $10^{-3}\; eV$ energy, whereas its momentum is lowered by $10^{-3}\; eV/c$. In the conventional scenario with exact Lorentz invariance, this is enough to allow the excited proton to decay into a proton or a neutron plus a pion, losing an important part of its energy. The small increase in the E/p ratio is enough to generate, in the final state, the increase of the nucleon mass term $m^2\, (2\, p)^{-1}$ (as momentum gets lower) as well as the pion mass term and the transverse energy of both particles. However, it can be checked [11] that in our scenario with LSV such a reaction is strictly forbidden, as the $\approx 2.10^{-23}$ increase of the E/p ratio cannot provide the energy required, due to the deformation term $\Delta\, E$, by the splitting of the incoming momentum. Elastic $p + \gamma$ scattering is permitted, but allows the proton to release only a small amount of its energy. Similar or more stringent bounds exist for channels involving lepton production. Obvious phase space limitations will also lower the collision rate, as compared to standard calculations using exact Lorentz invariance which predict photoproduction of real pions at such cosmic proton energies. The effect is strong enough to invalidate the GZK cutoff and explain the existence of the highest-energy cosmic-ray events. It will become more important at higher energies, as we get closer to the a^{-1} wavelength scale. If α is not universal, the particle with the highest value of α can always reach the earth, and other particles can below some energy above the GZK cutoff. Assuming exact universality of c, the situation for nuclei will crucially depend on the precise values of α for each nucleus and on the energy range. Models where c is not exactly universal are not ruled out, as the results from [24] do not apply [13], and deserve cosideration [18]. Our result is limited by the history of the Universe, as cosmic rays coming from distances closer and closer to horizon size will be older and older and, at early times, will have been confronted to rather different scenarios. But it is clear that DRK allows much older ultra-high energy cosmic rays, generated at much more remote sources, to reach the earth nowadays.

A previous attempt to explain the experimental absence of the predicted GZK cutoff by Lorentz symmetry violation at high energy [30] proposed an ansatz replacing relativistic kinematics by the relation:

$$E = m\, h\, (p^2\, E^{-2}) \qquad (8)$$

where the positive function h tends to $(1 - p^2\, E^{-2})^{-1/2}$ in the relativistic limit. These authors considered an expansion in powers of γ^4, where $\gamma = (1 - v^2 c^{-2})^{-1/2}$, v is the speed of the particle and the coefficient of the linear term in γ^4 had to be arbitrarily tuned to $\approx 10^{-44}$ in order to produce an effect of order 1 for a $10^{20}\; eV$ proton (leading to a potentially divergent expansion at higher energies). No such problems are encountered in our approach, where the required orders of magnitude come out naturally in terms of small perturbations.

LIFETIMES AT VERY HIGH ENERGY

In standard relativity, we can compute the lifetime of any unstable particle in its rest frame and, with the help of a Lorentz transformation, obtain the Lorentz-dilated lifetime for a particle moving at finite speed. The same procedure had been followed in previous estimates of the predictions of LLI breaking [27] for the decay of high-energy particles. This is no longer possible with the kinematics defined by (1), which explicitly incorporates LSV. Instead, two results are obtained [11,16]:

i) Assuming universal values of c and α , unstable particles with at least two massive particles in the final state of all their decay channels become stable at very high energy, as a consequence of the effect of LSV through (1). A typical order of magnitude for the energy E^{st} at which such a phenomenon occurs is:

$$E^{st} \approx c^{3/2} h^{1/2} (a\, m_2)^{-1/2} (m^2 - m_1^2 - m_2^2)^{1/2} \tag{9}$$

where: a) m is the mass of the decaying particle; b) we select the two heaviest particles of the final product of each decay channel, and m_2 is the mass of the lightest particle in this list; c) m_1 is the mass of the heaviest particle produced together with that of mass m_2 . With $a \approx 10^{-33}$ cm and $\alpha \approx 0.1$ for all particles, **the neutron would become stable** above $E \approx 10^{20}$ eV . **Some hadronic resonances** (e.g. the Δ^{++} , whose decay product contains a proton and a positron) **would become stable** above $E \approx 10^{21}$ eV . Similar considerations may apply to some supersymmetric particles. Most of these objects will decay before they can be accelerated to such energies, but they may result of a collision at very high energy or of the decay of a superluminal particle [10], The study of very high-energy cosmic rays can thus reveal as stable particles objects which would be unstable if produced at accelerators. If one of the light neutrinos (ν_e , ν_μ) has a mass in the ≈ 10 eV range, **the muon would become stable** at energies above $\approx 10^{22}$ eV . Weak neutrino mixing may restore muon decay, but with very long lifetime. **Similar considerations apply to the τ lepton**, which would become stable above $E \approx 10^{22}$ eV if the mass of the ν_τ is ≈ 100 eV but, again, a decay with very long lifetime can be restored by neutrino oscillations. For nuclei, the situation will depend on the details of DRK (basically, the value of α for each nucleus) and deserves further investigation using more precise theoretical models.

ii) With the same hypothesis as i), all unstable particles live longer than naively expected with exact Lorentz invariance and, at high enough energy, the effect becomes much stronger than previously estimated [27] ignoring the effects of DRK. At energies well below the stability region, partial decay rates are already modified by large factors leading to observable effects. Irrespectively of whether m_2 vanishes or not, the phenomenon occurs above $E \approx c^{3/2} h^{1/2} (m^2 - m_1^2)^{1/4} a^{-1/2}$ ($\approx 10^{18}$ eV for $\pi^+ \rightarrow \mu^+ + \nu_\mu$, if $a \approx 10^{-33}$ cm and $\alpha \approx 0.1$). The effect has a sudden, sharp rise, since a fourth power of the energy is involved in the calculation. In the LSV scenario, partial branching ratios become energy-dependent.

FINAL-STATE PHASE SPACE

No special constraint seems to arise from (2) if, in the VRF, two particles with equal, opposite momenta of modulus p with $\alpha \, (k \, a)^2 \ll 1$ collide to produce a multiparticle final state. But, as a consequence of LSV, the situation becomes fundamentally different at very high energy if one of the incoming particles is close to rest with respect to the VRF where formulae (1) - (3) apply. Assume a very high-energy particle (particle 1) with momentum $\vec{p}$, impinging on a particle at rest (particle 2). We take both particles to have mass m , and $p \gg mc$. In relativistic kinematics, we would have elastic final states where particle 1 has, with respect to the direction of $\vec{p}$, longitudinal momentum $p_{1,L} \gg mc$ and particle 2 has longitudinal momentum $p_{2,L} \gg mc$ with $p_{1,L} + p_{2,L} = p$. A total transverse energy $E_\perp \simeq mc^2$ would be left for the outgoing particles. But the balance is drastically modified by DRK if $\alpha \, (k \, a)^2 \, p$ becomes of the same order as $m \, c$ or larger. As the energy increases, stronger and stronger limitations of the available final-state phase space appear: the final-state configuration $p_{1,L} = p - p_{2,L} = (1 - \lambda) \, p$ becomes kinematically forbidden for $\alpha \, (k \, a)^2 \, p > 2 \, m \, c \, \lambda^{-1}(1 - \lambda)^{-1}/3$. Above $p \approx (m \, c \, a^{-2} \, h^2)^{1/3}$, "hard" interactions become severely limited by kinematical constraints. Similarly, with the same initial state, a multiperipheral final state configuration with N particles ($N > 2$) of mass m and longitudinal momenta $g^{i-1} \, p'_L$ ($i = 1, ..., N$, $g > 1$), where $p'_L = p \, (g - 1) \, (g^N - 1)^{-1}$, $g^N \gg 1$ and $p'_L \gg m \, c$, would have in standard relativity an allowed total transverse energy $E_\perp \, (N , g) \simeq m \, c^2 \, [1 - m \, c \, (2 \, p'_L)^{-1} \, (1 - g^{-1})^{-1}]$ which is positive definite. Again, using the new kinematics and the approximation (3), we find that such a longitudinal final state configuration is forbidden for values of the incoming momentum such that $\alpha \, (k \, a)^2 \, p \, c > 2 \, (3 \, g)^{-1} \, (1 + g + g^2) \, E_\perp \, (N , g)$. *Elastic, multiparticle and total cross sections will sharply fall at very high energy.*

For "soft" strong interactions, the approach were the two-body total cross section is the less sensitive to final-state phase space is, in principle, that based on dual resonance models and considering the imaginary part of the elastic amplitude as being dominated by the shadow of the production of pairs of very heavy resonances of masses M_1 and M_2 of order $\approx (p \, m \, c^3/2)^{1/2}$ in the direct channel [31,32]). Even in this scenario, we find important limitations to the allowed values of M_1 and M_2 , and to the two-resonance phase space, when $\alpha \, (k \, a)^2 \, p$ becomes $\approx m \, c$ or larger. Above $E \approx (2\pi)^{-2/3} \, (m \, \alpha^{-1} \, a^{-2} \, h^2 \, c^4)^{1/3}$, nonlocal effects play a crucial role and invalidate considerations based on Lorentz invariance and local field theory used to derive the Froissart bound [33], which seems not to be violated but ceases to be significant given the expected behaviour of total cross sections which, at very high-energy, seem to fall far below this bound. An updated study of noncausal dispersion relations [34,35], incorporating DRK and nonlocal dynamics, can possibly lead to new bounds. If the target is not at rest in the VRF, but its energy is small as compared to that of the incoming particle, its rest energy must be replaced by the actual target energy E_T in the above discussions. The absence of GZK cutoff is a particular application of this analysis, which has a much general validity.

EXPERIMENTAL IMPLICATIONS

With DRK, a small violation of the universality of c would not necessarily produce the Cherenkov effect in vacuum considered by Coleman and Glashow [24] for high-energy cosmic rays. If c and α are both universal and α is positive, all stable particles remain stable when accelerated to ultra-high energies and can reach any energy without spontaneously decaying or radiating in vacuum (the case $\alpha < 0$, not considered here, would lead to spontaneous decays and "Cherenkov" phenomena for particles at very high energy). If c is universal but α (positive) is not, there will in any case be at least one stable particle at ultra-high energy (that with the highest positive value of α). If none of the two constants is universal, any scenario is *a priori* possible. The mechanisms we just described compete with those considered in [24] and tend to compensate their effect: therefore, the bounds obtained by these authors do not apply to our ansatz where small violations of the universality of c can be compensated by the deformation term [13,18]. On the other hand, the discussion of velocity oscillations of neutrinos presented in [36] for the low-energy region is compatible with our theory. However, the universality of c seems natural in unified field theories (whereas that of the mass is naturally violated) and preserves the Poincaré relativity principle in the low-momentum limit. As previously stressed, deviations from the universality of c due to nucleon or nucleus structure are expected to be very small according to naive soliton models.

For ultra-high energy cosmic rays in the $E \approx 10^{20}$ eV region, we expect the most dramatic physical effects to be governed by the values of α for the particles considered. If Lorentz symmetry is broken and an absolute rest frame exists, high-energy particles are indeed different physical objects from low-energy particles, and high-energy tests of Lorentz symmetry are required. To reach direct comparison with a $\approx 3.10^{20}$ eV cosmic ray event, a $p - p$ collider with energy ≈ 400 TeV per beam would be required. Very-forward experiments at LHC and VLHC would be crucial steps in a systematic check of the validity of Lorentz symmetry, comparing their data with those of cosmic-ray events above $\approx 10^{16}$ eV. Evidence for Lorentz symmetry violation would no doubt be the most important physics outcome of particle physics experiments in the decades to come.

The possibility of taking the value of a^{-1} close to the wave vector of the highest-energy cosmic rays, i.e. $\approx 10^{26}$ cm^{-1}, was considered in [16] in connexion with a possible search for DRK effects through particle lifetimes. With $\alpha \approx 0.1$ and formulae (1)-(3), this would not be incompatible with low-energy bounds on LSV. But the value of E_{lim} would become too low leading to obvious incompatibilities with data in the very high-energy region (the fall of final-state phase space). New bounds on LSV thus emerge from high-energy data using the parametrization (1)-(3). Requiring that: a) cosmic rays with energies below $\approx 3.10^{20}$ eV deposit most of their energy in the atmosphere; b) the GZK cutoff is suppressed at energies above $\approx 10^{20}$ eV, leads in the DRK scenario to the constraint: 10^{-72} $cm^2 < \alpha\, a^2 < 10^{-61}$ cm^2, equivalent to $10^{-20} < \alpha < 10^{-9}$ for $a \approx 10^{-26}$ cm. Remarkably enough, assuming full-strength LSV forces a to be in the range 10^{-36} cm $<$ a $<$ 10^{-30} cm. Data on high-energy cosmic rays contain information relevant to the search for DRK signatures and should be carefully analyzed. The energy dependence of the π^0 lifetime above E $\approx$ 10^{18} eV can be a basic ingredient in generating the specific cascade development profile (e.g. electromagnetic showers versus hadronic showers and muons). Beyond DRK, strong signatures can be produced by other LSV effects: failure of the parton model for protons and nuclei (including its dual "soft" version [31,32]), substantially changing the multiplicities and event shape; strong deviations from the relativistic formulae for Lorentz contraction and time dilation leading to basic modifications of the dynamics... Because of its stability at very high energy, the neutron becomes a serious candidate to be a possible primary of the highest-energy cosmic-ray events.

Cosmic rays seem to indeed be able to test the predictions of (1) and set upper bounds on the fundamental length a, as well as constraints on α. Experiments like AUGER, OWL, AIRWATCH FROM SPACE and AMANDA present great potentialities. The study of early cascade development (perhaps with balloons installed in coincidence with the above experiments) will be crucial for the proposed investigation. Very high-energy data may even provide a way to measure neutrino masses and mixing, as well as other parameters related to phenomena beyond the standard model. If Lorentz symmetry were not violated, there would be no fundamental difference between the collision of a very high-energy cosmic ray and the "Lorentz equivalent" event at a collider. But, if Lorentz symmetry is violated, the study of the parameters governing LSV will provide us with a unique microscope directly focused on Planck-scale physics. Indeed, the E $\approx$ 10^{20} eV scale is closer, in order of magnitude, to the Planck scale than to the electroweak scale.

The possibility that above E_{lim} cosmic rays do not deposit all their energy in the atmosphere suggests to operate underground detectors in coincidence with air shower detectors, if ever feasible at the required scales. At E $\approx$ E_{lim}, the cosmic particle can still deposit enough energy in the atmosphere to produce a detectable air shower for satellite-based and balloon experiments.

REFERENCES

1. Poincaré, H., "A propos de la théorie de M. Larmor", *L'Eclairage électrique*, Vol. **5**, 5 (1895).

2. Poincaré, H., "Electricité et Optique: La lumière et les théories électriques", Ed. Gauthier-Villars, Paris 1901.

3. See, for instance, two essays by A.A. Logunov: "On the articles by Henri Poincaré on the dynamics of the electron", Ed. JINR, Dubna 1995, and "Relativistic theory of gravity and the Mach principle", Ed. JINR, Dubna 1997.

4. Feynman, R.P. et al., "The Feynman Lectures on Physics", Vol. I, Addison-Wesley 1964.

5. Poincaré, H., "Sur la dynamique de l'électron", *Comptes Rendus de l'Académie des Sciences*, Vol. **140**, p. 1504, June 5, 1905.

6. Einstein, A., *Annalen der Physik* **17**, 891 (1905).

7. Miller, A.I., in "Henri Poincaré, Science and Philosophy", Ed. Akademie Verlag, Berlin, and Albert Blanchard, Paris (1996).

8. Paty, M., same reference.

9. Feynman, R.P., quoted by A.A. Logunov in [3].

10. Gonzalez-Mestres, L., contribution to the 28th International Conference on High Energy Physics (ICHEP 96), Warsaw July 1996, paper hep-ph/9610474 of LANL (Los Alamos) electronic archive.

11. Gonzalez-Mestres, L., paper physics/9704017 of LANL electronic archive.

12. Gonzalez-Mestres, L., talk given at the International Conference on Relativistic Physics and some of its Applications, Athens June 1997, paper physics/9709006.

13. Gonzalez-Mestres, L., papers physics/9702026 and physics/9703020.

14. Gonzalez-Mestres, L., Proceedings of the 25th International Cosmic Ray Conference, Durban July-August 1997 (ICRC 97), Vol. **6** , p. 113 (1997), paper physics/9705032.

15. Gonzalez-Mestres, L., paper nucl-th/9708028 of LANL electonic archive.

16. Gonzalez-Mestres, L., ICRC Proceedings, Vol. **6** , p. 109 (1997).

17. Gonzalez-Mestres, L., contribution to the Europhysics International Conference on High-Energy Physics (HEP 97), Jerusalem August 1997, paper physics/9708028.

18. Gonzalez-Mestres, L., talk given at the Pre-Conference "Pierre Auger" Workshop of ICRC 97, paper physics/9706032.

19. Gonzalez-Mestres, L., paper physics/9706022 of LANL electonic archive.

20. Gonzalez-Mestres, L., talk given at the International Workshop on Topics on Astroparticle and Underground Physics (TAUP 97), paper physics/9712005.

21. Lamoreaux, S.K., Jacobs, J.P., Heckel, B.R., Raab, F.J. and Forston, E.N., *Phys. Rev. Lett.* **57** , 3125 (1986).

22. Hills, D. and Hall, J.L., *Phys. Rev. Lett.* **64** , 1697 (1990).

23. A. Einstein, "Geometrie und Erfahrung", *Preussische Akademie der Wissenchaften, Sitzungsberichte*, part **I**, p. 123 (1921).

24. Coleman, S. and Glashow, S.L., *Phys. Lett. B* **405**, 249 (1997).

25. See, for instance, Will, C., "Theory and Experiment in Gravitational Physics", Cambridge University Press (1993).

26. Rédei, L.B., *Phys. Rev.* **162** , 1299 (1967).

27. Anchordoqui, Dova, M.T., Gómez Dumm, D. and Lacentre, P., *Zeitschrift für Physik C* **73** , 465 (1997).

28. Greisen, K., *Phys. Rev. Lett.* **16** , 748 (1966).

29. Zatsepin, G.T. and Kuzmin, V.A., *Pisma Zh. Eksp. Teor. Fiz.* **4** , 114 (1966).

30. Kirzhnits, D.A., and Chechin, V.A., *Soviet Journal of Nuclear Physics*, **15** , 585 (1972).

31. Aurenche, P. and Gonzalez-Mestres, L., *Phys. Rev. D* **18** , 2995 (1978).

32. Aurenche, P. and Gonzalez-Mestres, L., *Zeitschrift für Physik C* **1** , 307 (1979).

33. Froissart, M., *Phys. Rev.* **123**, 1053 (1961)

34. Blokhintsev, D.I. and Kolerov, G.I., *Nuovo Cimento* 34 , **163** (1964).

35. Blokhintsev, D.I., *Sov. Phys. Usp.* **9** , 405 (1966).

36. Glashow, S., Halprin, A., Krastev, P.I., Leung, C.N. and Pantaleone, J., *Phys. Rev. D* **56** , 2433 (1997).

From Atmospheric Electricity to the 16-Joule Cosmic-Ray Air Shower

George W. Clark

Department of Physics and Center for Space Research
Massachusetts Institute of Technology
Cambridge, Massachusetts 02139

Abstract. The roots of cosmic-ray air shower research are traced from Benjamin Franklin to recent times, with a focus on the work of Bruno Rossi and post-war developments at MIT.

I   ATMOSPHERIC ELECTRICITY

Extensive showers of cosmic-ray particles in the open air were discovered by Bruno Rossi in 1934, and independently by Pierre Auger and Roland Maze several years later. But the historical roots of air shower research are in the investigations of atmospheric electricity that began with Benjamin Franklin. The eleventh edition of the Encyclopedia Brittanica, published in 1910, has a long article on that subject. It was written by Charles Chree, president of the Physical Society of London, after the discovery of X-rays, radioactivity, and the electron, but before the Rutherford nucleus, the Bohr atom, and cosmic rays. He begins with Franklin's perilous kite experiment which proved that there is, indeed, electricity in the air. Chree then skips a century to Linss's observation that a charged electroscope leaks with a decay constant of about one hour, regardless of how well it is insulated. [1] With the discovery of radioactivity and ionizing radiation just before the turn of the century, it seemed clear that electroscopes leak because they attract the ions produced by gamma rays from radioactive elements in the ground. In 10 pages of fine print and 14 tables of data Chree reviews the knowledge of atmospheric electricity and, in particular, the dependence of the leak rate on the seasons, time of the day, geographical position, temperature and humidity, the frequency of thunder storms, even on proximity to a water fall, but not on high altitude above the ground. He doesn't mention the very careful work of C. T. R. Wilson who wrote [44]

---

[1] Linss's observation can be checked with a bunch of Christmas tinsel (strips of aluminized mylar) knotted at one end and suspended by a silk thread. Stroked by a plastic pen that has been rubbed on hair or wool, the strips will flair out with a pretty symmetry, and gradually collapse in about an hour.

CP433, *Workshop on Observing Giant Air Showers from Space*
edited by J.F. Krizmanic et al.

Experiments were now carried out to test whether the continuous production of ions in dust-free air could be explained as being due to radiation from sources outside our atmosphere, possibly radiation like Röntgen rays or like cathode rays, but of enormously greater penetrating power. The experiments consisted in first observing the rate of leakage through the air in a closed vessel as before, the apparatus then being taken into an underground tunnel and the observations repeated there. If the ionization were due to such a cause, we should expect to observe a smaller leakage underground on account of absorption of the rays by the rocks above the tunnel.

His conclusion after the experiment:

There is thus no evidence of any falling off of the rate of production of ions in the vessel although there were many feet of solid rock overhead. It is unlikely, therefore, that the ionization is due to radiation which has traversed our atmosphere; it seems to be, as Geitel concludes, a property of the air itself.

II THE CLASSICAL ERA

Victor Hess sought the cause of atmospheric ionization in ten high-altitude flights in an open gondola lifted by a hydrogen-filled balloon. One flight was made during a solar eclipse. Another took place on August 7, 1912 when he ascended from the Vienna Prater with two companions and three ion chambers to an altitude of 5350 meters. He returned to Earth 230 km away with measurements showing that the rate of ionization in a sealed chamber had increased 16-fold at the top of his flight. He wrote [15]

The results of my observations are best explained by the assumption that a radiation of very great penetrating power enters our atmosphere from above.

He called it "Höhenstrahlung." Subsequent flights by W. Kolhörster confirmed Hess's discovery, including a heroic ascent to 9300 m where the ionization rate was measured at fifty times the ground value [17] [18].

The first World War began soon after Hess's discovery, and little progress was made toward understanding the nature of Höhenstrahlung till 1923. Then Robert Millikan became interested in it. Already a Nobel laureate for precise measurement of the electron's charge by the oil drop method invented by his graduate student [13], Millikan recognized the mysterious "radiation from above" as a possible new world of science to conquer. At first skeptical about the Hess discovery, he began a series of experiments that he would claim solved the problem of the nature and origin of the radiation which he renamed cosmic rays. He made precise measurements with ionization meters sent all over the world, up and down mountains, deep in lakes, and at high altitudes in pioneering experiments with unmanned balloons.

Before 1929, the most penetrating radiation known was gamma rays, and the only known interaction of high-energy photons with matter was Compton scattering from electrons for which the cross section was known to decrease with wavelength. Millikan concluded that his measurements of the attenuation curves of cosmic rays in air and water could be accounted for by assuming they are ultra short-wavelength gamma rays with discrete energies corresponding to the mass defects that would be released as energy in the instantaneous formation of helium, oxygen, nitrogen and silicon from hydrogen. He called the primary radiation the "birth cries" of atoms formed in interstellar space. Later, in the face of the evidence for cosmic rays with more energy than could possibly be derived from atom synthesis, Millikan changed his theory to the "atom annihilation hypothesis." Casting doubts on the work of Hess and Kolhörster, Millikan promoted his results as the real discovery of cosmic rays [22] [23].

Though Millikan's ideas seem strange, his measurements did convince the remaining skeptics of the existence of cosmic rays. And his wrong ideas persisted in the Oxford Unabridged Dictionary as late as the 1958 edition, which defines cosmic rays as:

Any of the rays of extremely high frequency and penetrating power produced, it is thought, beyond the earth's atmosphere, or nearly beyond it, by transmutations of atoms continually taking place in interstellar space. If they are waves, as R. A. Millikan thinks, their wavelength is shorter than that of any ray produced in laboratories, and their frequency is a thousand times greater than that of X-rays. They bombard the earth and are in small part responsible for the ionization in the earth's atmosphere.

The Nobel Committee awarded the 1936 Prize jointly to Hess "for his discovery of cosmic radiation", and to Carl Anderson, a former student of Millikan's, "for his discovery of the positron." Meanwhile, Millikan clung to his ideas. A 1949 issue of *Reviews of Modern Physics* contains a collection of papers from a symposium in honor of his eightieth birthday. Numerous topics of post-war cosmic-ray and high-energy physics are discussed, such as the heavy primaries, the pi meson, and other recent discoveries. The lead article by Millikan is entitled "The Present Status of the Evidence for the Atom-Annihilation Hypothesis" [24]. The last paragraph of Millikan's article begins with the sentence

No evidence has come in since then (1942) which is unfavorable to the theory, unless it be that which we ourselves obtained in August and September, 1947.

III MODERN TIMES: PENETRATING PARTICLES, THE SOFT COMPONENT, AND MULTIPLICATIVE SHOWERS

Modern cosmic ray physics began in 1927 with two discoveries. J. Clay discovered the latitude effect in a measurement of the change in cosmic-ray ionization between Holland and Java [12]. His results showed that the primary cosmic-rays consist, at least in part, of charged particles that are influenced by the geomagnetic field. D. V. Skobeltzyn, studying the energies of beta-decay electrons with a magnet cloud chamber that he built in the laboratory of Madame Curie, noticed the faint track of a particle with much less curvature, and therefore much more momentum, than any of the beta electrons [40]. That showed that some cosmic rays at ground level are also energetic charged particles. Two years later Skobeltzyn observed locally produced groups of up to four high-energy particles [41], and Bothe & Kolhörster used the new coincidence method, invented by Bothe, to detect particles that penetrated 4 cm of gold [6].

Bruno Rossi was beginning his first job as assistant at the Physics Institute of the University of Florence when he read the Bothe & Kolhörster paper. Acquainted with the well-known ideas of Millikan, Rossi was inspired by that paper to focus his research on what seemed to be a breakthrough into a whole new realm of physics. Among the first things he did was to invent the electronic triode coincidence circuit [26]. His circuit, precursor of the AND circuit of modern computers, achieved a great improvement in time resolution over the mechanical pen recorder that Bothe had used to record coincidences. It made possible a wide range of low counting-rate experiments that would otherwise be swamped by accidental coincidences.

In a summer visit to Bothe's Berlin laboratory, Rossi used his coincidence circuit with the local Geiger tubes to detect cosmic ray particles that penetrated 9.7 cm of lead. In his autobiography [38] Rossi describes how his own homemade tubes, constructed according to the published procedures of Geiger and Müller, had steel wires that had been dipped in nitric acid. He wrote

> I was puzzled about this matter until one day Bothe took me aside with a mysterious air and began: 'I will tell you a secret, but you must promise not to give it away to anyone'. After I had promised, he continued: 'my counters do not have a steel wire, as advertised; they have an aluminum wire'. To my shame, I must confess that, upon my return to Italy, I was not able to keep the secret of the aluminum wire from my friends in Florence and Rome. But I relieved my conscience by requiring of them the same oath of secrecy that Bothe had required of me.

In 1931 the Italian Royal Academy organized a Conference on Nuclear Physics attended by physics luminaries from around the world. Fermi invited Rossi to give the introductory talk on cosmic rays. In the presence of Millikan, he explained that charged particles capable of penetrating 9.7 cm of lead had more energy than the limit set by Millikan's atom birth-cry theory. Rossi wrote

Millikan clearly resented having his beloved theory torn to pieces by a mere youth, so much so that from that moment on he refused to recognize my existence.

The electronic age of nuclear physics was inaugurated in the laboratory on the hill of Arcetri overlooking Florence. From there the technique of electronic coincidence spread rapidly around the world. Rossi used it with three aligned Geiger tubes to detect "penetrating" particles that traversed 1 meter of lead [28]. With three Geiger tubes in a triangular configuration, he detected showers of particles produced in various thicknesses of lead. His observations of "multiplicative" showers and measurement of the "Rossi transition effect" were so unexpected at the time that his paper was rejected by one German journal, and accepted by another [29] only after Heisenberg vouched for his reliability. It was these measurements that made the first clear distinction between the rapidly absorbed "soft component" of secondary cosmic rays, soon to be recognized as electrons and photons, and the penetrating "hard component", that would later be identified as mesotrons, now called muons.

Rossi sent his first student, Giuseppe (Beppo) Occhialini, to Manchester where he developed a coincidence trigger for the cloud chamber of P. M. S. Blackett. The counter-controlled cloud chamber soon produced many pictures of locally-produced showers [5] which confirmed Rossi's results, and inspired the theories of electron-photon showers [4] [7].

Learning in Berlin of the work of Störmer on the motion of auroral particles in the geomagnetic field, Rossi conceived the idea of determining the sign of the charged primaries by observing the effects of the earth's magnetic field on the directional dependence of the cosmic ray intensity. He published the theory of the east-west effect in the *Physical Review* [27], and attempted unsuccessfully to measure it at the high geomagnetic latitude of Florence. After a four-year delay he was finally able to travel to the low geomagnetic latitude of Eritrea where he found the penetrating particle intensity at a zenith angle of 45 deg was 26% higher from the west than from the east. The delay cost him the priority of discovery because the effect had already been measured in two experiments at Mexico City [16] [1]. All three experiments proved the surprising result that most of the primaries are positive particles, not negative electrons as had been generally assumed.

Rossi's first report of his measurement of the east-west effect contains the discovery of extensive cosmic-ray air showers. To test his electronics for accidental coincidences he spread his Geiger tubes in a horizontal plane so that no single particle could trigger a coincidence. His translation of what he wrote [30] is

> It would seem therefore (since doubts about possible disturbances were ruled out by appropriate control experiments) that once in a while there arrive on the instruments very extended showers of particles which produce coincidences between counters even rather far from each other. Unfortunately I lacked the time to study more closely this phenomenon in order to establish with certainty the existence of the supposed corpuscular showers and investigate their origin.

Several years later Pierre Auger and Roland Maze rediscovered air showers in experiments aimed initially at studying locally produced multiplicative showers [2]. They carried out an extensive study of what came to be called "Auger showers" with trays of Geiger tubes and electronic circuits that registered the numbers of tubes simultaneously discharged by a shower. They concluded that the primary particles initiating the showers had energies up to at least 10^{15} eV.

Rossi's Rome talk had inspired Arthur Compton to became active in cosmic ray research. Rossi left Italy in 1938 after being removed by the Fascisti from his position as Director of the Physics Institute which he had established at Padua. After several months of cordial and productive visits to Bohr's Institute in Copenhagen and Blackett's laboratory at Manchester, Compton invited him to the University of Chicago. Within a few weeks he assembled a penetrating particle detector and took it to Mt. Evans in Colorado where, together with Norman Hilberry and J. B. Hoag, he measured the attenuations of mesotrons in air and in graphite. From the difference, caused by the disintegration of mesotrons in flight through the atmosphere, he derived the first certain experimental evidence for the decay of a subnuclear particle and a measure of the muon mean life [31].

In 1940 Rossi was appointed to the Cornell faculty. With his first American student, Kenneth Greisen, he wrote the review article which became the bible of cosmic ray theory and interpretation [32]. With new collaborators he measured the effect of relativistic time dilation on the survival of mesotrons in the atmosphere and verified quantitatively the predictions of the Einstein theory [33]. In 1942, with America at war, he made his second major electronic invention, the time-to amplitude converter (TAC). Rossi and Nereson used it to measure the mesotron mean life to an accuracy of ±0.4% [34] [25]. They did not describe the design of the TAC, now a standard instrument of nuclear physics laboratories, because they thought it might have military value. It was referred to as the "the timing circuit", and only described after the war [35].

IV POST-WAR AIR SHOWER RESEARCH AT MIT

After the war many new experiments with Geiger tubes and coincidence circuitry started up around the world. Large-area proportional detectors were not yet generally known. Rossi and Hans Staub, working at Los Alamos during the war, had developed large proportional pulse ion chambers [36] for use in the so-called RaLa (radioactive lanthanum) experiments which measured the compression achieved in implosion experiments in a canyon near the laboratory, and then in the experiment which measured the exponential increase in the intensity of gamma-rays emitted by the plutonium bomb during the first few microseconds of the Trinity test. As the war ended, Rossi turned his attention once more to the problems of cosmic rays.

In 1946 Rossi joined the physics faculty of MIT, and initiated a variety of projects aimed at exploring the composition and interactions of cosmic rays in the atmosphere, the nature of particles produced in high-energy interactions of cosmic rays, and the properties of the primary radiation. He conceived the method for measuring the size and core location of an air shower from measurements of the density of air shower particles over a widely-spread array of proportional counters. At his suggestion, Robert Williams, one of the several young Los Alamos scientists who had followed Rossi to MIT to complete their graduate studies, set up an array of pulse ion chambers at an altitude of 10,000 feet on Mt. Evans. Using a cloud chamber to determine the arrival directions, he measured the structure and size spectrum of showers generated by primaries with energies up to 10^{17} eV [43].

In 1952 Rossi suggested to Peter Bassi and me that we explore the use of the new liquid scintillators, with their short scintillation decay times, to measure the arrival directions of air showers by fast timing. With three liquid scintillation detectors set out in a line on the roof of MIT, distributed-line amplifiers, and a fast oscilloscope, we measured the relative arrival times of air shower particles and demonstrated that shower disks are, as one would imagine, thin and nearly flat, well suited for direction measurements [3].

The methods of density sampling and fast timing were soon incorporated in several air shower experiments around the world. At MIT we began with a square array of four detectors to study the celestial arrival directions of primaries with a mean energy of about 10^{15} eV. We found a disappointing isotropy in the celestial distribution of several thousand arrival directions [8]. Next, we set up an array of eleven 1-m^2 toluene-based liquid scintillators in the woods around the Agassiz station of the Harvard College Observatory. We encountered some unexpected problems: rabbits acquired a taste for the insulation of our coaxial cables, and one night lightening set a detector on fire. When the local fire company arrived they found a pillar of fire bursting from a mysterious contraption in the middle of the woods. Fortunately the woods were wet, and we didn't burn down the observatory. After tense discussions among high officials of MIT and Harvard, we were allowed to proceed with metal flame snuffers devised by John Linsley and suspended by fusible wires over each detector. Meanwhile, we set up a factory that turned out several tons of plastic scintillators to replace the liquid scintillators in the woods [9]. The Agassiz experiment extended the spectrum to 10^{18} eV [10].

The Bolivian Air Shower Joint Experiment, was carried out in a collaboration between the Institute for Nuclear Science in Tokyo, MIT, and the University of San Andres in La Paz. It aimed at the detection of "low-mu" air showers, i.e., air showers with very low ratios of muons to electrons. Low-mu showers are expected to be initiated by the high-energy gamma rays from the decay of neutral pions produced in the interactions of cosmic-ray nuclei with interstellar matter. That mechanism for production of cosmic gamma rays, suggested by Satio Hayakawa in 1952 [14], has provided the explanation for most of the diffuse galactic component of high-energy gamma rays that has been a major subject of space astronomy since it was first observed with the OSO-3 satellite [19]. The BASJE was designed to do gamma-ray astronomy at 10^{14} eV. A muon detector with 60 m^2 of plastic scintillators was developed by Koichi Suga during a visit to MIT. It was placed under a concrete roof supporting 200 tons of native galena, a remarkable structure built at 17,000 feet on Mt. Chacaltaya under the supervision of Ismael Escobar. An array of scintillation detectors, transported from the Agassiz experiment, was set up on platforms around the muon detector. The experiment detected showers with one-thousandth of the average density of muons (about one shower in a thousand). However, the distribution of their arrival directions showed no significant concentration toward the Milky Way where the production of gamma ray is expected to occur [11].

Finally, John Linsley and Livio Scarsi built a 2.5-km^2 air shower array at Volcano Ranch near Albuquerque, New Mexico, and extended the measurement of the primary spectrum to 10^{19} eV [20]. Linsley ultimately stretched the array to an area of 10 km^2 and detected a monster shower produced by a 16-Joule (10^{20} eV) primary particle [21]. The origin of that particle and others like it detected in the Haverah Park and Sydney experiments remains one of the most intriguing problems of astrophysics today .

V ACKNOWLEDGMENT

This historical sketch is an expansion of an after dinner talk given at the Workshop on the Highest Energy Cosmic Rays. The parts before 1950 are based in large measure on the collection of memoirs and essays edited by Sekido and Elliot [39], on the book and the autobiography of Rossi [37,38], and on a text by Stranathan [42]. The post-World War II part is narrowly focused on developments at MIT with no attempt to describe the important results obtained in other contemporary air shower experiments around the world.

REFERENCES

1. Alvarez, L. & Compton, A., *Phys. Rev.* **43**, 835 (1933).
2. Auger, P. & Maze, R., *Compt. Rend. Acad. Sci. (Paris)* **206**, 1721 (1938).
3. Bassi, P., Clark, G. W., & Rossi, B., *Phys. Rev* **92** 441, (1953).
4. Bhabha, H. & Heitler, W., *Proc. Roy. Soc.* **159**, 432 (1957).
5. Blackett, P. M. S. & Occhialini, G., *Proc. Roy. Soc.* **A149**, 699 (1933).
6. Bothe, W. & Kolhörster, W., *Z. Physik* **56**, 751 (1929).
7. Carlson, J. F., & Oppenheimer, J. R., *Phys. Rev.* **51**, 220 (1957).
8. Clark, G. W., *Phys. Rev.* **108**, 450 (1957).
9. Clark, G. W., Scherb, F., & Smith, W. B., *Rev. Sci. Inst.* **28**, 433 (1957).
10. Clark, G. W., Earl, J. Kraushaar, W. L., Linsley, J., Rossi, B. B., Scherb, F., and Scott, D., *Phys. Rev.* **122**, 637 (1961).
11. Clark, G. W., Suga, K., & Escobar, I., *Pont. Acad. Sci. Scripta Varia* **25**, 1 (1963).
12. Clay, J., *Proc. Acad. Wetensch., Amsterdam* **30**, 1115 (1927).
13. Fletcher, H., "My Work with Millikan on the Oil Drop Experiment", *Physics Today* **123**, 123 (1982).
14. Hayakawa, S., *Prog. Theoret. Phys.* **9**, 517 (1952).

15. Hess, V. F., *Wien. Ber.*, **121**, 2001 (1912).

16. Johnson, T. H., *Phys. Rev.* **43**, 834 (1934).

17. Kolhörster, W., *Physik. Z.* **14**, 1153 (1913).

18. Kolhörster, W., *Verhandl. d. Deutsch. Phys. Ges.* **16**, 719 (1914).

19. Kraushaar, W. L., Clark, G. W., Garmire, G. P., Borken, R., Higbie, P., Leong, C., & Thorsos, T., *Ap. J.* **177**, 341 (1972).

20. Linsley, J., Scarsi, L., & Rossi, B., *Phys. Rev. Letters* **6**, 485 (1961).

21. Linsley, J., *Scientific American* **239**, 48 (1978).

22. Millikan, R. A. & Cameron, G. H., *Nature* **121**, 19 (1928).

23. Millikan, R. A., *Science* **68**, 279 (1928).

24. Millikan, R. A., *Rev. Mod. Phys.* **91**, 1 (1949).

25. Nereson, N., & Rossi, B., *Phys. Rev.* **64**, 199 (1943).

26. Rossi, B., *Nature* **125**, 636 (1930).

27. Rossi, B., *Phys. Rev.* **36**, 606 (1930).

28. Rossi, B., *Nature* **20**, 65 (1932).

29. Rossi, B., *Phys. Zeitsch.* **33**, 304 (1932).

30. Rossi, B., *Ricerca Scient. V (1)*, 559 (1934).

31. Rossi, B., Hilberry, N., & Hoag, J. B., *Phys. Rev.* **56**, 837 (1939).

32. Rossi, B. & Greisen, K., *Rev. Mod. Phys.* **13**, 240 (1941).

33. Rossi, B., Greisen, K., Stearns, J. C., Froman, D. K., & Koontz, P. G., *Phys. Rev.* **61**, 675 (1942).

34. Rossi, B. & Nereson, N., *Phys. Rev.* **62**, 417 (1942).

35. Rossi, B. & Nereson, N., *Rev. Sci. Instr.* **17**, 65 (1946).

36. Rossi, B. & Staub, H., *Ionization Chambers and Counters*, McGraw-Hill Book Co., Inc. (1949).

37. Rossi B. B., *Cosmic Rays*, New York: McGraw-Hill, (1964).

38. Rossi, B. B., *Moments in the Life of a Scientist*, Cambridge Univ. Press (1990).

39. Sekido, Y. & Elliot, H., *Early History of Cosmic Ray Studies*, Reidel Publishing Company (1982).

40. Skobeltzyn, D., *Zs. Phys.* **43**, 354 (1927).

41. Skobeltzyn, D., *Zs. Phys.* **54**, 686 (1929).

42. Stranathan, J. D., *The Particles of Modern Physics*, Philadelphia: Blakiston (1942).

43. Williams, R., *Phys. Rev.* **74**, 1689 (1948).

44. Wilson, C. T. R., *Proc. Roy. Soc. London*, **A68**, 151 (1901).

Ultrahigh Energy Cosmic Rays from Topological Defects — Cosmic Strings, Monopoles, Necklaces, and All That

Pijushpani Bhattacharjee[1]

Laboratory for High Energy Astrophysics,
NASA/Goddard Space Flight Center, Code 661,
Greenbelt, MD 20771. USA.
and
Indian Institute of Astrophysics, Bangalore-560 034. INDIA.

Abstract. The topological defect scenario of origin of the observed highest energy cosmic rays is reviewed. Under a variety of circumstances, topological defects formed in the early Universe can be sources of very massive particles in the Universe today. The decay products of these massive particles may be responsible for the observed highest energy cosmic ray particles above 10^{20} eV. Some massive particle production processes involving cosmic strings and magnetic monopoles are discussed. We also discuss the implications of results of certain recent numerical simulations of evolution of cosmic strings. These results (which remain to be confirmed by independent simulations) seem to show that massive particle production may be a generic feature of cosmic strings, which would make cosmic strings an inevitable source of extremely high energy cosmic rays with potentially detectable flux. At the same time, cosmic strings are severely constrained by the observed cosmic ray flux above 10^{20} eV, if massive particle radiation is the dominant energy loss mechanism for cosmic strings.

INTRODUCTION

Cosmic Topological Defects (TDs) [1,2] such as magnetic monopoles, cosmic strings, domain walls, and various hybrid TD systems consisting of these basic kinds of TDs, are predicted to form in the early Universe as a result of symmetry-breaking phase transitions envisaged in unified theories of elementary particle interactions. Under a variety of circumstances these TDs can be sources of extremely massive unstable particles in the universe today [3–11]. The masses m_X of these so-called "X" particles (the quanta of the massive gauge- and higgs fields of the underlying spontaneously broken gauge theory) would typically be of order the symmetry-breaking energy scale at which the relevant TDs were formed, which, in Grand

---

[1] NAS/NRC Resident Senior Research Associate at NASA/GSFC on sabbatical leave from Indian Institute of Astrophysics, Bangalore-560 034. India.

CP433, *Workshop on Observing Giant Air Showers from Space*
edited by J.F. Krizmanic et al.

Unified Theories (GUTs), can be as large as $\sim 10^{16}$ GeV. The decay of these X particles can give rise to extremely energetic photons, neutrinos and nucleons with energy up to $\sim m_X$. If the X particle production rate from TDs is large enough, these extremely energetic particles may be detectable by ground-based as well as space-based large-area detectors being planned for detecting ultrahigh energy (UHE) (i.e., energy $\gtrsim 10^9$ GeV) cosmic rays. These cosmic ray detectors may thus provide us with a tool for studying the signature of TDs and thus of GUT scale physics.

There is currently much interest in the possibility that the Extremely High Energy (EHE) Cosmic Ray (EHECR) events — those with energies above 10^{11} GeV reported recently [12] — may be due to decays of massive X particles originating from TDs [13,10,14–17,11]. This possibility is of interest in view of the fact that the energies associated with the EHECR events are hard to obtain [18,13,19] within the standard diffusive shock acceleration mechanism [20] that involves first-order Fermi acceleration of charged particles at relativistic shocks associated with known powerful astrophysical objects; see, however, Ref. [21]. In addition, there is the problem of absence of any obviously identifiable sources for these EHECR events [22,13]. These problems are avoided in the TD scenario in a natural way. Firstly, no acceleration mechanism is needed: The decay products of the X particles have energies up to $\sim m_X$ which can be as large as, say, 10^{16} GeV. Secondly, the absence of obviously identifiable sources is not a problem because TDs need not necessarily be associated with any visible or otherwise active astrophysical objects such as AGNs or radio galaxies.

The basic ideas of the TD scenario of origin of EHECR have been reviewed in a number of discussions in the past; see, e.g., Refs. [7,23–25]. Detailed calculations of the predicted spectra of nucleons, photons, and neutrinos in the TD scenario have been done in the past several years [3,26,10,15,14,17,16,27]. Constraints on the TD scenario imposed by experimental data on EHECR and diffuse gamma ray background have also been discussed [28,15,17,16].

In what follows, I first discuss briefly some of the basic aspects of topological defects and their formation in the early Universe. The basic steps in the calculation of the 'observable' particle spectra resulting from the decay of the X particles is discussed. A simple benchmark calculation of the X particle production rate required to explain the observed EHECR particle flux is performed. I then discuss three specific X particle production mechanisms involving (1) "ordinary" cosmic strings, (2) monopolonia — metastable monopole-antimonopole bound states, and (3) cosmic "necklace" — a system of monopoles on strings. I also discuss the implications of the results of certain recent numerical simulations of evolution of cosmic strings in the Universe. These results, if confirmed by independent simulations, would imply that massive X-particle production and hence cosmic ray production might be a generic feature of cosmic strings, which would make cosmic strings an inevitable source of EHECR with potentially detectable flux. Indeed, in this case, we shall see that the measured EHECR flux already puts severe constraint on the energy scale of symmetry-breaking associated with any cosmic string forming phase transition in the early Universe.

I use natural units with $\hbar = c = k_B = 1$ throughout, unless otherwise stated.

TOPOLOGICAL DEFECTS AND X PARTICLE PRODUCTION: A BRIEF REVIEW

Topological Defects are sometimes characterized as "exotic". In actual fact, TDs are routinely seen, measured, and studied in condensed matter systems in laboratories. TDs form during phase transitions associated with the phenomenon of spontaneous symmetry breaking (SSB), which is a central concept in condensed matter physics as well as in the Standard Model of particle physics. Well-known examples of TDs in condensed matter systems are vortex lines in superfluid helium, magnetic flux tubes in type-II superconductors, disclination lines and 'hedgehogs' in nematic liquid crystals, and so on. Perhaps what is perceived as exotic is the existence of TDs in the cosmological context. However, it has been realized for quite some time now, particularly since the early seventies, that our Universe in its early stages must have behaved very much like condensed matter systems. Indeed, ideas of unified gauge theories of elementary particle interactions taken together with the hot big-bang model of the early Universe necessarily imply that our Universe in its early history has passed through a sequence of symmetry-breaking phase transitions as it expanded and cooled through certain critical temperatures. Depending on the symmetry breaking pattern, one or more kinds of the three basic kinds of topological defects — magnetic monopoles, cosmic strings and domain walls — could be formed during some of these phase transitions [1,2]. In fact, formation of magnetic monopoles, is *inevitable* in practically all Grand Unified Theories (GUTs) that provide a unified description of the electroweak and strong interactions[2]. The monopoles are analogous to the 'hedgehogs' in nematic liquid crystals and appear whenever the unbroken symmetry group possesses a local U(1) symmetry. The 'global' cosmic strings, which arise in breaking of global U(1) symmetry, are similar to vortex filaments in superfluid helium, and the 'local' or 'gauge' cosmic strings arising from breaking of a local U(1) symmetry are similar to the magnetic flux tubes in type-II superconductors. Cosmic domain walls appear whenever a discrete symmetry is spontaneously broken.

Interestingly, it has recently become possible to simulate the analogue of cosmic string formation in the early Universe by means of laboratory experiments [30] on vortex-filament formation in the superfluid transition of ^{3}He, which occurs at a temperature of a few millikelvin. The results of these experiments have provided striking confirmation of the basic Kibble-Zurek picture [1,2,31] of topological defect formation in general, which was initially developed within the context of defect formation in the early Universe [1].

---

[2] The inevitability of monopole formation leads to the well-known "monopole problem" of cosmology, which historically was one of the "problems" that motivated the idea of *inflationary cosmology*; for a review, see Ref. [29]

170

Symmetry Breaking Phase Transitions and Formation of Topological Defects

During a symmetry breaking phase transition, the system under consideration undergoes a transition from a state of higher symmetry to one of a lower (reduced) symmetry at a critical temperature during the cooling of the system. In spontaneous symmetry breaking (SSB), the system below the critical temperature possesses multiple degenerate ground states rather than a unique ground state. These degenerate ground states differ from each other by the 'phase angle' or some internal degrees of freedom of the "order parameter" field (OPF) whose absolute magnitude (which is same for all the ground states) is a measure of the order (or lack of symmetry) in the system. (In particle physics, the OPF is the "higgs" field.) The symmetry under consideration is invariance of the energy (or more precisely the Lagrangian for the OPF or the higgs field) under transformations that change the 'phase' of the OPF. The Lagrangian is always invariant under these transformations because, by construction, it depends only on the absolute magnitude of the OPF and not on its 'phase'. However, the ground states transform among each other under the action of these transformations, and so any chosen ground state is clearly not invariant under the transformations — the symmetry is spontaneously broken. The existence of multiple degenerate ground states is the defining characteristic of the phenomenon of spontaneous symmetry breaking. By convention, the absolute value of the OPF is taken to be zero in the high-temperature unbroken-symmetry phase and unity in the low-temperature broken symmetry phase.

Because of the availability of these multiple degenerate ground states in the low temperature phase, different parts of the system may choose to settle down to different ground states when making the transition to the low temperature phase, especially so if the transition happens in an out-of-equilibrium situation. Indeed, one expects that the choice of the phase of the OPF in regions separated by more than the correlation length of the thermal fluctuations of the OPF will be uncorrelated. The correlation length will always be finite in a finite physical system. In the context of the Universe as a whole, there is also an upper limit to the correlation length at any time t, namely, the causal horizon length $\sim ct$. Thus the choice of the phase of the OPF will be random and in general different in different parts of the Universe separated by more than the causal horizon length at the time of phase transition. This often leads to 'obstruction' in the way of uniform completion of the transition throughout the bulk of the system. Indeed, it is often the case that the random choice of different ground states in different regions leads to some regions being forced to remain in the unbroken symmetry phase. These regions are the 'topological defects'.

The topological nature of the defects becomes clear when one considers the configuration of the OPF in the low temperature phase at the end of the phase transition. The choice of the 'phase' of the OPF corresponding to different ground states in different parts of the system may be such that, in order to avoid energetically unfavored discontinuity in the spatial variation of the OPF, the magnitude of the OPF is forced to be zero on some geometrical points, lines, or surfaces, which define the 'center' of the defects. Actually, a defect has a finite size dictated by the need to minimize the overall energy of the system; the absolute value of the OPF increases *gradually* from zero at the center to its broken-symmetry-phase value at a finite distance from the center.

The topological stability of the defects is due to non-trivial topological 'winding' of the OPF configuration around the defect center. For example, in the case of the linear defect (the vortex filament in the superfluid helium, for example), the OPF (the wave function of the condensed helium atoms, for example) is a complex number whose phase turns by an integral multiple of 2π as one makes a complete circuit along a closed curve around any point on the defect line (the closed curve being on the normal plane cutting the defect line at the given point). Once formed, such a configuration cannot 'unwind' by itself and is thus topologically stable.

In the context of spontaneously broken gauge theories, explicit analytical and/or numerical finite-energy, extended, topologically stable solutions for the higgs- and gauge field configurations representing cosmic strings, monopoles and domain walls are known; see Ref. [2] for review. In the broken-symmetry phase, a higgs field responsible for spontaneous symmetry breaking of a local gauge theory is massive, as are the gauge bosons of the theory, the mass scale being set by the absolute magnitude of the higgs field (more precisely, its vacuum expectation value) in the broken symmetry phase. The size of the 'core' of a defect is of order m_X^{-1}, where X represents the higgs or the gauge field. Within the core, the symmetry remains unbroken, and the energy density (associated with the higgs and the gauge fields) is higher within the core than outside. The topological stability of the defect ensures the 'trapping' of the excess energy within the core of the defect, which is what makes a defect massive. It can be shown that the mass scale of a defect is fixed by the energy (or temperature) scale of the symmetry breaking phase transition at which the defect is formed. Thus, if we denote by T_c the critical temperature of the defect-forming phase transition in the early Universe, then the mass of a monopole formed at that phase transition is roughly of order T_c, the mass per unit length of a cosmic string is of order T_c^2, and the mass per unit area of a domain wall is of order T_c^3.

The X particles are the quanta of excitations of the higgs and gauge fields. In the broken symmetry phase, these quanta are massive, their mass is also roughly of order T_c. These massive quanta typically have very short life times, and so they all decayed away quickly soon after the phase transition in the early Universe, and none of those X particles survive in the present Universe. This is to say that outside of a defect in the low-energy Universe today, the higgs and the gauge fields are in their ground states and no excitations of the X particle quanta are present. However, inside the core of a defect, the symmetry is unbroken — the X particles are massless inside. Topological stability of the defect prevents this "piece of the early Universe" trapped inside the defect from decaying. If, however, there is a process which removes this topological protection, then the energy trapped inside the defect will

172

dissociate into quanta of the X particles which, being now in the broken-symmetry phase, will be massive and short lived, decaying into elementary particles such as quarks and leptons which, in turn, would eventually materialize into energetic nucleons, photons, neutrinos, etc., that might contribute to the observed EHECR flux.

Production of X particles from TDs may happen in a variety of ways directly or indirectly related to local removal of topological stability of (parts of) TDs. Examples include 'cusp' evaporation from cosmic strings [6], collapse of macroscopic cosmic string loops [8], shrinking of cosmic string loops to radii of order m_X^{-1} [9], annihilation of a monopole with an antimonopole [4,10], and so on. In the case of current-carrying superconducting cosmic strings [32,5], the charge carriers (which could be quanta of a superheavy fermion field trapped in 'zero mode' inside the string, or a charged scalar field living inside the string due to energetic reasons) are expelled from the string when the current on the string reaches a critical value — in the vacuum away from the string, these charge carriers act as the massive X particles. In the case of 'ordinary' cosmic strings, recent field theory simulations [33] of evolution of cosmic strings in the early Universe show that a cosmic string network loses energy directly into oscillations of the underlying gauge and higgs fields 'constituting' the string, which in quantum theory, correspond to quanta of massive X particles — a result which has important implications for both cosmic strings as well as for EHECR; this result, however, remains to be confirmed by independent simulations.

X-Particle Production Rate

The number density of X particles produced by TDs per unit time, dn_X/dt, can be generally written as [3]

$$\frac{dn_X}{dt}(t) = \frac{Q_0}{m_X} \left(\frac{t}{t_0}\right)^{-4+p} , \qquad (1)$$

where t_0 denotes the present epoch, and Q_0 is the rate of energy density injected in the form of X particles in the present epoch. The quantity Q_0 and the parameter p depend on the specific TD process under consideration. In writing Eq. (1) it is assumed that the only time scale in the problem is the hubble time t and that any other time scale involved can be expressed in terms of the hubble time. Similarly, we assume that any energy scale involved in the problem is expressible in terms of the energy scale η of the symmetry breaking at which the TDs under consideration are formed. (Note that m_X is fixed by η.) These assumptions are sometimes expressed by saying that the TDs under consideration evolve in a scale independent way. Indeed, it turns out that Eq. (1) is a phenomenologically good parametrization for the specific TD processes studied so far. For example, $p = 1$ for a process of X particle production from cosmic string loops [8], for a process involving collapsing monopole-antimonopole bound states [4,10], for a process of particle production from monopole-string systems called "necklaces" [11], and so on, while $p = 0$ for a process involving superconducting cosmic string loops [5].

Note that X particle production from TDs may occur continually at all epochs after the formation of the relevant TDs in the early Universe. However, only those X particles produced in the relatively recent epochs and at relatively close-by, non-cosmological distances ($\lesssim 100\,\mathrm{Mpc}$) are relevant for the question of EHECR. This is because, nucleons of energies above $10^{11}\,\mathrm{GeV}$ produced by the decay of X particles occurring at distances much larger than $\sim 50\,\mathrm{Mpc}$ suffer drastic energy loss during their propagation, due to photopion production on the the cosmic microwave background radiation (CMBR) fields (the so-called "GZK effect" [34]), and hence do not survive as EHECR particles today. Distance of sources of photons of energies above $\sim 10^{11}\,\mathrm{GeV}$ are also similarly restricted due to absorption through $e^+ e^-$ production on the radio background photons (see, e.g., [26]). The neutrinos, however, can survive from much earlier cosmological epochs, and may, if detected, prove to be the ultimate discriminant between a TD scenario and a more conventional scenario of origin of EHECR.

FROM X PARTICLES TO 'OBSERVABLE' PARTICLES

The X particles released from TDs would decay typically into quarks and leptons. The life-time τ is typically $\sim (\alpha m_X)^{-1}$ (where $\alpha \sim$ few $\times 10^{-2}$), which for $m_X \sim 10^{16}\,\mathrm{GeV}$ is $\sim 10^{-39}\,\mathrm{sec}$ or so. The decay is, therefore, essentially instantaneous at late cosmological epochs of interest to us. The quarks would hadronize (typically on a strong interaction time scale $\sim 10^{-23}\,\mathrm{sec}$, i.e., again practically instantaneously) by producing jets of hadrons, most of which would eventually be light mesons (pions) with a small admixture (typically $\lesssim 10\%$) of baryons and antibaryons (nucleons and antinucleons). The neutral pions decay to two photons, while the decay of charged pions gives rise to neutrinos. Some leptons (charged as well as neutral) could also be produced directly from the X particle decay. But by far the largest number of nucleons, photons, and neutrinos, would be produced through the hadronic channel. The spectra of produced particles are, therefore, essentially determined by the process of fragmentation of quarks/gluons into hadrons as described by QCD.

Quark → Hadron Fragmentation Spectrum

The exact process by which a single high energy quark gives rise to a jet of hadrons is not known; it involves some kind of non-perturbative physics that is not well understood. However, different semi-phenomenological approaches have been developed which describe the hadronic "fragmentation spectrum" of quarks/gluons that are in good agreement with the currently available experimental data on inclusive hadron spectra in quark/gluon jets in a variety of high energy processes.

In these approaches, the process of production of a jet containing a large number of hadrons, by a single high energy quark (or gluon), is 'factorized' into two stages. The first stage involves 'hard' processes involving large momentum transfers, whereby the initial high energy quark emits 'bremsstrahlung' gluons which themselves create more quarks and gluons through various QCD processes. These hard processes are well described by perturbative QCD. Thus a single high energy quark gives rise to a 'parton cascade' — a shower of quarks and gluons — which, due to the high energy nature of the process, is confined in a narrow cone or jet whose axis lies along the direction of propagation of the original quark. This first stage of the process, i.e., the parton cascade development, described by perturbative QCD, is terminated at a cut-off value, $\langle k_\perp^2 \rangle_{\text{cut-off}}^{1/2} \sim 1\,\text{GeV}$, of the typical transverse momentum. Thereafter, the second stage, involving the non-perturbative "confinement" process, takes over binding the quarks and gluons into color neutral hadrons. This second stage is usually described by one of the available phenomenological hadronization models such as the LUND string fragmentation model [35] or the cluster fragmentation model [36]. Detailed Monte Carlo numerical codes now exist [36–38] which incorporate the two stages outlined above. These codes provide a reasonably good description of a variety of relevant experimental data. Clearly, however, this is a numerically intensive approach.

Local Parton-Hadron Duality

There is an alternative approach that is essentially analytical and has proved very fruitful in terms of its ability to describe the gross features of hadronic jet systems, such as the inclusive spectra of particles, the particle multiplicities and their correlations, etc., reasonably well. This approach is based on the concept of "Local Parton Hadron Duality" (LPHD) [39]. Basically, in this approach, the second stage involving the non-perturbative hadronization process mentioned above is ignored, and the hadron spectrum is taken to be the same, up to an overall normalization constant, as the spectrum of partons (i.e., quarks and gluons) in the parton cascade after evolving the latter all the way down to a cut-off transverse momentum $\langle k_\perp^2 \rangle_{\text{cut-off}}^{1/2} \sim R^{-1} \sim$ few hundred MeV, where R is a typical hadronic size. At present the only justification for such an approach seems to be that it gives a remarkably good description of the experimental data including recent experimental results from LEP, HERA and TEVATRON [40]. Justification of LPHD at a more fundamental theoretical level, however, is not yet available. Nevertheless, it serves as a good phenomenological tool.

The main assumption in LPHD is that the actual hadronization process, i.e., the conversion of the quarks and gluons in the parton cascade into color neutral hadrons, occurs at a low virtuality scale of order of a typical hadron mass independently of the energy of the cascade initiating primary quark, and involves only low momentum transfers and local color 're-arrangement' which somehow does not drastically alter the form of the momentum spectrum of the particles in the parton cascade already determined by the 'hard' (i.e., large momentum transfer) perturbative QCD processes. Thus, the non-perturbative hadronization effects are lumped together in an 'unimportant' overall normalization constant which can be determined phenomenologically.

A good quantitative description of the perturbative QCD stage of the parton cascade evolution is provided by the so-called Modified Leading Logarithmic Approximation (MLLA) [41] of QCD, which allows the parton energy spectrum to be expressed analytically in terms of functions depending on two free parameters, namely, the effective QCD scale Λ_{eff} (which fixes the effective running QCD coupling strength $\alpha_s^{\text{eff}}(\tilde{Q}^2)$) and the transverse momentum cut-off $\tilde{Q}_0$. For the case $\tilde{Q}_0 = \Lambda_{\text{eff}}$, the analytical result simplifies considerably, and one gets what is referred to as the "limiting spectrum" [39,40]. For asymptotically high energies of interest, i.e., for energies E_{jet} of the original jet-initiating quark satisfying $E_{\text{jet}} \gg \Lambda_{\text{eff}}$, the limiting spectrum can be approximated by a Gaussian in the variable $\xi \equiv \ln(1/x)$, with $x \equiv E_{\text{parton}}/E_{\text{jet}}$, E_{parton} being the energy of a quark (parton) in the jet:

$$x \frac{dN_{\text{parton}}(Y, x)}{dx} \approx \frac{N_{\text{parton}}(Y)}{\sigma\sqrt{2\pi}} \exp\left[-\frac{(\xi - \bar{\xi})^2}{2\sigma^2}\right], \qquad (2)$$

where $Y \equiv \ln(E_{\text{jet}}/\Lambda_{\text{eff}})$, $\bar{\xi} \approx Y/2$, $2\sigma^2 = \left(\frac{bY^3}{36N_c}\right)^{1/2}$ with $b \equiv (11N_c - 2N_f)/3$, N_c being the number of colors and N_f the number of flavors of quarks involved, and $N_{\text{parton}}(Y) \sim \exp\{(16N_cY/b)^{1/2}\}$ is the average total multiplicity of the partons in the jet.

Eq. (2) gives us the spectrum of the partons in the jet. By LPHD hypothesis, the shape of the hadron spectrum, dN_h/dx (with $x = E_h/E_{\text{jet}}$, E_h being the energy of a hadron in the jet), is given by the same form as in Eq. (2), except for an overall normalization constant that takes account of the effect of conversion of partons into hadrons. Phenomenologically, for given values of Λ_{eff} and E_{jet}, the normalization constant can be determined simply from overall energy conservation, i.e., from the condition $\int_0^1 x\frac{dN_h(Y,x)}{dx}\,dx = 1$. The value of Λ_{eff} is not known a priori, but a fit to the inclusive charged particle spectrum in e^+e^- collisions at center-of-mass energy $E_{\text{cm}} = 2E_{\text{jet}} \sim 90\,\text{GeV}$ (Z-resonance) gives $\Lambda_{\text{eff}}^{\text{ch}} \sim 250\,\text{MeV}$.

Note that, within the LPHD picture, there is no way of distinguishing between various different species of hadrons. Phenomenologically, the experimental data can be fitted by using different values of Λ_{eff} for different species of particles depending on their masses. For our consideration of particles at EHECR energies, all particles under consideration will be extremely relativistic, and since, in our case, $E_{\text{jet}} \sim m_X/2 \gg \Lambda_{\text{eff}}$, the hadron spectrum will be relatively insensitive to the exact value of Λ_{eff}. Also, one can safely assume that at the asymptotically high energies of our interest, all hadrons — mesons as well as baryons — will have the same spectrum. However, the dominant species of particles in terms of their overall number will be the light mesons (pions); baryons typically constitute a fraction of $\lesssim (3 - 10)\%$ as indicated by existing collider data. For more details on various phenomenological aspects of the LPHD hypothesis, see the reviews [40].

Nucleon, Photon and Neutrino Injection Spectra

Using the MLLA + LPHD hadron spectrum discussed above, and assuming that each X particle on average undergoes N-body decay (typically $N \leq 3$) to N_q quarks (including antiquarks) and N_ℓ leptons (neutrinos and/or charged leptons), so that $N = N_q + N_\ell$, and assuming that the energy m_X is shared roughly equally by the N primary decay products of the X, the nucleon injection spectrum, $\Phi_N(E_i, t_i)$, from the decay of all X particles from TDs at any time t_i can be written as

$$\Phi_N(E_i, t_i) = \frac{dn_X(t_i)}{dt_i} N_q f_N \frac{1}{E_i} N_{\text{norm}} \frac{1}{\sigma\sqrt{2\pi}} \exp\left[-\ln^2\left(\frac{x_*}{x}\right)\right], \tag{3}$$

where E_i denotes the energy at injection, $\frac{dn_X}{dt_i}$ is the number of X-particle released per unit volume per unit time at time t_i, f_N is the nucleon fraction in the hadronic jet produced by a single quark, $x = NE_i/m_X$, $x_* = (N\Lambda_{\text{eff}}/m_X)^{1/2}$, and N_{norm} is the normalization constant defined by

$$N_{\text{norm}}(m_X) = \left(\int_0^1 dx \frac{1}{\sigma\sqrt{2\pi}} \exp\left[-\ln^2\left(\frac{x_*}{x}\right)\right]\right)^{-1}. \tag{4}$$

An important point about the nucleon injection spectrum given by Eq. (3) is that, unlike the spectrum predicted in the standard diffusive shock acceleration theory (see, e.g., Refs. [20,18,19]), the injection spectrum in the TD scenario (or, for that matter, in any non-acceleration scenario in which the energetic particles arise from decay of massive elementary particles), is not, in general, a power-law in energy. Although, in the energy regions of our interest, the spectrum (3) can be approximated [7] by power-law segments ($\propto E_i^{-\alpha}$), the power-law index α, in the energy regions of interest, is generally smaller than that in shock acceleration theories — the latter typically predict $\alpha \geq 2$. In other words, the injection spectrum in non-acceleration theories is generally harder (or flatter) compared to that in conventional acceleration theories. This fact has important consequences; it leads to the prediction of a pronounced "recovery" [3] of the evolved nucleon spectrum after a partial GZK "cut-off" and the consequent flattening of the spectrum above $\sim 10^{11}$ GeV. A relatively hard spectrum may also naturally give rise to a "gap" in the measured EHECR spectrum [14].

The photon injection spectrum from the decay of the neutral pions ($\pi^0 \to 2\gamma$) in the jets is given by

$$\Phi_\gamma(E_i, t_i) \simeq 2 \int_{E_i}^{m_X/N} \frac{dE}{E} \Phi_{\pi^0}(E), \tag{5}$$

where $\Phi_{\pi^0} \simeq \frac{1}{3}\frac{1-f_N}{f_N}\Phi_N$ is the neutral pion spectrum in the jet.

Similarly, the neutrino ($\nu_\mu + \bar{\nu}_\mu$) injection spectrum resulting from the charged pion decay [$\pi^\pm \to \mu^\pm \nu_\mu(\bar{\nu}_\mu)$] can be written as [42,3]

$$\Phi_{(\nu_\mu + \bar{\nu}_\mu)}(E_i) \simeq 2.34 \int_{2.34E_i}^{m_X/N} \frac{dE}{E} \Phi_{(\pi^+ + \pi^-)}(E) \,, \tag{6}$$

where $\Phi_{(\pi^+ + \pi^-)} \simeq \frac{2}{3}\frac{1-f_N}{f_N}\Phi_N$.

The decay of each muon (from the decay of a charged pion) produces two more neutrinos and an electron (or positron): $\mu^\pm \to e^\pm \nu_e(\bar{\nu}_e)\bar{\nu}_\mu(\nu_\mu)$. Thus each charged pion eventually gives rise to three neutrinos: one ν_μ, one $\bar{\nu}_\mu$ and one ν_e (or $\bar{\nu}_e$), all of roughly the same energy. So the total $\nu_\mu + \bar{\nu}_\mu$ injection spectrum will be roughly twice the spectrum given in Eq. (6), while the total $\nu_e + \bar{\nu}_e$ spectrum will be roughly same as that in Eq. (6).

Note that, if the hadron spectrum in the jet is generally approximated by a power-law in energy, then nucleon, photon and neutrino injection spectra will also have the same power-law form all with the same power-law index.

It is worth emphasizing here that while using the LPHD hadron spectrum in the analysis of the TD scenario of EHECR one should keep in mind that there is a great deal of uncertainty involved in extrapolating the QCD based hadron spectra — which have been tested so far only at relatively 'low' energies of ~ 100 GeV — to the extremely high energies of $\sim 10^{15}$ GeV or so.

Evolution of the Particle Spectra

Nucleons

The evolution of the nucleon spectrum is mainly governed by interactions of the nucleons with the CMBR. The relevant interactions are pair production by protons ($p\gamma_b \to pe^+e^-$), photoproduction of single or multiple pions by nucleons N ($N\gamma_b \to Nn\pi$, $n \geq 1$), and neutron decay. Here γ_b stands for a background photon, in this case, a CMBR photon. At EHECR energies, the photopion production is the dominant process. This is a drastic energy loss process for nucleons, and is the basis of the well-known prediction of the GZK cut-off [34] of the evolved EHECR nucleon spectrum at energies above $\sim 10^{11}$ GeV. This process also limits the distance of a possible source of the observed EHECR particles to distances less than ~ 50 Mpc [34,13,22].

γ-rays

The γ-rays at EHECR energies interact via pair production (PP: $\gamma\gamma_b \to e^+e^-$) and double pair production (DPP: $\gamma\gamma_b \to e^+e^-e^+e^-$), while the electrons (positrons) interact via inverse Compton scattering (ICS: $e\gamma_b \to e'\gamma$) and triplet pair production (TPP: $e\gamma_b \to ee^+e^-$). In addition, the electrons (positrons) suffer synchrotron energy loss in the extragalactic magnetic field (EGMF). The background photons (γ_b) involved in the PP process are mainly the universal radio background (URB) photons for γ-rays above $\sim 10^{19}$ eV, the CMBR photons for

γ-rays between $\sim 10^{14}\,\mathrm{eV}$ and $\sim 10^{19}\,\mathrm{eV}$, and the infrared and optical (IR/O) background photons for γ-rays below $\sim 10^{14}\,\mathrm{eV}$.

The evolution of the γ-ray spectrum is complicated due to the fact that PP and ICS processes together lead to development of electromagnetic (EM) cascades, whereby any electromagnetic energy in the form of γ-rays and/or electrons (positrons) released at an energy which is above the threshold for PP process on photons of a particular background, cascades down to progressively lower energies through a cycle of PP interactions (on the photons of progressively higher energy backgrounds) and ICS interactions (mainly on the CMBR). The cascading has the overall effect of increasing the γ-ray flux at EHECR energies because it causes an effective increase of the attenuation length of these γ-rays. Further cascading by any cascade photon stops either if the remaining path length of the cascade photon, as it propagates from its point of creation to the observation point, becomes less than the mean free path for the PP process on the relevant background photons, or if the energy of the propagating cascade photon falls below the threshold for the PP process. Thus, depending on the distance at which they are first injected, γ-rays of EHECR energies can, after propagation, give rise to a γ-ray spectrum that may span the energy range from (say) a few tens of MeV all the way up to EHECR energies. This, of course, means that any model of electromagnetic energy injection at EHECR energies has to meet the constraint that the resulting cascade γ-ray flux at lower energies should not exceed the measured flux at those energies. In the context of TD models of EHECR, this was first pointed out in Ref. [28].

The mean attenuation length of EHE γ-rays depends strongly on the density of URB and on the strength of the EGMF, both of which are uncertain at the present time. The EGMF typically inhibits cascade development because of synchrotron cooling of the e^-e^+ pairs produced in the PP process. Depending on the strength of the EGMF, the synchrotron cooling time scale may be shorter than the time scale of ICS, in which case the e^- or the e^+ under consideration loses energy through synchrotron radiation before it can undergo ICS, and thus cascade development stops. In this case the γ-ray flux is determined mainly by the "direct" γ-rays, i.e., the ones that originate at distances less than the absorption length due to PP process. The energy lost by synchrotron cooling does not, however, disappear — rather, it appears at a lower energy and can initiate fresh EM cascades by interacting with the photons of a higher energy background such as CMBR or IR/O depending on its energy. So the overall effect of a relatively strong EGMF is to deplete the γ-ray flux above some energy in the EHE region and increase the γ-ray flux below a corresponding energy in the 'low' (MeV – GeV) energy region, where it will have to meet the constraints imposed by the measured extragalactic diffuse γ-ray background [43,44].

In addition to uncertainties in the strength of EGMF and the URB, another major source of uncertainty in the cascade calculation is the poorly known IR/O background (see, e.g., Ref. [45] for a recent discussion of IR/O backgrounds). The latter strongly influences the cascade spectrum in the energy range from $\sim 10^{11}\,\mathrm{eV}$ to $\sim 10^{14}\,\mathrm{eV}$. Below this range, however, the cascade spectrum becomes relatively insensitive [46] to the model parameters that determine the IR/O background. This is a fortunate circumstance because this implies that the constraints on TD models derived [15–17,46] by comparing the measured 100 MeV – 10 GeV γ-ray background with TD model predictions are relatively insensitive to uncertainties in our precise knowledge of the IR/O background, and hence are fairly robust.

Neutrinos

EHE neutrinos suffer absorption through fermion-antifermion pair production on the thermal background neutrinos ($\nu + \bar{\nu}_b \to f\bar{f}$), where $f \equiv e, \mu, \tau, \nu, q$, and ν_b is a thermal background neutrino. Due to this absorption process, neutrinos of present observed energy $E_{\nu,0}$ would have to have been injected at redshifts less than $z_a(E_{\nu,0}) \simeq 3.5 \times 10^2 (10^{20}\,\mathrm{eV}/E_{\nu,0})^{2/7}$, for $E_{\nu,0} \gtrsim 3 \times 10^{14}\,\mathrm{eV}$ [3].

Actually, the μ's, τ's, and quarks created in the above absorption process generate further neutrinos through decay of the μ's and τ's and through decay of the charged pions created by quark fragmentation. This leads to a "neutrino cascade" [27], effectively increasing the size of the "neutrino horizon" of the Universe. This, in turn, has the effect of increasing the overall neutrino flux around 10^{20} eV by a factor of few relative to the case when the cascading effect is not taken into account.

Recently, it has been pointed out [47] that if neutrinos have a small mass m_ν in the eV range, then EHE neutrinos (antineutrinos) of energy $\sim 4\,(\mathrm{eV}/m_\nu) \times 10^{12}\,\mathrm{GeV}$ will annihilate on the relic antineutrinos (neutrinos) to produce the Z-boson with a resonant cross section of $\sim 10^{-32}\,\mathrm{cm}^2$. The hadronic decay of the Z will produce additional photons, neutrinos and nucleons, which will add to the photon and neutrino cascading processes mentioned above. This process may also leave specific signatures on the EHE neutrino spectrum, the detection of which may in the end provide an indirect signature of the neutrino mass and hence of dark matter. Detailed self-consistent calculation of nucleon, photon and neutrino fluxes, however, remain to be done.

Predicted Particle Fluxes and Constraints on TD models

As discussed above, the predicted particle fluxes depend on a number of parameters: the X particle mass m_X, the strength of EGMF, the URB, the injection spectra, and so on. Recently, detailed numerical calculation of the particle spectra, especially the spectrum of γ-rays in the entire energy range from $\sim 10^8$ eV to $\sim 10^{25}$ eV, have been done [15–17] taking into account the effects of electromagnetic cascading and EGMF. Fig. 1 shows the predicted diffuse particle fluxes for a representative set of values of various parameters involved. Here I discuss only the predicted diffuse fluxes of particles assuming uniform distribution of TD sources in the Universe. (Possible individual "bursting" sources of EHECR and the possible reconstruction of their "images" from data on arrival direction, arrival time and energy associated with individual events, leading to information about the source characteristics and, in particular about the EGMF, are discussed in the talk by G. Sigl; see this volume.)

Because the magnitude of X particle production rate is not known *a priori*, the best one can do at the present time is adopt a suitable normalization procedure for the absolute flux so as to be able to explain the EHECR data, and then check whether this normalization is consistent with all relevant observational data, not just on EHECR particles, but also on other relevant particle fluxes at lower energies, especially the diffuse gamma ray background measurements in the MeV – GeV region. The normalization of the absolute fluxes in Fig. 1 is optimized in the sense

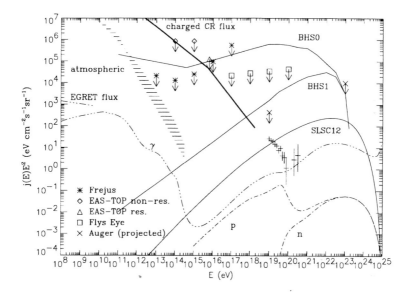

FIGURE 1. Predicted fluxes of γ-rays (dash - triple dotted line), protons and neutrons (dash-dotted lines) and $(\nu_\mu + \bar{\nu}_\mu)$ (solid lines) for TD models with $p = 1$, $m_X = 10^{16}\,\text{GeV}$ and EGMF of 10^{-12} Gauss. The neutrino curve marked "SLSC12" corresponds to the calculation of Ref. [16], while the curves marked "BHS0" and "BHS1" represent the neutrino fluxes from Ref. [3] and correspond respectively to $p = 0$ and the $p = 1$ models. Also shown are the estimates of the atmospheric neutrino background for different zenith angles [48] (hatched region marked "atmospheric"). Data points with error bars represent the combined cosmic ray data from the Fly's Eye and AGASA experiments [12] above 10^{19} eV, and the thick solid line represents piecewise power-law fit to the observed charged CR flux. The dash - triple dotted line on the left margin represents experimental upper limits on the diffuse γ-ray flux at 100 MeV – 5 GeV from EGRET data [43,44]. Points with arrows pointing downwards represent approximate upper limits on the diffuse neutrino flux from Frejus [49], the EAS-TOP [50], and the Fly's Eye [51] experiments, as indicated. The projected limit shown for the proposed Auger experiment assumes the acceptance estimated in Ref. [52] for non-detection over a five year period. (Courtesy G. Sigl)

that it has been determined by fitting the 'observable' particle (i.e., the combined nucleon and γ-ray) fluxes to the measured EHECR data by the maximum likelihood method [14] and corresponds to a likelihood significance of $> 50\%$.

It is clear that TD models, while potentially contributing dominantly to the particle fluxes above $\sim 10^{20}$ eV, make negligible contribution to cosmic ray flux below $\sim 10^{19}$ eV because of the relatively hard nature of the particle spectra in these models. Thus the flux below $\sim 2 \times 10^{19}$ eV is presumably due to a conventional acceleration scenario, and was not included in the fitting procedure.

Since pions are the most numerous particles in the jets, their decay products, i.e., photons and neutrinos dominate the number of particles at production. Since neutrinos suffer little attenuation and can come to us unattenuated from large cosmological distances (except for absorption due to e^+e^- pair production by interacting with the cosmic thermal neutrino background, the path length for which is $\gg 100$ Mpc), their fluxes, as expected, are the largest among all particles at the highest energies. However, their detection probability is much lower compared to those for protons and photons[3]. Photons also far outnumber nucleons at production, and depending on the level of the URB and EGMF, may dominate over the nucleon flux and thus dominate the 'observable' particle flux at EHECR energies.

An important point to note is that photons and neutrinos in the TD scenario are *primary* particles in that they are produced directly from the decay of the pions in the hadronic jets. In contrast, photons and neutrinos in conventional acceleration scenarios can be produced only through *secondary* processes — they are mainly produced by the decay of photoproduced pions resulting from the GZK interactions of primary HECR nucleons (produced by the acceleration process) with CMBR photons [42]. Of course, these secondary neutrinos and photons would also be there in the TD scenario, but their fluxes are sub-dominant to the primary ones at the highest energies.

The *shapes* of the EHE nucleon- and γ-ray spectra In the TD scenario are "universal" [3,26] in the sense that they are independent of specific TD process even though different TD sources evolve differently (as parametrized by the parameter p in the X particle production rate). This is because, at these energies, the attenuation lengths of nucleons and γ-rays are small ($\lesssim 100$ Mpc) compared to Hubble length and so the effects of unknown cosmological evolution of the TD sources are negligible compared to propagation effects. The universal shapes of the EHECR nucleon and γ-ray spectra reflect their injection spectra, which, as discussed earlier, are determined by QCD. Thus, large statistics measurement of the EHECR spectrum by future detectors may give us a probe of new physics, and in particular QCD, at energies not currently accessible in laboratory accelerators, *provided* a TD-like non-acceleration scenario of origin of EHECR is correct.

In contrast to the EHECR nucleons and γ-rays, the predicted γ-ray flux below

---

[3] The EHE neutrinos of TD origin would, however, be potentially detectable by the proposed space-based detectors like OWL and AIRWATCH, and ground based detectors like Auger, Telescope Array, and so on; see articles on these detector projects in this volume.

$\sim 10^{14}$ eV (the threshold for pair production on the photons of CMBR) and the predicted EHE neutrino flux depend on the total energy released integrated over redshift and hence are dependent on the specific TD model (i.e., specific value of p). In particular, the γ ray flux below $\sim 10^{11}$ eV scales as the total electromagnetic energy released from X particles integrated over all redshifts and increases with decreasing value of p. This has been used to constrain [53] TD models from considerations of CMBR distortions and from independent considerations of modified light element abundances due to ^{4}He photodisintegration; for example, this rules out [53] the $p = 0$ TD model.

The γ-ray flux in Fig. 1 is consistent with the estimates of the upper limit on the background in the 100 MeV to ~ 5 GeV from EGRET data [43]. Very recently, new estimates of the background flux have been presented [44] which now extend up to ~ 100 GeV with roughly the same power-law index as the earlier estimates [43]. If the EGMF is significantly larger than the value assumed ($\sim 10^{-12}$ Gauss) for the calculation of Fig. 1, then the electromagnetic cascade energy transferred to the relevant low energy region will be higher, and in this case the γ-ray flux in Fig. 1 will only be marginally consistent, or may even be inconsistent, with the experimental diffuse background flux estimates. Note, however, that the fluxes in Fig. 1 correspond to the case $m_X = 10^{16}$ GeV. Lower values of m_X are possible, which will give lower contribution to the low energy diffuse flux while at the same time producing enough energy in the form of X particles to explain the observed EHECR flux. This can be seen as follows:

As we shall discuss in more details in the next section, for a relatively hard power-law photon injection spectrum $\propto E^{-\alpha}$ with $\alpha < 2$, the energy injection rate Q_0 required to match a given differential EHECR flux at a given energy decreases with decreasing value of m_X (provided, of course, $m_X > 10^{11}$ GeV, the energy of the highest energy event). For example, it is easy to see (see next section) that for a photon injection spectrum $\propto E^{-1.5}$, the value of Q_0 required to explain the EHECR flux is roughly proportional to $m_X^{1/2}$. And since the cascade γ-ray flux in the $\lesssim 100$ GeV region is essentially directly proportional to Q_0 injected at EHECR energies, a reduction by a factor of 10, for example, of the predicted γ-ray flux in the $\lesssim 100$ GeV region is easily achieved by reducing m_X by about two orders of magnitude, i.e., for $m_X \lesssim 10^{14}$ GeV. And, of course, as already mentioned, the low energy γ-ray flux is also reduced if the EGMF strength is lower.

Thus, it seems that TD models of EHECR with $m_X \sim 10^{16}$ GeV and/or high EGMF ($\sim 10^{-9}$ Gauss) are somewhat difficult to simultaneously reconcile with EHECR data and the low energy diffuse γ-ray background. On the other hand, models with $m_X \lesssim 10^{14}$ GeV and low EGMF ($\lesssim 10^{-11}$ Gauss) can explain the EHECR data and at the same time be consistent with all existing data.

The neutrino flux indicated by curve marked "SLSC12" in Figure 1 corresponds to overall X particle production rate obtained by the maximum likelihood normalization of the 'observable' (i.e., photon plus nucleon) flux to the EHECR data for the $p = 1$ TD model. The *predicted* level of the neutrino flux, therefore, depends on the value of EGMF, the radio background, etc., which go into determining the photon flux. The earlier flux estimate (the curve marked "BHS1") in Fig.1 is higher simply because the overall X particle production rate was normalized to a higher value; that normalization was obtained by normalizing the predicted *proton* (as opposed to photon) flux with the measured flux at a lower energy $\sim 5 \times 10^{19}$ eV (where the measured cosmic ray flux is higher than that at EHECR energies) — the Fly's Eye and AGASA highest energy events beyond 10^{20} eV [12] were not yet discovered at that time! The $p = 0$ model is ruled out by the upper limits from Frejus as well as EAS-TOP experiments. (Actually, the $p = 0$ model is also ruled out from considerations [53] of ^{4}He photo-disintegration and CMBR distortions as mentioned above — it corresponds to unacceptably high rate of energy injection in the early cosmological epochs.) On the other hand, the new neutrino flux estimates for the $p = 1$ model are consistent with all existing experimental upper limits. At energies $\gtrsim 10^{20}$ eV, the predicted neutrino flux in the $p = 1$ TD model also dominates over the predicted flux from blazars/AGNs as well as over the predicted flux of "cosmogenic" neutrinos produced by interactions of UHE cosmic rays with CMBR [54] (not shown in Fig. 1). For more details on the detectability of neutrino fluxes in TD models, see, e.g., Ref. [16,27,55,56].

X-PARTICLE PRODUCTION RATE REQUIRED TO EXPLAIN THE EHECR FLUX: A BENCHMARK CALCULATION

To have an idea of the kind of numbers for the required X particle production rate, we can perform a simple (albeit crude) benchmark calculation as follows:

Since in TD models, photons are expected to dominate the observable EHECR flux, let us assume for simplicity that the highest energy events are due to photons. Let us assume a typical three-body decay mode of the X into a $q\bar{q}$ pair and a lepton: $X \to q\bar{q}\ell$. The two quarks will produce two hadronic jets. Let f_π denote the total pion fraction in a jet. Then the photons from the two jets carry a total energy $E_{\gamma,\text{Total}} \simeq \left(\frac{2}{3} \times 0.9 \times \frac{1}{3} \right) m_X(f_\pi/0.9) = 0.2 m_X(f_\pi/0.9)$. Let us assume a power-law hadronic fragmentation spectrum with index 1.5. Then the photon injection spectrum from a single X particle can be written as

$$\frac{dN_\gamma}{dE_\gamma} = \frac{3}{m_X} \times 0.3 \left(\frac{3E_\gamma}{m_X} \right)^{-1.5} \left(\frac{f_\pi}{0.9} \right), \tag{7}$$

which is properly normalized with the total photon energy $E_{\gamma,\text{Total}}$. We shall neglect cosmological evolution effects since photons of EHECR energies have a cosmologically negligible path length of only $\sim$ few tens of Mpc for absorption through pair production on the radio background.

With these assumptions, the photon flux $j_\gamma(E_\gamma)$ at the observed energy E_γ is simply given by

$$j_\gamma(E_\gamma) \simeq \frac{1}{4\pi} \lambda(E_\gamma) \frac{dn_X}{dt} \frac{dN_\gamma}{dE_\gamma}, \tag{8}$$

where $\lambda(E_\gamma)$ is the pair production absorption path length of a photon of energy E_γ.

Noting that $dn_X/dt = Q_0/m_X$, and normalizing the above flux to the measured EHECR flux corresponding to the highest energy event at $\sim 3 \times 10^{20}\,\text{eV}$, given by $j(3 \times 10^{20}\,\text{eV}) \approx 5.6 \times 10^{-41}\,\text{cm}^{-2}\,\text{eV}^{-1}\,\text{sec}^{-1}\,\text{sr}^{-1}$, we get

$$Q_{0,\text{required}} \approx 2.1 \times 10^{-21}\,\text{eV}\,\text{cm}^{-3}\,\text{sec}^{-1} \left(\frac{10\,\text{Mpc}}{\lambda_{\gamma,300}}\right) \left(\frac{m_X}{10^{16}\,\text{GeV}}\right)^{1/2}, \tag{9}$$

or

$$\left(\frac{dn_X}{dt}\right)_{0,\text{required}} \approx 2.1 \times 10^{-46}\,\text{cm}^{-3}\,\text{sec}^{-1} \left(\frac{10\,\text{Mpc}}{\lambda_{\gamma,300}}\right) \left(\frac{m_X}{10^{16}\,\text{GeV}}\right)^{-1/2}, \tag{10}$$

where $\lambda_{\gamma,300}$ is the absorption path length of a 300 EeV photon (1 EeV $\equiv 10^{18}\,\text{eV}$). The subscript 0 stands for the present epoch.

The above numbers are probably uncertain by up to an order of magnitude depending on the exact form of the injection spectrum, the absorption path length of EHECR photons, electromagnetic cascading effect, and so on. Indeed, since electromagnetic cascading effect (which we have neglected here) generally increases the photon flux, the above numbers are most likely overestimates. Nevertheless, they do serve as crude benchmark numbers. These numbers indicate that in order for TDs to explain the EHECR events, the X particles must be produced in the present epoch at a rate of $\sim 2 \times 10^{35}\,\text{Mpc}^{-3}\,\text{yr}^{-1}$, or in more "down-to-earth" units, about $\sim 23\text{Au}^{-3}\,\text{yr}^{-1}$, i.e., about 20 X particles within a solar system-size volume per year.

In the next section I discuss three plausible specific TD processes and examine their efficacies with regard to X particle production and EHECR keeping in mind the above rough estimates of the required X particle production rate.

X PARTICLE PRODUCTION FROM COSMIC STRINGS, MONOPOLES, AND NECKLACES

Cosmic Strings

Let us first recall the salient features of evolution of cosmic strings in the Universe; for a review, see Refs. [2]. Immediately after their formation, the strings would be in a random tangled configuration. One can define a characteristic length scale, ξ_s, of the initial string configuration in terms of the overall mass-energy density, ρ_s, of strings simply through the relation

$$\rho_s = \mu/\xi_s^2, \tag{11}$$

where μ denotes the string mass (energy) per unit length.

Initially, the strings find themselves in a dense medium, so they move under a strong frictional damping force. The damping remains significant until the temperature falls to $T \lesssim (G\mu)^{1/2}\eta$, where G is Newton's constant and η is the symmetry-breaking scale at which strings were formed. [Recall, for GUT scale cosmic strings, for example, $\eta \sim 10^{16}$ GeV, $\mu \sim \eta^2 \sim (10^{16}\,\text{GeV})^2$, and $G\mu \sim 10^{-6}$.] In the friction dominated epoch, a curved string segment of radius of curvature r quickly achieves a terminal velocity $\propto 1/r$. The small scale irregularities on the strings are, therefore, quickly smoothed out. As a result, the strings are straightened out and their total length shortened. The energy density in strings, therefore, decreases with time. This means that the characteristic length scale ξ_s describing the string configuration increases with time as the Universe expands. Eventually ξ_s becomes comparable to the causal horizon distance $\sim t$. At about this time, the ambient density of the Universe becomes dilute enough that damping becomes unimportant so that the strings start moving relativistically.

Beyond this point, there are two possibilities. Causality prevents the length scale ξ_s from growing faster than the horizon length. So, either (a) ξ_s keeps up with the horizon length, i.e., ξ_s/t becomes a constant, or (b) ξ_s increases less rapidly than t.

Let us consider the second possibility first. In this case, the string density falls less rapidly than t^{-2}. On the other hand, we know that the radiation density in the radiation-dominated epoch as well as matter density in the matter-dominated epoch both scale as t^{-2}. Clearly, therefore, the strings would come to dominate the density of the Universe at some point of time. It can be shown that this would happen quite early in the history of the Universe unless the strings are very light, much lighter than the GUT scale strings. A string dominated early Universe would be unacceptably inhomogeneous conflicting with the observed Universe[4].

The other possibility, which goes by the name of "scaling" hypothesis, seems to be more probable, as suggested by detailed numerical as well as analytical studies [2,33]. The numerical simulations generally find that the string density does reach the scaling regime given by $\rho_{s,\text{scaling}} \propto 1/t^2$, and then continues to be in this regime. It is, however, clear that in order for this to happen, strings must lose energy in some form at a certain rate. This is because, in absence of any energy loss, the string configuration would only be conformally stretched by the expansion of the Universe on scales larger than the horizon so that ξ_s would only scale as the scale factor $\propto t^{1/2}$ in the radiation dominated Universe, and $\propto t^{2/3}$ in the matter dominated Universe. In both cases, this would fail to keep the string density in the scaling regime, leading back to string domination. In order for the string density to

---

[4] However, a string dominated *recent* Universe — dominated by 'light' strings formed at a phase transition at about the electroweak symmetry breaking scale — is possible. Such a string dominated recent Universe may even have some desirable cosmological properties [57]. Such light strings are, however, not of interest to us in this discussion.

be maintained in the scaling regime, energy must be lost by the string configuration per unit proper volume at a rate $\dot{\rho}_{s,\text{loss}}$ satisfying the equation

$$\dot{\rho}_{s,\text{total}} = -2\frac{\dot{R}}{R}\rho_s + \dot{\rho}_{s,\text{loss}} , \tag{12}$$

where the first term on the right hand side is due to expansion of the Universe, R being the scale factor of the expanding Universe. In the scaling regime $\dot{\rho}_{s,\text{total}} = -2\rho_s/t$, which gives $\dot{\rho}_{s,\text{loss}} = -\rho_s/t$ in the radiation dominated Universe, and $\dot{\rho}_{s,\text{loss}} = -(2/3)\rho_s/t$ in the matter dominated Universe.

The important question is, in what form does the string configuration lose its energy so as to maintain itself in the scaling regime? One possible mechanism of energy loss from strings is formation of closed loops. Occasionally, a segment of string might self-intersect by curling up on itself. The intersecting segments may intercommute, i.e., "exchange partners" , leading to formation of a closed loop which pinches off the string. The closed loop would then oscillate and lose energy by emitting gravitational radiation and eventually disappear. It can be shown that this is indeed an efficient mechanism of extracting energy from strings and transferring it to other forms. The string energy loss rate estimated above indicates that scaling could be maintained by roughly of order one closed loop of horizon size ($\sim t$) formed in a horizon size volume ($\sim t^3$) in one hubble expansion time ($\sim t$) at any time t. In principle, as far as energetics is concerned, one can have the same effect if, instead of one or few large loops, a large number of smaller loops are formed. Which one may actually happen depends on the detailed dynamics of string evolution, and can only be decided by means of numerical simulations.

Early numerical simulations seemed to support the large (i.e., $\sim$ horizon size) loop formation picture. Subsequent simulations with improved resolution, however, found a lot of small-scale structure on strings, the latter presumably being due to kinks left on the strings after each crossing and intercommuting of string segments. Consequently, loops formed were found to be much smaller in size than horizon size and correspondingly larger in number. Further simulations showed that the loops tended to be formed predominantly on the scale of the cut-off length imposed for reasonable resolution of even the smallest size loops allowed by the given resolution scale of the simulation. It is, however, generally thought that the small-scale structure cannot continue to build up indefinitely, because the back-reaction of the kinky string's own gravitational field would eventually stabilize the small-scale structure at a scale of order $G\mu t$. The loops would, therefore, be expected to be formed predominantly of size $\sim G\mu t$, at any time t. Although much smaller than the horizon size, these loops would still be of 'macroscopic' size, much larger than the microscopic string width scale ($\sim \eta^{-1} \sim \mu^{-1/2}$). These loops would, therefore, also oscillate and eventually disappear by emitting gravitational radiation. Thus, according to above picture, the dominant mechanism of energy loss from strings responsible for maintaining the string density in the scaling regime would be formation of macroscopic-size ($\gg \eta^{-1}$) loops and emission of gravitational radiation by these loops. More details of this picture of cosmic string evolution can be found in Refs. [2].

How is the above picture of cosmic string evolution relevant for cosmic rays? How does one get X particles from cosmic strings?

One way of getting X particles from cosmic strings is through the so-called cusp-evaporation mechanism [6]. I will not discuss this mechanism here, but the resulting X particle production rate generally turns out to be too low to be of relevance to EHECR.

Another possibility arises as follows: As the closed loops oscillate and emit gravitational radiation, they lose energy and shrink. Eventually, when a loop's radius shrinks to a size of order the width of the string, the string unwinds and turns into X particles, which will then decay, producing high energy particles. However, each loop in the end only produces of the order of one X particle. The resulting cosmic ray flux is again too low to be observable [9].

Clearly, the only way cosmic strings may produce large number of X particles is if macroscopic lengths of string are involved in the X particle production process. One such mechanism was suggested in Refs. [7,8] based on the following arguments. The picture of gravitational radiation being the dominant energy loss mechanism for cosmic strings rests on the assumption that the loops themselves do not self-intersect frequently. Since the motion of a freely oscillating loop is periodic (with a period of $L/2$, L being the invariant length of the loop [58]), a loop formed in a self-intersecting configuration will undergo self-intersection within its first period of oscillation. If the loop does indeed self-intersect and break up into two smaller loops, and if the daughter loops again self-intersect breaking up into two even smaller loops, and so on, then one can see that a single initially large loop of size L can break up into a debris of tiny loops of size η^{-1} (thereby turning into X particles) within a time scale of $\sim L$. Since the largest loops are expected to be of size $< t$, one sees that the above time scale can be much smaller than one hubble time. In other words, one large loop can break up into a large number of tiny loops (X particles) within one hubble time. Such self-intersecting loops would not radiate much energy gravitationally because that would require many periods of oscillation. In other words, such repeatedly self-intersecting loops would be a channel through which energy contained in macroscopic lengths of string could go into X particles instead of into gravitational radiation. As shown in Ref. [58], some non-circular loops could also be in configurations which would collapse completely into double-line configurations and subsequently annihilate into X particles [59]. It has also been argued in Ref. [60] that the self-intersection probability of a loop increases exponentially with the number of small scale kinks on a loop.

At the time of work of Ref. [8], however, it was not clear as to what fraction of the string energy might go into X particles through processes discussed above. Treating this fraction to be a free parameter f, i.e., assuming that a fraction f of the energy extracted from the long strings per unit volume per unit time went into X particles, the authors of Ref. [8] found that the fraction f had to be rather small, $f \lesssim 7 \times 10^{-4}$; otherwise, the predicted cosmic ray flux from GUT scale cosmic strings would exceed the observed flux. Results of more recent calculation of the predicted observable particle flux [16] (see Fig. 1) correspond to an upper limit on f which is about two orders of magnitude lower[5].

The actual value of f is still unknown. But if gravitational radiation, and not massive particle radiation, is the dominant energy loss mechanism for cosmic strings, then the fraction f may actually be much smaller than the above upper limit, in which case the flux of cosmic rays produced by cosmic strings through X particles would be small and below the observed flux. However, rapid conversion of a significant fraction of the energy in macroscopically large loops into tiny loops (X particles) through repeated loop self-intersections due, for example, to the presence of large number of kinks on the loops [60], and consequent production of significant EHECR flux, cannot be ruled out.

Very recently, the notion gravitational radiation as the dominant energy loss mechanism for cosmic strings has been questioned by the results of new numerical simulations of evolution of cosmic strings [33]. Authors of Ref. [33] claim that if loop production is not artificially restricted by imposing a cutoff length for loop size in the simulation, then loops tend to be produced dominantly on the smallest allowed length scale in the problem, namely, on the scale of the width of the string. Such small loops promptly collapse into X particles. In other words, there is essentially no loop production at all — the string energy density is maintained in the scaling regime by energy loss from strings predominantly in the form of direct X particle emission, rather than by formation of large loops and their subsequent gravitational radiation. This new result, while subject to confirmation by independent simulations, obviously has important implications for EHECR. Indeed, in this case, the upper limit on the fraction f mentioned above implies severe constraint on GUT-scale cosmic strings. From the results of Ref. [33], the string energy density in the scaling regime is $\rho_{s,\text{scaling}} \simeq \mu/(0.3\,t)^2$. With X particle production as the dominant energy loss mechanism, we immediately see from Eq. (12) that rate of production of X particles from strings per unit volume must be $dn_X/dt \simeq 7.4(\mu/m_X)t^{-3}$ in the matter dominated era. Taking, for cosmic strings, $\mu \simeq \pi\eta^2$, and taking $m_X \sim 0.7\eta$, where η is the symmetry-breaking scale, we get by using the constraint imposed by Eq. (10), $\eta \lesssim 4.2\times10^{13}$ GeV. Thus the GUT scale cosmic strings with $\eta \sim 10^{16}$ GeV are ruled out by EHECR data if the results of Ref. [33] are correct. At the same time X particles from cosmic strings formed at a phase transition with $\eta \sim 10^{13}$ – 10^{14} GeV are able to explain the EHECR data. Cosmic strings may thus be a "natural" source of extremely high energy cosmic rays if massive particle radiation, and *not* gravitational radiation, is indeed their dominant energy loss mechanism.

Cosmic string formation with $\eta \sim 10^{14}$ GeV rather than at the GUT scale of $\sim 10^{16}$ GeV is not hard to envisage. For example, the symmetry breaking scheme $SO(10) \to SU(3) \times SU(2) \times U(1)_Y \times U(1)$ can take place at the GUT unification

---

[5] This is due to the fact that the 'observable' particle flux now includes the gamma ray flux in addition to the protons — in contrast to only the much lower proton flux considered in Ref. [8] — and also because of the maximum likelihood normalization of the predicted flux to the observed data.

scale $M_{\mathrm{GUT}} \sim 10^{16}\,\mathrm{GeV}$; with no U(1) subgroup broken, this phase transition produces no strings. However, the second U(1) can be subsequently broken with a second phase transition at a scale $\sim 10^{14}\,\mathrm{GeV}$ to yield the cosmic strings relevant for EHECR. Note that these strings would be too light to be relevant for structure formation in the Universe and their signature on the CMBR sky would also be too weak to be detectable. Instead, the extremely high energy end of the cosmic ray spectrum may offer a probing ground for signatures of these cosmic strings.

Monopoles

If monopoles were formed at a phase transition in the early Universe, then, as Hill [4] suggested in 1983, a metastable monopole-antimonopole bound state — "monopolonium" — is possible. At any temperature T, monopolonia would be formed with binding energy $E_b \gtrsim T$. The initial radius r_i of a monopolonium would be $r_i \sim \frac{1}{2}g_m^2/E_b$, where g_m is the magnetic charge (which is related to the electronic charge e through the Dirac quantization condition $eg_m = N/2$, N being the monopole's winding number). Classically, of course, the monopolonium is unstable. Quantum mechanically, the monopolonium can exist only in certain "stationary" states characterized by the principal quantum number n given by $r = n^2 a_m^{\mathrm{B}}$, where n is a positive integer, r is the instantaneous radius, and $a_m^{\mathrm{B}} = 8\alpha_e/m_M$ is the "magnetic" Bohr radius of the monopolonium. Here $\alpha_e = 1/137$ is the "electric" fine-structure constant, and m_M is the mass of a monopole. Since the Bohr radius of a monopolonium is much less than the Compton wavelength (size) of a monopole, i.e., $a_m^{\mathrm{B}} \ll m_M^{-1}$, the monopolonium does not exist in the ground ($n = 1$) state, because the monopole and the antimonopole would be overlapping, and so would annihilate each other. However, a monopolonium would initially be formed with $n \gg 1$; it would then undergo a series of transitions through a series of tighter and tighter bound states by emitting initially photons and subsequently gluons, Z bosons, and finally the GUT X bosons. Eventually, the cores of the monopole and the antimonopole would overlap, at which point the monopolonium would annihilate into X particles. Hill showed that the life time of a monopolonium is proportional to the cube of its initial radius. Depending on the epoch of formation, some of the monopolonia formed in the early Universe could be surviving in the Universe today, and some would have collapsed in the recent epochs. It can be shown [10] that monopolonia collapsing in the present epoch would have been formed in the early Universe at around the epoch of primordial nucleosynthesis.

The X particles produced by the collapsing monopolonia may give rise to EHECR. This possibility was studied in details in Ref. [10], who showed that this process, like cosmic strings, can also be described by an equation for X particle production rate described by Eq. (1) with $p = 1$. The efficacy of the process, however, depends on two parameters, namely, (a) the monopolonium-to-monopole fraction at formation (ξ_f) and (b) the monopole abundance. The latter is unknown, while the former is in principle calculable by using the classical Saha ionization formalism. However, phenomenologically, since a monopole mass can be typically $m_M \sim 40 m_X$ (so that each monopolonium collapse can release ~ 80 X particles), we see from Eq. (10) that one requires roughly (only!) about 3 monopolonium collapse per decade within a volume roughly of the size of the solar system. Whether or not this can happen depends, as already mentioned, on ξ_f as well as on the monopole abundance, the condition [10] being $(\Omega_M h_{100}^2) h_{100} \xi_f \simeq 1.7 \times 10^{-8} (m_X/10^{16}\,\text{GeV})^{1/2} (10\,\text{Mpc}/\lambda_{\gamma,300})$, where Ω_M is the mass density contributed by monopoles in units of closure density of the Universe, and h_{100} is the present hubble constant in units of $100\,\text{km sec}^{-1}\,\text{Mpc}^{-1}$. Thus, as expected, larger the monopole abundance, smaller is the monopolonium fraction ξ_f required to explain the EHECR flux. Note that, since ξ_f must always be less than unity, the above requirements can be satisfied as long as $(\Omega_M h_{100}^2) h_{100} > 1.7 \times 10^{-8} (m_X/10^{16}\,\text{GeV})^{1/2}$. Recall, in this context, that the most stringent bound on the monopole abundance is given by the Parkar bound [29], $(\Omega_M h_{100}^2)_{\text{Parkar}} \lesssim 4 \times 10^{-3} (m_M/10^{16}\,\text{GeV})^2$. The estimate of ξ_f obtained by using the Saha ionization formalism [4,10] shows that the resulting requirement on the monopole abundance (in order to explain the EHECR flux) is well within the Parkar bound mentioned above. The monopolonium collapse, therefore, seems to be an attractive scenario in this regard. A detailed numerical simulation of monopolonium formation to determine the monopolonium fraction at formation would, however, be useful in this context.

Necklaces

A cosmic necklace is a possible hybrid topological defect consisting of a closed cosmic string loop with monopole "beads" on them. Such a hybrid defect was first considered by Hindmarsh and Kibble [61]. Such defects could be formed in a two stage symmetry-breaking scheme such as $G \to H \times U(1) \to H \times Z_2$, where Z_2 is the discrete group $\{-1,1\}$ under multiplication. In such a symmetry breaking, monopoles are formed at the first step of the symmetry breaking if the group G is semisimple. In the second step, the so-called "Z_2" strings are formed and then each monopole gets attached to two strings, with the monopole magnetic flux channeled along the strings. Possible production of massive X particles from necklaces has been pointed out in Ref. [11].

The evolution of the necklace system is not well understood. The crucial quantity is the dimensionless ratio $r \equiv m_M/(\mu d)$, where m_M denotes the monopole mass, μ is the string mass per unit length, and d is the average separation between the monopoles. For $r \ll 1$, the monopoles play a subdominant role, and the evolution of the system is similar to that of ordinary cosmic strings. For $r \gg 1$, on the other hand, the monopoles determine the behavior of the system. The monopoles sitting on the strings tend to make the motion of the closed necklaces aperiodic. The authors of Ref. [11] assume that closed necklaces undergo frequent self-intersections, leading to monopole - antimonopole annihilation and, consequently, release of massive X particles. The X particle production rate for the necklace system can also be written in the form of Eq. (1) with $p = 1$. The efficacy of the process depends on free parameters, r, μ and m_M. For appropriate choice of the parameters, the required EHECR flux can be obtained. For more details see Ref. [11] and the article by Berezinsky in this volume.

Topological Defects in Supersymmmetric Theories

Recently, it has been pointed out [62] that in a wide class of supersymmetric unified theories, the higgs bosons associated with the gauge symmetry breaking can be 'light' — of mass of order the soft supersymmetry breaking scale $\sim$ TeV — even though the gauge symmetry breaking scale (and hence the mass of the gauge boson) itself may be much larger. The topological defects in these theories can, therefore, simultaneously be sources of $\sim$ TeV mass-scale higgs bosons as well as supermassive (mass up to $\sim 10^{16}$ GeV) gauge bosons. The decay of the TeV higgs may give a significant contribution to the observed diffuse γ-ray background above a few GeV, while the supermassive gauge boson decay may explain the EHECR. For more details, see Ref. [62].

CONCLUSION

There is no dearth of specific models of X particle producing processes involving topological defects. Almost all the "realistic" processes studied so far can be parametrized in the form of Eq. (1) with $p = 1$. The spectra of various kinds of particles produced for all these processes would essentially be similar to the ones shown in Fig. 1. Different processes might, however, contribute different amounts to the total flux. It is difficult to say which particular process may contribute most to the observed EHECR flux. However, in this respect, the cosmic string scenario seems to be relatively parameter free, especially if the new results of Ref. [33] are correct. Experimentally, it will be difficult, if not impossible, to tell which specific TD process, if at all, is responsible for the EHE cosmic rays. The best one can hope for is that proposed future experiments like OWL, Auger, and so on, may be able to tell us whether topological defects (or, for that matter, any *non-acceleration* mechanism in general) or some other completely different scenario is involved in producing the observed EHECR. In any event, the prospect of being able to look for signatures of new physics with the proposed EHECR experiments is certainly exciting, to say the least.

Note added: Recently, it has been pointed out [63] that if indeed decays of X particles from TD processes are responsible for the EHECR, then the same TD processes occurring in the early Universe may also have given rise to the observed baryon asymmetry of the Universe through CP- and Baryon number violating decays of the X particles.

ACKNOWLEDGMENTS

I am grateful to Günter Sigl for many helpful discussions, and in particular for providing the Figure. This work is supported by a NAS/NRC Resident Senior Research Associateship award at NASA/GSFC.

REFERENCES

1. T.W.B. Kibble, *J. Phys.* **A9**, 1387 (1976).
2. A. Vilenkin and E.P.S. Shellard, *Cosmic Strings and other Topological Defects* (Cambridge Univ. Press, Cambridge, 1994); M. Hindmarsh and T. W. B. Kibble, Rep. Prog. Phys. **58**, 477 (1995); T. W. B. Kibble, *Aust. J. Phys.* **50**, 697 (1997); T. Vachaspati, *Topological Defects in the Cosmos and Lab* (hep-ph/9802311) (to appear in *Contemporary Physics.*
3. P. Bhattacharjee, C.T. Hill, and D.N. Schramm, *Phys. Rev. Lett.* **69**, 567 (1992).
4. C.T. Hill, *Nucl. Phys.* **B 224**, 469 (1983).
5. C.T. Hill, D.N. Schramm, and T.P. Walker, *Phys. Rev.* **D 36**, 1007 (1987).
6. J. H. MacGibbon and R. H. Brandenberger, *Nucl. Phys.* **B 331**, 153 (1990); P. Bhattacharjee, *Phys. Rev.* **D 40**, 3968 (1989); M. Mohazzab and R. Brandenberger, *Int. Jour. Mod. Phys.* **D 2**, 183 (1993).
7. P. Bhattacharjee, in *Astrophysical Aspects of the Most Energetic Cosmic Rays*, eds. M. Nagano and F. Takahara (World Scientific, Singapore, 1991), *pp.* 382 – 399.
8. P. Bhattacharjee and N.C. Rana, *Phys. Lett.* **B 246**, 365 (1990).
9. A. J. Gill and T. W. B. Kibble, *Phys. Rev.* **D 50**, 3660 (1994).
10. P. Bhattacharjee and G. Sigl, *Phys. Rev.* **D 51**, 4079 (1995).
11. V. Berezinsky and A. Vilenkin, *Phys. Rev. Lett.* **79**, 5202 (1997).
12. D.J. Bird *et al*, *Phys. Rev. Lett.* **71**, 3401 (1993); *Astrophys. J.* **441**, 144 (1995); N. Hayashida *et al*, *Phys. Rev. Lett.* **73**, 3491 (1994); S. Yoshida *et al*, *Astropart. Phys.* **3**, 105 (1995).
13. G. Sigl, D.N. Schramm, and P. Bhattacharjee, *Astropart. Phys.* **2**, 401 (1994).
14. G. Sigl, S. Lee, D. N. Schramm, and P. Bhattacharjee, *Science*, **270**, 1977 (1995).
15. G. Sigl, S. Lee, and P. Coppi, *astro-ph/9604093*.
16. G. Sigl, S. Lee, D. N. Schramm, and P. Coppi, *Phys. Lett.* **B 392**, 129 (1997).
17. R.J. Protheroe and T. Stanev, *Phys. Rev. Lett.* **77**, 3708 (1996); **78**, 3420 (1997) (E).
18. A. M. Hillas, *Ann. Rev. Astron. Astrophys.* **22**, 425 (1984).
19. C.A. Norman, D.B. Melrose, and A. Achterberg, *Astrophys. J.* **454**, 60 (1995).
20. See, e.g., L. O'C Drury, *Rep. Prog. Phys.* **46**, 973 (1983); F. C. Jones and D. C. Ellison, *Sp. Sci. Rev.* **58**, 259 (1991).
21. J.P. Rachen and P.L. Biermann, *Astron. Astrophys.* **272**, 161 (1993); P. L. Biermann, in this volume.
22. J. W. Elbert, and P. Sommers, *Astrophys. J.* **441**, 151 (1995).
23. P. Bhattacharjee, in *Non-Accelerator Particle Physics*, Ed. R. Cowsik (World Scientific, Singapore, 1995).

24. P. Bhattacharjee, in *Extremely High Energy Cosmic Rays: Astrophysics and Future Observatories*, Ed. M. Nagano (ICRR, Univ. of Tokyo, 1996).

25. G. Sigl, *Sp. Sci. Rev.* **75**, 375 (1996).

26. F.A. Aharonian, P. Bhattacharjee, and D.N. Schramm, *Phys. Rev.* **D 46**, 4188 (1992).

27. S. Yoshida, *Astropart. Phys.* **2**, 187 (1994); S. Yoshida, H. Dai, C. Jui, and P. Sommers, *Astrophys. J.* **479**, 547 (1997).

28. X. Chi, C. Dahanayake, J. Wdowczyk, and A. W. Wolfendale, *Astropart. Phys.* **1**, 129 (1993); *ibid.* **1**, 239 (1993).

29. E.W. Kolb and M.S. Turner, *The Early Universe* (Addison-Wesley, Redwood City, California, 1990).

30. C. Bäuerle *et al*, *Nature* **382**, 332 (1996); V. Ruutu *et al*, *Nature* **382**, 334 (1996).

31. W.H. Zurek, *Nature* **317**, 505 (1985).

32. E. Witten, *Nucl. Phys.* **B 249**, 557 (1985).

33. G. Vincent, N.D. Antunes, and M. Hindmarsh, *Phys. Rev. Lett.* (to be published) (hep-ph/9708427); G. R. Vincent, M. Hindmarsh, and M. Sakellariadou, *Phys. Rev.* **D 56**, 637 (1997).

34. K. Greisen, *Phys. Rev. Lett.* **16**, 748 (1966); G. T. Zatsepin and V. A. Kuzmin, *Pisma Zh. Eksp. Teor. Fiz.* **4**, 114 (1966) [*JETP. Lett.* **4**, 78 (1966)]; F. W. Stecker, *Phys. Rev. Lett.* **21**, 1016 (1968); F. A. Aharonian and J. W. Cronin, *Phys. Rev.* **D 50**, 1892 (1994).

35. B. Andersson *et al*, *Phys. Rep.* **97**, 31 (1983).

36. G. Marchesini and B.R. Webber, *Nucl. Phys.* **B 310**, 461 (1988); G. Marchesini *et al*, *Comp. Phys. Comm.* **67**, 465 (1992); HERWIG version 5.9, hep-ph/9607393.

37. T. Sjöstrand and M. Bengtsson, *Comp. Phys. Comm.* **43**, 367 (1987).

38. L. Lönnblad, *Comp. Phys. Comm.* **71**, 15 (1992).

39. Ya. I. Azimov, Yu. L. Dokshitzer, V. A. Khoze, and S. I. Troyan, *Z. Phys.* **C 27**, 65 (1985); **C 31**, 213 (1986).

40. Yu. L. Dokshitzer, V. A. Khoze, A. H. Mueller, and S. I. Troyan, *Basics of perturbative QCD* (Editions Frontiers, Saclay, 1991); R. K. Ellis, W. J. Stirling, and B. R. Webber, *QCD and Collider Physics* (Cambridge Univ. Press, Cambridge, England, 1996); V. A. Khoze and W. Ochs, *Int. J. Mod. Phys.* (To be published) (hep-ph/9701421).

41. A.H. Mueller, *Nucl. Phys.* **B213**, 85 (1983); *ibid.* **B241**, 141 (1984) (E).

42. F.W. Stecker, *Astrophys. J.* **228**, 919 (1979).

43. A. Chen, J. Dwyer, and P. Kaaret, *Astrophys. J.* **463**, 169 (1996).

44. P. Sreekumar *et al*, *Astrophys. J.* **494**, 523 (1998); P. Sreekumar, F. W. Stecker, and S. C. Kappadath, in *Proc. Fourth Compton Symp.*, Eds. C. D. Dermer, M. S. Strickman, and J. D. Kurfess (AIP Conf. Proc. 410, 1997).

45. M.H. Salamon and F.W. Stecker, *Astrophys. J.* **493**, 547 (1998).

46. P. S. Coppi and F. A. Aharonian, *Astrophys. J.* **487**, L9 (1997).

47. T.J. Weiler, *E-print* hep-ph/9710431.

48. P. Lipari, *Astropart. Phys.* **1**, 195 (1993).

49. W. Rhode *et al.*, *Astropart. Phys.* **4**, 217 (1996).

50. M. Aglietta *et al*, in *Proc. 24th Int. Cosmic Ray Conf.* **1**, 638 (1995).

51. R.M. Baltrusaitis *et al*, *Astrophys. J.* **281**, L9 (1984); *Phys. Rev.* **D 31**, 2192 (1985).

52. G. Parente and E. Zas, *E-print* astro-ph/9606091.

53. G. Sigl, K. Jedamzik, D. N. Schramm, and V. Berezinsky, *Phys. Rev.* **D 52**, 6682 (1995).

54. F.W. Stecker, C. Done, M. H. Salamon, and P. Sommers, *Phys. Rev. Lett.* **66**, 2697 (1991); **69**, 2738 (1992) (E).

55. R. Gandhi, C. Quigg, M.H. Reno, and I. Sarcevic, *Astropart. Phys.* **5**, 81 (1995).

56. J.J. Blanco-Pillado, R.A. Vazquez, and E. Zas, *Phys. Rev. Lett.* **78**, 3614 (1997).

57. A. Vilenkin, *Phys. Rev. Lett.* **53**, 1016 (1984); D. Spergel and Ue-Li Pen, *Astrophys. J.* **491**, L67 (1997).

58. T.W.B. Kibble and N. Turok, *Phys. Lett.* **B 116**, 141 (1982).

59. P. Bhattacharjee, T.W.B. Kibble and N. Turok, *Phys. Lett.* **B 119**, 95 (1982).

60. X.A. Siemens and T.W.B. Kibble, *Nucl. Phys.* **B 438**, 307 (1995).

61. M.B. Hindmarsh and T.W.B. Kibble, *Phys. Rev. Lett.* **55**, 2398 (1985).

62. P. Bhattacharjee, Q. Shafi, and F.W. Stecker, *Phys. Rev. Lett.* (1998) (in press) (*E-print* hep-ph/9710533).

63. P. Bhattacharjee, *E-print* hep-ph/9803223.

Galactic and Extragalactic Magnetic Fields in the Local Universe: An Overview

Philipp P. Kronberg

University of Toronto, Department of Astronomy,
60 St. George St. Toronto, M1S 3H8 Canada

ABSTRACT

The methods of detecting and measuring interstellar and intergalactic magnetic fields are reviewed and briefly discussed. Recent results are surveyed for magnetic fields in the Milky Way, its halo, galaxy halos, galaxy clusters, and the intergalactic medium. Preliminary evidence for the magnetic field strength in the circumgalactic halo of our galaxy is discussed, and suggests that there are significant contrasts in magnetic field strength in and around galaxies such as ours, and between galaxy clusters and the general intergalactic medium. The general trend of recent results indicates that, wherever we detect intergalactic hot gas and galaxies, we also find magnetic fields at levels of $\approx 10^{-7}$ G, or higher. The hitherto undetected, weaker fields in the general intergalactic medium outside of clusters and in large intergalactic voids might, in future, be measurable through observations of γ-rays and/or cosmic ray nuclei. There appear to be close, though so far incompletely understood connections between ultra-high energy cosmic rays, high energy photons, diffuse hot gas, and magnetic fields in interstellar and intergalactic space.

1. INTRODUCTION

Observations over the last ten years have produced magnetic field detections not only in galaxy disks, but also in galaxy halos, and clusters of galaxies. There is also evidence for widespread magnetic field on supra-galaxy cluster scales in the "nearby" (≤ 100 Mpc) universe, that is currently regarded as the "horizon" for ultra high energy cosmic rays (UHECR). Additionally, some very distant galaxy systems, up to redshifts of at least 2, are found to contain magnetic field strengths of at least a few μG. At the current stage it seems that the more we look for extragalactic magnetic fields, the more ubiquitous we find them to be. Magnetic fields in diffuse astrophysical plasmas can be observationally detected, or measured by the following "traditional" means:

(1) Through the detection of synchrotron radiation, which requires both cosmic ray electrons and a magnetic field. The emissivity, in units of radiated power per unit volume, can be expressed as

$$\varepsilon(\nu) \propto n_{cr0} \cdot \Xi(\gamma) \cdot (B\sin\varphi)^{(\gamma+1)/2} \nu^{(1-\gamma)/2} \tag{1}$$

CP433, *Workshop on Observing Giant Air Showers from Space*
edited by J.F. Krizmanic et al.

196

where the energy distribution of the cosmic ray electrons is given by $n_{er}(E) = n_{er0} E^{-\gamma}$. φ is the average electron pitch angle with respect to the local field direction, B is the magnetic field strength, ν the frequency of observation, and $\Xi(\gamma)$ is a slowly varying function of γ.

(2) By the detection of Zeeman Effect splitting of a radio transition ν_{mn} in the interstellar magnetic field.

$$\nu = \nu_{mn} \pm eB(4\pi mc)^{-1} \text{ Hz} \tag{2}$$

(3) By measuring the induced polarization of optical starlight (intrinsically unpolarized) due to intervening dust grains that align themselves with the interstellar magnetic field and selectively scatter or absorb according to polarization angle.

(4) Faraday rotation due to a magnetic field in the presence of non-relativistic electrons: The rotation measure from a "cloud" at a redshift z, having an electron column density N_e cm^{-2} and electron density-weighted line-of-sight magnetic field strength $< B_{||} >$ G, is (cf. Kronberg and Perry, 1982(1)).

$$RM = 2.63\times10^{-13}(1+z)^{-2} N_e(z) < B_{||}(z) > \quad \text{rad m}^{-2}. \tag{3}$$

Or, in terms of the local electron density along the column, $n_e(l)$,

$$RM = \Delta\chi/(\Delta\lambda^2) = 8.1\times10^5 \int n_e(l) B_{||} \, dl \quad \text{rad m}^{-2} \tag{4}$$

where l is in pc, n_e in cm^{-3}, and B in Gauss.

(5) RM's of pulsars in our Galaxy can similarly be obtained from multi-frequency observations of their linear polarization, averaged over the pulses. Because the refractive index of the ionized interstellar gas varies significantly over the frequency ranges at which pulsars are usually detected (VHF to UHF), the relative pulse delay vs. frequency relation, gives additionally a *Dispersion Measure*

$$DM \propto \int n_e \, dl \quad \text{cm}^{-3}\text{pc}. \tag{5}$$

Comparison of (5) and (4) shows that the ratio RM/DM = $< B_{||} >$, an electron density-weighted average of the line-of-sight component of the field along the sight line to the pulsar. This type of magnetic field measurement was first performed by Lyne and Smith (1968)(2). Unfortunately, pulsars are so far too faint to observe in external galaxies, so that this method has not given measurements of extragalactic field strengths. However, the efficiency of measuring of pulsar RM's is improving rapidly so that pulsars will in future become increasingly important probes of the interstellar magnetic field. Pulsars provide the further advantage that the DM itself gives an independent estimate of the distance to the pulsar (within our Galaxy). This is in contrast to extragalactic polarized sources, whose radiation traverses the entire "local" galactic sightline from the direction of the source in question.

Unfortunately, Zeeman splitting (2) is only just detectable in our Galaxy at the level of a few µG. Its advantage is, by analogy with pulsars, that the galactic sightline can be dynamically estimated from the Doppler shift of the line system in question. Its disadvantage lies in the general difficulty of making the measurements. Also, comparably direct field strength measurements for external galaxies have not been possible, much less in hot intergalactic gas, because of the Doppler smearing of the radio lines due to turbulence and thermal line broadening. Optical polarization of starlight (3) is, up to now only practical for nearby, relatively face-on galaxies (where the optical dust extinction is low) and in any case it does not reveal the field sign, nor measure its strength.

Diffuse synchrotron radiation (1) and Faraday rotation (4) are the most practical methods for *detecting* magnetic fields in most galaxy systems and intergalactic space. Measurements of the field *strength* are not so easy, however. This is because, although diffuse synchrotron radiation is relatively easy to detect, especially at the lower (VHF and UHF) radio frequencies, the measured emissivity $\varepsilon(\nu)$ is weighted by the relativistic electron density (eq.(1)) which cannot be measured independently (except by direct cosmic ray measurements in the vicinity of the Earth). In the case of Faraday rotation (4), the observed RM (eqs. (3,4)) is scaled by the co-spatial, non-relativistic free electron density (n_e), and also by $\sqrt{N}$, N being the number of reversals along the line of sight. Both n_e and N must, and sometimes can be independently estimated from observation.

Faraday rotation of a polarized radio source can occur in the following places: (i) In the Earth's ionosphere, which normally produces less than 1 - 2 rad/m^2. (ii) In the Milky Way, where it varies from $\approx$ 2 rad/m^2 at high galactic latitudes to 500 rad/m^2 near the galactic plane. (iii) An intervening galaxy cluster can Faraday rotate a background radio source by typically 20 to 100 rad/m^2, and some dense, ``cooling flow'' galaxy clusters have been known to produce up to 10,000 rad/m^2 in large ($\geq$ 10 kpc) zones near the cluster core (Taylor and Perley (1993)(3)). However such large RM's are rare, and most radio galaxies and quasars have integrated RM's less than $\approx$ 30 rad/m^2. (iv) *Within* the large, galaxy - size outer radio lobes of extended radio galaxies. (v) Finally, on the largest known scales, any widespread, all-pervading intergalactic magnetic field could exist at some level, but it has not yet been detected. Any cumulative extragalactic RM due to a widespread intergalactic field up to a redshift of 2.5 is $\leq$ 3 rad/m^2.

In some instances the magnetic field *strength* in the system causing the excess RM can be further studied by other observations that can provide n_e, or the column density, N_e, of electrons that are associated with the system in question (*cf.* equation (3)). Examples are: (a) The X-ray – determined profile of n_e within galaxy clusters, (b) column density estimates of absorption lines in the system that causes an excess RM, and (c) measurment of the frequency-dependent delay of galactic pulsar pulses, *i.e.* the interstellar plasma Dispersion Measure (DM) as discussed above.

2. INTERSTELLAR MAGNETIC FIELDS IN THE MILKY WAY

Evidence for a widespread interstellar magnetic field has existed since the early days of radio astronomy, when it was discovered that the galactic disk is pervaded by diffuse synchrotron emission — the dominant signal in low resolution maps of the radio sky at frequencies below 500 MHz.

2.1 The Magnetic Field in the Disk of the Milky Way

Since the 1960's, Faraday RM's from linear polarization measurements of ever larger numbers of polarized extragalactic radio sources has led to a more refined modelling of the large scale Galactic magnetic field structure (Simard-Normandin and Kronberg 1979(4),1980(5), Sofue, Fujimoto, and Wielebinski 1986(6), Vallée *et al.* 1988(7), Clegg *et al.* 1992(8), Rand and Lyne 1994(9)). The Galactic magnetic field appears organized on a grand scale, and also has some large scale field reversal(s). Figure 1 shows a recent plot of the integrated Faraday rotation measures on an equal area projection in galactic coordinates (l,b). The systematic variations of RM on large angular scales are caused primarily by the Milky Way's interstellar magnetic field. Although most RM's above $|b| \approx$ 28° are small, a minority of high latitude sources have (for their (l,b), location) an anomalously high RM that is uncorrelated with their galactic position. The latter sources indicate an additional extragalactic origin of the RM, -- either in an unusually dense galaxy-cluster environment around the radio galaxy or quasar, or a cluster or intervening single - galaxy system seen in projection in front of a distant quasar (*cf.* Kronberg and Perry, 1982(1)). Nearly all significantly "excess" RM's of extragalactic origin apply to sources in the distant extragalactic universe which, for the purposes of this conference, I shall define as >100Mpc.

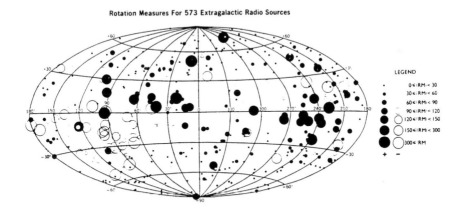

Rotation Measures For 573 Extragalactic Radio Sources

Figure 1. The Faraday Rotation measures of a sample of extraglactic radio sources projected onto galactic coordinates (l,b), where $(0,0)$ is the direction the galactic centre, and l increases to the left. Solid circles denote positive RM's, and open circles show negative values. The diameters increase with increasing |RM|.

Models based on data such as in Figure 1, and incorporating pulsar RM's and DM's indicate that, within our 2 - 3 kpc vicinity at 9 kpc from the galactic center, the galactic disk has an ordered component of magnetic field, of *ca.* 2 μG strength, directed toward *l* ≈ 90°. There is persistent evidence for at least one large scale reversal near the Sagittarius Arm, a little inside of our galactocentric radius (Simard-Normandin and Kronberg 1979(4), 1980(5), Vallée 1991(10), Rand and Lyne 1994(9)). Whether there are large scale reversals in the outer part of the galactic disk is currently disputed, and must await additional Faraday RM's from pulsars and extragalactic soures in that region of the galactic sky. The random component of the interstellar field within the galactic disk appears roughly comparable to the uniform component. The galactic field strength could be as high as 4 - 5 μG toward the center of the galactic plane. Local variations in |**B**|, and firm measurements of |<**B**$_r$>| / |**B**$_0$| are difficult to specify at this stage.

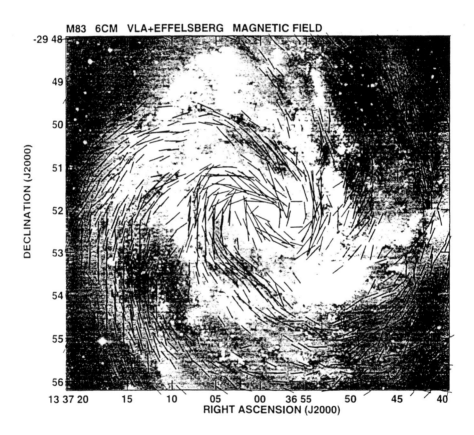

Figure 2. Polarization structure of the diffuse synchrotron radiation in the spiral galaxy M83. (Personal Communication: Previously unpublished image from S. Sukumar, R. Beck, *et al.*, paper in preparation. Printed here with the kind permission of R. Beck. Copyright 1998)

2.2 Evidence for Magnetic Fields in the Disks of Other Nearby Spiral Galaxies

Images of the polarized radio synchrotron emission have been made for more than 20 nearby spiral galaxies over the past decade. Although few of these results give corresponding images of the Faraday rotation hence the field sense, they do indicate large scale field ordering, consistent with what we conclude for the Milky Way (*cf.* Figure 2, and Beck *et al.* 1996(11) for a recent review). The ordered field component usually follows the spiral arm pattern. It often appears more ordered and, for ill-understood reasons, sometimes even stronger *between* the spiral arms (*cf.* Beck and Hoernes 1996(12)). The extent to which there are large scale reversals of field *sign* in external "quiescent" spiral galaxies (as we observe in the Milky Way) is not yet clear, since full Faraday rotation images of studies of external galaxies are only beginning to be made. What these results conclusively show is that all spiral-like galaxies with interstellar cosmic rays and thermal gas possess an interstellar magnetic field (*cf.* Wielebinski 1990(13)).

2.3 Evidence for Disk – to – Off – Disk Field Strength Contrast in |B|

Firm, quantitative measurements of the magnetic field strength and for the galactic *halo* have so far been elusive -- both around our own Galaxy, and around other "quiescent" disk galaxies—although not for strong outflow galaxies, as I will show in § 3. In the following I review the scant evidence so far for the galactic *vs.* near-extragalactic *contrast* in field strength, as inferred from the scale height (Z_0) of the mean interstellar magnetic field strength. The question is important to answer, since it would establish a range for CR proton gyroradii, hence the propagation characteristics of cosmic ray particles in the outer halo of our Galaxy, and within the Local Group of galaxies.

Figure 3 illustrates one attempt to quantify the diminution of the RM with galactic latitude in our galactic vicinity. Simard-Normandin and Kronberg (1980)(5) estimated a full width in RM. *i.e.* $n_e B_{||}$, of ca. 1.4 kpc, *i.e.* $2Z_0^{RM}$. This can be compared with another measured scale height, namely that of the ionized gas layer by imaging the diffuse optical H_a emission, giving a scale height of ≈ 1.1 kpc (Rand *et al.* 1990(14), Dettmar 1990(15)). Yet another "handle" on the $|B|(Z)$ relation can be gleaned from Breuermann, Kanbach and Berkhuijsen's (1985)(16) determinations of the scale height of the Galaxy's continuum synchrotron emission, Z_0^{SYNCHR}, based on the Haslam *et al.* (17,18) 408 MHz all sky survey. Breuermann *et al.*(16) estimate that $Z_0^{SYNCHR} \approx 1$ kpc at a galactocentric radius of 3kpc, increasing to ≈ 3 kpc at 12 kpc galactocentric radius. (I assume a galactocentric radius for the Sun, $R_\odot = 8$ kpc in distinction to 10 kpc used by Breuermann *et al.* (16)).

The form of the adopted functions is variously defined in the literature (exponential, sech(Z/Z_0), and polynomial), and we shall not attempt distinctions here. For our purpose of estimating how $<|B|(Z)>$ declines above the galactic plane, we need to "unfold" $|B|$ from $n_e B_{||}$ (eq.(4)) for the RM analysis, and from $n_{er0}.(B\sin\varphi)^{(\gamma+1)/2}$ (eq. (1)) for the synchrotron emissivity analysis.

Ignoring the question of the exact form of $|B|(Z)$, simple assumptions would lead to $Z_0^B \approx 2Z_0^{RM}, \approx 2Z_0^{SYNCHR}, \approx 2Z_0^{ne}$. Based on the above results, I estimate that $|B|(Z)$ decreases by nearly an order of magnitude to a few $\times 10^{-7}$ G at $Z \approx 4$ kpc above the plane in the vicinity of the sun. Evidence for a thickening of the galactic synchrotron halo at large galactocentric distances (cf. Phillipps et al. 1981(19)) suggests a correspondingly larger Z_0^B at, say 1.5 solar circle radii. However, since the synchrotron emissivity at $Z=0$ is lower in the outer galaxy, the absolute magnetic field strength at $Z = 4$ kpc and $R=1.5 \times R_\odot$ may well be comparable to that at $Z = 4$ kpc and $R_\odot$.

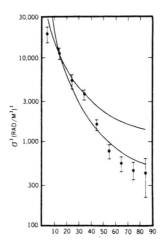

Figure 3. The variance of our Galaxy's rotation measure, calculated for 10° wide rings in galactic latitude (b). The upper and lower curves are from models assuming (i) completely random, and (ii) completely field geometries, respectively. This demonstrates that the galaxy's magneto-ionic heght is not very sensitive to assumptions about $|<B_r>| / |B_0|$. (Source: Simard-Normandin and Kronberg 1980 (5)).

We may ask if the out-of-plane field strength falloff can be clarified by observations of some of similar, nearby "quiescent" galaxies. Unfortunately, the halo field structure and strength is difficult to measure where the halo synchrotron emissivity is low (characteristic of "quiescent" spiral galaxies). Also, any halo Faraday rotation usually gets overwhelmed by the RM in the galaxy disks-- which are seen in projection along the same line-of-sight columns for any galaxy that is tilted toward our line of sight. Exceptions are the very few nearby galaxies that are seen edge-on. For two such galaxies, NGC891, and NGC 4631, Hummel et al. (1991)(20) have analysed the synchrotron halo emission, and the depolarization due to ionized gas mixed in with the synchrotron-emitting cosmic ray gas in the halo of both of these galaxies. They conclude a magnetic field scale height $Z_0^B \approx 4$ kpc for NGC891 and ≈ 8 kpc for NGC4631 (see also figure below). Scaling these numbers to my estimates above, namely where $|B|$ declines by a factor of 5 to 10 from the disk value, their values suggest significantly higher scale heights. This is most notably the case for NGC4631. The higher Z_0^B values for these galaxies are most likely due to their relatively more vigorous star-formation activity.

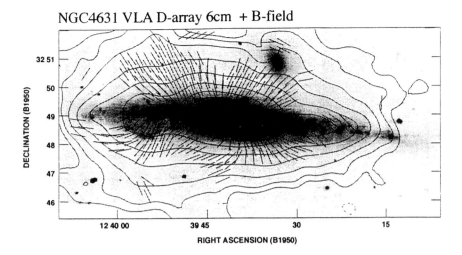

NGC4631 VLA D-array 6cm + B-field

Figure 4 An overlay of the 5GHz (λ6cm) radio emission (total intensity contours + normals to the de-Faraday rotated linear polarization orientations) of the edge-on galaxy NGC4631, and an image in the optical R-band. (*source*: Golla and Hummel 1994 (21)).

3. MAGNETIC FIELDS IN THE HALOS OF "ACTIVE" STARFORMING GALAXIES, AND RADIO GALAXIES

3.1 Active and Starbursting Galaxies

Other galaxies that are less quiesent have larger halos -- in synchrotron radiation, X-rays, and optical thermal emission. There seems to be a progression of activity level, in the sense that, for example, NG253 a nearby galaxy seen nearly edge-on has a radio halo extending to ≈ 9kpc (Carilli *et al.* 1992)(22). NGC4631, also just discussed, has a relatively thick halo, associated with stronger outflow due to an enhanced rate of star formation (Figure 4 above).

The inferred magnetic field strengths in these out-of-plane halos is a few microgauss, *i.e.* comparable with that in the denser disk. As the recent example of NGC4631 from Golla and Hummel (1994)(21) shows (*cf.* Fig. 4), such out-of plane fields can be remarkably organized, and appear to be directed out of the plane if the star formation activity (hence outflow wind) is sufficiently intense. These magnetized halos can also be related to the early origins of galactic magnetic fields, in that this magnetized halo gas, will have gradually leaked into the i.g.m. from galactic winds at early cosmological epochs (see Kronberg and Lesch (1996)(23)).

More extreme examples of star formation-driven outflow are illustrated by the nearby starburst galaxy M82, whose halo has recently revealed a large-scale organized magnetic field, as illustrated in Fig. 5 below. M82's halo has also been detected in X-rays, H_α emission, and in other emission lines. The origin of M82's outflow appears localized to the $\approx$ 700pc-wide starburst region, and the outflow velocity is significantly above M82's escape velocity. Since the origin of the outflow is driven by supernovae, enhanced star formation in any galaxy is able, in principle, to eject both cosmic rays and magnetic field into the i.g.m.

Another astrophysical source of large, magnetized cosmic ray gas are the well-known diffuse, strongly emitting radio lobes of some radio galaxies and quasars. The supra-galaxy scale morphology of double radio sources, their associated large energies (Burbidge 1956(24)), and the high collimation of their radiating jets make them a remarkable phemonenon of the extragalactic universe. Large scale jets and associated radio lobes source structure are described in a previous issue of this *Conference Series* by Bridle (1991)(25). Jets also occur on stellar scales, *e.g.* in binary star systems and protostellar objects, where a magnetized accretion disk produces a highly collimated outflow (*cf.* Mestel 1972(26), Lovelace, Berk and Contopoulos 1991(27)).

Figure 5. Lines show the projected orientation of the de-Faraday rotated polarization of synchrotron emission from the halo of cosmic ray electrons and magnetic field around the starburst galaxy M82. *Source:* Reuter *et al.* (1994)(28).

3.2 Radio Galaxies

Figure 6 shows a 1.4 GHz image of the double radio source, Cygnus A (3C405), one of the first images in which the filamentary structure of the outer lobes was revealed, thanks to the combination of resolution, sensitivity and imaging dynamic range of the NRAO Very Large Array (VLA). The origin of these rope-like, enhanced - $|B|$ zones has also been reproduced in 3D MHD simulations -- *e.g.* by D.A. Clark (1993)(29). Thus, we are now beginning to understand the field amplification mechanisms which can ``inflate'' significant volumes of intergalactic space with magnetic fields—which produce the (easily detectable) synchrotron radiation. Systems such as this have been suggested as possible sites of acceleration of CR nuclei to very high energies – see Peter Biermann's article (30) in this issue, and references therein. The closest, radio-powerful such system is Virgo A (3C274) in the Virgo cluster of galaxies at *ca.* 11 Mpc.

GALAXY NUCLEUS

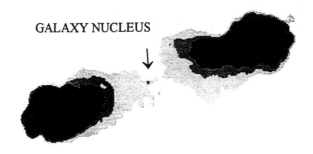

Figure 6.   Image of the radio galaxy Cygnus A (3C405) at 1.5″ resolution, at a frequency of 1.4 GHz The observations were made with the NRAO VLA. (*Source*: R.A. Perley,   personal communication. Printed here with the kind permission of R.A. Perley. Copyright 1998)

4. MAGNETIC FIELDS IN GALAXY CLUSTERS

Intergalactic gas in galaxy clusters has a typical electron density of 10^{-4} - 10^{-2} cm^{-3}, temperature in the range 10^7 to 10^8 K, and an extent of 1 Mpc for the cluster core. At this temperature range, which makes clusters significant X-ray sources ($L_x \approx 10^{43}$ - 10^{45} erg/s), the ion sound speed is comparable to the galaxies' velocity dispersion in the cluster, which is 400 - 1200 km s^{-1}. A minority of clusters such as the Coma cluster, that are in a merging phase, contain an enhanced population of cosmic ray electrons which, over dimensions comparable to that of the hot gas, emit a diffuse ``halo'' of synchrotron radiation, thus revealing an intracluster magnetic field. Kim *et al.* (1990)(31) and more recently Feretti *et al.* (1995)(32) estimate magnetic field strengths in the range 2 – 6 microgauss (scaled to H_0=50 km/sec/Mpc^3) in the core of the Coma cluster (see Fig. 7).

The existence of a strong radio halo does not seem to much affect intracluster magnetic field measurements, given that Kim, Tribble and Kronberg (1991)(33) find, in a statistical RM analysis for about 50 Abell clusters that $|\mathbf{B}| \approx 2 \ \mu G$, or perhaps more is common in the i.g.m. of the core areas of galaxy clusters. These results, in part, and the Coma cluster $|\mathbf{B}|$ estimates used n_e estimates from the diffuse X-ray bremsstrahlung in the clusters. A more extensive review of magnetic fields in galaxy clusters the reader is given in Kronberg (1994)(34).

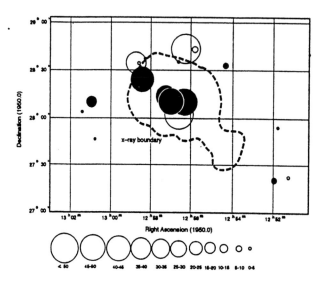

Figure 7 RM distribution of 18 background radio sources relative to the X-ray boundary of the Coma Cluster of galaxies, which shows a clear excess RM added by the intracluster medium (source Kim et al. 1990)(30). Filled and open circles show positive, and negative RM's respectively. The dashed line shows the ROSAT X-ray boundary at approx. 0.12 counts/(400 arcsec^2) for the ROSAT PSPC detector, in the range 0.5 - 2.4 keV (from Briel, Henry, and Böhringer 1992 (35)).

5. MAGNETIC FIELDS ON SUPRACLUSTER SCALES (> 1Mpc)

Cosmic rays serve as particularly effective "illuminators" of intergalactic magnetic fields at lower radio frequencies. This is because of the relatively high spectral density of synchrotron radiation at low frequencies, reflecting the high value of γ, typically 2.4 - 3, which defines the power law distribution of cosmic ray electron energies: $N(E) \propto E^{-\gamma}$ (cf. eq. 1). The critical frequency (near which most of the synchrotron radiation is emitted) is related to E by $\nu_C = (3eB \ \sin\varphi/4\pi \ m^3c^5)E^2$ (Pacholczyk 1970(36)). An advantage of observing at the lowest possible radio frequencies is that we preferentially detect the lowest energy CR electrons, which survive the longest to "keep illuminating" the associated magnetic field. To calculate the longest possible loss time for a CR electron, we define a "cosmic background- equivalent" magnetic field, B_{bge}, for which a CR electron's energy loss rate by synchrotron radiation ($dE/dt \propto |\mathbf{B}|^2E^2$) equals that due to inverse

Compton scattering off the microwave background radiation (dE/dt) $\propto \varepsilon_{bg} E^2$, where $\varepsilon_{bg} = 4.8.10^{-13} (1+z)^4$ erg cm^{-3}.

For $z \ll 1$, to which we restrict the present discussion, $B_{bge} \approx 3.10^{-6}$ G. Larger magnetic field strengths will cause shorter, synchrotron radiation-dominated lifetimes, whereas for B < B_{bge}, a CR electron's loss rate, being dominated by inverse Compton scattering, will depend only on its energy and ε_{bg} (cf. Rees 1967(37), Pacholczyk 1970(36) for a discussion of the basic radiation processes). The *inverse Compton-dominated* lifetime, τ_{max}, for synchrotron-radiating CR electrons can be written as

$$\tau_{max} \approx 3.10^8 (\nu/327\text{MHz})^{-0.5} (|B|/|B_{bge}|)^{0.5} \text{ years} \qquad (6)$$

where ν is the frequency of observation, and we express the electron energy in terms of ν and B. $|B|/|B_{bge}| \leq 1$, but the factor $|B|/|B_{bge}|^{0.5}$ is probably not far below unity (*cf.* Kronberg (1994)(34)) as suggested by recently measured i.c.m. field values. Thus we see that in these circumstances, the maximum "fossil lifetime" of relativistic electrons scales to first order by $\nu^{-0.5}$. This illustrates why we want to image around galaxy clusters and superclusters at the lowest possible frequency, while still preserving an acceptable angular resolution.

Observing frequencies of interest are between approximately 20 MHz and 400 MHz: Below 20 MHz, refraction and absorption by the ionized component of the interstellar medium and/or the ionosphere become important. Above 400 MHz, the emissivity becomes too low for the most sensitive detection of diffuse intergalactic synchrotron emission. In addition, the higher energy electrons, which radiate on average at higher frequencies, are less likely to have propagated to large distances from their acceleration sites because of their shorter loss times.

Until recently, the low available resolution, difficulties with earthbound and solar interference, and ionospheric phase fluctuations have prevented accurate imaging at frequencies below $\approx$ 300MHz. To distinguish true diffuse intergalactic synchrotron emission from a blend of discrete extragalactic radio sources, an angular resolution of 1' or better is required. This, at ν = 75 MHz for example, requires a well-filled antenna array (to achieve the required image quality), which extends to $\approx$30 km or more (to achieve the required resolution). Furthermore, interference suppression and phase irregularities in the ionosphere need to be eliminated, which require computer-intensive data processing techniques. These requirements are being approximately fulfilled for the first time -- by the NRAO VLA whose frequency range will be extended downward to 75 MHz for the first time in 1998.

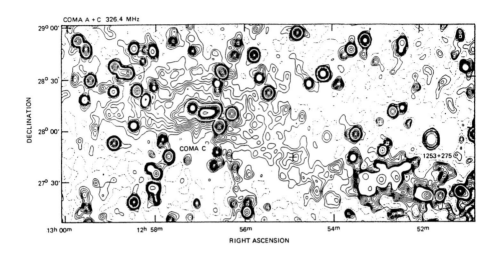

Figure 8 Radio image showing 326 MHz emission along in intergalactic ''bridge'', which appears to connect the Coma cluster of galaxies (Coma C) with the Coma A complex of radio emission elsewhere in the Coma supercluster (Note that the declination scale is compressed relative to the R. A. scale).

[38] Reprinted by permission from *Nature,* **341** 720 K.- T. Kim et al, ©1989 Macmillan Magazines Ltd.

These instrumental developments open up new possibilities for the detection of large scale i.g. magnetic fields. Relation (5) shows that radio images at $\nu = 30$ MHz could image diffuse, magnetized intergalactic gas in the local universe which was produced $\approx 10^9$ years ago, on the assumption that $|B|$ is not much different from $|B_{bge}|$. Such observations could give clues from the local Universe to the evolution of magnetic field strength over a significant period of cosmic time in such systems, e.g. in old, extended radio source lobes. An important complement observation will be that of sensitive imaging of diffuse X-ray or extreme-UV emission which is produced by either inverse Compton or thermal bremsstrahlung. Figure 8 shows a WSRT image published by Kim *et al.* (1989)(38) of the 326 MHz intergalactic synchrotron emission surrounding the Coma cluster of galaxies, which may be prototypical of the type of radiation which could be imaged in future at much lower frequencies. They found emission extending beyond the Coma cluster, indicating an extended, magnetized region on a supracluster scale. Measurements of this type do not, of course, yield the field strength directly, but the assumption of equipartition between particle and magnetic energy density gives field strengths of ca. $2 \cdot 10^{-7}(1+k)^{2/7}$ G in the region in Kim et al.'s (1989)(38) observations, which extend well outside of the cluster core. (k is the ratio of CR proton to electron energies, usually estimated at between 10 and 100.)

These first sensitive low frequency images suggest that future lower frequency images of high sensitivity and resolution will yield a great deal of new information on the distribution of intergalactic magnetic fields on large scales. Low frequency radio astronomy will thus play an important role in the direct detection of the stronger intergalactic magnetic fields. It likewise provides a direct way of tracing intergalactic cosmic ray electrons. If, as suggested by Biermann (30), cosmic ray electrons in some intergalactic systems in effect trace either present or past acceleration sites of UHECR nuclei, then low frequency extragalactic radio astronomy may in future also provide important clues to the acceleration and distribution of UHECR's.

6. POSSIBLE FUTURE USE OF UHECR'S AND γ RAYS TO INFER THE STRENGTH AND STRUCTURE OF INTERGALACTIC MAGNETIC FIELDS

The observed cosmic ray (CR) spectrum of protons, is found to extend to $ca. 10^{20}$ eV and beyond. If the Universe is pervaded with these most energetic CR particles, then the gyroradius, r_g, of a 10^{20} eV proton in a 10^{-9} gauss field is comparable with the typical distance between acceleration sites, if luminous radio sources like Cygnus A are the origin of these CR's(*cf.* Rachen and Biermann, 1993 (39))

In the nearby Universe, 10^{20} eV protons will travel in nearly straight lines for $B_{i.g.} \ll 10^{-9}$ gauss, which may be characteristic of the interiors (voids) of the large, ≈ 50 Mpc bubbles that define the large scale distribution of optically visible matter. Around the periphery of the voids, i.e. the bubble ``surfaces'' where the galaxies lie, they will be substantially deflected if $B_{i.g} = 10^{-7 \to -6}$ G, and possibly diffused in galaxy clusters and superclusters.

Of particular interest will be measurements of the much lower field strengths expected within the intergalactic voids. Here, we have few prospects using the "traditional" methods in §1 of estimating or limiting the magnetic field strengths, since the likely field levels below 10^{-8} G are generally beyond the reach of Faraday RM measurements. There is hope, however that future observations of high energy photons and high energy cosmic ray nuclei may provide intergalactic magnetic field measurements in these weaker régimes.

For example, a UHECR nucleus with known mass and charge has a Larmor radius given by

$$R_L \approx 100 \; E_{20} \; Z_{-1} \; B_{-9} \; \text{Mpc}. \tag{7}$$

If the origin of the CR nucleus is identified with an extragalactic system of known distance, a magnetic field deflection, hence field strength estimate can be obtained from the measured direction of arrival. Alternatively, slightly different arrival directions from multiple CR events originating at a common, known source could in future serve to distinguish different high energy CR nuclei (*cf.* Cronin, 1997(40)). Of course, **B** variations along the path of propagation make estimates of $|\mathbf{B}|$ somewhat less simple than equation (6) implies.

Another method would be provided by the simultaneous emission of an extragalactic γ-ray burster and a $\approx 10^{20}$ eV CR proton. In this case the measured difference in arrival times might be used to infer a weak field at the level of *ca.* $10^{-10} \cdot$G (*cf.* Biermann 1997(41)) that may be appropriate in intergalactic voids. Plaga (1995)(42) has proposed an analysis of the distribution in *time* of secondary γ rays due to e^+e^- - photon cascades from very high energy primary γ-ray events. This measurement would be sensitive to i.g. magnetic field strengths of $\approx 10^{-12}$ G and below. The forgoing methods that use UHECR's and γ rays to probe B $_{i.g}$ are described here only as examples of what will be possible in future. They also serve to illustrate interconnections that are yet to be explored and understood, between UHECR's, high energy γ-rays, the ambient intergalactic radiation field, and magnetic fields.

ACKNOWLEDGEMENTS

I am grateful to the Natural Sciences and Engineering Research Council of Canada for a research grant, the support of a Research Award of the Alexander von Humboldt Foundation in Germany, and to Richard Wielebinski for his hospitality at the Max-Planck-Institut für Astronomie. I also thank R.A. Perley and Rainer Beck for kindly permitting the presentation here of previously unpublished results, Elly Berkhuijsen and Peter Biermann for informative discussions, and John Bulicz for his assistance in the preparation of this article.

REFERENCES

1. Kronberg, P.P., and Perry, J.J. 1982. *Astrophys J.* **263**, 518
2. Lyne, A.G. and Smith F.G., 1968. *Nature* **218**, 124
3. Taylor, G.B., and Perley, R.A. 1993, *Astrophys J.* **416** , 554
4. Simard-Normandin M., and Kronberg, P.P., 1979, *Nature* **279**, 115
5. Simard-Normandin M., and Kronberg, P.P. 1980, *Astrophys J.* **242**, 74
6. Sofue Y. Fujimoto M., and Wielebinski, R. 1986, *Ann Rev. Astron. Astrophys* **24**, 22, 459
7. Vallée, J.P., Simard-Normandin M., and Bignell, J.P. 1988 *Astrophys J.* **331**, 321
8. Clegg A.W., Cordes, J.M., Simonetti, J.H., and Kulkarni, S.R. 1992, *Astrophys J.* **386**, 43
9. Rand, R.J., and Lyne, A.G., 1994 *Mon. Not. Roy. Astron. Soc.* **268**, 497
10. Vallée, J.P. 1991. *Astrophys J.* **366**, 450
11. Beck, R., Brandenburg, A., Moss, D., Shukurov, A., and Sokoloff, D. 1996. *Ann Rev Astron. & Astrophys.* 34, 155
12. Beck R., and Hoernes P. 1996. *Nature*, **379**, 47
13. Wielebinski, R., 1990 *The Interstellar Medium in Galaxies*, (Dordrecht: Kluwer), p.349
14. Rand, R.J., Kulkarni, S. and Hester, J.J. 1990 *Astrophys. J. (Lett)* **352**, L1
15. Dettmar, R.-J. 1990. *Astronomy & Astrophysics* **232**, L15
16. Breuermann, K., Kanbach G., and Berkhuijsen, E.M. 1985 *Astronomy & Astrophysics* **153**, 17

17. Haslam, C.G.T., Salter, C.J., Stoffel, H., Wilson, W.E., Cleary, M.N., Cooke D.J., and Thomasson, P. 1981 *Astronomy & Astrophysics.* **100**, 209
18. Haslam, C.G.T., Salter, C.J., Stoffel, H., Wilson, W.E. 1981 *Astronomy & Astrophysics.* Suppl. **47**, 1
19. Phillipps, S., Kearsey, S., Osborne, J.L., Haslam, C.G.T.,and Stoffel, H. 1981., *Astronomy & Astrophysics* **103**, 405
20. Hummel. E., Beck, R., and Dahlem, M. 1991. *Astronomy & Astrophysics* **248**, 23
21. Golla, G. and Hummel. E. 1994. *Astronomy & Astrophysics* **284**, 777
22. Carilli, C.L., Holdaway, M.A., Ho, P.T.P., and Depree, G.C., 1992. *Astrophys J.* **399**, L59
23. Kronberg, P.P. and Lesch, H. 1997. in *The Physics of Galactic Halos* ed. Lesch *et al.* Berlin: Academie Verlag,. 175
24. Burbidge, G.R., 1956. *Astrophys J.* **124**, 416
25. Bridle, A.H.,1991. *Am. Inst. of Phys. Conf. Series*, **254**, 386
26. Mestel, L., 1972 in *Stellar Evolution* ed. H-Y Chiu and A Muriel (Cambridge MA: MIT Press), p.643
27. Lovelace, R.V.E., Berk, H.L., and Contopoulos J. 1991. *Astrophys J.* **379**, 696
28. Reuter, H.-P., Klein, U., Lesch, H., Wielebinski,R. and Kronberg, P.P. 1994. *Astronomy & Astrophysics.* **282**, 724
29. Clarke, D.A. 1993, in *Jets in Extragalactic Radio Sources* (Ringberg Workshop Proc.) *ed.* H.-J. Röser and K. Meisenheimer (Berlin:Springer) p.243
30. Biermann, P.L. 1998. "Powerful Radio Galaxies as the Sources of High Energy Cosmic Rays" *Am. Inst. of Phys. Conf. Series*, **this issue**
31. Kim, K.-T., Kronberg, P.P., Dewdney, P.E., and Landecker, T.L. 1990. *Astrophys J.* **355**, 29
32. Feretti, L., Dallacasa, D., Giovannini, G., and Tagliani, A. 1995. *Astronomy & Astrophysics.* **302**, 680
33. Kim, K.T., Tribble, P, and Kronberg, P.P. 1991, *Astrophys J.* **379**, 80
34. Kronberg, P.P. 1994. *Rep. Prog. Phys.* **57**, 325
35. Briel, U., Henry, J.P., and Böhringer, H. 1992. *Astronomy & Astrophysics.* **259**, L31
36. Pacholczyk, A., 1970. *Radio Astrophysics* (San Francisco: Freeman)
37. Rees, M.J., 1967. *Mon. Not. Roy. Astron. Soc.* **137**, 429
38. Kim, K.-T., Kronberg, P.P., Giovannini, G., and Venturi, T. 1989. *Nature* **341**, 720
39. Rachen, J.P., and Biermann, P.L. 1993 *Astronomy & Astrophysics* **272**, 161
40. Cronin, J.W., 1997 preprint.
41. Biermann, P.L. 1997. *Journal of Physics G* **23**, 1
42. Plaga, R. 1995. *Nature* **374**, 430

Intergalactic Propagation of Heavy Nuclei - Are They the Trans-Greisen Events?

F. W. Stecker

Laboratory for High Energy Astrophysics, Code 661, NASA/Goddard Space Flight Center, Greenbelt, MD 20771, USA.

Abstract. The critical interactions for energy loss and photodisintegration of ultra-high energy cosmic-ray (UHCR) nuclei occur with photons of the infrared background radiation (IBR). I present here the results of a new estimation of the photodisintegration and propagation of (UHCR) nuclei in intergalactic space. I have reexamined this problem making use of a new determination of the IBR based on empirical data, primarily from IRAS galaxies, and also collateral information from TeV γ-ray observations of two nearby BL Lac objects. The results of this new calculation indicate that a 200 EeV nucleus which started out as Fe can have propagated ∼ 100 Mpc though the IBR. I argue that it is possible that the highest energy cosmic rays observed may be nuclei rather than protons.

Shortly after the discovery of the cosmic microwave background radiation (CBR), it was shown that cosmic rays above ∼100 EeV (10^{20}eV) should be attenuated by photomeson interactions with CBR photons [1]. It was later calculated that heavier nuclei with similar *total* energies would also be attenuated, but by a different process, *viz.*, photodisintegration interactions with IBR photons [2]. The particular detections of two events with energies well above these expected cutoffs, one at ∼ 200 EeV [3] and one at ∼ 300 EeV [4] have provided a double problem for cosmic-ray physicists. How does nature accelerate particles to these extreme energies and how do they get here from extragalactic sources [5]? To answer these questions, new physics has been invoked, physics involving the formation and annihilation of topological defects (TDs) which may have been produced in the very earliest stages of the big bang, perhaps as a result of grand unified theories. TD annihilation has unique observational consequences, such as the copious production of UHCR neutrinos and γ-rays [6]. A new ground based detector array experiment named after Pierre Auger [7] and an interesting satellite experiment called *OWL* have been proposed to look for such consequences [8]. Another idea recently proposed puts forth the speculation that the highest energy air-shower events may have been produced by a supersymmetric baryon designated by S^0 which is composed of

CP433, *Workshop on Observing Giant Air Showers from Space*
edited by J.F. Krizmanic et al.

the quark-gluino combination $uds\tilde{g}$ [9]. However, I will reexamine here the more conventional scenario by which UHCRs are accelerated at extragalactic sites.

Although cosmic acceleration to energies above 100 EeV pushes our present theoretical ideas to their extreme, it has been suggested that such acceleration may occur in hot spots in the lobes of radio galaxies [10]. I will assume here that such acceleration processes can occur in nature. We now turn specifically to the propagation problem. A UHCR proton of energy ~ 200 EeV has a lifetime against photomeson losses of $\sim 3 \times 10^{15}$s; one of energy 300 EeV has a lifetime of about half that [11]. These values correspond to linear propagation distances of ~ 30 and 15 Mpc respectively. Even shorter lifetimes were calculated for Fe nuclei, based on photodisintegration off the IBR [2]. Recent estimates of the lifetimes of UHCR γ-rays against electron-positron pair production interactions with background radio photons give values below 10^{15}s [12]. Within such distances, it is difficult to find candidate sources for UHCRs of such energies.

I present here the results of a new estimation of the photodisintegration and propagation of UHCR nuclei through the IBR in intergalactic space. I have reexamined this problem making use of a new determination of the IBR based on empirical data, primarily from IRAS galaxies, recently calculated by Malkan and Stecker [13]. Malkan and Stecker calculated the intensity and spectral energy distribution (SED) of the IBR based on empirical data, some of which was obtained for almost 3000 IRAS galaxies. It is these sources which produce the IBR. The data used for this calculation included (1) the luminosity dependent SEDs of these galaxies, (2) the 60 μm luminosity function for these galaxies, and (3) the redshift distribution of these galaxies. The magnitude of the IBR flux derived by these authors is approximately an order of magnitude lower that that used by Puget, Stecker and Bredekamp [2] in their extensive examination of photodisintegration of UHCR nuclei. This determination of a lower value for the magnitude of the IBR is also indicated by the observed lack of strong absorption in the multi-TeV γ-ray spectra of the active galaxies known as the BL Lac objects Mrk 421 [14] and Mrk 501 [15]. The lack of an obvious absorption feature up to an energy greater than ~ 5-10 TeV is consistent with the new, lower value for the IBR [16].

The Malkan-Stecker SED of the IBR has a similar shape to the one labeled "HIR" in the paper of Puget, et al. [2] in the mid-IR and far-IR range. However, it is approximately an order of magnitude lower in intensity. Therefore, one may replace the lifetimes given in ref. [2] by lifetimes which are longer by a factor of 10-20. Here I conservatively assume a factor of 10. Figure 1 is adapted from Fig. 14 of Puget, et al [2]. It indicates how a flux of UHCR Fe nuclei with an initial E^{-3} differential power-law spectrum will develop a cutoff at a critical energy, E_c which has an inverse dependence on the propagation time. In fact, for energies in the range near 200 EeV, $E_c \simeq 150(ct/100Mpc)^{-1/2}$ EeV.

A rescaling of Figure 13 of Puget, et al. [2] indicates that a nucleus which started out as Fe, reaching Earth from a distance of ~ 100 Mpc with a final energy of ~ 300 EeV would have an *average* value of $A \simeq 4$, corresponding to a helium nucleus. Such a nucleus would still have a mean-free path against both pair-production and photomeson production off the 2.7 K cosmic background radiation which would be greater than 100 Mpc.

I conclude from this that the highest energy CR induced air-showers could have been produced by UHCR nuclei propagating from a distance of the order of 100 Mpc! Stanev, Biermann and Lloyd-Evans [17] have examined the arrival directions of the highest energy air-shower events. These authors have pointed out that the 200 EeV event [3] is within 10° of the direction of the powerful radio galaxy NGC315 and the 300 EeV event is within 12° of the powerful radio galaxy 3C134. NGC315 lies at a distance of only ~ 60 Mpc from us. The distance to 3C134 is unfortunately unknown because its location behind a dense molecular cloud in our Galaxy obscures the spectral lines required for a measurement of its redshift.

Assuming that the two highest energy air showers observed were produced by helium nuclei (see above), taking the average magnetic field in our galaxy to be ~ 3μG, their gyroradii would be ~ 50 kpc. If they traveled ~ 500 pc through the galactic disk, with the thickness of the disk being ~ 100 pc, they would suffer a deflection of ~ 6°, allowing them to be observed within 10° of the source. (An intergalactic magnetic field might produce produce a larger deflection. However, both the strength and length scales of intergalactic magnetic fields are unknown.)

It should be also pointed out that it is reasonable to expect that the highest energy cosmic rays may be nuclei. This is because the maximum energy to which a particle can be accelerated in a source of a given size and magnetic field strength is proportional to its charge, Ze. That charge is 26 times larger for Fe than it is for protons. I conclude that is indeed possible that the highest energy cosmic rays which have been observed are heavy nuclei from radio galaxies which lie at distances within 100 Mpc from Earth.

A test for this heavy nucleus origin hypothesis would be to look for an event of energy greater than ~ 500 EeV. At these energies, even for heavy nuclei, photodisintegration off the 2.7 K background radiation should reduce the propagation length to less than 100 Mpc. Thus, the observation of events at energies greater than 500 EeV, perhaps by satellites [8], would force one to consider more exotic physical mechanisms to be the cause [6] [9].

REFERENCES

1. K. Greisen, Phys. Rev. Letters **16**, 748 (1966); G.T. Zatsepin and V.A. Kuz'min, Zh. Experim. i Teor. Fiz. - Pis'ma Redakt. **4**, 114 (1966).
2. J.L. Puget, F.W. Stecker and J.H. Bredekamp, Astrophys. J. **205**, 638 (1976).
3. N. Hayashida, *et al.*, Phys. Rev. Letters **73**, 3491 (1994).
4. D.G. Bird, *et al.*, Astrophys. J. **441**, 144 (1995).
5. J.W. Elbert and P. Sommers, Astrophys. J. **441**, 151 (1995).
6. G. Sigl, Space Sci. Rev. **75**, 375 (1996) and refs. therein; P. Bhattacharjee, Q. Shafi and F.W. Stecker, e-print hep-ph/9710533.
7. J.W. Cronin, Nucl. Phys. b (Proc. Suppl.) **28B**, 213 (1992).

8. J.F. Ormes, *et al.* in Proc. 25th Intl. Cosmic Ray Conf. (Durban, S.A.) eds. M.S. Potgieter, *et al.* **5**, 273 (1997).

9. G. R. Farrar, these proceedings.

10. P.L. Biermann and P.A. Strittmatter, Astrophys. J. **322**, 643 (1987); F. Takahara, Prog. Theor. Phys. (Japan) **83**, 1071 (1990).

11. F.W. Stecker, Phys. Rev. Letters **21**, 1016 (1968).

12. R.J. Protheroe and P.L. Biermann, Astroparticle Phys. **6**, 45 (1996).

13. M.A. Malkan and F.W. Stecker, Astrophys. J. **496**, in press (1998), preprint astro-ph/9710072.

14. J.E. McEnery, *et al.* in Proc. 25th Intl. Cosmic Ray Conf. (Durban, S.A.) eds. M.S. Potgieter, *et al.* **3**, 257 (1997).

15. F. Aharonian, *et al.*, Astron. and Astrophys., in press, preprint astro-ph/9706019 (1997).

16. F.W. Stecker and O.C. De Jager in Proc. Kruger Natl. Park Intl. Workshop on TeV Gamma-Ray Astrophysics, in press (1997), preprint astro-ph/9710145; F.W. Stecker and O.C. De Jager, to be published.

17. T. Stanev, P.L. Biermann and J. Lloyd-Evans, Phys. Rev. Letters **75**, 3056 (1995).

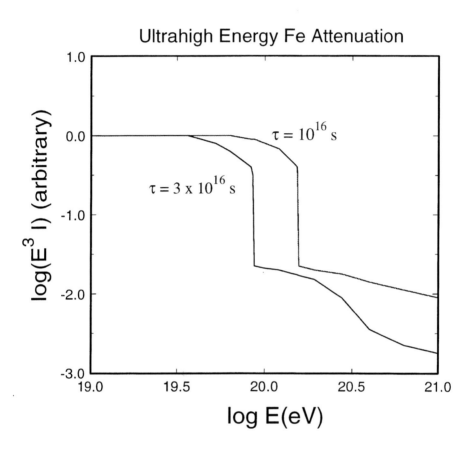

FIGURE 1. Attenuated E^{-3} differential spectrum for ultrahigh energy Fe nuclei for propagation times as indicated.

Propagation of Cosmic Rays and Neutrinos Through Space

Shigeru Yoshida

Institute for Cosmic Ray Research, University of Tokyo
Tanashi, Tokyo 188, Japan

Abstract.
We discuss the propagation of Extremely High Energy (EHE) cosmic ray protons and neutrinos. Their extremely high energies involve the interactions with the cosmological relic radiations of photons and neutrinos, which result in the modifications of the energy spectra of EHE particles after their propagation: The GZK cutoff for protons and the dip and steepening for neutrinos. Although the GZK cutoff has been recognized as one of the key features of extragalactic cosmic rays, it is also very important to recognize that the effects of the relic neutrino background to the EHE neutrino spectrum might provide the observational signatures of existence of the background neutrinos. Especially the massive neutrino case ($m_\nu \sim$ eV) would encourage the modification effects because the Z_0 resonance effect appears at around $10^{21} \sim 10^{22}$ eV, only one or two orders of magnitude higher than the highest energies of cosmic rays we have detected. Even for the massless neutrino case, the Topological Defects model would lead to appearance of the sizable effects of the relic neutrinos. Measurements of these phenomena should be considered as a main target of physics in any future experiment to observe EHE particles from space.

EHE PARTICLES AND THE COSMIC RELIC BACKGROUND

The study of the origin of Extremely High Energy (EHE) cosmic rays has been always coupled with the effects of their interactions with the cosmic microwave background, because their main effect, the Greisen-Zatsepin-Kuzmin cutoff (GZK cutoff) [1,2] is the centerpiece of the evidence to show the travel of cosmic rays from outside the Galaxy. Detailed understanding of the effects by the propagation of EHE cosmic rays is important because they are *tools* to identify the origin of cosmic rays. It is also a less remarked fact that the collisions of EHE particles with the relic low energy background could be a key in physics of EHE *neutrinos*. Neutrinos can usually penetrate cosmological distances without interacting with the low energy background and nobody would have to care about details of their propagation. This is not always true, however, if the neutrino energies are extremely

CP433, *Workshop on Observing Giant Air Showers from Space*
edited by J.F. Krizmanic et al.

high because the collision energy would reach to the mass of the Z_0. Especially when EHE neutrinos travel in high redshift epochs, which should happen in the cosmic ray production model involving the topological defects [3] or very distant quasars, the interactions with the cosmic neutrino background are not negligible. The effects of the interactions could not be *tools* to identify the source of EHE neutrinos, but would be interesting because the Big Bang scenario do predict the existence of the relic neutrino background which can constitute a part of the dark matter [4] but nobody has ever detected it. The collisions of cosmic ray particles with the low energy backgrounds would put the Extremely High Energy astrophysics to the very unique position. This is the motivation and the objectives of this paper.

THE GZK MECHANISM

The energy loss processes by pair creation and photopion production between EHE cosmic ray nucleons and 2.7 K background photons have been studied extensively. The studies in early days were based on analytical/numerical calculations of the transport equations [5,6] which have been followed up by numerical calculations with the Monte Carlo method (For examples, [7–9]) which took into account the broadening of the energy distribution of the recoil particles produced by the photopion production. Although the cosmological evolution of the background photon field leads to the dependence of the energy spectral shape of cosmic rays on the redshift of the sources, the energy where the cutoff starts tuned out to be universal values of $4 \sim 5 \times 10^{19}$ eV regardless of details in the source models, if the sources are distributed isotropically in universe, which is highly reasonable at least to a first approximation.

The remarkable products of the GZK mechanism are the secondary neutrinos which produced by decay of the photopions. Figure 1 shows the expected flux of

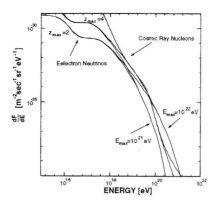

FIGURE 1. The spectrum of the secondary neutrinos produced by the GZK mechanism.

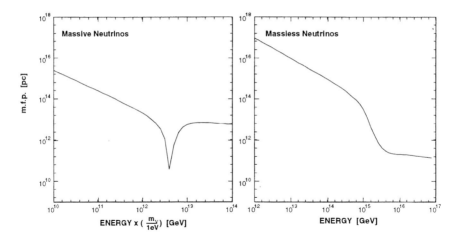

FIGURE 2. The mean free paths of EHE electron neutrinos for collisions with the cosmological background neutrinos in the present universe.

the GZK neutrinos [7]. The flux below 10^{19} eV depends on the activity of the cosmic ray emissions in earlier epochs which in this figure is defined by the parameter z_{max}, the turn-on time of the cosmic ray production, while the flux above 10^{19} eV depends strongly on the maximum accelerated energy of EHE cosmic rays E_{max} at the sources. Thus the intensity of the GZK neutrinos should be a good probe to understand the EHE cosmic rays emissions. This is because the neutrinos can propagate cosmological distances. However, the universe could become opaque even for neutrinos if energies of neutrinos reach to $\sim 10^{22}$ eV or greater and EHE neutrinos are emitted in very high redshift era. We discuss this interesting possibility in the next section.

EHE NEUTRINOS AND THE COSMOLOGICAL NEUTRINO BACKGROUND

The cosmological neutrino background predicted by the Big Bang model might play an important role in propagation of EHE neutrinos. Figure 2 shows the mean free paths of EHE neutrinos for their collisions with the background neutrinos [10,11]. The effect of the Z_0 resonance decreases the mean free path $\sim 10^{11}$ pc when the collision energy reaches to the mass of the Z_0. This path is much longer than the radius of the present Universe, however, the evolution of the neutrino background field shortens the mean free path. For example, at the epoch when the redshift $z \simeq 100$, the mean free path of EHE neutrinos becomes $\sim 100 kpc$ at 10^{21} eV. This suggests that the EHE neutrinos emitted from sources at high redshift epochs form the neutrino *cascade* during the propagation. Some channels in the

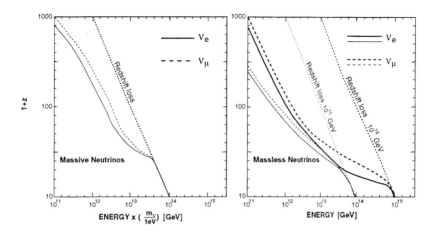

FIGURE 3. The mean free paths of EHE electron neutrinos for collisions with the cosmological background neutrinos in the present universe. The primary energies are 10^{15} GeV (thin lines) and 10^{16} GeV (thick lines). The solid lines show electron-neutrinos and the dashed lines show muon-neutrinos. The dotted lines show the upper bounds of the horizon taking into account the energy loss due to redshift.

interactions with the neutrino background lead to the absorption of EHE neutrinos (for examples, [12,13]) while the secondary products of muons, tauons, and pions emit neutrinos through their decay processes and further contribute to the neutrino cascade.

The effect of the neutrino cascade on the propagation of EHE neutrinos would be easily illustrated by considering the horizon of the universe for neutrinos. Figure 3 shows the curves of the the maximum redshift up to which neutrinos are not attenuated in their propagation [10,11]. The upper bounds of the horizon determined from the redshift energy loss are also shown for comparison. It is found that the energy loss effect due to the interactions with the black-body neutrinos contracts the horizon. This means that the distant sources at very high redshift epoch do not contribute to the bulk of EHE neutrinos. These EHE neutrinos are *absorbed*. However, the bulk of the secondary neutrinos moderate the absorption effect for the propagation from a high redshift epoch when the neutrino cascade develops significantly. As seen in figure 3, the cascading effect expands the horizon for neutrinos below 10^{12} GeV compared with that obtained by extrapolation from the higher energy region in which the cascade does not develop significantly during propagation. Twice as many muon neutrinos are produced as electron neutrinos by the pion decay processes, and the cascade expands the horizon for muon neutrinos further. The absorption effect due to the energy loss and the cascade development causes sizable modifications on the energy spectrum shape of EHE neutrinos if their energies extend up to $10^{21} \sim 10^{25}$ eV depending on the mass of neutrinos.

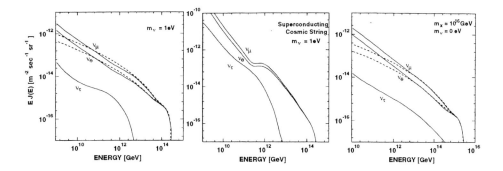

FIGURE 4. The differential flux of the EHE neutrinos from the collapse of the TDs. The dashed curves show the results calculated if only the energy loss due to the expansion of the Universe is considered.

The topological defects (TDs) scenario [3,14] actually provides the situation involving the neutrino cascade because the energies of EHE neutrinos produced by the annihilation of TDs range up to the GUT energy scale ($\sim 10^{16}~GeV$) and the TDs annihilation rate is high at early times (high redshift epochs). Figure 4 shows the expected spectrum [11]. It is found that the secondary neutrinos produced in the neutrino cascade enhance the intensity below $\sim 10^{20}~eV$. For the massive neutrino case, the spectrum has the dip structure at around the Z_0 resonance energy especially for case of the strong evolution (the superconducting string).

PHOTONS AND PROTONS PRODUCED BY EHE NEUTRINOS

We saw the effects of the secondary neutrinos produced in the neutrino cascade in the previous section. It should be noted that the channels of the hadronic decay in the cascading produce not only neutrinos but photons and nucleons. This mechanism might be one of the *sources* to produce the EHE cosmic rays we are now observing and detection of these secondary cosmic rays would provide another signature of the massive neutrino background. Weiler pointed out that collisions of EHE neutrinos with possibly clustered massive neutrinos in our galaxy might be responsible for the super GZK events detected by AGASA and Fly's Eye [15]. Even if the massive neutrinos are not clustered, these *secondary* photons and nucleons can be significant component of the observed EHE cosmic rays if very powerful astronomical objects such as AGNs are much more luminous in neutrinos than protons/gamma rays with their energies up to the Z_0 resonance energy. The number of produced gamma rays per unit length per unit energy is approximately given by

$$\frac{dN_\gamma}{dE_\gamma dL} = (1+z)^3 n_0 \int dE_\nu \frac{dN_\nu}{dE_\nu} \frac{dF}{dLdE_\gamma} \quad (1)$$

$$\frac{dF}{dLdE_{rec}} \simeq \int dE_{rec} \frac{d\sigma}{dE_{rec}}\Big|_{s=2m_\nu E_\nu} \frac{1}{E_{rec}} \frac{4}{3} \int_{E_\gamma/E_{rec}}^{1} \frac{dx}{x} \frac{dn_h}{dx} \tag{2}$$

where E_{rec} is the recoil energy of the collision particles, dn_h/dx is the pion fragmentation spectrum, dN_ν/E_ν is the differential flux of EHE neutrinos, n_0 is the number density of the background neutrinos in the present universe, z is the redshift. Since the dominant channel of the neutrino-neutrino interactions is the s-channel Z_0 exchange in this case, the cross section is written as

$$\frac{d\sigma}{dE_{rec}} = \frac{1}{E_\nu} \frac{2G^2}{\pi} M_z^2 \frac{s}{(s-M_z)^2 + M_z^2\Gamma_z^2} \left[g_L^2\eta^2 + g_R^2(1-\eta)^2\right] \tag{3}$$

where $\eta = E_{rec}/E_\nu$, g_L and g_R are the left-handed and right-handed coefficients respectively. The fragmentation spectrum approximately follows a power low formula in the region we are interested in here ($10^{-3} \leq x \leq 10^{-1}$). Using the fact that the power low index of the fragmentation spectrum δ is close to that of the EHE neutrino spectrum γ and that g_L is by far larger than g_R for hadronic Z_0 decay, all these integrals can be analytically calculated to give the following approximated formula:

$$\frac{dN_\gamma}{dE_\gamma dL} \simeq \frac{4}{\delta(\delta+2)} \sqrt{2}\pi \frac{\Gamma_z}{M_z} \frac{(1+z)^3}{\lambda_z} \frac{dN_\nu}{dE_\nu}\Big|_{E_\nu=E_{res}} \cdot \frac{dn_h}{dx}\Big|_{x=E_\gamma/E_{res}} \tag{4}$$

$\lambda_z \simeq 38$ Gpc is the mean free path of EHE neutrinos at the Z_0 resonance energy

$$E_{res} = 4 \times 10^{21} \left(\frac{m_\nu}{1eV}\right)^{-1} \quad eV. \tag{5}$$

When the source distance is much shorter than λ_z, the secondary gamma ray spectrum is given by

$$\frac{dN_\gamma}{dE_\gamma} = \int dL \frac{dN_\gamma}{dE_\gamma dL} = \frac{4}{\delta(\delta+2)} \sqrt{2}\pi \frac{\Gamma_z}{M_z} \frac{1}{\lambda_z}$$
$$\times \frac{c}{H_0} \frac{2}{3}\left[(1+z_s)^{\frac{3}{2}} - 1\right] \frac{dN_\nu}{dE_\nu}\Big|_{E_\nu=E_{res}} \frac{dn_h}{dx}\Big|_{x=E_\gamma/E_{res}} \tag{6}$$

Here H_0 is the Hubble constant and z_s is the source redshift. Then one can estimate the EHE neutrino flux required to give the secondary gamma ray intensity which is comparable to the observed cosmic ray flux at 3×10^{19} eV, $1.3 \times 10^{-34}m^{-2}\,sec^{-1}\,sr^{-1}eV^{-1}$ (measured by AGASA). As an example, let us consider Markarian 421, the closest Blazer, to represent the possible sources. For $dn_h/dx \sim ax^{-\delta} \sim 0.4x^{-1.4}$, $h = 0.5$ and $z_s = 0.0308$ (redshift of Markarian 421), we get

$$\frac{dN_\nu}{dE_\nu} = 2.5 \times 10^{-27} \left(\frac{m_\nu}{1eV}\right)^{0.4} \left(\frac{E_\nu}{10^{15}eV}\right)^{-1.0} \quad m^{-2}\,sec^{-1}\,sr^{-1}eV^{-1} \quad (\gamma = 1) \tag{7}$$

222

$$\frac{dN_\nu}{dE_\nu} = 5.0 \times 10^{-24} \left(\frac{m_\nu}{1eV}\right)^{-0.1} \left(\frac{E_\nu}{10^{15}eV}\right)^{-1.5} \quad m^{-2}\sec^{-1}sr^{-1}eV^{-1} \quad (\gamma = 1.5)$$

$$(8)$$

Our assumption of the hard spectrum of the neutrinos ($\gamma \sim 1 \sim 1.5$) is generally expected in the AGN neutrino model where the photopion production is the mechanism to produce neutrinos ([16]). The obtained intensities at around 10^{15} eV are comparable to the prediction range of the different models of AGN neutrino production (e.g. [18,17]). All the requirement is extending of the neutrino spectrum up to $E_{res} \sim 10^{21}$ eV with intensity of $\sim 10^{-33}m^{-2}\sec^{-1}sr^{-1}eV^{-1}$ at E_{res}. Because the neutrino has no energy loss process at the acceleration site, this is not totally unreasonable. Using the effective number of blazers with Markarian 421 flux, $\sim 3 \times 130sr^{-1}$ ([16]), we can calculate the neutrino luminosity for Markarian 421 in this model and find that

$$L_\nu \sim 2.6 \times 10^{7\sim9} \left(\frac{m_\nu}{1eV}\right)^{-0.6} m^{-2}\sec^{-1}eV \sim 10^{1\sim3}L_{\gamma,Mkn421}|_{TeV}. \quad (9)$$

This might be still conservative for several reasons. The observed gamma ray luminosity may have been reduced by absorption in the source or in the extragalactic space. Since neutrinos can travel cosmological distances, additional neutrino sources from which EGRET cannot detect their photon emission are likely to exist. Emission from distant sources would develop more neutrino cascading because of the evolution of the neutrino background.

It should be remarked, however, that the calculations discussed above give us only rough estimations about the required luminosity because the produced gamma rays (and protons) also lose their energy by the GZK mechanism, which we did not include in our estimation. It is absolutely necessary to carry out the detail numerical calculations based on the unified treatment of the propagation of neutrinos, photons and nucleons in the low energy background fields in order to check the feasibility of this hypothesis. The study of the propagation by the unified calculations is in progress and will be presented elsewhere.

SUMMARY

We have discussed the propagation of EHE cosmic ray protons and neutrinos mainly focusing on the effects of their interactions in the cosmological relic background field, one of the unique features of EHE astrophysics. The GZK cutoff is very the centerpiece among them now, but the others especially on the neutrinos would bring us new understandings of the universe.

Figure 5 shows the expected spectra of cosmic rays including the GZK neutrinos and those expected in the TDs scenario. The effects of the cosmological background field determine the spectral shape as well as their intensities at least to some extent. Measurements of these curves and their implications from aspects of physics of

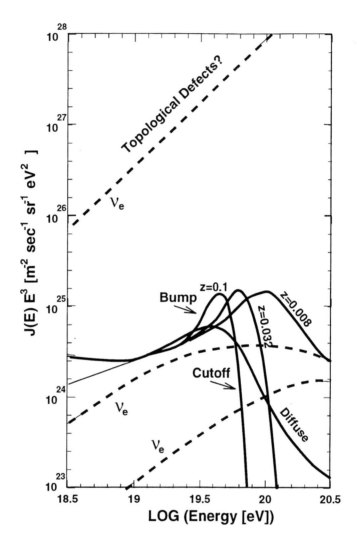

FIGURE 5. The EHE cosmic ray spectra. The lower two dashed curves show optimistic and conservative fluxes of the GZK neutrinos depending on the assumptions about the cosmic ray radiations in high redshift epochs. "Diffuse" means the case of the isotropic source distribution.

the cosmological relic backgrounds are our motivations to build the cosmic ray observatories in space.

ACKNOWLEDGMENTS

The author wishes to thank the workshop organizers for their warm hospitality at this meeting. He also wishes to acknowledge very valuable discussions with Dr. G. Sigl. He is grateful to Prof. J. W. Cronin for his encouragement in my early studies concerning the problems of EHE neutrino propagation.

REFERENCES

1. Greisen, K., *Phys. Rev. Lett.* **16**, 748 (1966).
2. Zatsepin, G. T., and Kuzmin. V. A., *Sov.Phys. JEPT Lett.* **4**, 78 (1966).
3. Bhattacharjee, P., Hill, C.T., and Schramm, D.N., *Phys.Rev.Lett.* **69**, 567 (1992).
4. Primack, J.R., Holtzman, J., Klypin, A., and Caldwell, D.O., *Phys.Rev.Lett.* **74**, 2160 (1995).
5. Hill, C. T., and Schramm, D. N., *Phys.Rev.* **D31**, 564 (1985).
6. Berezinsky, V. S., and Grigor'eva, S. I., *Astron.& Astrophys.* **199**, 1 (1988).
7. Yoshida, S., and Teshima, M., *Prog. Theor. Phys.* **89**, 833 (1993).
8. Protheroe, R.J., and Johnson, P., *Astropart. Phys.* **4**, 253 (1996).
9. Lee, S, *astro-ph/9604098, submitted to Phys.Rev.D* (1996).
10. Yoshida, S., *Astropart.Phys.* **2**, 187 (1994).
11. Yoshida, S., et al., *Astrophys. J.* **479**, 547 (1997).
12. Roulet, E., *Phys.Rev.* **D47**, 5247 (1993).
13. Weiler, T. J., *Astrophys.J.* **285**,495 (1984).
14. Sigl, G., Lee, S., and Schramm, D. N., *Phys.Lett.* **392 B**, 129 (1997).
15. Weiler, T. J., *VAND-TH-97-8, hep-ph/9710431* and *this conference* (1997).
16. Halzen, F., and Zas, E., *Astrophys. J.* **488**, 669 (1997).
17. Mannheim, K., *Astropart.Phys.* **3**, 295 (1995).
18. Protheroe, R.J., *ADP-AT-96-7* (1996).

Can Ultra High Energy Cosmic Rays be Evidence for New Particle Physics?

Glennys R. Farrar

Department of Physics and Astronomy
Rutgers University, Piscataway, NJ 08855, USA

Abstract. Candidate astrophysical acceleration sites capable of producing the highest energy cosmic rays ($E > 10^{19.5}$ eV) appear to be at far greater distances than is compatible with their being known particles. The properties of a new particle which can account for observations are discussed and found to be tightly constrained. In order to travel 100's or 1000's of Mpc through the cosmic microwave background radiation without severe energy loss and yet produce a shower in Earth's atmosphere which is consistent with observations, it must be a hadron with mass of order a few GeV and lifetime greater than about 1 week. A particle with the required properties was identified years ago in the context of supersymmetric theories with a very light gluino. Laboratory experiments do not exclude it, as is discussed briefly.

INTRODUCTION

Properties of ultra-high-energy cosmic rays (UHECRs) present a paradox within the standard astrophysical and particle physics framework. According to the Greisen-Zatsepin-Kuzmin bound [1], the spectrum of cosmic rays should cut off at about $10^{19.5}$ eV unless the source is closer than of order tens of megaparsecs. Yet many cosmic ray events have been observed at higher energy than this, and suitable cosmic accelerators within the GKZ range have not been found [2–5]. Some have considered the possibility of exotic relatively nearby sources, e.g., decay of super long-lived relics of the big bang [6,7] or associated production in gamma ray bursts [8]. Another possibility is that the UHECRs are exotic particles which can be transmitted from cosmological distances of order 100-1000's Mpc, where suitable conventional accelerators are found [9,10]. It turns out that it is difficult for a particle to simultaneously have the properties necessary to evade the GZK bound and also interact like an ordinary hadron in the atmosphere, as the UHECRs do [10]. However, somewhat miraculously, a particle with the correct properties to account for the UHECRs is automatically present in an interesting class of supersymmetric theories. The properties of this particle were delineated before the observation of the UHECRs and are not subject to extensive tuning [11,9].

CP433, *Workshop on Observing Giant Air Showers from Space*
edited by J.F. Krizmanic et al.
© 1998 The American Institute of Physics 1-56396-766-9/98/$15.00

BRIEF REVIEW OF LIGHT GLUINO
PHENOMENOLOGY

A supersymmetric version of a locally gauge invariant theory such as the standard model necessarily has massless fermions – the superpartners of the gauge bosons – called gauginos. Therefore the theory has an "accidental"[1] chiral symmetry which I will generically call R-invariance below. When the gauge invariance and supersymmetry are spontaneously broken, the R-invariance may or may not break.

In supersymmetry (SUSY) breaking scenarios which do not break R-invariance at all, every gaugino is massless at tree level[2] and R-parity is conserved. Such SUSY breaking has several attractive theoretical consequences such as the absence of the "SUSY CP problem" [12,9]. Gauginos get calculable masses through radiative corrections from electroweak (gaugino/higgsino-Higgs/gauge boson) and top-stop loops [13]. Evaluating these loops within the allowed parameter space leads to a gluino mass range $m_{\tilde{g}} \sim \frac{1}{10} - \frac{1}{2}$ GeV [12,9], while analysis of the η' mass and properties narrows this to $m(\tilde{g}) \approx 120 \pm 40$ MeV [14]. The photino mass range depends on more unknowns than the gluino mass, such as the higgs and higgsino sectors, but can be estimated to be $m_{\tilde{\gamma}} \sim \frac{1}{10} - 1\frac{1}{2}$ GeV [12].

In other SUSY breaking scenarios some or all gauginos are massive, with masses of order the squark or slepton masses. For purposes of understanding ultra-high-energy cosmic rays, we will be interested in models in which at least the gluino is massless or light at tree level. Such models have been constructed in the gauge-mediated framework in refs. [15,16]. In Raby's model the gluino mass can be tuned over a range ~ 100 GeV, and the gluino can be the lightest supersymmetric particle.

If the gluino lifetime is long compared with the strong interaction time scale, 10^{-23} sec, it binds with quarks, antiquarks and/or gluons to make color-singlet hadrons (generically called R-hadrons [17]). The lightest of these is expected to be the gluino-gluon bound state, designated R^0. It is predicted to have a mass in the range $1.3 - 2.2$ GeV [18,9], approximately degenerate with the lightest glueball (0^{++}) and "gluinoball" (0^{-+}, $\tilde{g}\tilde{g}$), for a gluino mass of ≈ 100 MeV [12,14]. The existance of an "extra" isosinglet pseudoscalar meson, $\eta(1410)$, which is difficult to accomodate in standard QCD but which matches nicely the mass and properties predicted for the pseudoscalar $\tilde{g}\tilde{g}$, is encouraging for the light gluino ansatz [19,9].

The lightest R-hadron with non-zero baryon number is the $uds\tilde{g}$ bound state designated S^0 [11]. As we shall see in the next section, the very highest energy cosmic rays reaching Earth [9] could well be S^0's. On account of the very strong hyperfine attraction among the quarks in the flavor-singlet channel, the S^0 mass is about 210 ± 20 MeV lower than that of the lightest R-nucleons [11,34]. If we knew the mass of the crypto-exotic flavor singlet baryon uds-gluon, we could place

---

[1] A global symmetry which is not imposed by hand but is the inevitable consequence of the gauge symmetries of the theory, like baryon and lepton number in the standard model, is called an accidental symmetry.

[2] In a typical supergravity case, all dimension-3 SUSY breaking terms are absent so that in conventional notation, m_1, m_2, m_3, $A \approx 0$.

the S^0 mass to within a couple of hundred MeV by analogy[3]. The $1/2^-$ baryon $\Lambda(1405)$ could be this $uds\tilde{g}$ state [9], in which case we see the S^0 mass could be as low as about 1.5 GeV. We can obtain an estimate of the maximum possible mass of the $uds\tilde{g}$ as follows. Add the mass of a Λ (to account for the effective mass of the quarks in a color singlet state, in the presence of chiral symmetry breaking) to the mass of a glueball (to account for the mass associated with combining a pair of color octets whose short-distance mass is zero) plus a bare gluino mass of 120 MeV: $m(S^0) < 1120 + 1600 + 120 = 2740$ MeV. This estimate does not account for the hyperfine attraction due to the uds in an S^0 being in a flavor singlet state whereas the uds in a Λ is in a flavor octet state [11,34]. A more realistic estimate would replace the Λ mass of 1120 MeV with the nucleon mass minus 210 MeV, leading to $m(S^0) \approx 940 - 210 + 1600 + 120 = 2450$ MeV.

If new gluino-containing hadrons have lifetimes shorter than about 10^{-10} sec, they can be discovered through missing energy or beam dump experiments [17]. However if the gluino is nearly massless it is long enough lived that the standard techniques are inapplicable [11]. The non-negligible radiative mass for the photino compared to the R^0, leads to an R^0 lifetime in the range $10^{-10} - 10^{-5}$ sec [12,9] if the photino mass is radiative. The dominant decay mode is $R^0 \to \pi^+\pi^-\tilde{\gamma}$ [9]. If neutralinos have tree level masses large compared to the gluino's, the R^0 would be stable or very long-lived, depending on the mass of the gravitino. When the R^0 is stable or long-lived, the usual SUSY signatures relying on prompt neutralino or goldstino (gravitino) production [17] are not useful [11]. As a consequence, gluino masses less than about $\frac{1}{2}$ GeV are largely unconstrained [18][4]. Proposals for direct searches for hadrons containing gluinos, via their decays in K^0 beams and otherwise, are given in Refs. [18,9]. For the moment, the experimental cuts preclude investigating the parameter ranges of theoretical interest, but part of the parameter space relevant when photinos provide dark matter should be amenable to study[5] In the course of the next two years it should be possible to exclude the all-gauginos-light scenario if it is not correct [25]. For a recent detailed survey of the experimental constraints on light gaugino scenarios, see [22].

An attractive feature of models with all gauginos massless or extremely light at tree level is that relic photinos naturally provide the correct abundance of dark matter [26,27]. By contrast, finding a good explanation for dark matter is a problem

---

[3] Due to the heavy mass of squarks, no rigorous supersymmetry argument equates the $uds\tilde{g}$ and $uds g$ masses in the massless gluino limit, unlike the case for glueball and glueballino in quenched approximation.

[4] The ALPEH claim to exclude light gluinos [20] assigns a 1σ theoretical systematic error based on varying the renormalization scale over a small range. Taking a more generally accepted range of scale variation and accounting for the large sensitivity to hadronization model, the ALEPH systematic uncertainty is comparable to that of other experiments and does not exclude light gluinos [21,22]. The claim of Nagy and Troscsanyi [23], that use of R_4 allows a 95% cl exclusion, has even worse problems. In addition to scale sensitivity, their result relies on using the central value of α_s. When the error bars on α_s are included, their limit is reduced to 1σ, even without considering the uncertainty due to scale and resummation scheme sensitivity.

[5] See [24] for a first experimental effort at placing a limit on R^0 production and decay via $\pi^+\pi^-\tilde{\gamma}$.

if the gluino is the only light gaugino. In this case the lightest neutralino would not be stable on cosmological time scales and thus could not be the dark matter particle. The gravitino is not a satisfactory dark matter candidate, if that is the lightest susymmetric particle, due to structure formation considerations [30]. Nor can the lightest gluino-containing hadrons provide sufficient relic dark matter density, even if absolutely stable as in the model of [16], because they annihilate too efficiently [28,27][6]. However new types of matter with conserved quantum numbers can be present in the theory, so the absence of a neutralino dark matter candidate may not be an insurmountable problem in models in which there is a light gluino but no light photino.

Now let us turn to the lifetime of the S^0, recalling the relevant mass estimates above for a gluino mass of 120 MeV: $m(R^0) = 1.3 - 2.2$ GeV, $m(S^0) = 1.4 - 2.7$ GeV, and $m_{\tilde{\gamma}}$ must lie in the range $\sim 0.9 - 1.7$ GeV if photinos account for the relic dark matter [27]. Thus the strong-interaction decay $S^0 \rightarrow \Lambda R^0$ is unlikely to be kinematically allowed, nor does the weak-interaction decay $S^0 \rightarrow n R^0$ seem likely. Even if kinematically allowed, the decay $S^0 \rightarrow n\tilde{\gamma}$ would lead to a long S^0 lifetime since it involves a flavor-changing-neutral-weak transition mediated by squarks. However the S^0 may be kinematically unable to decay since the mass estimates above are compatible with $m(S^0) < m(p) + m(e^-) + m_{\tilde{\gamma}}$ [9]. Requiring the S^0 to be stable or very long lived leads to the favored mass range $1.5 \lesssim m(S^0) \lesssim 2.6$ GeV.

If a gluino-containing hadron is absolutely stable, the most important consideration is whether it binds to nucleons to produce new stable nuclei which accumulate near Earth [11]. If so, limits on exotic isotopes give stringent limits on their abundance [11,18,28,29,31]. The S^0 is not expected to bind to nuclei [18]. The large ($\gtrsim 400$ MeV) energy gap to intermediate states accessible by pion exchange implies the effective nuclear potential seen by an S^0 is too shallow to support a bound state, except conceivably for very heavy nuclei[7].

To summarize, the gluino may be extremely light ($m_{\tilde{g}} \approx 120$ MeV) and give rise to new hadrons with masses below 3 GeV. Such a scenario predicts a particle like the $\eta(1410)$ and thus resolves the mystery of the existance of this state. As we shall see in the next section, the lightest R-baryon can naturally account for the ultra-high-energy cosmic rays which have been observed above the GKZ bound. The other gauginos may also be light. Such scenarios are very attractive because the photino naturally accounts for relic dark matter and there is no SUSY-CP problem. The all-gauginos-light scenario should be excludable within a year or so via LEP experiments [25]. The only-gluino-light scenario will be more difficult to exclude because that will require better theoretical control of perturbative and non-perturbative effects in QCD [22].

---

[6] In order to produce an interesting dark matter density, the gluino mass must be so large that it is inconsistent with properties of our galaxy [29].

[7] See [31] for an approach to estimating the nuclear-size dependence of the effective potential. It is less clear that a stable R^0 or $u\bar{d}\tilde{g}$ would not bind to nuclei and therefore be excluded; see [28,16,31] for a discussion of some of the issues.

ULTRA-HIGH-ENERGY COSMIC RAYS

If the light gaugino scenario is correct, the lightest R-baryon, $S^0 \equiv uds\tilde{g}$ [11], may be responsible for the very highest energy cosmic rays reaching Earth [9]. As is well-known to this audience, the observation of several events with energies $\gtrsim 2\ 10^{20}$ eV [2,3] presents a severe puzzle for astrophysics[8]. Protons with such high energies have a large scattering cross section on the cosmic microwave background photons, because E_{cm} can be sufficient to excite the $\Delta(1230)$ resonance [1]. Consequently the scattering length of ultra-high-energy protons is of order 10 Mpc or less. The upper bound on the energy of cosmic rays which could have originated in the local cluster, $\sim 10^{19.5}$ eV, is called the Greisen-Zatsepin-Kuzmin (GZK) bound.

Two of the highest energy cosmic ray events come from the same direction in the sky [2,3]; the geometrical random probability for this is $\sim 10^{-3}$ using a 1-sigma error box[9]. The nearest plausible source in that direction is the Seyfert galaxy MCG 8-11-11 (aka UGC 03374), but it is 62-124 Mpc away [4]. An even more attractive source is the AGN 3C 147, but its distance is at least 1200 Mpc [4]. The solid curves in Fig. 1, reproduced from ref. [10], show the spectrum of high energy protons as a function of their initial distance, for several different values of the injection energy. Compton scattering and photoproduction, as well as redshift effects, have been included [10]. It is evidently highly unlikely that the highest energy cosmic ray events can be due to protons from MCG 8-11-11, and even more unlikely that two or three high energy protons could penetrate such distances or originate from 3C 147.

It is also unlikely that the UHECR primaries are photons. First of all, photons of these energies have a scattering length, 6.6 Mpc, comparable to that of protons when account is taken of scattering from radio as well as CMBR photons [4]. Secondly, the atmospheric showers appear to be hadronic rather than electromagnetic: the UHECRs observed via extensive air shower detectors have the large muon content characteristic of a hadronic primary and the shower development of the 3.2×10^{20} eV Fly's Eye event has been found to be incompatible with that of a photon primary [33].

However the ground-state R-baryon, the flavor singlet scalar $uds\tilde{g}$ bound state denoted S^0, could explain these ultra-high-energy events [9,10]. The S^0 lifetime is plausibly longer than $\sim 10^5$ sec, the proper time required for a few 10^{20} GeV particle of mass 2 GeV to travel 100 Mpc. Furthermore, the GZK bound for the S^0 is several times higher than for protons. Three effects contribute to this: (a) The S^0 is neutral, so its interactions with photons cancel at leading order and are only present due to inhomogeneities in its quark substructure. (b) The S^0 is heavier than the proton. (c) The mass splitting between the S^0 and the lowest lying resonances which can be reached in a γS^0 collision (mass $\equiv M^*$) is larger than the proton-$\Delta(1230)$ splitting.

---

[8] For an introduction and references see [32] and talks at this conference.

[9] A third event above the GZK bound in the same direction has also been identified, as well as another triplet and two other pairs of UHE events [2].

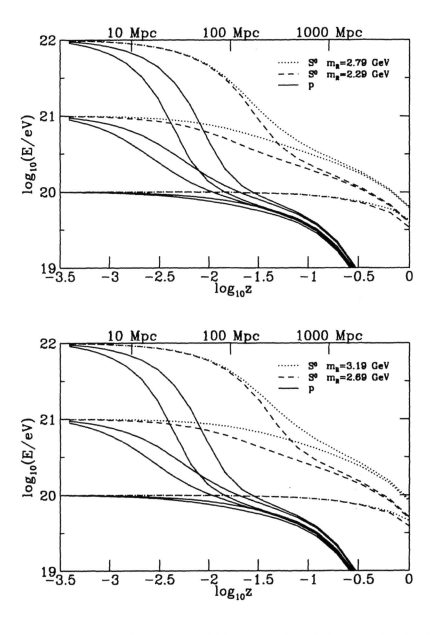

FIGURE 1. The figures show the primary particle's energy as it would be observed on Earth today if it were injected with various energies ($10^{22}eV$ eV, 10^{21} eV, and 10^{20} eV) at various redshifts. The distances correspond to luminosity distances. The mass of S^0 is 1.9GeV in the upper plot while it is 2.3GeV in the lower plot. Here, the Hubble constant has been set to 50 km sec^{-1} Mpc^{-1}.

The threshold energy for exciting the resonances in γS^0 collisions is larger than in γp collisions by the factor [9] $\frac{m_{S^0}}{m_p} \frac{(M^* - M_{S^0})}{(M_\Delta - m_p)}$, where $M^* - M_{S^0}$ is the mass splitting between the S^0 and the lowest mass resonance excited in γS^0 scattering. Since the photon couples as a flavor octet, the resonances excited in $S^0 \gamma$ collisions are flavor octets. Since the S^0 has spin-0 and the photon has helicity ± 1, only a spin-1 R_Λ or R_Σ can be produced in the intermediate state. There are two R-baryon flavor octets with $J = 1$, one with total quark spin 3/2 and the other with total quark spin 1/2, like the S^0. Neglecting the mixing between these states which is small, their masses are about 385-460 and 815-890 MeV heavier than the S^0, respectively [34]. This is a much larger splitting than $M_\Delta - m_p = 290$ MeV Thus one expects that the GZK bound is a factor of 2.7 - 7.5 higher for S^0's than for p's, depending on which R-hyperons are strongly coupled to the γS^0 system [9]. A detailed calculation of S^0 scattering on microwave photons, including $e^+ e^-$ pair production and redshift effects can be found in [10]. The results for a typical choice of parameters are shown in Fig. 1, confirming the rough estimate of ref. [9].

However as the above discussion makes clear, any neutral stable hadron with mass larger than a few times the proton mass will have a long enough mean free path in the CMBR to evade the GZK bound. Many extensions of the standard model contain stable colored particles besides quarks and gluons which, due to confinement, will be "clothed" with quarks and gluons to form new stable hadrons. However as pointed out in [10], there is also an upper bound on the mass of an acceptable UHECR primary (uhecron). This comes about because the fractional energy loss per collision with atmospheric nuclei is of order $(1 \text{ GeV})/m_U$, where m_U is the mass of the uhecron. But if the energy loss per collision is too small, the uhecron shower development does not resemble that of a nucleon. Detailed Monte Carlo simulation is necessary to pin down the maximum acceptable mass [35], but it seems unlikely to exceed of order ten GeV. Therefore new heavy colored particles whose masses are $\gtrsim 100$ GeV could not be the UHECR primaries even if they were not excluded otherwise. Supersymmetry breaking schemes such as [16] which allow parameters to be adjusted to make the gluino stable but do not require it to be nearly massless must be fine tuned to account for the observed UHECR showers. It is remarkable that the mass of the S^0 and its excitations in the nearly-massless gluino scenario fortuitously falls in the rather narrow range required to explain the UHECR's and yet be consistent with present laboratory and astrophysical constraints.

The question of production/acceleration of UHECR's is a difficult one, even if the UHECR primary could be a proton. Mechanisms for accelerating protons are reviewed in ref. [5]. Most of the mechanisms proposed for protons have variants which work for S^0's [10]. Indirect production via decay of defects or long-lived relics of the big bang proceeds by production of extremely high energy quarks (or gluinos). Since all baryons and R-baryons eventually decay to protons and S^0's respectively, the relative probability that a quark or gluino fragments into an S^0 compared to a proton can be expected to be of order $10^{-1} - 10^{-2}$. This estimate incorporates the difficulty of forming hadrons with increasingly large numbers of constituents, as reflected in the baryon to meson ratio in quark fragmentation which is typically of order 1:10. To be conservative, an additional possible suppression of up to a factor of 10 is included because the typical mass of R-baryons is greater than that of baryons.

Mechanisms which accelerate protons also produce high energy S^0's, via the production of R_p's ($uud\tilde{g}$ bound states) in pN collisions [10]. A problem with some proton acceleration mechanisms which is overcome with S^0's is that astrophysical accelerators capable of producing ultra-high-energy protons may have such large densities that the protons are unlikely to escape without colliding and losing energy. In the scenario at hand, a high proton collision rate is actually advantageous for producing R_p's. These R_p's decay to $S^0\pi^+$ via a weak interaction, with lifetime estimated to be $2 \cdot 10^{-11} - 2 \cdot 10^{-10}$ sec [9]. The S^0N cross section is likely to be smaller than the NN cross section by up to a factor of 10 [9]. Furthermore the S^0 interaction with electrons and photons is negligible.

It may be significant that the predicted time-dilated lifetime of an R_p of energy $\approx 3 \ 10^{20}$ eV, is of order seconds – a characteristic timescale for Gamma Ray Bursts. Mechanisms for producing ultra-high-energy protons in GRB's would translate to the production of R_p's [8].

Laboratory experiments can be used to get upper bounds on the production of R_p's, which may be helpful in deciding whether the S^0 production mechanisms discussed above is plausible. The E761 collaboration at Fermilab searched for evidence of $R_p \rightarrow S^0\pi^+$ [36]. Their result is shown in Fig. 2. If the lifetime of the R_p is of order nanoseconds and $m(S^0) \lesssim 2.1$ GeV (so $m(R_p) \lesssim 2.3$ GeV), these limits would make it difficult to produce sufficient high energy S^0's via R_p's. But for a lifetime of order $2 \ 10^{-1} - 2 \ 10^{-2}$ ns as estimated in [9] and the favored S^0 mass of ≈ 2.4 GeV, the E761 limits are too weak to be a constraint. As detailed in [22], a second generation experiment of this type would be very valuable.

SUMMARY

Cosmic ray events with energies above the GZK bound may be due to a quasistable hadron containing a very light gluino, such as the $uds\tilde{g}$ bound state called S^0. When the gluino mass arises only radiatively, due to the spontaneous breaking of electroweak symmetry, its mass is about 100 MeV. This implies a favored S^0 mass range of $1.5 - 2.6$ GeV. The GZK cutoff for S^0's occurs at higher energy than for protons, and S^0's of 3×10^{20} eV can come from cosmological distances where appropriate accelerators are found. The atmospheric shower of a high energy S^0 is similar to that of a nucleon.

The S^0's are not deflected by electric or magnetic fields and therefore should accurately point to their sources[10]. That means that if ultra-high-energy cosmic rays originate in persistent astrophysical objects such as AGN's and if S^0's are the

---

[10] Since the S^0 is a neutral spin-0 particle, even its magnetic dipole moment vanishes.

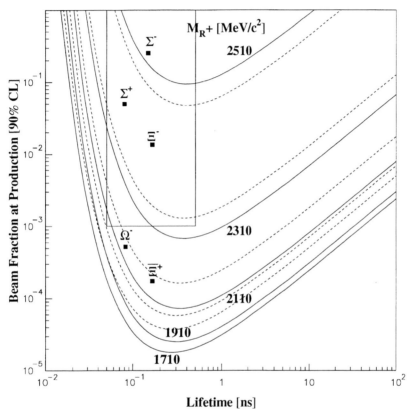

FIGURE 2. E761 limits vs. $\tau(R_p)$ [36]. Solid contours give limits for various values of the R_p mass, which is about 210 MeV above the S^0 mass. Ignore the box and dotted contours.

primaries, the UHECR events will cluster about certain directions in the sky. It should also be possible to identify a source "behind" each UHE event. In fact, four different "clusters" – two pairs and two triplets – have been identified among the highest energy events. The events in each of these clusters is consistent with pointing directly to the same location. A candidate astrophysical source, 3C 147, has been identified for the triplet containing two of the highest energy events. At well over a gigaparsec, it is near enough for S^0's to have arrived with little energy loss.

With the large sample of ultra-high-energy cosmic rays which hopefully will be obtained by HiRes and the Auger Project, the question of clustering will be settled. Whether the UHECR events cluster or not, they may be due to S^0's. If so, the prediction of a "GZK" cutoff in the spectrum, but shifted to higher energy, can be tested.

Laboratory experiments presently lack sufficient sensitivity to exclude the possibility of a very light gluino, but that should change within a year or two if the photino is light enough to account for dark matter.

REFERENCES

1. K. Greisen. *Phys. Rev. Lett.*, 16:748, 1966; G. T. Zatsepin and V. A. Kuzmin. *Sov. Phys.-JETP Lett.*, 4:78, 1966.
2. N. Hayashida. *Phys. Rev. Lett.*, 73:3491, 1994;
3. D. J. Bird et al. *Ap. J.*, 441:144, 1995.
4. J. Elbert and P. Sommers. *Ap. J.*, 441:151, 1995.
5. P. Biermann. *J. Phys. G: Nucl. Part. Phys.*, 23:1-27, 1997.
6. V. A. Kuzmin and V. A. Rubakov, astro-ph/9709178.
7. V. Berezinsky and M. Kachelreiss, hep-ph/9709485.
8. *Phys. Rev. Lett.*, 75:386, 1995.
9. G. R. Farrar. *Phys. Rev. Lett.*, 76:4111, 1996.
10. D. J. Chung, G. R. Farrar, and E. W. Kolb. Technical Report RU-97-14 and astro-ph/9707036, FNAL and Rutgers Univ, 1997.
11. G. R. Farrar. *Phys. Rev. Lett.*, 53:1029–1033, 1984.
12. G. R. Farrar. Technical Reports RU-95-17 (hep-ph/9504295) and RU-95-25 (hep-ph/9508291), Rutgers Univ., 1995. Invited talk SUSY95, Paris, May 1995, RU-95-73 (hep-ph/9704310).
13. R. Barbieri and L. Maiani, *Nucl. Phys.*, B243:429, 1984; G. R. Farrar and A. Masiero,Technical Report RU-94-38, hep-ph/9410401, Rutgers Univ., 1994; D. Pierce and A. Papadopoulos, *Nucl. Phys.*, B430:278, 1994; G. R. Farrar, Technical Report RU-95-26 (hep-ph/9508292), Rutgers Univ., 1995.
14. G. R. Farrar and G. Gabadadze. *Phys. Lett.*, 397B:104, 1997.
15. R. Mohapatra and S. Nandi. UMD-PP-97-082 (hep-ph/9702291) 1997. Z. Chacko et al. UMD-PP-97-102 (hep-ph/9704307) 1997.
16. S. Raby. OHSTPY-HEP-T-97-002 (hep-ph/9702299) 1997.
17. G. R. Farrar and P. Fayet. *Phys. Lett.*, 76B:575–579, 1978. *ibid* 79B:442–446, 1978.

18. G. R. Farrar. *Phys. Rev.*, D51:3904, 1995.

19. F. E. Close, G. R. Farrar, and Z. P. Li. *Phys. Rev.*, D55:5749, 1997.

20. The Aleph Collaboration. *Z. Phys.*, C76:191-199, 1997.

21. B. Gary, CTEQ Workshop, FNAL, Nov. 1996; G. R. Farrar, Rencontres de la Valee d'Aoste, La Thuile, Feb. 1997, RU-97-22 (hep-ph/9707467).

22. G. R. Farrar, Invited Review SUSY97, Philadelphia, May 1997, RU-97-79 (hep-ph/9710277).

23. Z. Nagy and Z. Trocsanyi, hep-ph/9708343.

24. The KTeV Collaboration. *Phys. Rev. Lett.*, 79:4083, 1997.

25. G. R. Farrar. RU-97-20 (hep-ph/9706393) Rutgers Univ., 1997. Phys. Lett. B, in press.

26. G. R. Farrar and E. W. Kolb. *Phys. Rev.*, D53:2990, 1996.

27. D. J. Chung, G. R. Farrar, and E. W. Kolb. Technical Report FERMILAB-Pub-96-097-A, RU-97-13 and astro-ph/9703145, FNAL and Rutgers Univ, 1997. Phys. Rev. D. to be published.

28. R. Plaga. *Phys. Rev.* D51:6504, 1995.

29. E. Nardi and E. Roulet *Phys. Lett.*, 245B:105, 1990.

30. E. Pierpaoli et al.,astro-ph/9709047.

31. S. Nussinov, hep-ph/9610236.

32. J. Lloyd-Evans and A. Watson. *Phys. World*, 9:47, 1996.

33. F. Halzen et al. *Astroparticle Phys.*, 3:151, 1995.

34. F. Bucella, G. R. Farrar, and A. Pugliese. *Phys. Lett.*, B153:311–314, 1985.

35. I. F. Albuquerque, G. R. Farrar, and E. W. Kolb. Work in progress.

36. I. F. Albuquerque and others (E761 Collaboration). *Phys. Rev. Lett.*, 78:3252, 1997.

Angle-Time-Energy Images of Ultra-High Energy Cosmic Ray Sources

Günter Sigl*

Department of Astronomy & Astrophysics
Enrico Fermi Institute, The University of Chicago, Chicago, IL 60637-1433

Abstract. Substantial amount of information both on the source and on characteristics of intercepting magnetic fields is encoded in the distribution in arrival times, directions, and energies of charged ultra-high energy cosmic rays from discrete sources. We present a numerical approach that allows to extract such information from data from next generation experiments.

INTRODUCTION

The origin of ultra-high energy cosmic rays is still a major unresolved mystery in astrophysics. It is hard to imagine a mechanism producing particles of energy up to several 100 EeV ($= 10^{20}$ eV). In addition, sources must be closer than $\simeq 50$ Mpc because of the limited range of nucleons due to photo-pion production at these energies. No obvious astrophysical sources have been found within this distance [1,2]. Although deflection of charged primaries can be strong in the direction of strong magnetic fields along large mass agglomerations such as the supergalactic plane [3], a deflection of several degrees at the most is expected along most other lines of sight for nucleons above 10^{20} eV, due to the Faraday rotation limit on the large-scale field. For next generation experiments with their anticipated much improved exposure, this opens up the possibility to do "particle astronomy" and pinpoint sources along the arrival directions.

It has been noted in that respect that a sub class of events above 4×10^{19} eV seems to cluster in arrival directions [4]. If these clusters originated in discrete sources, some interesting qualitative consequences result already, such as a limit on the intercepted magnetic fields that is comparable to the Faraday rotation limit [5]. Next generation experiments should in this case see clusters of several tens or even hundreds of events at these energies in case of the Pierre Auger Project [6] and the Orbital Wide-angle Light Collector (OWL) [8], respectively. This just follows from scaling to the relevant expected exposures. A data pool of arrival directions, times, and energies of that size contains a substantial amount of information on both the source of a given cluster of events and magnetic fields intercepting the line of sight.

CP433, *Workshop on Observing Giant Air Showers from Space*
edited by J.F. Krizmanic et al.
© 1998 The American Institute of Physics 1-56396-766-9/98/$15.00

This motivated us to conduct a detailed feasibility study for the potential of future experiments to reconstruct certain parameters that characterize the source mechanism and the large-scale magnetic field both of which are poorly known at present. We first briefly describe our method and then summarize our results and give some examples.

DESCRIPTION OF APPROACH

Here we describe the essential ingredients of our numerical approach; more details can be found in Refs. [9,10].

The propagation of nucleons through extragalactic space is simulated using the Monte Carlo technique: First, a magnetic field realization is set up on a grid via Fast Fourier Transformation by sampling a power spectrum of the form $\langle B^2(k) \rangle \propto k^{n_B}$ for wavenumbers $k < 2\pi/l_c$ and 0 otherwise, where l_c characterizes the coherence scale and n_B the magnetic field power spectrum. As results are quite insensitive to n_B, we assume $n_B = 0$ if not stated otherwise. More sophisticated models for the magnetic field including the role of the large scale structure of galaxies, for example, along the lines discussed in Ref. [3], may be implemented in the future. Many nucleon trajectories are then calculated between a given source and observer by sampling direction of emission, injection energy and the stochastic pion production loss that becomes important above the Greisen-Zatsepin-Kuzmin (GZK) cut-off [11] at a few 10^{19} eV. Pair production by protons has been incorporated as a continuous energy loss. One of the main problems that has to be solved when images of discrete sources are discussed, has not been considered in other work on propagation and deflection [12] and consists of the fact that different trajectories not only originate at the same source, but also have to reach the same observer.

From the injection energies, direction, time, and energy of arrival recorded for the trajectories we then calculate histograms for the distribution in these quantities by convolving with the injection spectrum (typically a power law with index γ for the differential spectrum) in energy and with a timescale T_S that characterizes the emission timescale. Histograms are also smeared out in energy to account for finite energy resolution (typically $\Delta E/E \simeq 0.14$, a value expected for future detectors) and are proportional to the source fluence N_0. We also use the parameter

$$\tau_E \simeq 2.0 \left(\frac{D}{30\,\mathrm{Mpc}}\right)^2 \left(\frac{E}{100\,\mathrm{EeV}}\right)^{-2} \left(\frac{B_{\mathrm{rms}}}{10^{-11}\,\mathrm{G}}\right)^2 \left(\frac{l_c}{1\,\mathrm{Mpc}}\right) \mathrm{yr}. \qquad (1)$$

which is the average time delay for a proton of energy E over a distance D in a field of r.m.s. strength B_{rms} when energy loss is negligible, and $\tau_E \ll D$ [13]. It is related to the average deflection angle θ_E by

$$\theta_E \simeq 0.02° \left(\frac{D}{10\,\mathrm{Mpc}}\right)^{-1/2} \left(\frac{\tau_E}{1\,\mathrm{yr}}\right)^{1/2}. \qquad (2)$$

238

The subscript E is given in EeV in the following.

Clusters of events are then obtained by sampling the histogram with Poisson statistics over a time window of width T_{obs} which constitutes the experimental lifetime, at a random position. Conversely, for a given cluster of events, a likelihood can be calculated for a given histogram that corresponds to certain values of the physical parameters described above. Averaging over different observational window positions and different realizations of the magnetic field for the same parameters yields the likelihood function $\mathcal{L}(\tau_{100}, T_S, D, \gamma, N_0, l_c, n_B)$. Marginalization over part of these parameters, using priors that account for certain constraints and other available information, can be used to reduce the parameter space.

RESULTS AND EXAMPLES

We first give a brief outline of the main features of the angle-time-energy images of clusters of ultra-high energy nucleons which have been described in detail in Ref. [14].

If both $T_S < \tau_{100}$, and τ_{100} is small compared to T_{obs}, arrival time and energy are correlated according to $\tau_E \propto E^{-2}$; see Eq. (1). The angular image can not be resolved in this case.

A source, such that $\tau_{100} \gg T_S$ and $\tau_{100} \gg T_{\text{obs}}$, can be seen only in a limited range of energies, at a given time, as first pointed out in Ref. [13], and demonstrated in Fig. 1. Below the GZK cut-off, the width of this stripe, in the time-energy plane and within the observational window of length T_{obs}, is then governed by the ratio $D\theta_E/l_c$ for the energy at which events are observed: If this ratio is much smaller than 1, all nucleons have experienced the same magnetic field structure during their propagation and the width is very small in the absence of pion production; in the opposite case the width is expected to be $\Delta\tau_E/\tau_E \sim 60\%$, even for negligible energy loss. Furthermore, the angular image is point-like or diffuse, with θ_E describing the systematic off-set from the direction to the source, or the angular extent of the diffuse image that is centered on the source, respectively, in these two cases. If $D\theta_E/l_c \sim 1$, several images of the source can result [15].

For a source emitting continuously at all energies of interest here, $i.e.$ with $T_S \gg \tau_{30}$ and $T_S \gg T_{\text{obs}}$, events of any energy can be recorded at any time. Whereas the above remarks on the angular image now apply for all energies (see Fig. 2), the distribution of arrival time $vs.$ energy is now uniform.

Finally, for a source, such that $\tau_{100} < T_S$ and $\tau_{30} > T_S$, together with $T_S \gg T_{\text{obs}}$, there exists an energy E_C, such that $\tau_{E_C} = T_S$. In this case, protons with $E < E_C$ are not detected, as they could not have reached us within T_{obs}. However, protons with $E > E_C$ are detected as for a continuously emitting source, $i.e.$ with a uniform distribution of arrival times $vs.$ energy (see, e.g., Fig. 3).

We now summarize results on the potential to reconstruct the parameters τ_{100}, T_S, D, γ, N_0, l_c, and n_B in these scenarios. Details have been presented in Ref. [10].

The likelihood presents different degeneracies between different parameters, which complicates the analysis. As an example, the likelihood is degenerate in the ratios N_0/T_S, or $N_0/\Delta\tau_{100}$, with N_0 the total fluence, and $\Delta\tau_{100}$ the spread in arrival time: these ratios represent rates of detection. Another example is given by the degeneracy between the distance D and the injection energy spectrum index γ. Yet another is the ratio $D\theta_E/l_c \propto (D\tau_E)^{1/2}/l_c$, that controls the size of the scatter around the mean of the $\tau_E - E$ correlation. Therefore, in most general cases, values for the different parameters cannot be pinned down, and generally, only domains of validity are found. We remark, however, that the generic scenarios discussed above are, in general, easy to distinguish from the likelihood function (see, e.g., Fig. 2).

We find that the distance to the source is obtained from the pion production signature, above the GZK cut-off, when the emission timescale of the source dominates over the time delay. The lower the minimal energy above which the source appears as emitting continuously, the higher the accuracy on the distance D. The error on D is, in the best case, typically a factor 2, for one cluster of $\simeq 40$ events. In this case, where the emission timescale dominates over the time delay at all observable energies, information on the magnetic field is only contained in the angular image. Qualitatively, the size of the angular image is proportional to $B_{\mathrm{rms}}(Dl_c)^{1/2}/E$, whereas the structure of the image, *i.e.* the number of separate images, is controlled by the ratio $D\theta_E/l_c \propto D^{3/2}B_{\mathrm{rms}}/El_c^{1/2}$. Finally, the case where the time delay dominates over the emission timescale, with a time delay shorter than the lifetime of the experiment, also allows to estimate the distance with a reasonable accuracy.

The injection spectrum index γ can be measured provided ultra-high energy cosmic rays are recorded over a bandpass in energy that is sufficiently broad. In general, it is comparably easy to rule out a hard injection spectrum if the actual $\gamma \gtrsim 2.0$, but it is much harder to distinguish between $\gamma = 2.0$ and 2.5.

The strength of the magnetic field can only be obtained from the time-energy image in this latter case because the angular image will not be resolvable. When the time delay dominates over the emission timescale, and is, at the same time, larger than the lifetime T_{obs} of the experiment, only a lower limit corresponding to T_{obs}, can be placed on the time delay, hence on the strength of the magnetic field. When combined with the Faraday rotation upper limit, this would nonetheless allow to bracket the r.m.s. magnetic field strength within a few orders of magnitude. Here as well, significant information is contained in the angular image.

The coherence length enters the ratio $(D\tau_E)^{1/2}/l_c$ that controls the scatter around the mean of the $\tau_E - E$ correlation in the time-energy image. It can therefore be estimated from the width of this image, provided the emission timescale is dominated by τ_E (otherwise the correlation would not be seen), and some prior information on D and τ_E is available. If the source appears continuous and the time delay is large enough to resolve the angular image, l_c can be constrained or even estimated from the fact that $D\theta_E/l_c$ passes through 1 at the energy where the scatter $\Delta\theta_E/\theta_E$ becomes comparable to 1 (it is much smaller at energies that are higher but still below the GZK cut-off; see Fig. 2). Our simulations showed no sensitivity to the magnetic field power spectrum characterized by n_B.

An emission timescale much larger than the experimental lifetime may be estimated if a lower cut-off in the spectrum is observable at an energy E_C, indicating that $T_S \simeq \tau_{E_C}$. The latter may, in turn, be estimated from the angular image size via Eq. (2), where the distance can be estimated from the spectrum visible above the GZK cut-off, as discussed above. An example for this scenario is shown in Fig. 3. For angular resolutions $\Delta\theta$, timescales in the range

$$3 \times 10^3 \left(\frac{\Delta\theta}{1°}\right)^2 \left(\frac{D}{10\,\text{Mpc}}\right) \text{yr} \lesssim T_S \simeq \tau_E \lesssim 10^4 \cdots 10^7 \left(\frac{E}{100\,\text{EeV}}\right)^{-2} \text{yr} \qquad (3)$$

could be probed. The lower limit follows from the requirement that it should be possible to estimate τ_E from θ_E, using Eq. (2), otherwise only an upper limit on T_S, corresponding to this same number, would apply. The upper bound in Eq. (3) comes from constraints on maximal time delays in cosmic magnetic fields, such as the Faraday rotation limit in case of the cosmological large-scale field (smaller number) and knowledge on stronger fields associated with the large-scale galaxy structure (larger number). Eq. (3) constitutes an interesting range of emission timescales for many conceivable scenarios of ultra-high energy cosmic rays. For example, the hot spots in certain powerful radio galaxies that have been suggested as ultra-high energy cosmic ray sources [16], have a size of only several kpc and could have an episodic activity on timescales of $\sim 10^6$ yr.

CONCLUSIONS

A wealth of information on both the production mechanism of highest energy cosmic rays and on the structure of large-scale magnetic fields is encoded in angle-time-energy images of discrete sources. If the clustering suggested by AGASA is real, tens (for the Pierre Auger Project) to hundreds (for the OWL Project) of events above a few 10^{19} eV can be expected from individual sources alone. With resolutions of 10-20% in energy and fractions of a degree in angle, next generation experiments should be able to exploit this information.

Special thanks go to Martin Lemoine for an ongoing extensive collaboration on this subject. I also thank Angela Olinto and David Schramm for collaboration in earlier stages. This work was supported, in part, by the DoE, NSF, and NASA at the University of Chicago.

REFERENCES

1. G. Sigl, D. N. Schramm, and P. Bhattacharjee, *Astropart. Phys.* **2**, 401 (1994).
2. J. W. Elbert, and P. Sommers, *Astrophys. J.* **441**, 151 (1995).
3. for a discussion of this case see, e.g., P. L. Biermann, H. Kang, J. P. Rachen, and D. Ryu, e-print astro-ph/9709252, to appear in Proc. Moriond Meeting on High Energy Phenomena, Jan. 1997, Les Arcs.

4. N. Hayashida et al., *Phys. Rev. Lett.* **77**, 1000 (1996).

5. G. Sigl, D. N. Schramm, S. Lee, and C. T. Hill, *Proc. Natl. Acad. Sci. USA* **94**, 10501 (1997).

6. J. W. Cronin, *Nucl. Phys. B (Proc. Suppl.)* **28B**, 213 (1992); The Pierre Auger Observatory Design Report (2nd ed.) 14 March 1997.

7. Proc. of *International Symposium on Extremely High Energy Cosmic Rays: Astrophysics and Future Observatories* (Institute for Cosmic Ray Research, Tokyo, 1996).

8. J. F. Ormes et al., in *Proc. 25th International Cosmic Ray Conference* (Durban, 1997), eds.: M. S. Potgieter et al., 5, 273; Y. Takahashi et al., in [7], p. 310.

9. G. Sigl, M. Lemoine, and A.V. Olinto, *Phys. Rev. D* **56**, 4470 (1997).

10. G. Sigl and M. Lemoine, e-print astro-ph/9711060, submitted to *Astropart. Phys.*

11. K. Greisen, *Phys. Rev. Lett.* **16**, 748 (1966); G. T. Zatsepin and V. A. Kuzmin, *Pis'ma Zh. Eksp. Teor. Fiz.* **4**, 114 (1966) [*JETP. Lett.* **4**, 78 (1966)].

12. see, e.g., G. A. Medina Tanco, E. M. de Gouveia Dal Pino, and J. E. Horvath, *Astropart. Phys.* **6**, 337 (1997); R. Lampard, R. W. Clay, and B. R. Dawson, *Astropart. Phys* **7**, 213 (1997).

13. E. Waxman and J. Miralda-Escudé, *Astrophys. J.* **472**, L89 (1996).

14. M. Lemoine, G. Sigl, A. V. Olinto, and D.N. Schramm, *Astrophys. J.* **486**, L115 (1997).

15. M. Lemoine and G. Sigl, work in progress.

16. J. P. Rachen, and P. L. Biermann, *Astron. Astrophys.* **272**, 161 (1993).

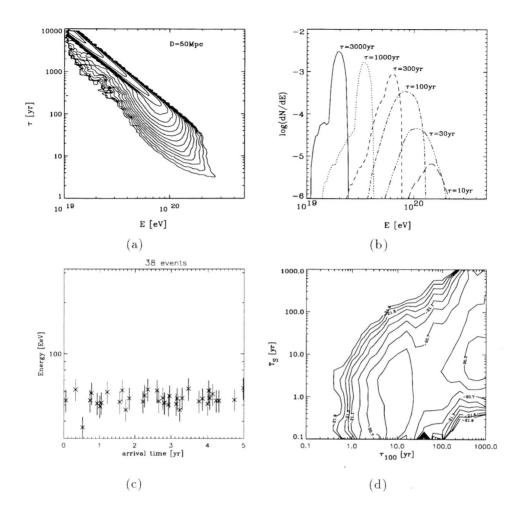

FIGURE 1. (a) An arrival time-energy histogram for $\gamma = 2.0$, $\tau_{100} = 100\,\mathrm{yr}$, $T_S \ll \tau_{100}$, $l_c \simeq 1\,\mathrm{Mpc}$, $D = 50\,\mathrm{Mpc}$, corresponding to $B_{\mathrm{rms}} \simeq 4 \times 10^{-11}\,\mathrm{G}$. Contours are in steps of a factor $10^{0.4} = 2.51$; (b) Observable energy spectrum for several positions of the observational window in the histogram in (a); (c) Example for a cluster in the arrival time-energy plane resulting from one of the cuts shown in (b); (d) The likelihood function, marginalized over N_0 and γ, for $D = 50\,\mathrm{Mpc}$, $l_c = 0.25\,\mathrm{Mpc}$, for the cluster shown in (c), in the $T_S - \tau_{100}$ plane. The contours shown go from the maximum down to about 0.01 of the maximum in steps of a factor $10^{0.2} = 1.58$. The fall-off at $\tau_{100} \gtrsim 50\,\mathrm{yr}$ and $T_S \lesssim 3\,\mathrm{yr}$ is a numerical artifact due to limited statistics. The true parameters are reasonably well reconstructed.

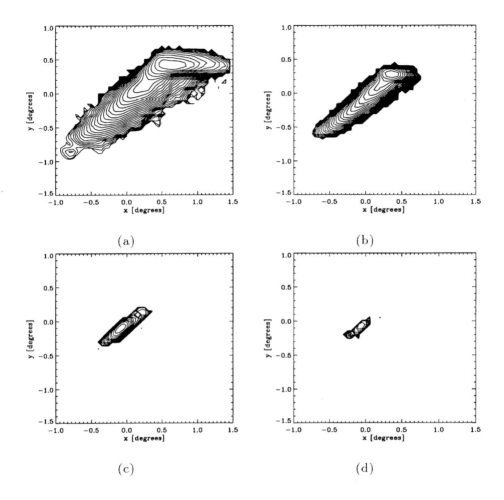

FIGURE 2. An angle-histogram for $\gamma = 2.0$, $\tau_{100} = 10^4$ yr, $T_S \gg \tau_{100}$, $l_c \simeq 1$ Mpc, $D = 50$ Mpc, corresponding to $B_{\rm rms} \simeq 4 \times 10^{-10}$ G. An angular resolution of $0.05°$ was assumed. The point $x = y = 0$ is the source position and the contours decrease in steps of 0.1 in the logarithm to base 10. (a) Image integrated over all energies $E > 30$ EeV. Two partially blended, elongated images at $x \simeq 0.2°$, $y \simeq 0.2°$ and at $x \simeq 0.7°$, $y \simeq 0.4°$ are clearly visible, the second one being more luminous by about a factor 4; (b) Same for $E > 100$ EeV. The two images are now closer to the source position; (c) Same for $E > 200$ EeV. The second image has almost disappeared; (d) Same for $E > 300$ EeV. As a consequence, $D\theta_E/l_c \simeq 1$ for $E \simeq 100$EeV. If D can be estimated from the energy spectrum, an estimate for l_c results.

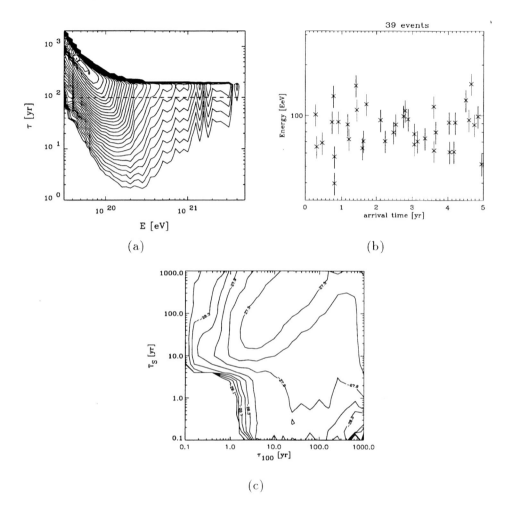

FIGURE 3. (a) An arrival time-energy histogram for $\gamma = 2.0$, $\tau_{100} = 50\,\mathrm{yr}$, $T_S = 200\,\mathrm{yr}$, $l_c \simeq 1\,\mathrm{Mpc}$, $D = 50\,\mathrm{Mpc}$, corresponding to $B_{\mathrm{rms}} \simeq 3 \times 10^{-11}\,\mathrm{G}$. Contours are in steps of a factor $10^{0.4} = 2.51$; (b) Example for a cluster in the arrival time-energy plane resulting from the cut indicated in (a) by the dashed line at $\tau \simeq 100\,\mathrm{yr}$; (c) Same as Fig. 1 (d), but for the cluster shown in (b). Note that the likelihood clearly favors $T_S \simeq \tau_{50}$. For τ_{100} large enough to be estimated from the angular image size, $T_S \gg T_{\mathrm{obs}}$ can, therefore, be estimated as well.

Relic Neutrinos, Monopoles, and Cosmic Rays above $\sim 10^{20}$ eV

Thomas J. Weiler[1]

Department of Physics & Astronomy, Vanderbilt University
Nashville TN 37235
email: weilertj@ctral1.vanderbilt.edu

Abstract. The observation of cosmic ray events above the Greisen–Kuzmin–Zatsepin (GZK) cut–off of 5×10^{19} eV offers an enormous opportunity for the discovery of new physics. We explore two possible origins for these super–GZK events. The first example uses Standard Model (SM) physics augmented only by $\stackrel{<}{\sim}$ eV neutrino masses as suggested by solar, atmospheric, and terrestrial neutrino detection, and by the comological need for a hot dark matter component. In this example, cosmic ray neutrinos from distant, highest energy sources annihilate relatively nearby on the relic neutrino background to produce "Z-bursts", highly collimated, highly boosted ($\gamma_Z \sim 10^{11}$) hadronic jets. The SM and hot Big Bang cosmology give the probability for each neutrino flavor at its resonant energy to annihilate within the halo of our galactic supercluster as likely within an order of magnitude of 1%. The kinematics are completely determined by the neutrino masses and the properties of the Z boson. The burst energy is $E_R = 4 \ (\text{eV}/m_\nu) \times 10^{21}$ eV, and the burst content includes, on average, thirty photons and 2.7 nucleons with super–GZK energies. The second example goes beyond SM physics to invoke relativistic magnetic monopoles as the cosmic ray primaries. Motivations for this hypothesis are twofold: (i) conventional primaries are problematic, while monopoles are naturally accelerated to $E \sim 10^{20}$ eV by galactic magnetic fields; (ii) the observed highest energy cosmic ray flux is just a few orders of magnitude below the Parker flux limit for monopoles. By matching the cosmic monopole production mechanism to the observed highest energy cosmic ray flux we estimate the monopole mass to be $\stackrel{<}{\sim} 10^{10}$ GeV. Several tests of the neutrino annihilation and monopole hypotheses are indicated.

THE COSMIC RAY PUZZLE ABOVE 100 EEV

The recent discoveries by the AGASA [1], Fly's Eye [2], Haverah Park [3], and Yakutsk [4] collaborations of air shower events with energies near and above 10^{20} eV present an outstanding puzzle in high energy cosmic ray physics. It was anticipated that the highest energy cosmic primaries would be protons from outside the galaxy, perhaps produced in active galactic nuclei (AGNs) [5]. It was also anticipated that

---

[1] This work is supported in part by the Department of Energy grant DE–FG05–85ER40226.

CP433, *Workshop on Observing Giant Air Showers from Space*
edited by J.F. Krizmanic et al.

the highest energies for protons arriving at earth would be $\sim 5 \times 10^{19}$ eV. The origin of this Greisen–Kuzmin–Zatsepin (GZK) cut–off [6] is degradation of the proton energy by the resonant scattering process $p + \gamma_{2.7K} \rightarrow \Delta^* \rightarrow N + \pi$ when the proton is above the resonant threshold for Δ^* production; $\gamma_{2.7K}$ denotes a photon in the 2.7K cosmic background radiation. A proton produced at its cosmic source with an initial energy E_p will on average arrive at earth with only a fraction $\sim (0.8)^{D/6\,\mathrm{Mpc}}$ of its original energy. Proton energy is not lost significantly only if the highest energy protons come from rather nearby sources, $\stackrel{<}{\sim} 50$ to 100 Mpc [7]. However, no AGN sources are known to exist within 100 Mpc of earth. Hence, the observation of air shower events above 5×10^{19} eV challenges standard theory. [2]

A primary nucleus mitigates the cut–off problem (energy per nucleon is reduced by $1/A$), but has additional problems: above $\sim 10^{19}$ eV nuclei may be photo–dissociated by the 2.7K background [9], and possibly disintegrated by the particle density ambient at the astrophysical source. Gamma–rays and neutrinos are other possible primary candidates for the highest energy events. The gamma–ray possibility appears inconsistent with the time–development of the Fly's Eye event, but is not ruled out for this event [10]. However, the mean free path for a $\sim 10^{20}$ eV photon to annihilate on the radio background to $e^+ e^-$ is believed to be only 10 to 40 Mpc [10], and the density profile of the Yakutsk event [4] showed a large number of muons which argues against gamma–ray initiation. Concerning the neutrino possibility, the Fly's Eye event occured high in the atmosphere, whereas the expected event rate for early development of a neutrino–induced air shower is down from that of an electromagnetic or hadronic interaction by six orders of magnitude [10].

New ideas are welcomed. Here we summarize two. The first one [11] uses standard model (SM) particle physics, the relic neutrino background predicted by Big Bang cosmology, and neutrino mass values suggested by oscillation data. The second one [12] postulates the existence of a small flux of relativistic magnetic monopoles. Each of these ideas offers an enromous discovery potential.

NEUTRINO–RELIC NEUTRINO ANNIHILATION AND Z–BURSTS

It was noted some time ago [13] that the mean free path $\lambda_j = [n_{\nu_j} \, \sigma_{ann}(\nu_j + \bar{\nu}_j \rightarrow Z)]^{-1}$ for a cosmic ray neutrino to annihilate at the Z resonance on a background of nonrelativistic relic antineutrinos (and vice versa) having mass m_j and density n_{ν_j} is only slightly larger than the Hubble size of the Universe, $D_H \equiv c \, H_0^{-1} = 0.9 \, h_{100}^{-1} \times 10^{28}$ cm. This means that the annihilation probability per cosmic distance of travel may be significant.

---

[2] The suggestion has been made that hot spots of radio galaxies in the supergalactic plane at distances of tens of megaparsecs may be the sources of the super–GZK primaries [8]. Statistical support for this hypothesis is weak at present.

The invariant, energy–averaged annihilation cross section for the process $\nu_j + \bar{\nu}_j \to Z$ is given by the integral over the Z pole: $<\sigma_{\mathrm{ann}}> \equiv \int \frac{ds}{M_Z^2} \sigma_{ann}(s)$, with s the square of the energy in the center of momentum frame. The standard model value for this cross section is $<\sigma_{\mathrm{ann}}> = 4\pi G_F/\sqrt{2} = 4.2 \times 10^{-32}\mathrm{cm}^2$ for each neutrino type j (flavor or mass basis), independent of any neutrino mixing–angles since the annihilation mechanism is a neutral current process. The energy of the neutrino annihilating at the peak of the Z–pole is $E_{\nu_j}^R = M_Z^2/2m_{\nu_j} = 4\,(\mathrm{eV}/m_{\nu_j}) \times 10^{21}$ eV. The energy–averaged annihilation cross section $<\sigma_{\mathrm{ann}}>$ is the effective cross section for all neutrinos within $\frac{1}{2}\delta E_R/E_R = \Gamma_Z/M_Z = 3\%$ of their peak annihilation energy. We will refer to neutrinos with energy in the range 0.97 E_R to 1.03 E_R as being at the resonant energy. (We will sometimes use E_R generically for resonant energy, as we do here, with the understanding that there are really three different resonant energies, one for each neutrino mass.) At a given resonant energy $E_{\nu_j}^R$, only relic neutrinos with the jth mass m_j may annihilate.

Each resonant neutrino annihilation produces a Z boson which immediately decays (its lifetime is 3×10^{-25} s in its rest frame). 70% of these decays are hadronic, consisting of a particle burst known to include on average fifteen neutral pions and 1.35 baryon–antibaryon pairs [14]. The fifteen π^0's decay to produce thirty high energy photons. We will refer to the end product of this Z production and hadronic decay as a "Z–burst." The nucleons and photons in the Z–burst are candidates for the primary particles with energy above the GZK–cutoff [11,15].

Let us call the distance over which a stable particle can propagate without losing more than an order of magnitude of its energy the GZK distance. For a photon it is 10 to 40 Mpc, with the exact number depending on the strengths of the diffuse radio and infrared background. For a proton it is $D_{\mathrm{GZK}} \sim -[\ln(1/10)/\ln(0.8)] \times 6$ Mpc $\sim$ 50 to 100 Mpc. If the Z–burst points in the direction of earth and occurs within the GZK distance, then one or more of the photons and nucleons in the burst may initiate a super–GZK air shower at earth. The mean multiplicity in Z decay is about 40 [14]. This dilutes the energy per hadron somewhat compared to $E_{\nu_j}^R$, but also provides a larger flux of photons (from π^0–decays) and a few baryons per burst. Shown in an attached Figure is a schematic of the Z–burst mechanism within the GZK zone.

The annihilation/Z–burst rate depends on the relic neutrino density. The mean neutrino density of the universe is predicted by hot Big Bang cosmology. The density of neutrinos with mass below an MeV (the neutrino decoupling temperature) is given by a relativistic Fermi-Dirac distribution characterized by a single temperature parameter. The distribution is that of relativisitic neutrinos, even though the neutrinos are nonrelativistic today, because the distribution is determined at the decoupling epoch. As a result of photon reheating from the era of $e^+e^- \to \gamma\gamma$ annihilation, the neutrino temperature $T_\nu \sim 1.95\mathrm{K}$ turns out to be a factor of $(4/11)^{1/3}$ less than that of the photon temperature, $T_\gamma = 2.73\mathrm{K}$. The resulting mean neutrino number density is $<n_{\nu_j}> = (3\zeta(3)/4\pi^2)T_\nu^3 = 54\,\mathrm{cm}^{-3}$ for each light flavor j with an equal number density for each antineutrino flavor. [3] Note that the predicted relic neutrino density is normalized via the temperature relation to the relic photon density which is measured. Consequently, the predicted mean density of $<n_{\nu_j}> = 54\,\mathrm{cm}^{-3}$ must be considered firm.

Probability for Annihilating to a Z–burst

In a universe where the neutrinos are nonrelativistic and uniformly distributed, the mean annihilation length for neutrinos at their resonant energy would be $\lambda = (<\sigma_{\mathrm{ann}}><n_{\nu_j}>)^{-1} = 4.4 \times 10^{29}$ cm, which is $50\,h_{100}$ times the Hubble distance. A cosmic ray neutrino arriving at earth from a cosmically distant source will have traversed approximately a Hubble distance of space, and so its annihilation probability on the relic neutrino sea is roughly $D_H/\lambda_{ann} = 2.0\,h_{100}^{-1}\,\%$ (neglecting cosmic expansion).

For a more careful derivation, we let $F_{\nu_j}(E_\nu, x)$ denote the flux of the jth neutrino flavor, as would be measured at a distance x from the source, with energy within dE of E_ν. The units of $F_{\nu_j}(E_\nu, x)$ are neutrinos/energy/area/time/solid angle. This flux may be quasi–isotropic ("diffuse"), as might arise from a sum over cosmically–distant sources such as AGNs; or it may be highly directional, perhaps pointing back to sources within our supergalactic plane. The production rate of Z's with energy within dE of E_ν, per unit length, per area, time, and solid angle, is therefore $dF_{\nu_j}(E_\nu, x)/dx = \sigma_{ann}(E_\nu)n_{\nu_j}(x)F_{\nu_j}(E_\nu, x)$. Integrating this equation over the distance D from the emission site to earth, and integrating over neutrino energy then gives the total rate of resonant annihilation, $i.e.\ Z$–burst production (in the narrow resonance approxiamtion), within the distance D:

$$\delta F_{\nu_j}(D) = E_R F_{\nu_j}(E_R, 0) \int \frac{ds}{M_Z^2}\left[1 - \exp(-\sigma_{ann}(s)\,S_j(D))\right],\ \ S(D) \equiv \int_0^D dx\,n_{\nu_j}(x).$$

$S(D)$ is the neutrino column density from earth to the distance D. If $\sigma_{ann}(s)S(D)$ is small compared to one, then

$$\delta F_{\nu_j}(D) \approx E_R <\sigma_{\mathrm{ann}}> S_j(D)F_{\nu_j}(E_R, 0).$$

For $S(D) \ll 1/\sigma_{ann}(s)$, the rate for Z–burst generation depends linearly on the relic neutrino column density. Writing $S(D)$ as $<n_{\nu_j}> \times D$, we make contact with our previous simple estimate.

For each 50 Mpc of travel through the mean neutrino density, the probability for a neutrino with resonant energy to annihilate to a Z–boson is 3.6×10^{-4}. Since the branching fraction for a Z to decay to hadrons is 70%, one part in 4000 of the resonant neutrino flux will be converted into a Z–burst containing ultrahigh energy photons and nucleons within the 50 Mpc GZK–distance of earth in this unclustered

---

[3] We do not consider here the possibility of unequal numbers of relic neutrinos and antineutrinos. For unpolarized Majorana neutrinos, the two numbers must be equal.

universe. We live in a matter–rich portion of the universe. Consequently, it is expected that our local relic neutrino density is somewhat enhanced compared to the universal average, due the local potential well. Therefore, the value of 0.025% for the probability of a resonant neutrino creating a Z–burst within the GZK zone, obtained without neutrino clustering, is the <u>absolute</u> <u>minimum</u> annihilation probability in a Big Bang universe.

The annihilation probability within the GZK volume is enhanced due to neutrino trapping by the factor $\xi \equiv < n_\nu(< D_{\text{GZK}}) > / < n_\nu >$, where $< n_\nu >$ is the average value for the neutrino density throughout the entire Universe, and $< n_\nu(< D_{\text{GZK}}) >$ is the average value within the GZK zone. How large is the local relic neutrino enhancement expected to be? The mean baryon density of our Galaxy (within a sphere of radius ~ 15 kpc) compared to the baryon density averaged over the visible universe is enhanced by about 10^6. However, one expects the relic neutrino density to be much less enhanced than the baryon density for several reasons. First of all, neutrinos do not dissipate energy as easily as electrically charged baryons and elctrons. Secondly, since gravity couples universally to energy and is blind to particle quantum numbers, it is unlikely that the mass fraction of neutrinos in any halo is larger than the universal mass fraction, $f_\nu \sim \Omega_\nu = \sum_j m_{\nu_j}/(92h_{100}^2\,\text{eV})$, which is of the order of a per cent for $m_\nu \sim 1$ eV. Thirdly, Pauli blocking presents a significant barrier to clustering of light–mass fermions on the scale of our Galactic halo. As a crude estimate of Pauli blocking, one may use the zero temperature Fermi gas as a model of the gravitationally bound halo neutrinos. Requiring that the Fermi momentum of the neutrinos not exceed the virial velocity σ of the Galaxy, one gets $\xi = n_{\nu_j}/54\,\text{cm}^{-3} \lesssim 10^3(m_{\nu_j}/\text{eV})^3(\sigma/220\text{kms}^{-1})^3$. Finally, neutrinos have a much larger Jeans ("free–streaming") length than do baryons at the crucial time when galaxies start to grow nonlinearly. This free–streaming length of $\lambda_{FS} \sim 100\,(10\text{eV}/m_\nu)$ Mpc is the minimum size of an overdensity which can gravitationally contain the collisionless neutrinos, and should roughly correspond to the expected size of a neutrino halo today. Within the uncertainties of the estimate, this length is the size of our local supergalactic cluster. Taking 20 Mpc as the supercluster size, and 100 Mpc as the characteristic distance between superclusters, one gets a density enhancement of $\sim 5^3 \sim 100$. If neutrinos manage to cluster on the smaller galactic or galactic cluster scales, then their density may be larger by another factor of 10 or so, but probably not more, on these smaller scales. We multiply the probability above for converting a resonant neutrino into a hadronic Z–burst within 50 Mpc of earth, obtained for a smooth relic backgroung, by the density enhancement factor ~ 100 to get our best conversion estimate of about 2%.

This is our main result, which we repeat for emphasis: *the probability for neutrinos at their resonant energy to annihilate within the halo of our Super Galactic Cluster is likely within an order of magnitude of 1%, with the exact value depending on unknown aspects of relic neutrino clustering.*

Note that two crucial elements are required for this mechanism to produce super–GZK air showers. They are the existence of a neutrino flux at $\gtrsim 10^{21}$ eV, and the existence of a neutrino mass in the 0.1 to 10 eV range. Concerning the flux, it is not unlikely that whatever mechanism produces the most energetic hadrons also produces charged pions of comparable energy. Thus, one may expect neutrino production at ultrahigh energy, coming from pion decay and subsequent muon decay [16]. The opacity difference between neutrino and proton in dense sources such as AGNs makes credible the possibility of a neutrino flux considerably above the proton flux at highest energies. [4] Concerning possible neutrino masses, the simplest explanation for the anomalous atmospheric–neutrino flavor–ratio [18] is neutrino oscillations driven by a mass–squared difference of $\sim 10^{-3}$ to $10^{-2}\mathrm{eV}^2$ [19], which implies a mass of at least 0.03 eV. Also, the recent LSND measurement is claimed [20] to indicate a neutrino mass of 1 to 2 eV. Furthermore, according to Big Bang cosmology, the fraction of closure density provided by possible neutrino masses is $\Omega_\nu = \sum_j m_{\nu_j}/(92\,h_{100}^2\,\mathrm{eV})$, with $0.5 \leq h_{100} \equiv H_0/(100\mathrm{km/s/Mpc}) \leq 1$. One sees that $\sim$ eV neutrino masses are consistent, and even required if neutrino hot dark matter is to contribute in any significant way to the evolution of large–scale structures.

Before considering specific signatures of the annihilation mechanism for generating 100 EeV particles, we briefly mention two subtleties:

As their momenta red–shifted in the expanding universe, the relic neutrinos evolved to the unpolarized nonrelativistic state which they occupy today. As a result, if the neutrino is a Dirac particle, then the sterile right–handed neutrino and the sterile left–handed antineutrino fields are populated equally with the two active fields. Therefore, for Dirac neutrinos the active densities available for annihilation with the incident high energy neutrino are half of the total densities, and the Z–burst production probability is half of what we quote in this article. For Majorana neutrinos, there are no sterile fields and the total densities are active in annihilation. Majorana neutrinos are favored over Dirac neutrinos in currently popular theoretical models with nonzero neutrino mass [21].

It is possible, in fact it is probable, that massive neutrinos exhibit mixing in analogy to the quark sector of the standard model of particle physics. In the mixed case, the flavor states are unitary mixtures of the mass states. Letting $\alpha = e, \mu, \tau$ label flavor and $j = 1, 2, 3$ label mass, one writes $|\nu_\alpha> = \sum_j U_{\alpha j} |\nu_j >$. Then each neutrino flavor at the resonant energy of a given mass state has a nonzero probability to annihilate, but with an extra probability factor of $|U_{\alpha j}|^2$. For example, the ν_μ's and ν_e's from pion and mu decay will annihilate at the resonant energy of m_2 with the probability factors $|U_{\mu 2}|^2$ and $|U_{e2}|^2$, respectively, times what we calculate below. If there is mixing, the factors $|U_{\alpha j}|^2$ can be easily multiplied in.

We note that the phenomenon of neutrino oscillations is irrelevant in our context of annihilation, because the $\nu - \bar{\nu}$ annihilation process requires just a single transformation from flavor to mass basis, so the phase differences induced between

---

[4] There is also the possibility that the highest energy neutrinos originate in quark jets produced when some supermassive relic particles decay, in which case the neutrino flux greatly exceeds the proton flux [17].

mass states are not observed.

The p, n, γ Flux Above 100 EeV from Z–bursts

The decay products of the Z are well–known from the millions of Z's produced at LEP and at the SLC. [14]. The respective branching fractions for Z–decay into hadrons, neutrino–antineutrino pairs, and charged lepton pairs are 70%, 20%, and 10%. The mean multiplicity $< N >$ in hadronic Z decay is about 40 particles, of which, on average, 17 are charged pions, 9 are neutral pions, and importantly, 1.35 are baryon–antibaryon pairs which become 2.7 nucleons and antinucleons. Unstable hadrons decay to produce more pions. We estiamte that the total pion and nucleon count for the Z–burst is then, 15 π^0's, 28 $\pi^\pm$'s, and 2.7 nucleons (we now mean "nucleons" to include the antinucleons as well). What is of major interest are the 15 π^0's and the 2.7 nucleons. The 2.7 nucleons may be protons or neutrons. Although the lifetime of a neutron is only ten minutes in its rest frame, in the lab frame its lifetime is enhanced by the boost factor $\gamma_N \sim E_R/ < N > m_N \sim (\text{eV}/m_\nu) \times 10^{11}$. The neutron will typically travel $c\tau_n \gamma_N \sim (\text{eV}/m_\nu)$ Mpc in free space before decaying. Each time the neutron scatters on the 2.7K photon background (mean free path of 6 Mpc), it's clock for decay is reset to zero, so energetic neutrons from great distances may still reach the earth. (In fact, a nucleon above the GZK cut–off energy will on average spend half of its transit time as a neutron, and half as a proton.)

Among the 15 π^0's, 9 are produced directly in Z–decay, while the other 6 arise from decays of various hadronic resonances. A comparison of the data [14] for the momentum spectra for direct pions and for protons produced in Z–decay reveals that the boosted mean energy per proton is larger than that of a direct pion by a factor ~ 3.5. We expect the energy of the six π^0's produced through resonance decays to be softer yet, by another factor of ~ 3. Weighting the direct and secondary pions appropriately then, we arrive at a factor of about six for the softness of the mean pion energy compared to the mean nucleon energy. Since the photon on average carries half of the parent pion energy, the mean energy of a photon in a Z–burst is expected to be less than that of a nucleon by an order of magnitude.

The 30 photons and the 2.7 nucleons are the candidate primary particles for inducing super–GZK air showers in the earth's atmosphere. The a priori photon to nucleon ratio is about 30. However, the hardness of the nucleon spectrum compared to the photon spectrum mitigates this ratio if a selection is made for the very highest energy particles. Of course, these average values for the multiplicities and energies must be used with some caution, since fluctuations in multiplicity and particle-types per event, and in energy per individual particle, are large.

Further Signatures from Z–bursts

The particle spectrum in Z–decay and the Lorentz factor of the Z, $\gamma_Z = E_R/M_Z = M_Z/2m_\nu = 0.9\,(m_\nu/\text{eV})^{-1} \times 10^{11}$, determine the possible signatures of Z–bursts. We comment on some of the possible Z–burst observables, beyond the photon– and nucleon–initiated super–GZK air showers:

(i) The Z–decay products which in the Z rest frame lie within the forward hemisphere are boosted into a highly–collimated lab–frame cone of half–angle $1/\gamma_Z = 2(m_{\nu_j}/\text{eV}) \times 10^{-11}$ radians. Z–bursts originating within $20(\text{eV}/m_{\nu_j})$ parsecs of earth, if directed toward the earth, arrive with a transverse spatial spread of less than one earth diameter. It is possible for the decay products of a single Z–burst to initiate multiple air showers. A large area surface array (e.g. the Auger project) or an orbiting all–earth observing satellite (e.g. the OWL or AIRWATCH proposals) [22] could search for these nearly coincident showers.

(ii) The mean number of baryon–antibaryon pairs per hadronic Z–decay is 1.35. Baryon number conservation requires each hadronic Z decay to contain an integer number (possibly zero) of baryon–antibaryon pairs. If the number of baryon pairs per hadronic shower is governed by Poisson statistics, then the probabilities for 0, 1, 2, 3, 4, and 5 pairs are 26%, 35%, 24%, 10%, 4%, and 1%, respectively.

(iii) The energy of the Z–bursts are fixed at $4\,(\text{eV}/m_{\nu_j}) \times 10^{21}$ eV by the neutrino mass(es). The energy of individual particles produced in the burst can approach this value but cannot exceed it. This may serve to distinguish the Z–burst hypothesis from some recent speculations for super–GZK events based on SUSY or GUT–scale physics [17], in which cut–off energies are expected to be much higher.

(iv) From the highest super–GZK event energy E^{max}, one can deduce an upper bound on the neutrino mass of $m_\nu < M_Z^2/2E^{\text{max}} = 4\,(10^{21}\text{eV}/E^{\text{max}})$ eV. Similarly, from the mean energy $<E>$ of super–GZK events one can estimate the mass of the participating neutrino flavor via $m_\nu = M_Z^2/2E_R \sim M_Z^2/(2<N><E>) \sim 0.5\,(10^{20}\text{eV}/<E>)$ eV; if there is a selection bias toward events at higher energy, then this formula gives a lower bound on the neutrino mass.

(v) The most significant time–scale for the Z–burst is the hadronization time in the lab frame, which is $\gamma_Z \times \text{fm}/c \sim 10^{-12}$ s. However, some hard photons will arrive "late." These are the photons arising from pions produced in kaon decays (with rest frame lifetimes of 0.9×10^{-10} s, 1.2×10^{-8} s, and 5×10^{-8} s, respectively, for the K_S, $K^\pm$, and K_L.) These particles contribute 3 π^0's per Z–burst. There are also 0.25 π^0's from Λ decay and 0.08 π^0's from $\Sigma^\pm$ decay. Λ and $\Sigma^\pm$ lifetimes in the rest frame are $\mathcal{O}(10^{-10})$ s. In the boosted lab frame, the kaon and strange baryon lifetimes will be $\mathcal{O}(20 \text{ to } 10^4)$ s. We therefore expect ~ 7 hard photons per event to straggle by this amount of time. Long time–scales are also available from the decays in flight of the charged pions and muons, and from the inverse Compton and synchrotron processes of the $e^\pm$'s. (Recall that there are about 28 $\pi^\pm$'s per Z–burst, which cascade through 28 $\mu^\pm$'s to 28 $e^\pm$'s.) It is not clear to us whether these processes can generate an observable photon yield.

(vi) There could be a "neutrino pile–up" at two to three decades of energy below

E_R. These pile–up neutrinos are the result of the hadronic decay chain which includes $Z \rightarrow \sim 28\,(\pi^{\pm} \rightarrow \nu_{\mu} + \mu^{\pm} \rightarrow \nu_{\mu} + \bar{\nu}_{\mu} + \nu_e/\bar{\nu}_e + e^{\pm})$; *i.e.* each of the 70% of the resonant neutrino interactions which yield hadrons produces about 85 neutrinos with mean energy $\sim E_{\pi}/4 \sim E_R/160$. These neutrinos are in addition to the neutrinos piling up from the decay of pions photo–produced by any super–GZK nucleons scattering on the 2.7K background. [23]

(vii) A very interesting issue is to what extent the boosted Z–decay products will contain a copious amount of observable gamma–rays produced by internal brehmsstrahlung during the decay process. It seems possible that Z–bursts may generate short duration gamma–ray bursts observable at earth. Even extreme infrared radiation becomes observable after boosting by $\gamma_Z \sim 10^{11}$. A 10^{-5} radio photon becomes an MeV gamma–ray, and a 10^{-2} eV infrared photon becomes a hard GeV gamma–ray. The glib statement that "the 1/E brehmsstrahlung singularity produces photons with such low energy that they cannot be observed" may not hold for Z–bursts. As we have stated, the hadronization time when boosted to the lab frame, $\sim 10^{-12}$ s, sets the basic time–scale for particle emission and brehmsstrahlung.

There are a few more observations that should be made concerning the hypothesis under discussion:

(i) If the highest energy neutrino cosmic ray flux points back to discrete sources of origin, then the super–GZK event arrival directions should point back to these same sources. However, if the highest energy neutrino flux is diffuse, then the super–GZK event directions should correlate with the spatial distribution of the relic neutrino density. Perhaps the angular distribution of super–GZK events can be used to perform halo tomography of our supergalactic cluster.

(ii) If the super–GZK events are due to neutrino annihilation on relics as hypothesized here, and if the high energy neutrino flux is eventually measured, then an estimate of the relic neutrino column density out to $D_{\mathrm{GZK}} \sim 50$ Mpc may be made. Let $\mathcal{L}(> E_{\mathrm{GZK}})$ be the luminosity of super–GZK events (in units of events/area/time/solid angle). Then the column density of the annihilating neutrino flavor out to D_{GZK} is $S^{\nu_j}_{\mathrm{GZK}} \sim \mathcal{L}(> E_{\mathrm{GZK}})/ < \sigma_{\mathrm{ann}} > F_{\nu_j}(E_R)\,\delta E_R = 4.5 \times 10^{32}[\mathcal{L}(> E_{\mathrm{GZK}})/E_R F_{\nu_j}(E_R)]\,\mathrm{cm}^{-2}$. If $F_{\nu_j}(E)$ is measured below the resonant energy, an estimate of the neutrino column density can still be made by extrapolating the flux to E_R. For example, if a power law is assumed with a spectral index α, then $F_{\nu_j}(E_R) = (E/E_R)^{\alpha}\,F_{\nu_j}(E)$.

MAGNETIC MONOPOLES AS THE HIGHEST ENERGY PRIMARY PARTICLES

The idea [24] that the primary particles of the highest energy cosmic rays may be magnetic monopoles has been resuscitated recently [12]. Two "coincidences" in the data support this hypothesis. The first is that the energies above the cut-off are naturally attained by monopoles when accelerated by known cosmic magnetic fields. The second is that the observed cosmic ray flux above the cut–off is not far below the theoretically allowed "Parker limit" monopole flux.

To impart its kinetic energy to the induced air–shower, the monopole must be relativistic. Let us define the monopole's inelasticity factor η as the fraction of monopole energy transferred to the air shower in the mean. Then a relativistic monopole primary must satisfy $M \ll E_M = \eta^{-1} 10^{20}$ eV. The Kibble mechanism [25] for monopole generation in an early–universe phase transition establishes a monotonic relationship between the monopoles' flux and mass. We will see that there results from this a second upper bound on the monopole mass, which turns out to be very similar the first. The consistency of these two bounds is a third "coincidence." Thus, we arrive at a flux of monopoles of mass $M \stackrel{<}{\sim} 10^{10}$ GeV as a viable explanation of the highest energy cosmic ray data.

Monopole Kinetic Energies

The kinetic energy of cosmic monopoles is easily obtained. As pointed out by Dirac, the minimum charge for a monopole is $q_M = e/2\alpha$ (which implies $\alpha_M = 1/4\alpha$). In the local interstellar medium, the magnetic field B is approximately 3×10^{-6} gauss ($\equiv B_{-6}$) with a coherence length $L \sim 300$ pc ($\equiv L_{300}$). Thus, a galactic monopole will typically have kinetic energy:

$$E_K \sim q_M B L \sqrt{N}$$
$$\simeq 6 \times 10^{20} (\frac{B}{B_{-6}}) \, (\frac{L}{L_{300}})^{1/2} \, (\frac{R_M}{R_{30}})^{1/2} \text{ eV},$$

where $N \sim R_M/L \sim 100 \, (R_M/R_{30})/(L/L_{300})$ is the number of magnetic domains encountered by a typical monopole as it traverses the galactic magnetic field region of size $R_M \equiv 30 \, R_{30}$ kpc. Note that this energy is above the GZK cut–off. Thus, the "acceleration problem" for $E \stackrel{>}{\sim} 10^{20}$ eV primaries is naturally solved in the monopole hypothesis.

Monopole Flux

To obtain the theoretically predicted monopole flux, it is worthwhile to review how and when a monopole is generated in a phase transition [25,26]. The topological requirement for monopole production is that a semisimple gauge group changes so that a $U(1)$ factor becomes unbroken. If the mass or temperature scale at which the symmetry changes is Λ, then the monopoles appear as topological defects, with mass $M \sim \alpha^{-1}\Lambda$. We use $M \sim 100 \, \Lambda$ in the estimates to follow. All that is necessary to ensure that the monopoles are relativistic today, i.e. $M \stackrel{<}{\sim} 10^{10}$ GeV, and so produce relativistic air showers, is to require this symmetry breaking scale associated with the production of monopoles be at or below $\sim 10^8$ GeV.

At the time of the phase transition, roughly one monopole or antimonopole is produced per correlated volume [25]. The resulting monopole number density today is

$$n_M \sim 0.1\,(\Lambda/10^{17}\mathrm{GeV})^3(l_H/\xi_c)^3\mathrm{cm}^{-3}, \tag{1}$$

where ξ_c is the phase transition correlation length, bounded from above by the horizon size l_H at the time of the phase transition, or equivalently, at the Ginsburg temperature T_G of the phase transition. The correlation length may be comparable to the horizon size (second order or weakly first order phase transition) or considerably smaller than the horizon size (strongly first order transition). From Eq. (1), the general expression for the relativistic monopole flux may be written

$$F_M = c\,n_M/4\pi \sim 0.2\,(M/10^{16}\mathrm{GeV})^3(l_H/\xi_c)^3 \tag{2}$$

per cm^2·sec·sr.

The "Parker limit" on the galactic monopole flux [27] is $F_M^{PL} \leq 10^{-15}/\mathrm{cm}^2/\mathrm{sec}/\mathrm{sr}$. It is derived by requiring that the measured galactic magnetic fields not be depleted (by accelerating monopoles) faster than the fields can be regenerated by galactic magnetohydrodynamics. Comparing this Parker limit with the general monopole flux in Eq. (2), we see that the Parker bound is satisfied if $M \stackrel{<}{\sim} 10^{11}(\xi_c/l_H)$ GeV.

The $M \stackrel{<}{\sim} 10^{10}$ GeV restriction also serves to ameliorate possible overclosure of the universe by an excessive monopole mass density. The resulting monopole mass density today relative to the closure value is

$$\Omega_M \sim 0.1\,(M/10^{13}\mathrm{GeV})^4(l_H/\xi_c)^3. \tag{3}$$

Monopoles less massive than $\sim 10^{13}(\xi_c/l_H)^{3/4}$ GeV do not overclose the universe. From Eqs. (3) and (2) we may also write for the *relativistic* monopole closure density

$$\Omega_{RM} \sim 10^{-8}(\langle E_M \rangle/10^{20}\mathrm{eV})(F_M/F_M^{PL}),$$

which shows that the hypothesized monopole flux does not close the universe regardless of the nature of the monopole–creating phase transition (parameterized by ξ_c/l_H) as long as the monopole flux is not orders of magnitude in violation of the Parker limit.

There is no obvious reason why monopoles accelerated by cosmic magnetic fields should have a falling spectrum, or even a broad spectrum. So we assume that the monopole spectrum is peaked in the energy half–decade 1 to 5×10^{20} eV. With this assumption, the monopole differential flux obtained from Eq. (2) is

$$\frac{dF_M}{dE} \sim 4 \times 10^{-40}(\frac{M}{10^{10}\mathrm{GeV}})^3(\frac{l_H}{\xi_c})^3$$

per cm^2·sec·sr·eV. Comparing this monopole flux to the measured differential flux $(dF/dE)_{Exp} \sim 10^{-38\pm2}$ per cm^2·sec·sr·eV above 10^{20} eV (summarized in [10]), we

256

infer $M \sim (\xi_c/l_H) \times 10^{10\pm1}$ GeV. We note that the monopole mass derived here from the flux requirement is remarkably consistent with the prior mass requirement arising from the requirement that the $E \sim 10^{20}$ eV monopoles be relativistic. We also note that the observed highest energy cosmic ray flux lies just three to four orders of magnitude below the Parker limit for monopole flux. A much larger observed flux would have exceeded the Parker limit, while a slightly lower flux would not have been observed. If the monopole hypothesis is correct, it is possible that we are seeing evidence for some dynamical reason forcing the monopole flux to roughly saturate the Parker bound.

Monopole–induced Air Showers

Any proposed primary candidate must be able to reproduce the observed shower evolution of the 3×10^{20} eV Fly's Eye event, which is marginally consistent with that expected in a proton–initiated shower. Does a monopole–induced air shower fit the Fly's Eye event profile? We do not know. The hadronic component of the monopole shower is likely to be complicated. The interior of the monopole is symmetric vacuum, in which all the fermion, Yang–Mills, and Higgs fields of the grand unified theory coexist. Thus, even though the Compton size of the monopole is incredibly tiny, its strong interaction size is the usual confinement radius of $\sim$ 1 fm, and its strong interaction cross–section is indeed strong, $\sim 10^{-26}$cm^2, and possibly growing with energy like other hadronic cross–sections. Furthermore, a number of unusual monopole–nucleus interactions may take place, including enhanced monopole–catalyzed baryon–violating processes with a strong cross–section $\sim 10^{-27}$cm^2 [28]; catalyzation of the inverse process $e^- + M \rightarrow M + \pi + (\bar{p}$ or $\bar{n})$, followed by pion/antibaryon initiation of a hadronic shower; binding of one or more nucleons by the monopole [29], in which case the monopole–air interaction may resemble a a relativistic nucleus–nucleus collision; strong polarization of the air nuclei due to magnetic interaction with the individual nucleon magnetic moments and electric $(\vec{E} = \gamma_M e/2\alpha r^2 \hat{\phi})$ interaction with the proton constituents, possibly causing fragmentation [29]; hard elastic magnetic scattering of ionized nuclei (in the rest frame of the monopole the charged nucleus will see the monopole as a reflecting magnetic mirror); and possible electroweak–scale sphaleron processes [30] at the large Q–value of the monopole–air nucleus interaction $(\sim \gamma_M Am_N \sim$ TeV). It has been pointed out that a strong cross–section is not enough; the energy transfer (inelasticity) must also be large, which may be difficult to achieve for a massive particle like the monopole [31]. Clearly, more theoretical work is required to understand a monopole's air shower development.

On the other hand, the monopole's electromagnetic showering properties are straightforward. A magnetic monopole has a rest–frame magnetic field $B_{RF} = q_M\hat{r}/r^2$. When boosted to a velocity $\vec{\beta}_M$, an electric field $\vec{E}_M = \gamma_M \vec{\beta}_M \times \vec{B}_{RF}$ is generated, leading to a "dual Lorentz" force acting on the charged constituents of air atoms. The electromagnetic energy loss of a relativistic monopole traveling through matter is very similar to that of a heavy nucleus with similar γ–factor and charge $Z = q_M/e = 1/2\alpha = 137/2$. One result is a $\sim 6\,\mathrm{GeV}/(\mathrm{g\,cm^{-2}})$ "minimum–ionizing monopole" electromagnetic energy loss. Integrated through the atmosphere, the total electromagnetic energy loss is therefore $\sim (6.2/\cos\theta_z)$ TeV, for zenith angle $\theta_z \stackrel{<}{\sim} 60°$. For a horizontal shower the integrated energy loss is ~ 240 TeV. A second electromagnetic prediction is Cerenkov radiation at the usual angle but enhanced by $(137/2)^2 \sim 4700$ compared to a proton primary. This enhanced Cerenkov radiation may help in the identification of the monopole primary.

The smaller energy transfer per collision for a massive monopole as compared to that of the usual primary candidates may constitute a signature for heavy monopole primaries. Moreover, the back–scattered atmospheric particles in the center–of–mass system (which is roughly half of the scattered particles) are forward–scattered in the lab frame into a cone of half–angle $1/\gamma_M$; at the given energy of $E \sim 10^{20}$ eV, this angle will be large for a heavy monopole primary compared to the angle for a usual primary particle, possibly offering another monopole signature. A further signature may be that monopole primaries will arrive anisotropically distributed on the sky, showing a preference for the direction of the local galactic magnetic field.

CONCLUSIONS FOR NEUTRINO–ANNIHILATION AND MONOPOLE HYPOTHESES

In summary, if one or more neutrino mass is within about an order of magnitude of an eV, and if there is a sufficient flux of cosmic ray neutrinos at $\stackrel{>}{\sim} 10^{21}$ eV, then $\nu_{\mathrm{cr}} + \bar{\nu}_{\mathrm{relic}}$ (or vice versa) $\rightarrow Z \rightarrow hadrons \rightarrow nucleons\ and\ photons$ within the GZK volume ($\sim (50\mathrm{Mpc})^3$ of earth may be the origin of air shower events observed above the GZK cut–off. Possible signatures to validate or invalidate this hypothesis are abundant. If the hypothesis is correct, then air shower observations may show the existence of the relic neutrino gas liberated from the primordial early–universe plasma when the universe was only one second old!

On the other hand, the primary particles initiating the highest energy cosmic ray events may be relativistic magnetic monopoles of mass $M \stackrel{>}{\sim} 10^{10}$ GeV. Energies of $\sim 10^{20}$ eV can easily be attained via acceleration in a typical galactic magnetic field, and the observed highest energy cosmic ray flux (not far below the Parker limit) can be explained by the Kibble mechanism of early–universe monopole production. Again there are an abundant number of identifying signatures available.

There are good prospects for more cosmic ray data at the highest energies. Present cosmic ray detection efforts are ongoing, and the "Hi–Res", "Telescope Array", "Auger", "OWL", and "Airwatch" projects has been formed to coordinate international efforts to collect air–shower data from ever–larger parts of our sky.

REFERENCES

1. S.Yoshida, et al., (AGASA Collab.) *Astropart. Phys.* **3**, 105 (1995); N. Hayashida et al., *Phys. Rev. Lett.* **73**, 3491 (1994).

2. D. J. Bird et al., (Fly's Eye Collab.) *Phys. Rev. Lett.* **71**, 3401 (1993); *Astrophys. J.* **424**, 491 (1994); *ibid.* **441**, 144 (1995).

3. G. Brooke et al. (Haverah Park Collab.), Proc. 19th Intl. Cosmic Ray Conf. (La Jolla) **2**, 150 (1985); reported in M. A. Lawrence, R. J. O. Reid, and A. A. Watson (Haverah Park Collab.), *J. Phys.* **G 17**, 733 (1991).

4. N. N. Efimov et al., (Yakutsk Collab.) ICRR Symposium on Astrophysical Aspects of the Most Energetic Cosmic Rays, ed. N. Nagano and F. Takahara, World Scientific pub. (1991); and Proc. 22nd ICRC, Dublin (1991).

5. P. L. Biermann and P. A. Strittmatter, *Astrophys. J.* **322**, 643 (1987); J. P. Rachen and P. L. Biermann, *Astron. & Astrophys.* **272**, 161 (1993); G. Sigl, D. N. Schramm, and P. Bhattacharjee, *Astropart. Phys.* **2**, 401 (1994); an excellent overview of models proposed to generate ultrahigh energy primaries is given by P. L. Biermann, *J. Phys. G:* **23**, 1 (1997).

6. K. Greisen, *Phys. Rev. Lett.* **16**, 748 (1966); G. T. Zatsepin and V. A. Kuzmin, *Pisma Zh. Eksp. Teor. Fiz.* **4**, 114 (1966); F. W. Stecker, *Phys. Rev. Lett.* **21**, 1016 (1968); J. L. Puget, F. W. Stecker and J. H. Bredekamp, *Astrophys. J.*, **205**, 638 (1976); V. S. Berezinsky and S. I. Grigoreva, *Astron. & Astrophys.*, **199**, 1 (1988).

7. S. Yoshida and M. Teshima, *Prog. Theor. Phys.* **89**, 833 (1993); F. A. Aharonian and J. W. Cronin, *Phys. Rev.* **D50**, 1892 (1994); G. Sigl, D. N. Schramm, and P. Bhattacharjee, *Astropart. Phys.* **2**, 401 (1994); J. W. Elbert and P. Sommers, *Astrophys. J.* **441**, 151 (1995);

8. T. Stanev, P. Biermann, J. Lloyd-Evans, J. Rachen and A. Watson, *Phys. Rev. Lett.* **75**, 3056 (1995).

9. F. W. Stecker, *Phys. Rev.* **180**, 1264 (1969); astro–ph/9710353.

10. F. Halzen, R. A. Vazquez, T. Stanev, and V. P. Vankov, *Astropart. Phys.*, **3**, 151 (1995).

11. T. J. Weiler, hep–ph/9710431.

12. T. W. Kephart and T. J. Weiler, *Astropart. Phys.* **4**, 271 (1996); T. J. Weiler and T. W. Kephart, *Nucl. Phys. Proc. Suppl.* **B51**, 218 (1996);

13. T. J. Weiler, *Phys. Rev. Lett.* **49**, 234 (1982); *Astrophys. J.* **285**, 495 (1984); and in *High Energy Neutrino Astrophysics*, ed. V. J. Stenger, J. G. Learned, S. Pakvasa, and X. Tata, Honolulu HI, 1992, pub. World Scientific.

14. Particle Data Group, *Phys. Rev.* **D54**, pp. 187–8 (1996).

15. E. Roulet, *Phys. Rev.* **D47**, 5247 (1993); P. Gondolo, G. Gelmini, and S. Sarkar, *Nucl. Phys.* **B393**, 111 (1993); S. Yoshida, H. Dai, C. Jui, and P. Sommers, *Astrophys. J.* **479**, 547 (1997); D. Fargion, B. Mele and A. Salis, astro-ph/9710029.

16. K. Mannheim and P. L. Biermann, *Astron. & Astrophys.* **221**, 211 (1989); R. J. Protheroe and A. P. Szabo, *Phys. Rev. Lett.* **69**, 2885 (1992); *Astropart. Phys.* **2**, 375 (1994); K. Mannheim, *Astropart. Phys.* **3**, 295 (1995); F. W. Stecker and M. H. Salamon, *Space Sci. Rev.* **75**, 341 (1996); R. J. Protheroe, in *Towards the Millennium in Astrophysics: Problems and Prospects*, Erice 1996, eds. M.M. Shapiro and J.P.

Wefel, pub. World Scientific (astro-ph/9612213), and in *Accretion Phenomena and Related Outflows*, ed. D. Wickramashinghe et al., 1996 (astro-ph/9607165); F. Halzen and E. Zas, astro–ph/9702193.

17. P. Bhattacharjee, C. T. Hill, and D. N. Schramm,' *Phys. Rev. Lett.* **69**, 567 (1992); P. Bhattacharjee and G. Sigl, *Phys. Rev.* **D51**, 407 (1995); G. Sigl, S. Lee, D. N. Schramm, and P. Coppi, *Phys. Lett.* **B392**, 129 (1997); V. Berezinsky, X. Martin and A. Vilenkin, astro-ph/9703077; V.Berczinsky and Λ. Vilcnkin, astro ph/9704257; V. Berezinsky, M. Kachelriess, and A.Vilenkin, *Phys. Rev. lett.* **79**, 4302 (1997) and astro-ph/9708217; V. Berezinsky and M. Kachelriess, hep-ph/9709485; P. H. Frampton, B. Keszthelyi, Y. J. Ng, astro-ph/9709080; V. A. Kuzmin and V. A. Rubakov, astro-ph/9709187; P. Bhattachatjee, Q. Shafi, and F. Stecker, hep–ph/9710533.

18. K. S. Hirata et al. (Kamiokande Collaboration), *Phys. Lett.* **B205**, 416 (1988); ibid, **B280**, 146 (1992); R. Becker–Szendy et al. (IMB Collaboration), *Phys. Rev.* **D46**, 3720 (1992); W. W. M. Allison et al. (Soudan Collaboration), *Phys. Lett.* **B391**, 491 (1997); Y. Totsuka (SuperKamiokande Collaboration), Plenary Talk at the International Conference on Lepton–Photon Interactions, July 28–Aug.1, 1997, Hamburg (to appear in Proc.).

19. J. G. Learned, S. Pakvasa, and T. J. Weiler, *Phys. Lett.* **B207**, 79 (1988); V. Barger and K. Whisnant, *Phys. Lett.* **B209**, 365 (1988); K. Hidaka, M. Honda, and S. Midorikawa, *Phys. Rev. Lett.* **61**, 1537 (1988).

20. C. Athanassopoulos et al. (LSND Collaboration), *Phys. Rev. Lett.* **77**, 3082 (1996); *Phys. Rev.* **C54**, 2685 (1997); nucl–ex/9709006 (1997);

21. Particle physics possibilities for neutrino mass, and the phenomenological implications of nonzero mass for terrestrial experiments and for cosmology are reviewed in G. Gelmini and E. Roulet, *Rept. Prog. Phys.* **58**, 1207 (1995).

22. Presentations by the "Telescope Array", "Hi–Res", "Auger", "OWL" and "AIR-WATCH" are available in these proceedings. In addition, the Auger Project has a home page at
http://www–td–auger.fanl.gov:82/; and the "Orbiting Wide–angle Light–collectors" collaboration has a homepage at
http://lheawww.gsfc.nasa.gov/docs/gamcosray/hecr/OWL/.

23. C. T. Hill and D. N. Schramm, *Phys. Rev.* **D31** , 564 (1985); see also F. A. Aharonian and J. W. Cronin in ref. [7].

24. N. A. Porter, *Nuovo Cim.* **16**, 958 (1960); E. Goto, *Prog. Theo. Phys.* **30**, 700 (1963).

25. T. W. B. Kibble, *J. Phys.* **A9**, 1387 (1976), and *Phys. Rept.* **67**, 183 (1980); M. B. Einhorn, D. L. Stein, and D. Toussaint, *Phys. Rev.* **D21**, 3295 (1980); A. H. Guth and E. J. Weinberg, *Phys. Rev.* **D23**, 876 (1981).

26. E. W. Kolb and M. S. Turner, "The Early Universe," Addison-Wesley pub., NY (1991).

27. E. N. Parker, *Astrophys. J.* **160**, 383 (1970); *ibid.* , **163**, 225 (1971); *ibid.* ,**166**, 295 (1971); M. S. Turner, E. N. Parker, and T. Bogdan, *Phys. Rev.* **D26**, 1296 (1982).

28. V. Rubakov, *JETP Lett.* **33**, 644 (1981); *Nucl. Phys.* **B203**, 311 (1982); C. G. Callan, Jr., *Phys. Rev.* **25**, 2141 (1982).

29. G. Giacomelli, in "Theory and Detection of Magnetic Monopoles in Gauge Theories", ed. N. Craigie, World Scientific pub., 1986; K. Olaussen, H. A. Olsen, P. Osland,

and I. Overbo, *Phys. Rev. Lett.* **52**, 325 (1984);

30. A sphaleron is the minimum–energy baryon– and lepton–number violating classical field configuration of the standard model. An overview of sphaleron physics can be found in ref. [26].
31. R. N. Mohapatra and S. Nussinov, hep–ph/9708497.

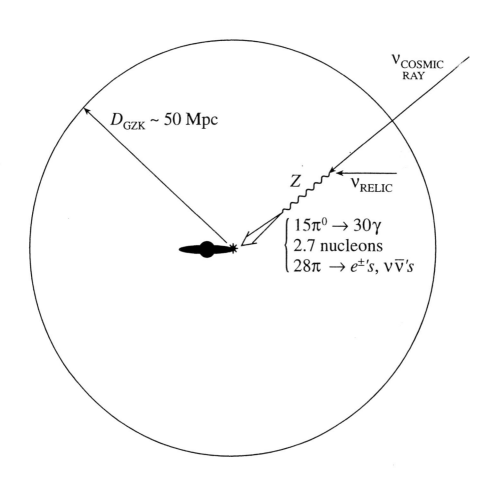

FIGURE 1. A schematic of the Z-burst mechanism within the GZK zone.

COMMENTS ON THE PHYSICS POTENTIAL OF ULTRA-HIGH-ENERGY NEUTRINO INTERACTIONS WITH OWL

David B. Cline

Department of Physics and Astronomy, Box 951547
University of California Los Angeles
Los Angeles, CA 90095-1547, USA

Abstract. The study of neutrino interactions and sources for energies $\geq 10^{19}$ eV is discussed. The OWL NASA project could play a very important role in this field. We make a simple estimate of the neutrino rate in OWL and find a small value. It is essential to design OWL with either a larger target volume or a planned upgrade to larger target volume.

INTRODUCTION

The OWL project could be of great importance to detect ultra-high-energy (UHE) neutrinos from new physics processes in the Universe. However, one should be somewhat realistic about the potential rates for the detector and the impact of this on the design of OWL ($\sim 10^{13}$ tons) and even of the future staging of OWL to go to much larger ($\sim 10^{14} - 10^{15}$ tons) mass.

In my opinion, one key idea would be the possibility of separating ν_e, ν_μ, and ν_τ interactions to help identify the source of the UHE neutrinos if they are detected by OWL.

POSSIBLE FLUX OF NEUTRINOS FOR OWL

We rely on the work in Ref. 1 for many of our flux estimates in this brief report. In this calculation (and in the references of this paper), one finds that the estimated neutrino fluxes vary from 6×10^{-8} cm^{-2} y^{-1} to 6×10^{-9} cm^{-2} y^{-1}. The latter is the level of neutron flux that could come from a pileup of events at the Greisen-Zepsin. The higher fluxes come from speculative processes like monopole annihilation. One might think that a flux of $\sim 10^{-7}$ cm^{-2} y^{-1} could be a reasonable upper limit for the neutrino flux in many models.

CP433, *Workshop on Observing Giant Air Showers from Space*
edited by J.F. Krizmanic et al.
© 1998 The American Institute of Physics 1-56396-766-9/98/$15.00

RATIO OF v_τ TO OTHER NEUTRINOS

In most models, such as the pileup at the Greisen cutoff, we expect the ratio of $\dot v_\tau$ to other neutrinos to be very small. There are two possible exceptions to this:

1. If neutrino oscillations are important (*i.e.*, v_μ to v_τ),

2. New physics could give a larger fraction of v_τ in the flux.

Therefore, it could be that the detection of v_τ interactions in OWL could be a very important method to observe new physics. Many years ago, I suggested the same ideas for the DUMAND detector.

NEUTRINO EVENT RATE IN OWL

We estimate the rate for $E_v \sim 10^{20}$ eV. The neutrino interaction cross sections have recently been recalculated.[2] Using the results of GQRS97 found in Ref. 2, we use the total cross section of 3×10^{-32} cm^{-2} for $E_v \sim 10^{20}$ eV. Using 10^{13} tons for a target mass and a detector efficiency of 1, we would get 100 events per year for a flux of 6×10^9 cm^{-2} y^{-1}. However, the detection of these UHE showers will require a cut on the data to a very small solid angle with a loss of at least 10^{-2}. Other factors could further reduce the rate so that a 10^{13} ton target may only give one detected event per year unless the neutrino flux is very much larger than that from the Greisen pileup. Our conclusion is that ultimately a larger OWL target, perhaps $10^{14} - 10^{15}$ tons, may be required for the neutrino physics to be done.

POSSIBLE DETECTION OF v_τ EVENTS IN OWL

At the incredible energies of 10^{20} eV, a τ particle can travel a large distance before decaying ($\ell_{decay} \sim 100$ km). On the other hand, an electron will produce a short distance shower, and an UHE muon will produce a longer track with many small bursts of hadronic energy. In principle, these different behaviors could lead to a signature that would allow separation of the different neutrinos. We suggest a detailed study of this by Monte Carlo simulation in OWL (3).

SUMMARY

In conclusion, the OWL project could be one of the most interesting for UHE particle physics in the next century, provided the detector is large enough to detect neutrinos.

REFERENCES

1. For information and further references, see Yoshida, S., Dai, H., Jui, C.C.H., and Sommers, P., *Ap. J.* **479**, 547-559 (1997).
2. The most recent work is Quigg, C., *et al.*, FNAL report GQRS97. Previous work is found in Quigg, C., Reno, M.H., and Walker, T.P., *Phys. Rev. Lett.* **57**, 774-782 (1986); also McKay, D.W. and Ralstan, J.P., *Phys. Lett. B* **167**, 103-107 (1986).
3. Various OWL talks, these proceedings.

Large Natural Cherenkov Detectors: Water and Ice

Francis Halzen

Physics Department, University of Wisconsin, Madison, WI 53706, USA

Abstract. In this review we first address 2 questions:

- why do we need kilometer-scale muon and neutrino detectors?

- what do we learn from the operating Baikal and AMANDA detectors about the construction of kilometer-scale detectors?

I will subsequently discuss the challenges for building the next-generation detectors. The main message is that these are different, in fact less ominous, than for commissioning the present, relatively small, detectors which must reconstruct events far outside their instrumented volume in order to achieve large effective telescope area.

I WHY KILOMETER-SCALE DETECTORS?

High energy neutrino telescopes are multi-purpose instruments; their science mission covers astronomy and astrophysics, cosmology, particle physics and cosmic ray physics. Their deployment creates new opportunities for glaciology and oceanography, possibly geology [1]. The observations of astronomers span 19 decades in energy or wavelength, from radio-waves to the high energy gamma rays detected with satellite-borne detectors [2]. Major discoveries have been historically associated with the introduction of techniques for exploring new wavelengths. The important discoveries were surprises. In this spirit, the primary motivation for commissioning neutrino telescopes is to cover uncharted territory: wavelengths smaller than 10^{-14} cm, or energies in excess of 10 GeV. This exploration has already been launched by truly pioneering observations using air Cherenkov telescopes [3]. Larger cosmic ray arrays with sensitivity above 10^7 TeV, an energy where charged protons may point back at their sources with minimal deflection by the galactic magnetic field, will be pursuing similar goals [4]. Could the high energy skies be devoid of particles? No, cosmic rays with energies exceeding 10^8 TeV have been recorded [4]. Between GeV gamma rays and the most energetic cosmic rays, there is uncharted territory spanning some eight orders of magnitude in wavelength. Exploring this energy region with neutrinos does have the definite advantage that they can, unlike

CP433, *Workshop on Observing Giant Air Showers from Space*
edited by J.F. Krizmanic et al.

high energy photons and nuclei, reach us, essentially without attenuation in flux, from the largest red-shifts.

The challenge is that neutrinos are difficult to detect: the small interaction cross sections that enable them to travel without attenuation over a Hubble radius, are also the reason why kilometer-scale detectors are required in order to capture them in sufficient numbers to do astronomy [5]. There is nothing magical about this result — I will explain this next.

Cosmic neutrinos, just like accelerator neutrinos, are made in beam dumps. A beam of accelerated protons is dumped into a target where they produce pions in collisions with nuclei. Neutral pions decay into gamma rays and charged pions into muons and neutrinos. All this is standard particle physics and, in the end, roughly equal numbers of secondary gamma rays and neutrinos emerge from the dump. In man-made beam dumps the photons are absorbed in the dense target; this may not be the case in an astrophysical system where the target material can be more tenuous. Also, the target material may be light rather than nuclei. For instance, with an ambient photon density a million times larger than the sun, approximately 10^{14} per cm^3, particles accelerated in the superluminal jets associated with active galactic nuclei (AGN), may meet more photons than nuclei when losing energy. Examples of cosmic beam dumps are tabulated in Table 1. They fall into two categories. Neutrinos produced by the cosmic ray beam are, of course, guaranteed and calculable. We know the properties of the beam and the various targets: the atmosphere, the hydrogen in the galactic plane and the CMBR background. Neutrinos from AGN and GRBs (gamma ray bursts) are not guaranteed, though both represent good candidate sites for the acceleration of the highest energy cosmic rays. That they are also the sources of the highest energy photons reinforces this association.

TABLE 1. Cosmic Beam Dumps

Beam	Target
cosmic rays	atmosphere
cosmic rays	galactic disk
cosmic rays	CMBR
AGN jets	ambient light, UV
shocked protons	GRB photons

In astrophysical beam dumps, like AGN and GRBs, there is typically one neutrino and photon produced per accelerated proton [1]. The accelerated protons and photons are, however, most likely to suffer attenuation in the source before they can escape. So, a hierarchy of particle fluxes emerges with protons < photons < neutrinos. A generic neutrino flux can be obtained from this relation by, conservatively, equating the neutrino with the observed cosmic ray flux. The detector size can now be determined [6] by taking into account the detection efficiency for neutrinos which is much reduced compared to protons.

Neutrino telescopes are conventional particle detectors which use natural and clear water and ice as the Cherenkov medium. A three dimensional grid of photo-multiplier tubes maps the Cherenkov cone radiated by a muon of neutrino origin. Nanosecond timing provides degree resolution of the muon track which is, at high energy, aligned with the neutrino direction. The probability to detect a TeV neutrino is roughly 10^{-6} [1]. It is easily computed from the requirement that, in order to be detected, the neutrino has to interact within a distance of the detector which is shorter than the range of the muon it produces. In other words, in order for the neutrino to be detected, the produced muon has to reach the detector. Therefore,

$$P_{\nu \to \mu} \simeq \frac{R_\mu}{\lambda_{\text{int}}} \simeq A E_\nu^n , \qquad (1)$$

where R_μ is the muon range and λ_{int} the neutrino interaction length. For energies below 1 TeV, where both the range and cross section depend linearly on energy, $n = 2$. Between TeV and PeV energies $n = 0.8$ and $A = 10^{-6}$, with E in TeV units. For EeV energies $n = 0.47$, $A = 10^{-2}$ with E in EeV.

At PeV energy the cosmic ray flux is of order 1 per m^2 per year and the probability to detect a neutrino of this energy is of order 10^{-3}. A neutrino flux equal to the cosmic ray flux will therefore yield only a few events per day in a kilometer squared detector. At EeV energy the situation is worse. With a rate of 1 per km ^{2} per year and a detection probability of 0.1, one can still detect several events per year in a kilometer squared detector provided the neutrino flux exceeds the proton flux by 2 orders of magnitude or more. For the neutrino flux generated by cosmic rays interacting with CMBR photons and such sources as AGN and topological defects [7], this is indeed the case. All above estimates are conservative and the rates should be higher because the neutrinos escape the source with a flatter energy spectrum than the protons [6]. In summary, the cosmic ray flux and the neutrino detection efficiency define the size of a neutrino telescope. Needless to say that a telescope with kilometer squared effective area represents a neutrino detector of kilometer cubed volume.

II BAIKAL AND THE MEDITERRANEAN

First generation neutrino detectors are designed to reach a relatively large telescope area and detection volume for a neutrino threshold of tens of GeV, not higher and, possibly, lower. This relatively low threshold permits calibration of the novel instrument on the known flux of atmospheric neutrinos. Its architecture is optimized for reconstructing the Cherenkov light front radiated by an up-going, neutrino-induced muon rather than for detecting signals of TeV energy and above. Up-going muons are to be identified in a background of down-going, cosmic ray muons which are more than 10^6 times more frequent for a depth of 1 kilometer.

The "landscape" of neutrino astronomy is sketched in Table 2. With the termination of the pioneering DUMAND experiment, the efforts in water are, at present,

TABLE 2.

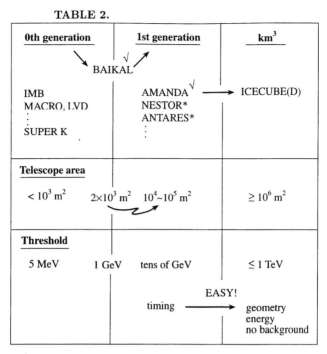

0th generation	1st generation	km³

BAIKAL √ ↗ ↘

IMB	AMANDA √ ——→ ICECUBE(D)
MACRO, LVD	NESTOR*
⋮	ANTARES*
SUPER K	⋮

Telescope area

< 10³ m² 2×10³ m² 10⁴~10⁵ m²	≥ 10⁶ m²

$< 10^3 \, \mathrm{m}^2 \quad 2\times10^3 \, \mathrm{m}^2 \quad 10^4 \text{\textasciitilde} 10^5 \, \mathrm{m}^2 \qquad \geq 10^6 \, \mathrm{m}^2$

Threshold

5 MeV   1 GeV   tens of GeV   ≤ 1 TeV

EASY!

timing ——→ geometry
energy
no background

√ ν-candidates
* R & D

spearheaded by the Baikal experiment [8]. Operating with 144 optical modules (OM) since April 1997, the *NT-200* detector will be completed by April 1998. The Baikal detector is well understood and the first atmospheric neutrinos have been identified; we will discuss this in more detail further on. The Baikal site is competitive with deep oceans although the smaller absorption length requires somewhat denser spacing of the OMs. This does however result in a lower threshold which is a definite advantage, for instance in WIMP searches. They have shown that their shallow depth of 1 kilometer does not represent a serious drawback. By far the most significant advantage is the site with a seasonal ice cover which allows reliable and inexpensive deployment and repair of detector elements.

In the following years, *NT-200* will be operated as a neutrino telescope with an effective area between $10^3 \sim 5 \times 10^3$ m², depending on the energy. Presumably too small to detect neutrinos from AGN and other extraterrestrial sources, *NT-200* will serve as the prototype for a larger telescope. For instance, with 2000 OMs, a threshold of $10 \sim 20$ GeV and an effective area of $5 \times 10^4 \sim 10^5$ m², an expanded Baikal telescope would fill the gap between present underground detectors and planned high threshold detectors of cube kilometer size. Its key advantage would be low threshold.

The Baikal experiment represents a proof of concept for deep ocean projects. These should have the advantage of larger depth and optically superior water. Their challenge is to design a reliable and affordable technology. Three groups are confronting the problem: NESTOR and Antares in the Mediterranean and a group at LBL, Berkeley. The latter has made seminal contributions to the design of digital OMs. Most of them have recently joined the AMANDA experiment and, as far as deployments are concerned, their effort will proceed in ice.

The NESTOR collaboration [9], as part of an ongoing series of technology tests, has recently deployed two aluminum "floors", 34 m in diameter, to a depth of 2600 m. Mechanical robustness was demonstrated by towing the structure, submerged below 2000 m, from shore to the site and back. The detector will consist of 12 six-legged floors separated by 30 m.

The Antares collaboration [10] is in the process of determining the critical detector parameters at a 2000 m deep, Mediterranean site off Toulon, France. A deliberate development effort will lead to the construction of a demonstration project consisting of 3 strings with a total of 200 OMs.

For neutrino astronomy to become a viable science several of these, or other, projects will have to succeed. Astronomy, whether in the optical or in any other wave-band, thrives on a diversity of complementary instruments, not on "a single best instrument". When the Soviet government tried out the latter method by creating a national large mirror project, it virtually annihilated the field.

III   FIRST NEUTRINOS FROM BAIKAL

The Baikal Neutrino Telescope is being deployed in Lake Baikal, Siberia, 3.6 km from shore at a depth of 1.1 km. An umbrella-like frame holds 8 strings, each instrumented with 24 pairs of 37-cm diameter $QUASAR$ photomultiplier tubes (PMT). Two PMTs in a pair are switched in coincidence in order to suppress background from bioluminescence and PMT noise.

They have analysed 212 days of data taken in 94-95 with 36 OMs. Upward-going muon candidates were selected from about 10^8 events in which more than 3 pairs of PMTs triggered. After quality cuts and χ^2 fitting of the tracks, a sample of 17 up-going events remained. These are not generated by neutrinos passing the earth below the detector, but by showers from down-going muons originating below the array. In a small detector such events are expected. In 2 events however the light does not decrease from bottom to top, as expected from invisible showering muons below the detector. A detailed analysis [11] yields a fake probability of 2% for both events.

After the deployment of 96 OMs in the spring of 96, three neutrino candidates have been found in a sample collected over 18 days. This is in agreement with the expected number of approximately 2.3. One of the events is displayed in Fig. 1. In this analysis the most effective quality cuts are the traditional χ^2 cut and a cut on the probability of non-reporting channels not to be hit, and reporting channels

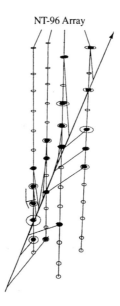

NT-96 Array

FIGURE 1. Candidate neutrino event from NT-96 in Lake Baikal.

to be hit ($P_{\text{no hit}}$ and P_{hit}, respectively). To guarantee a minimum lever arm for track fitting, they were forced to reject events with a projection of the most distant channels on the track smaller than 35 meters. This does, of course, result in a loss of threshold.

IV   THE AMANDA SOUTH POLE NEUTRINO DETECTOR

Construction of the first-generation AMANDA detector [12] was completed in the austral summer 96–97. It consists of 300 optical modules deployed at a depth of 1500–2000 m; see Fig. 2. An optical module consists of an 8 inch photomultiplier tube and nothing else. Calibration of this detector is in progress, although data has been taken with 80 OM's which were deployed one year earlier in order to verify the optical properties of the ice (AMANDA-80).

The performance of the AMANDA detector is encapsulated in the event shown in Fig. 3. Coincident events between AMANDA-80 and four shallow strings with 80 OM's (see Fig. 2), have been triggered for one year at a rate of 0.1 Hz. Every 10 seconds a cosmic ray muon is tracked over 1.2 kilometer. The contrast in detector response between the strings near 1 and 2 km depths is dramatic: while the Cherenkov photons diffuse on remnant bubbles in the shallow ice, a straight track with velocity c is registered in the deeper ice. The optical quality of the deep ice can be assessed by viewing the OM signals from a single muon triggering 2

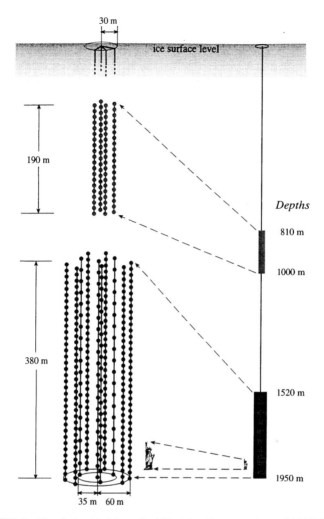

FIGURE 2. The Antarctic Muon And Neutrino Detector Array (AMANDA).

strings separated by 79.5 m; see Fig. 3b. The separation of the photons along the Cherenkov cone is well over 100 m, yet, despite some evidence of scattering, the speed-of-light propagation of the track can be readily identified.

The optical properties of the ice are quantified by studying the propagation in the ice of pulses of laser light of nanosecond duration. The arrival times of the photons after 20 m and 40 m are shown in Fig. 4 for the shallow and deep ice [13]. The distributions have been normalized to equal areas; in reality, the probability that a photon travels 70 m in the deep ice is $\sim 10^7$ times larger. There is no diffusion resulting in loss of information on the geometry of the Cherenkov cone in the deep ice.

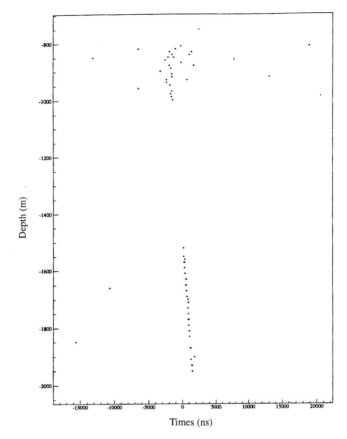

FIGURE 3a. Cosmic ray muon track triggered by both shallow and deep AMANDA OM's. Trigger times of the optical modules are shown as a function of depth. The diagram shows the diffusion of the track by bubbles above 1 km depth. Early and late hits, not associated with the track, are photomultiplier noise.

V   INTERMEZZO: AMANDA BEFORE AND AFTER

The AMANDA detector was antecedently proposed on the premise that inferior properties of ice as a particle detector with respect to water could be compensated by additional optical modules. The technique was supposed to be a factor 5~10 more cost-effective and, therefore, competitive. The design was based on then current information:

- the absorption length at 370 nm, the wavelength where photomultipliers are maximally efficient, had been measured to be 8 m,

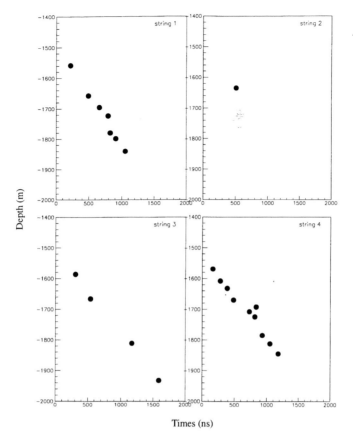

FIGURE 3b. Cosmic ray muon track triggered by both shallow and deep AMANDA OM's. Trigger times are shown separately for each string in the deep detector. In this event the muon mostly triggers OM's on strings 1 and 4 which are separated by 79.5 m.

- the scattering length was unknown,
- the AMANDA strategy was to use a large number of closely spaced OM's to overcome the short absorption length. Muon tracks triggering 6 or more OM's are reconstructed with degree accuracy. Taking data with a simple majority trigger of 6 OM's or more at 100 Hz yields an average effective area of 10^4 m^2.

The reality is that:

- the absorption length is 100 m or more, depending on depth [14],
- the scattering length is 25~30 m (preliminary),
- because of the large absorption length OM spacings are similar, actually larger, than that those of proposed water detectors. Also, a typical event triggers 20

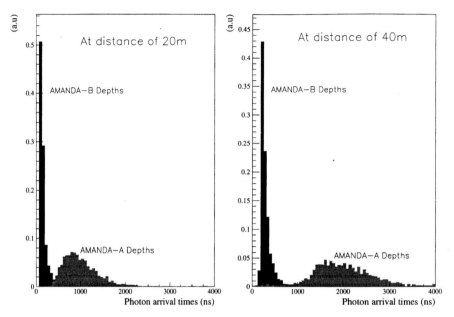

FIGURE 4. Propagation of 510 nm photons indicate bubble-free ice below 1500 m, in contrast with ice with some remnant bubbles above 1 km.

OM's, not 6. Of these more than 5 photons are not scattered. In the end, reconstruction is therefore as before, although additional information can be extracted from scattered photons by minimizing a likelihood function which matches measured and expected delays [15].

The measured arrival directions of background cosmic ray muon tracks, reconstructed with 5 or more unscattered photons, are confronted with their known angular distribution in Fig. 5. The agreement with Monte Carlo simulation is adequate. Less than one in 10^5 tracks is misreconstructed as originating below the detector [13]. Visual inspection reveals that the remaining misreconstructed tracks are mostly showers, radiated by muons or initiated by electron neutrinos, which are reconstructed as up-going tracks of muon neutrino origin. At the 10^{-6} level of the background, candidate events can be identified; see Fig. 6. This exercise establishes that AMANDA-80 can be operated as a neutrino detector; misreconstructed showers can be readily eliminated on the basis of the additional information on the amplitude of OM signals. Monte Carlo simulation, based on this exercise confirms, that AMANDA-300 is a 10^4 m^2 detector (somewhat smaller for atmospheric neutrinos and significantly larger for high energy signals) with 2.5 degrees mean angular resolution [15]. We have verified the angular resolution of AMANDA-80 by reconstructing muon tracks registered in coincidence with a surface air shower array [16].

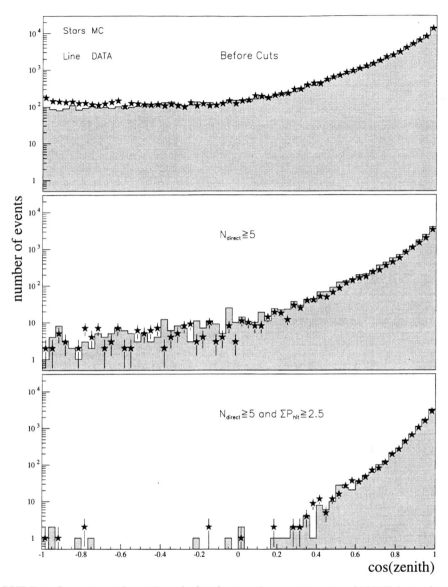

FIGURE 5. Reconstructed zenith angle distribution of muons triggering AMANDA-80: data and Monte Carlo. The relative normalization has not been adjusted at any level. The plot demonstrates a rejection of cosmic ray muons at a level of 10^{-5} which is reached by only 2 cuts: a cut on the number of unscattered photons and on P_{hit}, a quantity introduced in the context of the Baikal experiment.

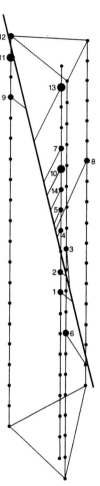

FIGURE 6. A candidate up-going, neutrino-induced muon in the AMANDA-80 data. The numbers indicate the time sequence of triggered OMs, the size of the dots the relative amplitude of the signal.

VI TOWARDS ICE CUBE(D)

A strawman detector with effective area in excess of 1 km^2 consists of 4800 OM's: 80 strings spaced by $\sim$ 100 m, each instrumented with 60 OM's spaced by 15 m. A cube with a side of 0.8 km is thus instrumented and a through-going muon can be visualized by doubling the length of the lower track in Fig. 3a. It is straightforward to convince oneself that a muon of TeV energy and above, which generates single photoelectron signals within 50 m of the muon track, can be reconstructed by geometry only. The spatial positions of the triggered OM's allow a geometric track reconstruction with a precision in zenith angle of:

$$\text{angular resolution} \simeq \frac{\text{OM spacing}}{\text{length of the track}} \simeq 15\,\text{m}/800\,\text{m} \simeq 1\,\text{degree};\qquad(2)$$

no timing information is really required. Timing is still necessary to establish whether a track is up- or down-going, not a challenge given that the transit time of the muon exceeds 2 microseconds. Using the events shown in Fig. 3, we have, in fact, already demonstrated that we can reject background cosmic ray muons. Once ICE CUBE(D) has been built, it can be used as a veto for AMANDA and its threshold lowered to GeV energy.

With half the number of OM's and half the price tag of the Superkamiokande and SNO solar neutrino detectors, the plan to commission such a detector over 5 years is not unrealistic. The price tag of the default technology used in AMANDA-300 is $6000 per OM, including cables and DAQ electronics. This signal can be transmitted to the surface by fiber optic cable without loss of information. Given the scientific range and promise of such an instrument, a kilometer-scale neutrino detector must be one of the best motivated scientific endeavors ever.

VII   ABOUT WATER AND ICE

The optical requirements of the detector medium can be readily evaluated, at least to first order, by noting that string spacings determine the cost of the detector. The attenuation length is the relevant quantity because it determines how far the light travels, irrespective of whether the photons are lost by scattering or absorption. Remember that, even in the absence of timing, hit geometry yields degree zenith angle resolution. Near the peak efficiency of the OM's the attenuation length is 25–30 m, larger in deep ice than in water below 4 km. The advantage of ice is that, unlike for water, its transparency is not degraded for blue Cherenkov light of lower wavelength, a property we hope to take further advantage of by using wavelength-shifter in future deployments.

The AMANDA approach to neutrino astronomy was initially motivated by the low noise of sterile ice and the cost-effective detector technology. These advantages remain, even though we know now that water and ice are competitive as a detector medium. They are, in fact, complementary. Water and ice seem to have similar attenuation length, with the role of scattering and absorption reversed. As demonstrated with the shallow AMANDA strings [17], scattering can be exploited to range out the light and perform calorimetry of showers produced by electron-neutrinos and showering muons. Long scattering lengths in water may result in superior angular resolution, especially for the smaller, first-generation detectors. This can be exploited to reconstruct events far outside the detector in order to increase its effective volume.

ACKNOWLEDGEMENTS

This research was supported in part by the U.S. Department of Energy under Grant No. DE-FG02-95ER40896 and in part by the University of Wisconsin Research Committee with funds granted by the Wisconsin Alumni Research Foundation.

REFERENCES

1. For a review, see T.K. Gaisser, F. Halzen and T. Stanev, *Phys. Rep.* **258**(3), 173 (1995); R. Gandhi, C. Quigg, M. H. Reno and I. Sarcevic, *Astropart. Phys.*, **5**, 81 (1996).
2. M. T. Ressel and M. S. Turner, *Comments Astrophys.* **14** (1990) 323.
3. M. Punch *et al.*, *Nature* **358**, 477–478 (1992); J. Quinn et al., *Ap. J.* **456**, L83 (1995); Schubnell et al., astro-ph/9602068, Ap. J. (1997 in press).
4. The Pierre Auger Project Design Report, Fermilab report (Feb. 1997) and references therein.
5. F. Halzen, *The case for a kilometer-scale neutrino detector*, in Nuclear and Particle Astrophysics and Cosmology, Proceedings of Snowmass 94, R. Kolb and R. Peccei, eds.
6. F. Halzen and E. Zas, Ap. J. **488**, 669 (1997).
7. G. Sigl, D.N. Schramm and P. Bhattacharjee, astro-ph/9403039, *Astropart. Phys.* **2**, 401 (1994) and references therein.
8. G. V. Domogatsky, these proceedings.
9. L. Trascatti, these proceedings.
10. F. Feinstein, these proceedings.
11. I.A.Belolaptikov et al., Proceedings of the 25th International Cosmic Ray Conference, Durban, South Africa (astroph/9705245).
12. S. W. Barwick *et al.*, *The status of the AMANDA high-energy neutrino detector*, in Proceedings of the 25th International Cosmic Ray Conference, Durban, South Africa (1997).
13. S. Tilav *et al.*, *First look at AMANDA-B data*, in Proceedings of the 25th International Cosmic Ray Conference, Durban, South Africa (1997).
14. The AMANDA collaboration, *Science* **267**, 1147 (1995).
15. C. Wiebusch *et al.*, *Muon reconstruction with AMANDA-B*, in Proceedings of the 25th International Cosmic Ray Conference, Durban, South Africa (1997).
16. T. Miller *et al.*, *Analysis of SPASE-AMANDA coincidence events*, in Proceedings of the 25th International Cosmic Ray Conference, Durban, South Africa (1997).
17. R. Porrata *et al.*, *Analysis of cascades in AMANDA-A*, in Proceedings of the 25th International Cosmic Ray Conference, Durban, South Africa (1997).

ULTRA HIGH ENERGY COSMIC RAYS FROM DECAYING SUPERHEAVY PARTICLES [1]

V.Berezinsky

INFN, Laboratori Nazionali del Gran Sasso, 67010 Assergi, Italy

Abstract.
  Decaying superheavy particles can be produced by Topological Defects or, in case they are quasi-stable, as relics from the early Universe. The decays of these particles can be the sources of observed Ultra High Energy Cosmic Rays ($E \sim 10^{10} - 10^{12}\ GeV$). The Topological Defects as the UHE CR sources are critically reviewed and cosmic necklaces and monopole-antiminopole pairs are identified as most plausible sources. The relic superheavy particles are shown to be clustering in the halo and their decays produce UHE CR without GZK cutoff. The Lightest Supersymmetric Particles with Ultra High Energies are naturally produced in the cascades accompanying the decays of superheavy particles. These particles are discussed as UHE carriers in the Universe.

I INTRODUCTION

The observation of cosmic ray particles with energies higher than $10^{11}\ GeV$ [1] gives a serious challenge to the known mechanisms of acceleration. The shock acceleration in different astrophysical objects typically gives maximal energy of accelerated protons less than $(1-3) \cdot 10^{10}\ GeV$ [2]. The unipolar induction can provide the maximal energy $1 \cdot 10^{11}\ GeV$ only for the extreme values of the parameters [3]. Much attention has recently been given to acceleration by ultrarelativistic shocks [4], [5]. The particles here can gain a tremendous increase in energy, equal to Γ^2, at a single reflection, where Γ is the Lorentz factor of the shock. However, it is known (see e.g. the simulation for pulsar relativistic wind in [6]) that particles entering the shock region are captured there or at least have a small probability to escape.

Topological defects, TD, (for a review see [7]) can naturally produce particles of ultrahigh energies (UHE). The pioneering observation of this possibility was made by Hill, Schramm and Walker [8] (for a general analysis of TD as UHE CR sources see [9] and for a review [10]).

---

[1] talk given at the workshop "Observing the Highest Energy Particles from Space", University of Maryland , 13 - 15 November, 1997

CP433, *Workshop on Observing Giant Air Showers from Space*
edited by J.F. Krizmanic et al.

In many cases TD become unstable and decompose to constituent fields, super-heavy gauge and Higgs bosons (X-particles), which then decay producing UHE CR. It could happen, for example, when two segments of ordinary string, or monopole and antimonopole touch each other, when electrical current in superconducting string reaches the critical value and in some other cases.

In most cases the problem with UHE CR from TD is not the maximal energy, but the fluxes. One very general reason for the low fluxes consists in the large distance between TD. A dimension scale for this distance is the Hubble distance H_0^{-1}. However, in some rather exceptional cases this dimensional scale is multiplied to a small dimensionless value r. If a distance between TD is larger than UHE proton attenuation length (due to the GZK effect [11]), then the flux at UHE is typically exponential suppressed.

Ordinary cosmic strings can produce particles when a loop annihilate into double line [12]. The produced UHE CR flux is strongly reduced due to the fact that a loop oscillates, and in the process of a collapse the two incoming parts of a loop touch each other in one point producing thus the smaller loops, instead of two-line annihilation. However, this idea was recently revived due to recent work [13]. It is argued there that the energy loss of the long strings is dominated by production of very small loops with the size down to the width of a string, which immediately annihilate into superheavy particles. A problem with this scenario is too large distance between strings (of order of the Hubble distance). For a distance between an observer and a string being the same, the observed spectrum of UHE CR has an exponential cutoff at energy $E \sim 3 \cdot 10^{19} \ eV$.

Superheavy particles can be also produced when two segments of string come into close contact, as in *cusp* events [14]. This process was studied later by Gill and Kibble [15], and they concluded that the resulting cosmic ray flux is far too small. An interesting possibility suggested by Brandenberger [14] is the *cusp* "evaporation" on cosmic strings. When the distance between two segments of the cusp becomes of the order of the string width, the cusp may"annihilate" turning into high energy particles, which are boosted by a very large Lorentz factor of the cusp [14]. However, the resulting UHE CR flux is considerably smaller than one observed [16].

Superconducting strings [17] appear to be much better suited for particle produc-tion. Moving through cosmic magnetic fields, such strings develop electric currents and copiously produce charged heavy particles when the current reaches certain critical value. The CR flux produced by superconducting strings is affected by some model-dependent string parameters and by the history and spatial distribu-tion of cosmic magnetic fields. Models considered so far failed to account for the observed flux [18].

Monopole-antimonopole pairs $(M\bar{M})$ can form bound states and eventually an-nihilate into UHE particles [19], [20]. For an appropriate choice of the monopole density n_M, this model is consistent with observations; however, the required (low) value of n_M implies fine-tuning. In the first phase transition $G \rightarrow H \times U(1)$ in the early Universe the monopoles are produced with too high density. It must then be diluted by inflation to very low density, precisely tuned to the observed UHE CR flux.

Monopole-string network can be formed in the early Universe in the sequence of symmetry breaking

$$G \to H \times U(1) \to H \times Z_N. \qquad (1)$$

For $N \geq 3$ an infinite network of monopoles connected by strings is formed. The magnetic fluxes of monopoles in the network are channeled into into the strings that connect them. The monopoles typically have additional unconfined magnetic and chromo-magnetic charges. When strings shrink the monopoles are pulled by them and are accelerated. The accelerated monopoles produce extremely high energy gluons, which then fragment into UHE hadrons [21]. The produced flux is too small to explain UHE CR observation [22].

Cosmic necklaces are TD which are formed in a sequence of symmetry breaking given by Eq.(1) when $N = 2$. The first phase transition produces monopoles, and at the second phase transition each monopole gets attached to two strings, with its magnetic flux channeled along the strings. The resulting necklaces resemble "ordinary" cosmic strings with monopoles playing the role of beads. Necklaces can evolve in such way that a distance between monopoles diminishes and in the end all monopoles annihilate with the neighboring antimonopoles [23]. The produced UHE CR flux [23] is close to the observed one and the shape of the spectrum resembles one observed. The distance between necklaces can be much smaller than attenuation length of UHE protons.

Superheavy relic particles can be sources of UHE CR [24,25]. In this scenario Cold Dark Matter (CDM) have a small admixture of long-lived superheavy particles. These particles must be heavy, $m_X > 10^{12}\ GeV$, long-lived $\tau_X > t_0$, where t_0 is the age of the Universe, and weakly interacting. The required life-time can be provided if this particle has (almost) conserved quantum number broken very weakly due to warmhole [25] or instanton [24] effects. Several mechanisms for production of such particles in the early Universe were identified. Like other forms of non-dissipative CDM , X-particles must accumulate in the halo of our Galaxy [25] and thus they produce UHE CR without GZK cutoff and without appreciable anisotropy.

The UHE carriers produced at the decay of superheavy relic particles or from TD, can be be nucleons, photons and neutrinos or neutralinos [28]. Production of neutralinos occurs in particle cascade, which originates at the decay of superheavy X-particle in close analogy to QCD cascade [28]. Though flux of UHE neutralino is of the same order as neutrino flux, its detection is more problematic because of smaller cross-section. The other particles discussed as carrier of UHE signal are gluino [26–28], in case it is Lightest Supersymmetric Particle (LSP), and heavy monopole [29,27].

In this paper I will present the results obtained in our joint works with M.Kachelriess and A.Vilenkin about necklaces and superheavy relic particles as possible sources of UHE CR. Neutralino and gluino as the carriers of UHE signal will be also shortly discussed.

II NECKLACES

Necklaces produced in a sequence of symmetry breaking $G \to H \times U(1) \to H \times Z_2$ form the infinite necklaces having the shape or random walks and a distribution of closed loops. Each monopole in a necklace is attached to two strings.

The monopole mass m and the string tension μ are determined by the corresponding symmetry breaking scales, η_s and η_m ($\eta_m > \eta_s$): $m \sim 4\pi\eta_m/e$, $\mu \sim 2\pi\eta_s^2$. Here, e is the gauge coupling. The mass per unit length of string is equal to its tension, μ. Each string attached to a monopole pulls it with a force $F \sim \mu$ in the direction of the string. The monopole radius δ_m and the string thickness δ_s are typically of the order $\delta_m \sim (e\eta_m)^{-1}$, $\delta_s \sim (e\eta_s)^{-1}$.

An important quantity for the necklace evolution is the dimensionless ratio

$$r = m/\mu d, \tag{2}$$

We expect the necklaces to evolve in a scaling regime. If ξ is the characteristic length scale of the network, equal to the typical separation of long strings and to their characteristic curvature radius, then the force per unit length of string is $f \sim \mu/\xi$, and the acceleration is $a \sim (r+1)^{-1}\xi^{-1}$. We assume that ξ changes on a Hubble time scale $\sim t$. Then the typical distance travelled by long strings in time t should be $\sim \xi$, so that the strings have enough time to intercommute in a Hubble time. This gives $at^2 \sim \xi$, or

$$\xi \sim (r+1)^{-1/2}t. \tag{3}$$

The typical string velocity is $v \sim (r+1)^{-1/2}$.

It is argued in Ref.([23]) that $r(t)$ is driven towards large value $r \gg 1$. However, for $r \geq 10^6$ the characteristic velocity of the network falls down below the virial velocity, and the necklaces will be trapped by gravitational clustering of the matter. This may change dramatically the evolution of network. One possible interesting effect for UHE CR can be enhancement of necklace space density within Local Supercluster – a desirable effect as far absence of the GZK cutoff is concerned. However, we restrict our consideration by the case $r < 10^6$. The distance between necklaces is still small enough, $\xi \gtrsim 3$ Mpc, to assume their uniform distribution, when calculating the UHE CR flux.

Self-intersections of long necklaces result in copious production of closed loops. For $r \gtrsim 1$ the motion of loops is not periodic, so loop self-intersections should be frequent and their fragmentation into smaller loops very efficient. A loop of size ℓ typically disintegrates on a timescale $\tau \sim r^{-1/2}\ell$. All monopoles trapped in the loop must, of course, annihilate in the end.

Annihilating $M\bar{M}$ pairs decay into Higgs and gauge bosons, which we shall refer to collectively as X-particles. The rate of X-particle production is easy to estimate if we note that infinite necklaces lose a substantial fraction of their length to closed loops in a Hubble time. The string length per unit volume is $\sim \xi^{-2}$, and the

282

monopole rest energy released per unit volume per unit time is $r\mu/\xi^2 t$. Hence, we can write

$$\dot{n}_X \sim r^2\mu/(t^3 m_X), \tag{4}$$

where $m_X \sim e\eta_m$ is the X-particle mass.

X-particles emitted by annihilating monopoles decay into hadrons, photons and neutrinos, the latter two components are produced through decays of pions.

The diffuse flux of ultra-high energy protons can be evaluated as

$$I_p(E) = \frac{c\dot{n}_X}{4\pi m_X} \int_0^{t_0} dt\ W_N(m_X, x_g) \frac{dE_g(E, t)}{dE} \tag{5}$$

where dn_X/dt is given by Eq.(4), E is an energy of proton at observation and $E_g(E, t)$ is its energy at generation at cosmological epoch t, $x_g = E_g/E$ and $W_N(m_X, x)$ is the fragmentation function of X-particle into nucleons of energy $E = x m_X$. The value of dE_g/dE can be calculated from the energy losses of a proton on microwave background radiation (e.g. see [3]). In Eq.(5) the recoil protons are taken into account, while in Ref.([23]) their contribution was neglected.

The fragmentation function $W_N(m_X, x)$ is calculated using the decay of X-particle into QCD partons (quark, gluons and their supersymmetric partners) with the consequent development of the parton cascade. The cascade in this case is identical to one initiated by e^+e^- -annihilation. We have used the fragmentation function in the gaussian form as obtained in MLLA approximation in [34] and [35].

In our calculations the UHE proton flux is fully determined by only two parameters, $r^2\mu$ and m_X. The former is restricted by low energy diffuse gamma-radiation. It results from e-m cascades initiated by high energy photons and electrons produced in the decays of X-particles.

The cascade energy density predicted in our model is

$$\omega_{cas} = \frac{1}{2} f_\pi r^2 \mu \int_0^{t_0} \frac{dt}{t^3} \frac{1}{(1+z)^4} = \frac{3}{4} f_\pi r^2 \frac{\mu}{t_0^2}, \tag{6}$$

where t_0 is the age of the Universe (here and below we use $h = 0.75$), z is the redshift and $f_\pi \sim 1$ is the fraction of energy transferred to pions. In Eq.(6) we took into account that half of the energy of pions is transferred to photons and electrons. The observational bound on the cascade density, for the kind of sources we are considering here, is [36] $\omega_{cas} \lesssim 10^{-5}\ eV/cm^3$. This gives a bound on the parameter $r^2\mu$.

In numerical calculations we used $r^2\mu = 1 \times 10^{28}\ GeV^2$, which results in $\omega_{cas} = 5.6 \cdot 10^{-6}\ eV/cm^3$, somewhat below the observational limit. Now we are left with one free parameter, m_X, which we fix at $1 \cdot 10^{14}\ GeV$. The maximum energy of protons is then $E_{max} \sim 10^{13}\ GeV$. The calculated proton flux is presented in Fig.1, together with a summary of observational data taken from ref. [37].

Let us now turn to the calculations of UHE gamma-ray flux from the decays of X-particles. The dominant channel is given by the decays of neutral pions. The flux can be readily calculated as

$$I_\gamma(E) = \frac{1}{4\pi}\dot{n}_X \lambda_\gamma(E) N_\gamma(E), \qquad (7)$$

where $\dot{n}_X$ is given by Eq.(4), $\lambda_\gamma(E)$ is the absorption length of a photon with energy E due to e^+e^- pair production on background radiation and $N_\gamma(E)$ is the number of photons with energy E produced per one decay of X-particle. The latter is given by

$$N_\gamma(E) = \frac{2}{m_X} \int_{E/m_X}^1 \frac{dx}{x} W_{\pi^0}(m_X, x) \qquad (8)$$

where $W_{\pi^0}(m_X, x)$ is the fragmentation function of X-particles into π^0 pions.

At energy $E > 1 \cdot 10^{10}$ GeV the dominant contribution to the gamma-ray absorption comes from the radio background. The significance of this process was first noticed in [38](see also book [3]). New calculations for this absorption were recently done [39]. We have used the absorption lengths from this work.

When evaluating the flux 7 at $E > 1 \cdot 10^{10}$ GeV we neglected cascading of a primary photon, because pair production and inverse compton scattering occur at these energies on radio background, and thus at each collision the energy of a cascade particle is halved. Moreover, assuming an intergalactic magnetic field $H \geq 1 \cdot 10^{-9}$, the secondary electrons and positrons loose their energy mainly due to synchrotron radiation and the emitted photons escape from the considered energy interval [40].

The calculated flux of gamma radiation is presented in Fig.1 by the curve labelled γ. One can see that at $E \sim 1 \cdot 10^{11}$ GeV the gamma ray flux is considerably lower than that of protons. This is mainly due to the difference in the attenuation lengths for protons (110 Mpc) and photons (2.6 Mpc [39] and 2.2 Mpc [38]). At higher energy the attenuation length for protons dramatically decreases (13.4 Mpc at $E = 1 \cdot 10^{12}$ GeV) and the fluxes of protons and photons become comparable.

A requirement for the models explaining the observed UHE events is that the distance between sources must be smaller than the attenuation length. Otherwise the flux at the corresponding energy would be exponentially suppressed. This imposes a severe constraint on the possible sources. For example, in the case of protons with energy $E \sim (2-3) \cdot 10^{11}$ GeV the proton attenuation length is 19 Mpc. If protons propagate rectilinearly, there should be several sources inside this radius; otherwise all particles would arrive from the same direction. If particles are strongly deflected in extragalactic magnetic fields, the distance to the source should be even smaller. Therefore, the sources of the observed events at the highest energy must be at a distance $R \lesssim 15$ Mpc in the case or protons.

In our model the distance between sources, given by Eq.(3), satisfies this condition for $r > 3 \cdot 10^4$. This is in contrast to other potential sources, including supeconducting cosmic strings and powerful astronomical sources such as AGN, for which this condition imposes severe restrictions.

The difficulty is even more pronounced in the case of UHE photons. These particles propagate rectilinearly and their absorption length is shorter: $2 - 4\ Mpc$ at $E \sim 3 \cdot 10^{11}\ GeV$. It is rather unrealistic to expect several powerful astronomical sources at such short distances. This condition is very restrictive for topological defects as well. The necklace model is rather exceptional regarding this aspect.

III UHE CR FROM RELIC QUASISTABLE PARTICLES

This possibility was recognized recently in Refs([24,25]).

Our main assumption is that Cold Dark Matter (CDM) has a small admixture of long-lived supermassive X-particles. Since, apart from very small scales, fluctuations grow identically in all components of CDM, the fraction of X-particles, ξ_X, is expected to be the same in all structures. In particular, ξ_X is the same in the halo of our Galaxy and in the extragalactic space. Thus the halo density of X-particles is enhanced in comparison with the extragalactic density. The decays of these particles produce UHE CR, whose flux is dominated by the halo component, and therefore has no GZK cutoff. Moreover, the potentially dangerous e-m cascade radiation is suppressed.

First, we address the elementary-particle and cosmological aspects of a super-heavy long-living particle. Can the relic density of such particles be as high as required by observations of UHE CR? And can they have a lifetime comparable or larger than the age of the Universe?

Let us assume that X-particle is a neutral fermion which belongs to a representation of the $SU(2) \times U(1)$ group. We assume also that the stability of X-particles is protected by a discrete symmetry which is the remnant of a gauge symmetry and is respected by all interactions except quantum gravity through wormhole effects. In other words, our particle is very similar to a very heavy neutralino with a conserved quantum number, R', being the direct analogue of R-parity (see [30] and the references therein). Thus, one can assume that the decay of X-particle occurs due to dimension 5 operators, inversely proportional to the Planck mass m_{Pl} and additionally suppressed by a factor $\exp(-S)$, where S is the action of a wormhole which absorbs R'-charge. As an example one can consider a term

$$\mathcal{L} \sim \frac{1}{m_{Pl}} \bar{\Psi} \nu \phi \phi \exp(-S), \tag{9}$$

where Ψ describes X-particle, and ϕ is a $SU(2)$ scalar with vacuum expectation value $v_{EW} = 250$ GeV. After spontaneous symmetry breaking the term (9) results in the mixing of X-particle and neutrino, and the lifetime due to $X \rightarrow \nu + q + \bar{q}$, e.g., is given by

$$\tau_X \sim \frac{192(2\pi)^3}{(G_F v_{EW}^2)^2} \frac{m_{\rm Pl}^2}{m_X^3} e^{2S}, \tag{10}$$

where G_F is the Fermi constant. The lifetime $\tau_X > t_0$ for X-particle with $m_X \geq 10^{13}$ GeV needs $S > 44$. This value is within the range of the allowed values as discussed in Ref. [31].

Let us now turn to the cosmological production of X-particles with $m_X \geq 10^{13}$ GeV. Several mechanisms were identified in [25], including thermal production at the reheating stage, production through the decay of inflaton field at the end of the "pre-heating" period following inflation, and through the decay of hybrid topological defects, such as monopoles connected by strings or walls bounded by strings.

For the thermal production, temperatures comparable to m_X are needed. In the case of a heavy decaying gravitino, the reheating temperature T_R (which is the highest temperature relevant for our problem) is severely limited to value below $10^8 - 10^{10}$ GeV, depending on the gravitino mass (see Ref. [32] and references therein). On the other hand, in models with dynamically broken supersymmetry, the lightest supersymmetric particle is the gravitino. Gravitinos with mass $m_{3/2} \leq 1$ keV interact relatively strongly with the thermal bath, thus decoupling relatively late, and can be the CDM particle [33]. In this scenario all phenomenological constraints on T_R (including the decay of the second lightest supersymmetric particle) disappear and one can assume $T_R \sim 10^{11} - 10^{12}$ GeV. In this range of temperatures, X-particles are not in thermal equilibrium. If $T_R < m_X$, the density n_X of X-particles produced during the reheating phase at time t_R due to $a + \bar{a} \to X + \bar{X}$ is easily estimated as

$$n_X(t_R) \sim N_a n_a^2 \sigma_X t_R \exp(-2m_X/T_R), \tag{11}$$

where N_a is the number of flavors which participate in the production of X-particles, n_a is the density of a-particles and σ_X is the production cross-section. The density of X-particles at the present epoch can be found by the standard procedure of calculating the ratio n_X/s, where s is the entropy density. Then for $m_X = 1 \cdot 10^{13}$ GeV and ξ_X in the wide range of values $10^{-8} - 10^{-4}$, the required reheating temperature is $T_R \sim 3 \cdot 10^{11}$ GeV.

In the second scenario mentioned above, non-equilibrium inflaton decay, X-particles are usually overproduced and a second period of inflation is needed to suppress their density.

Finally, X-particles could be produced by TD such as strings or textures. Particle production occurs at string intersections or in collapsing texture knots. The evolution of defects is scale invariant, and roughly a constant number of particles ν is produced per horizon volume t^3 per Hubble time t. ($\nu \sim 1$ for textures and $\nu \gg 1$ for strings.) The main contribution to to the X-particle density is given by the earliest epoch, soon after defect formation, and we find $\xi_X \sim 10^{-6} \nu (m_X/10^{13}\ GeV)(T_f/10^{10}\ GeV)^3$, where T_f is the defect formation temperature. Defects of energy scale $\eta \gtrsim m_X$ could be formed at a phase transition at

286

or slightly before the end of inflation. In the former case, $T_f \sim T_R$, while in the latter case defects should be considered as "formed" when their typical separation becomes smaller than t (hence $T_f < T_R$). It should be noted that early evolution of defects may be affected by friction; our estimate of ξ_X will then have to be modified. X particles can also be produced by hybrid topological defects: monopoles connected by strings or walls bound by strings. The required values of n_X/s can be obtained for a wide range of defect parameters.

The decays of X-particles result in the production of nucleons with a spectrum $W_N(m_X, x)$, where m_X is the mass of the X-particle and $x = E/m_X$. The flux of nucleons $(p, \bar{p}, n, \bar{n})$ from the halo and extragalactic space can be calculated as

$$I_N^i(E) = \frac{1}{4\pi} \frac{n_X^i}{\tau_X} R_i \frac{1}{m_X} W_N(m_X, x), \tag{12}$$

where index i runs through h (halo) and ex (extragalactic), R_i is the size of the halo R_h, or the attenuation length of UHE protons due to their collisions with microwave photons, $\lambda_p(E)$, for the halo case and extragalactic case, respectively. We shall assume $m_X n_X^h = \xi_X \rho_{CDM}^h$ and $m_X n_X^{ex} = \xi_X \Omega_{CDM} \rho_{cr}$, where ξ_X describes the fraction of X-particles in CDM, Ω_{CDM} is the CDM density in units of the critical density ρ_{cr}, and $\rho_{CDM}^h \approx 0.3 \; GeV/cv^3$ is the CDM density in the halo. We shall use the following values for these parameters: a large DM halo with $R_h = 100$ kpc (a smaller halo with $R_h = 50$ kpc is possible, too), $\Omega_{CDM} h^2 = 0.2$, the mass of X-particle in the range 10^{13} GeV $< m_X < 10^{16}$ GeV, the fraction of X-particles $\xi_X \ll 1$ and $\tau_X \gg t_0$, where t_0 is the age of the Universe. The two last parameters are convolved in the flux calculations in a single parameter $r_X = \xi_X t_0/\tau_X$. For $W_N(m_X, x)$ we shall use like in the previous section the QCD fragmentation function in MLLA approximation. For the attenuation length of UHE protons due to their interactions with microwave photons, we use the values given in the book [3].

The high energy photon flux is produced mainly due to decays of neutral pions and can be calculated for the halo case as

$$I_\gamma^h(E) = \frac{1}{4\pi} \frac{n_X}{\tau_X} R_h N_\gamma(E), \tag{13}$$

where $N_\gamma(E)$ is the number of photons with energy E produced per decay of one X-particle, which is given by Eq.(8)

For the calculation of the extragalactic gamma-ray flux, it is enough to replace the size of the halo, R_h, by the absorption length of a photon, $\lambda_\gamma(E)$. The main photon absorption process is $e^+ e^-$-production on background radiation and, at $E > 1 \cdot 10^{10}$ GeV, on the radio background. The neutrino flux calculation is similar.

Before discussing the obtained results, we consider the astrophysical constraints.

The most stringent constraint comes from electromagnetic cascade radiation, discussed in the previous section. In the present case this constraint is weaker, because the low-energy extragalactic nucleon flux is ~ 4 times smaller than that one from the Galactic halo (see Fig. 2). Thus the cascade radiation is suppressed by the same factor.

The cascade energy density calculated by integration over cosmological epochs (with the dominant contribution given by the present epoch $z = 0$) yields in our case

$$\omega_{cas} = \frac{1}{5} r_X \frac{\Omega_{CDM} \rho_{cr}}{H_0 t_0} = 6.3 \cdot 10^2 r_X f_\pi \ \text{eV/cm}^3. \tag{14}$$

To fit the UHE CR observational data by nucleons from halo, we need $r_X = 5 \cdot 10^{-11}$. Thus the cascade energy density is $\omega_{cas} = 3.2 \cdot 10^{-8} f_\pi \ \text{eV/cm}^3$, well below the observational bound.

Let us now discuss the obtained results. The fluxes shown in Fig. 2 are obtained for $R_h = 100$ kpc, $m_X = 1 \cdot 10^{13}$ GeV and $r_X = \xi_X t_0 / \tau_X = 5 \cdot 10^{-11}$. This ratio r_X allows very small ξ_X and $\tau_X > t_0$. The fluxes near the maximum energy $E_{max} = 5 \cdot 10^{12}$ GeV were only roughly estimated (dotted lines on the graph).

It is easy to verify that the extragalactic nucleon flux at $E \leq 3 \cdot 10^9$ GeV is suppressed by a factor ~ 4 and by a much larger factor at higher energies due to energy losses. The flux of extragalactic photons is suppressed even stronger, because the attenuation length for photons (due to absorption on radio-radiation) is much smaller than for nucleons (see Ref. [39]). This flux is not shown in the graph. The flux of high energy gamma-radiation from the halo is by a factor 7 higher than that of nucleons and the neutrino flux, given in the Fig.2 as the sum of the dominant halo component and subdominant extragalactic one, is twice higher than the gamma-ray flux.

The spectrum of the observed EAS is formed due to fluxes of gamma-rays and nucleons. The gamma-ray contribution to this spectrum is rather complicated. In contrast to low energies, the photon-induced showers at $E > 10^9$ GeV have the low-energy muon component as abundant as that for nucleon-induced showers [42]. However, the shower production by the photons is suppressed by the LPM effect [43] and by absorption in geomagnetic field (for recent calculations and discussion see [41,44] and references therein).

We wish to note that the excess of the gamma-ray flux over the nucleon flux from the halo is an unavoidable feature of this model. It follows from the more effective production of pions than nucleons in the QCD cascades from the decay of X-particle.

The signature of our model might be the signal from the Virgo cluster. The virial mass of the Virgo cluster is $M_{\text{Virgo}} \sim 1 \cdot 10^{15} M_\odot$ and the distance to it $R = 20$ Mpc. If UHE protons (and antiprotons) propagate rectilinearly from this source (which could be the case for $E_p \sim 10^{11} - 10^{12}$ GeV), their flux is given by

$$F_{p,\bar{p}}^{\text{Virgo}} = r_X \frac{M_{\text{Virgo}}}{t_0 R^2 m_X^2} W_N(m_X, x). \tag{15}$$

The ratio of this flux to the diffuse flux from the half hemisphere is $6.4 \cdot 10^{-3}$. This signature becomes less pronounced at smaller energies, when protons can be strongly deflected by intergalactic magnetic fields.

IV LSP IS UHE CARRIER

LSP is the Lightest Supersymmetric Particle. It can be stable if R-parity is strictly conserved or unstable if R-parity is violated. To be able to reach the Earth from most remote regions in the Universe, the LSP must have lifetime longer than $\tau_{LSP} \gtrsim t_0/\Gamma$, where t_0 is the age of the Universe and $\Gamma = E/m_{LSP}$ is the Lorentz-factor of the LSP. In case $m_{LSP} \sim 100\ GeV$, $\tau_{LSP} > 1\ yr$.

Theoretically the best motivated candidates for LSP are the neutralino and gravitino. We shall not consider the latter, because it is practically undetectable as UHE particle.

In all elaborated SUSY models the gluino is not the LSP. Only, if the dimension-three SUSY breaking terms are set to zero by hand, gluino with mass $m_{\tilde{g}} = \mathcal{O}(1\ GeV)$ can be the LSP [45]. There is some controversy if the low-mass window $1\ GeV \lesssim m_{\tilde{g}} \lesssim 4\ GeV$ for the gluino is still allowed [46,47]. Nevertheless, we shall study the production of high-energy gluinos and their interaction with matter being inspired by the recent suggestion [26] (see also [27]), that the atmospheric showers observed at the highest energies can be produced by colorless hadrons containing gluinos. We shall refer to any of such hadron as $\tilde{g}$-hadron. Light gluinos as UHE particles with energy $E \gtrsim 10^{16}$ eV were considered in some detail in the literature in connection with Cyg X-3 [48,49]. Additionally, we consider heavy gluinos with $m_{\tilde{g}} \gtrsim 150$ GeV [27].

UHE LSP are most naturally produced at the decays of unstable superheavy particles, either from TD or as the relic ones [28].

The QCD parton cascade is not a unique cascade process. A cascade multiplication of partons at the decay of superheavy particle appears whenever a probability of production of extra parton has the terms $\alpha \ln Q^2$ or $\alpha \ln^2 Q^2$, where Q is a maximum of parton transverse momentum, i.e. m_X in our case. Regardless of smallness of α, the cascade develops as long as $\alpha \ln Q^2 \gtrsim 1$. Therefore, for extremely large Q^2 we are interested in, a cascade develops due to parton multiplication through $SU(2) \times U(1)$ interactions as well. Like in QCD, the account of diagrams with $\alpha \ln Q^2$ gives the Leading Logarithm Approximation to the cascade fragmentation function.

For each next generation of cascade particles the virtuality of partons q^2 diminishes. When $q^2 \gg m_{SUSY}^2$, where m_{SUSY} is a typical mass of supersymmetric particles, the number of supersymmetric partons in the cascade is the same as their ordinary partners. At $q^2 < m_{SUSY}^2$ the supersymmetric particles are not produced any more and the remaining particles decay producing the LSP. In Ref.([28]) a simple Monte Carlo simulation for SUSU cascading was performed and the spectrum of emitted LSP was calculated. LSP take away a considerable fraction of the total energy ($\sim 40\%$).

The fluxes of UHE LSP are shown in Fig. 3 for the case of their production in cosmic necklaces (see section II). When the LSP is neutralino, the flux is somewhat lower than neutrino flux. The neutralino-nucleon cross- section, $\sigma_{\chi N}$, is also smaller than that for neutrino. For the theoretically favorable masses of supersymmetric particles, $\sigma_{\chi N} \sim 10^{-34} \ cm^2$ at extremely high energies. If the the masses of squarks are near their experimental bound, $M_{L,R} \sim 180 \ GeV$, the cross-section is 60 times higher.

Gluino as the LSP is another phenomenological option. Let us discuss shortly the status of the gluino as LSP.

In all elaborated SUSY models the gluino is not LSP, and this possibility is considered on purely phenomenological basis. Accelerator experiments give the lower limit on the gluino mass as $m_{\tilde{g}} \gtrsim 150$ GeV [46]. The upper limit of the gluino mass is given by cosmological and astrophysical constraints, as was recently discussed in [27]. In this work it was shown that if the gluino provides the dark matter observed in our galaxy, the signal from gluino annihilation and the abundance of anomalous heavy nuclei is too high. Since we are not interested in the case when gluino is DM particle, we can use these arguments to obtain an upper limit for the gluino mass. Calculating the relic density of gluinos (similar as in [27]) and using the condition $\Omega_{\tilde{g}} \ll \Omega_{\rm CDM}$, we obtained $m_{\tilde{g}} \ll 9$ TeV.

Now we come to very interesting argument against existence of a light stable or quasistable gluino [50]. It is plausible that the *glueballino* ($\tilde{g}g$) is the lightest hadronic state of gluino [48,49]. However, *gluebarino*, i.e. the bound state of gluino and three quarks, is almost stable because baryon number is extremely weakly violated. In Ref. [50] it is argued that the lightest gluebarino is the neutral state ($\tilde{g}uud$). These charged gluebarinos are produced by cosmic rays in the earth atmosphere [50], and light gluino as LSP is excluded by the search for heavy hydrogen or by proton decay experiments (in case of quasistable gluino). In the case that the lightest gluebarino is neutral, see [45], the arguments of [50] still work if a neutral gluebarino forms a bound state with the nuclei. Thus, a light gluino is disfavored.

The situation is different if the gluino is heavy, $m_{\tilde{g}} \gtrsim 150 \ GeV$. This gluino can be unstable due to weak R-parity violation [30] and have a lifetime $\tau_{\tilde{g}} \gtrsim 1$ yr, *i.e.* long enough to be UHE carrier (see beginning of this section). Then the calculated relic density at the time of decay is not in conflict with the cascade nucleosynthesis and all cosmologically produced $\tilde{g}$-hadrons decayed up to the present time. Moreover, the production of these gluinos by cosmic rays in the atmosphere is ineffective because of their large mass.

Glueballino, or more generally $\tilde{g}$-hadron, looses its energy while propagating from a source to the Earth. The dominant energy loss of the $\tilde{g}$-hadron is due to pion production in collisions with microwave photons. Pion production effectively starts at the same Lorentz-factor as in the case of the proton. This implies that the energy of the GZK cutoff is a factor $m_{\tilde{g}}/m_p$ higher than in case of the proton. The attenuation length also increases because the fraction of energy lost near the threshold is small, $\mu/m_{\tilde{g}}$, where μ is a pion mass. Therefore, even for light $\tilde{g}$-

hadrons, $m_{\tilde{g}} \gtrsim 2 \; GeV$, the steepening of the spectrum is less pronounced than for protons.

The spectrum of $\tilde{g}$-hadrons from the cosmic necklaces accounted for absorption in intergalactic space, is shown in Fig. 3.

A very light UHE $\tilde{g}$-hadron interacts with the nucleons in the atmosphere similarly to UHE proton. The cross-section is reduced only due to the radius of $\tilde{g}$-hadron and is of order of $\sim 1 \; mb$ [49]. In case of very heavy $\tilde{g}$-hadron the total cross-section can be of the same order of magnitude, but the cross-section with the large energy transfer, relevant for the detection in the atmosphere, is very small [28]. This is due to the fact that interaction of gluino in case of large energy transfer is characterized by large Q^2 and thus interaction is a deep inelastic QCD scattering.

Thus, only UHE gluino from low-mass window $1 \; GeV \leq m_{\tilde{g}} \leq 4 \; GeV$ could be a candidate for observed UHE particles, but it is disfavored by the arguments given above.

V   CONCLUSIONS

Topological Defects naturally produce particles with extremely high energies, much in excess of what is presently observed. However, the fluxes from most known TD are too small. So far only necklaces [23] and monopole-antimonopole pairs [20] can provide the observed flux of UHE CR.

Another promising sources of UHE CR are relic superheavy particles [24,25]. These particles should be clustering in the halo of our Galaxy [25], and thus UHE CR produced at their decays do not have the GZK cutoff. The signatures of this model are dominance of photons in the primary flux and Virgo cluster as a possible discrete source.

Apart from protons, photons and neutrinos the UHE carriers can be neutralinos [28], gluino [26–28] and monopoles [29,27]. While neutralino is a natural candidate for the Lightest Supersymmetric Particle (LSP) in SUSY models, gluino can be considered as LSP only phenomenologically. LSP are naturally produced in the parton cascade at the decay of superheavy X-particles. In case of neutralino both fluxes and cross-sections for interaction is somewhat lower than for neutrino. In case of gluinos the fluxes are comparable with that of neutralinos, but cross-sections for the production of observed extensive air showers are large enough only for light gluinos. These are disfavored, especially if the charged gluebarino is lighter than the neutral one [50].

ACKNOWLEDGEMENTS

This report is based on my recent works with Michael Kachelriess and Alex Vilenkin [23,25,28]. I am grateful to my co-authors for pleasant and useful cooperation and for many discussions.

Many thanks are to the organizers of the workshop for the most efficient work. I am especially grateful to Jonathan Ormes for all his efforts as the Chairman of the Organizing Committee and for inviting me to this most interesting meeting.

REFERENCES

1. N.Hayashida et al., Phys.Rev.Lett. **73**, 3491, (1994),
   D.J.Bird et al., Ap.J., **424**, 491, (1994).
2. C.T.Norman, D.B. Melrose and A. Achterberg, Ap.J **454**, 60, (1995).
3. V.S.Berezinsky, S.V.Bulanov, V.A. Dogiel, V.L.Ginzburg, and V.S.Ptuskin, "Astrophysics of Cosmic Rays", chapter 4, ELSEVIER, 1990.
4. M. Vietri, Ap.J **453**, 863 (1995).
5. E.Waxman, Phys. Rev. Lett., **75**, 386 (1995).
6. M.Hoshino, J. Arons, Y.A.Gallant and A.B.Langdon, Ap.J,**390**, 454 (1992).
7. A. Vilenkin and E.P.S. Shellard, Cosmic Strings and Other Topological Defects, Cambridge University Press, Cambridge, 1994; M.B. Hindmarsh and T.W.B. Kibble, Rep. Prog. Phys. **55**, 478 (1995).
8. C.T. Hill, D.N. Schramm and T.P. Walker, Phys. Rev. D36 (1987) 1007;
9. P. Bhattacharjee, C.T. Hill and D.N. Schramm, Phys. Rev. Lett. 69 (1992) 567; G. Sigl, D.N. Schramm and P. Bhattacharjee, Astropart. Phys. 2 (1994) 401;
10. G.Sigl, astro-ph/9611190.
11. K.Greisen, Phys. Rev. Lett. **16**, 748, (1966); G.T.Zatsepin and V.A.Kuzmin, Pisma Zh. Exp. Teor. Fiz. **4**, 114 (1996).
12. P.Bhattacharjee and N.C.Rana, Phys. Lett. **B 246**, 365 (1990).
13. G.Vincent, N.Antunes and M.Hindmarsh, hep-ph/9708427.
14. R.Brandenberger, Nuclear Physics, **B 293**, 812 (1987).
15. A.J. Gill and T.W.B. Kibble, Phys. Rev. D50 (1994) 3660.
16. J.H. MacGibbon and R.H. Brandenberger, Nucl. Phys. **B331**, 153 (1990); P. Bhattacharjee, Phys. Rev. **D40**, 3968 (1989).
17. E. Witten, Nucl. Phys. **B249**, 557 (1985).
18. V.Berezinsky and A.Vilenkin, in preparation.
19. C.T. Hill, Nucl. Phys. **B224**, 469 (1983).
20. P. Bhattacharjee and G. Sigl, Phys. Rev. **D51**, 4079 (1995).
21. V. Berezinsky, X. Martin and A. Vilenkin, Phys. Rev **D 56**, 2024 (1997) .
22. V.Berezinsky, P.Blasi and A.Vilenkin, in preparation.
23. V.Berezinsky and A.Vilenkin, astro-ph 9704257, to be published in PRL.
24. V.A.Kuzmin and V.A.Rubakov , Talk at the Workshop "Beyond the Desert", Castle Rindberg 1997, astro-ph/9709187.
25. V.Berezinsky, M.Kachelriess and A.Vilenkin, astro-ph/9708217, to be published in PRL.
26. D.J.H. Chung, G.R.Farrar and E.W.Kolb, astro-ph/9707036.
27. R.N.Mohapatra and S.Nussinov, hep-ph/9708497.
28. V.Berezinsky and M.Kachelriess, hep-ph/9709485.
29. T. Kephart and T.Weiler, Astrop. Phys. **4**, 271 (1996).

30. V. Berezinsky, A. S. Joshipura and J. W. F. Valle , hep-ph/9608307, to be published in PR D.

31. R. Kallosh, A. Linde, D. Linde, and L. Susskind, Phys.Rev **D52**, 912 (1995).

32. J. Ellis, J. E. Kim, and D. V. Nanopoulos, Phys. Lett. **B145**, 181, (1984); J. Ellis, G. B. Gelmini, C.Jarlskog, G.G.Ross and J.W.F.Valle, Phys. Lett. **B150**, 142 (1985); S. Sarkar, Rep. Prog. Phys. **59**, 1493 (1996).

33. T. Gherghetta, Nucl.Phys. **B485**, 25 (1997).

34. Yu.L.Dokshitzer, V.A,Khose, A.H.Mueller and S.I.Troyan, "Basics of Perturbative QCD", Editions Frontiers, 1991.

35. R.K.Ellis, W.J.Stirling and B.R.Webber, "QCD and Collider Physics", in preparation, 1997.

36. V.Berezinsky, Nucl. Phys. **B380**, 478, (1992).

37. S.Yoshida et al., Astrop. Phys., **3**, 105, (1995).

38. V.S.Berezinsky, Soviet Phys. Nucl. Phys., **11**, 399 (1970).

39. R.J.Protheroe and P.L.Biermann, Astroparticle Physics, **6**, 45, (1996)

40. V.S.Berezinsky, S.I.Grigorieva and O.F.Prilutsky, Proc. of 13th ICRC, Denver, **54**, 479, (1973).

41. R. Protheroe and T. Stanev, Phys. Rev. Lett. **77**, 3708 (1996) and erratum.

42. F. A. Aharonian, B. L. Kanevsky and V. A. Sahakian, J. Phys. **G17**, 1909 (1991).

43. L. D. Landau and I. Pomeranchuk, Dokl. Akad. Nauk SSSR, **92**, 535 (1953); A. B. Migdal, Phys. Rev.,**103**, 1811 (1956).

44. K. Kasahara, Proc. of Int. Symp. "Extremely High Energy Cosmic Rays" (ed. M.Nagano), Tokyo, Sept. 25-28, p.221, (1996).

45. G. R. Farrar, Phys. Rev. Lett. **76** (1996) 4111 and references therein cited.

46. Particle Data Group, Phys. Rev. **D54** (1996) 1.

47. Aleph collaboration, CERN-PPE-97/002, to be published in Z. Phys. C; G. R. Farrar, hep-ph/9707467.

48. G. Auriemma, L. Maiani and S. Petrarca, Phys. Lett. **B164** (1985) 179.

49. V. S. Berezinskii and B. L. Ioffe, Sov. Phys. JETP **63** (1986) 920.

50. M. B. Voloshin and L. B. Okun, Sov. J. Nucl. Phys. 43 (1986) 495.

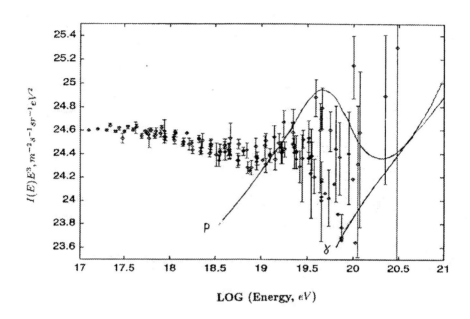

FIGURE 1. Predicted proton (p) and gamma-ray (γ) fluxes from necklaces in comparison with experimental data.

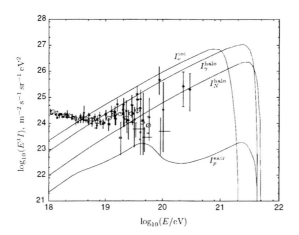

FIG. 2: Predicted fluxes from decaying X-particles: nucleons $(p, \bar{p}, n, \bar{n})$ from the halo (curve I_N^{halo}), extragalactic protons (curve I_p^{extr}), photons from the halo (curve I_γ^{halo}), and neutrinos from the halo and the extragalactic space (curve I_ν^{tot}).

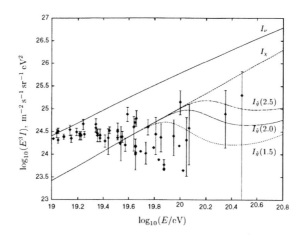

FIG. 3: Predicted fluxes from cosmic necklaces with $r^2\mu = 2 \cdot 10^{27}$ GeV^2: neutrinos (curve I_ν), neutralinos (curve I_χ) and $\tilde{g}$-hadrons with mass $m_{\hat{G}} = 2.5$ GeV, $m_{\hat{G}} = 2.0$ GeV and $m_{\hat{G}} = 1.5$ GeV.

Air Shower Modelling[1]

Thomas K. Gaisser

Bartol Research Institute
University of Delaware
Newark, DE 19716

Abstract. Simulation of showers requires extrapolation of models of hadronic interactions beyond the range of accelerator measurements. After briefly reviewing the problems following from this fact, I comment on the proton-proton cross section at air shower energies and on the use of a simplified scheme for efficient simulation of longitudinal shower profiles.

INTRODUCTION

New and proposed experiments to measure air showers with energies of 10^{20} eV and more [1–5] place severe demands on models of hadronic interactions. Interpretation of the cascades (for example, to determine the nature of the primary cosmic-rays that initiate the showers) will require extrapolation to $\sqrt{s} > 400$ TeV in the nucleon-nucleon center-of-mass system, or more than two hundred times the energy of the highest energy collider experiments. The problem is actually much worse because cascade development depends primarily on the forward fragmentation region (particles that carry a significant fraction of the beam momentum), whereas collider measurements concentrate on the central region which accounts for most of the multiplicity but only a small fraction of the energy of the interaction. In addition, the showers involve interactions with nuclei (both as targets and projectiles). Models thus depend in an important way on accelerator experiments at much lower energy (several hundred GeV lab energy) to constrain the properties of collisions of hadrons and nuclei over all of phase space.

As an illustration, consider the numbers in table 1, which contains a particular extrapolation of the nucleon-nucleon cross section [6]. The third and fourth columns of the table list respectively the corresponding inelastic p-air cross section and the number of wounded nucleons as estimated estimated from the simple geometric formula, $N_W = A\,\sigma_{pp}^{inel}/\sigma_{p-air}^{inel}$. Since the extrapolation of the pp cross section is itself uncertain, so is N_W.

---

[1] Research supported in part by the U.S. Department of Energy under Grant Number DE FG02 01 ER 40626

CP433, *Workshop on Observing Giant Air Showers from Space*
edited by J.F. Krizmanic et al.
© 1998 The American Institute of Physics 1-56396-766-9/98/$15.00

TABLE 1. Set of parameters for inelastic interactions of protons in nitrogen nuclei (the atmosphere).

$\sqrt{s}$	σ_{pp}^{inel} (mb)	σ_{p-air}^{inel} (mb)	N_W
53 GeV	33	285	1.7
200	40	315	1.8
546	50	345	2.1
900	54	365	2.2
1800	62	397	2.3
40 TeV	110	525	3.0
400 TeV	145	620	3.4

Although the extrapolation of the proton-proton cross section is uncertain, it is clear that any reasonably increasing cross section will lead to a significantly increased number of participating nucleons in nuclei at the ultra-high energies sought by these groups. The treatment of nuclei is therefore a critical component of event generators to be used for interpretation of giant air showers. It is interesting to note that most of the models of hadronic interactions in use to simulate air showers at the highest energy (e.g. Refs. [6–9]) share a common feature in their treatment of interactions on nuclear targets. This is the assumption that the energetic valence constituents of a projectile nucleon interact inelastically only once in passing through a target nucleus. When more than one target nucleon participates, all but one involve interactions with the less energetic sea of the projectile. This leads to hadron-nucleus collisions that are relatively more elastic than in a simple geometrical model in which the projectile loses large amounts of energy in successive collisions inside a nucleus. The evidence in favor of the relatively elastic model of hadron-nucleus collisions has been reviewed recently in this context in Ref. [10].

THE PROTON-PROTON CROSS SECTION

As Knapp [11] has emphasized, the cross section extrapolations used in different simulations vary widely. For example, the value of σ_{p-air}^{inel} at 10^{20} eV varies from 450 mb [9] to 600 mb [6]. These numbers are based on extrapolations of the underlying nucleon-nucleon cross sections together with the Glauber multiple scattering formalism [12], which relates hadronic and nuclear cross sections. The relation between the hadron-nucleon cross section and the corresponding hadron-nucleus cross section depends significantly also on the elastic slope parameter,

$$B(s) = \frac{d}{dt}\left[\ln\left(\frac{d\sigma_{el}}{dt}\right)\right]_{t=0}. \tag{1}$$

This relation is discussed in the context of cosmic-ray cascades in Ref. [13]. Qualitatively, the relation is such that for a given value of σ_{pp}^{total}, a larger value of the slope parameter corresponds to a larger proton-air cross section.

Conversely, since the only data above accelerator energies comes from cosmic-ray measurements, one can start from a measurement that is sensitive to σ_{p-air}^{inel} and try to infer the corresponding value of σ_{pp}^{tot}. There are some complications having to do with quasielastic processes, but these can be handled in a straightforward way. [13] As an example, consider the Fly's Eye value of $\sigma_{pp}^{tot} = 120 \pm 15$ mb at $\sqrt{s} = 30$ TeV [14]. This value is obtained using a geometrical scaling assumption [15] to extrapolate the slope parameter to this energy. This results in a large value of $B > 30$ GeV^{-2} and hence (for a measured value of $\sigma_{p-air}^{inel} \approx 540 \pm 50$ mb) a small value of σ_{pp}^{tot}. The original Fly's Eye value of $\sigma_{pp}^{tot} = 120 \pm 15$ mb is consistent with the extrapolated Donnachie & Landshoff parameterization [16], as shown in *Review of Particle Physics* [17]. Using a different model for the slope parameter [18], however, as advocated in the review article of Block & Cahn [19], leads to a slower increase in $B(s)$ and to a larger value of $\sigma_{pp}^{tot} \approx 175^{+40}_{-30}$ mb [13].

In addition to uncertainties in converting from σ_{p-air}^{inel} to σ_{pp}^{tot}, there are significant uncertainties in the determination of σ_{p-air}^{inel} itself. Extracting a p-air cross section from an air shower measurement depends on the primary composition (the proton cross section cannot be measured if there are no protons in the primary cosmic rays) and on the model of hadronic interactions used to interpret the air shower data. Both at Akeno [20] and at Fly's Eye [14], the approach is to look at the attenuation length (Λ) for deeply penetrating showers, on the assumption that, for a given energy, the most deeply penetrating showers are mostly protons. The model-dependence then is compressed into a parameter $a > 1$ in the relation

$$\Lambda = a \times \lambda = a \times \frac{14.5\, m_p}{\sigma_{p-air}^{inel}}. \tag{2}$$

Here λ is the interaction length of protons in air, which has a mean atomic mass of 14.5. The effective value of a depends both on the hadronic interaction model and on the way in which the fluctuations are affected by a contamination of helium and heavier nuclei [21].

An analysis of the tail of the X_{max} distribution of the Fly's Eye stereo data [22] is consistent with a large range of $\sigma_{p-air}^{inel} \sim 530 \pm 100$ mb at $\sqrt{s} \sim 30$ TeV. Assuming $B \approx 20$ GeV^{-2} at this energy, this would allow almost any value of $\sigma_{pp}^{tot} > 100$ mb. (See Fig. 7 of Ref. [13].)

LONGITUDINAL PROFILES OF BIG SHOWERS

A simple hybrid technique with a threshold $E_{th} \leq 10^{-3} \times E_0$, where E_0 is the primary energy per nucleon, is sufficient for simulating longitudinal profiles of high energy showers. In such a scheme [23] one uses a full Monte Carlo for all interactions, with $E_{int} > E_{th}$. Below this energy, parameterizations of longitudinal development of subshowers are used. The parameterizations, of the general form of Ref. [24], are constructed in a bootstrap approach, starting with a full Monte Carlo at low energy and working to higher energy.

The reason such a scheme works can be understood by looking at the fraction of shower energy transferred into the electromagnetic component by interactions with energy $E_{int} > z_{min} \times E_0$ as a function of z_{min}. This value exceeds $\sim 80\%$ for $z_{min} < 10^{-3}$.

In Ref. [23] we calculated $\langle X_{max} \rangle$ as a function of energy for protons and for iron using the SIBYLL [6] interaction model. The solid lines in Fig. 1 show the results as compared to data and to other calculations. Differences among models [11,25] arise primarily from the treatment of the central region and are therefore of greater importance for calculation of low energy muons, where this simple hybrid method is not adequate in any case. The KNP model [26] used in Ref. [27] has a naive treatment of propagation of the projectile through the nucleus, leading to a large value of the inelasticity. In addition, the KNP calculations have been corrected for experimental acceptance. Despite the uncertainties, the comparison still shows some evidence for a transition from heavier toward lighter composition in the energy range above 10^{18} eV. Study of the shape of the X_{max} distribution [22] shows that its breadth requires a mixture of components in this energy region, a result which is less dependent on models than the interpretation of the magnitude of $\langle X_{max} \rangle$. The AGASA group has measured the muon to electron ratio in order to determine the composition in the same energy range as the Fly's Eye result [28]. From their analysis they concluded that there was no change in composition. Recently, however, Dawson, Meyhandan and Simpson [29] have done an analysis of both the Fly's Eye depth of maximum measurements and the AGASA measurements. They conclude that both consistently show evidence for a transition from nearly pure iron around 10^{17} eV to a lighter composition above 10^{18} eV.

Finally, in Fig. 2 [33] I show a comparison of the profile of the 3×10^{20} eV Fly's Eye event [34] to individual simulated shower profiles, ten protons (solid lines) and ten iron (dotted lines). The striking thing about this event is that there is nothing unusual about it except for its size ($\sim 2 \times 10^{11}$ particles at maximum) and energy. It looks like a proton or nucleus.

ACKNOWLEDGEMENT. I am grateful to Ralph Engel, Paolo Lipari and Todor Stanev for collaboration on this work.

REFERENCES

1. T. Abu-Zayyad et al. (Hi-Res Collaboration) Proc. 25th Int. Cosmic Ray Conf. (Durban) 5, 321 (1997).
2. Cosmic Rays Above 10^{19} eV (ed. M. Boratav et al.), Nucl. Phys. Proc. Suppl. 28B (1992); The Pierre Auger Project (Design Report), The Auger Collaboration (1995).
3. M. Teshima et al. (Telescope Array Collaboration) Proc. 25th Int. Cosmic Ray Conf. (Durban) 5, 369 (1997).
4. J. Ormes et al., http://lheawww.gsfc.nasa.gov/docs/gamcosray/hecr/owl.
5. J. Linsley et al., Proc. 25th Int. Cosmic Ray Conf. (Durban) 5 381, 385 (1997).

6. R.S. Fletcher, T.K. Gaisser, Paolo Lipari & Todor Stanev, *Phys. Rev. D* **50**, 5710 (1994).

7. N.N. Kalmykov, S.S. Ostapchenko & A.I. Pavlov, *Nucl. Phys. B (Proc. Suppl.)* **52B**, 17 (1997).

8. J. Ranft, INFN/AE-97/45 *DPMJET version II.3 and II.4* (1997).

9. K. Werner, *Phys. Reports* **232**, 87 (1993).

10. G.M. Frichter, T.K. Gaisser & Todor Stanev, *Phys. Rev. D* **56**, 3135 (1997).

11. J. Knapp, Rapporteur talk at 25th Int. Cosmic Ray Conf. (Durban) (1997) (astro-ph/9710277).

12. R.J. Glauber & G. Matthiae, *Nucl. Phys.* **B21**, 135 (1970).

13. T.K. Gaisser, U.P. Sukhatme & G.B. Yodh, *Phys. Rev. D* **36**, 1350 (1987).

14. R.M. Baltrusaitis *et al. Phys. Rev. Letters* **52**, 1380 (1984).

15. J. Dias de Deus & P. Kroll, *Acta Phys. Pol. B* **9**, 159 (1978). See also A.J. Buras & J. Dias de Deus, *Nucl. Phys.* **B71**, 481 (1974).

16. A. Donnachie & P.V. Landshoff, *Phys. Letters* **B296**, 227 (1992).

17. *Review of Particle Physics, Phys. Rev. D* **54**, 193 (1996).

18. T. Chou & C.N. Yang in *Proc. 2nd Int. Conf. on High Energy Physics and Nuclear Structure* (Rehovot, ed. G. Alexander, North Holland, 1967). See also C. Bourrely, J. Soffer & T.T. Wu, *Phys. Rev. Letters* **54**, 757 (1985).

19. M.M. Block & R.N. Cahn, *Rev. Mod. Phys.* **57**, 563 (1985).

20. M. Honda, *et al.*, *Phys. Rev. Letters* **70**, 525 (1993).

21. Todor Stanev, in *Proc. Int. Conf. on Physics Simulations at High Energy* (Madison) ed. V. Barger, T. Gottschalk & F. Halzen (World Scientific, Singapore), 141 (1986).

22. T.K. Gaisser *et al.*, *Phys. Rev.* **D47**, 1919 (1993).

23. T.K. Gaisser, Paolo Lipari & Todor Stanev, *Proc. 25th Int. Cosmic Ray Conf.* (Durban) **6**, 281 (1997).

24. T.K. Gaisser & A.M. Hillas, *Proc. 15th Int. Cosmic Ray Conf.* (Plovdiv), **EA**, Volume 8, 353 (1977).

25. H. Klages, Highlight talk at 25th Int. Cosmic Ray Conf. (Durban) (1997).

26. B.Z. Kopeliovich, N.N. Nikolaev & I.K. Potashnikova, *Phys. Rev. D* **39**, 769 (1989).

27. D. Bird *et al. Phys. Rev. Letters* **71**, 3401 (1993).

28. N. Hayashida *et al. J. Phys. G* **21**, 1101 (1995).

29. B.R. Dawson, R. Meyhandan & K.M. Simpson, *Proc. 25th Int. Cosmic Ray Conf.* (Durban) **4**, 25 (1997).

30. R. Boothby *et al.* (DICE Collaboration) astro-ph/9710168 (1997).

31. J. Cortina *et al.* (HEGRA Collaboration) *Proc. 25th Int. Cosmic Ray Conf.* **4**, 69 (1997).

32. N.N. Kalmykov & G.B. Khristiansen, *J. Phys. G* **21**, 1279.

33. T.K. Gaisser in *Proc. Escuela Mexicana de Astrofísica Nuclear* (ed. Jorge Hirsch, to be published by Cambridge University Press, 1998).

34. D.J. Bird *et al. Ap.J.* **441**, 144 (1995).

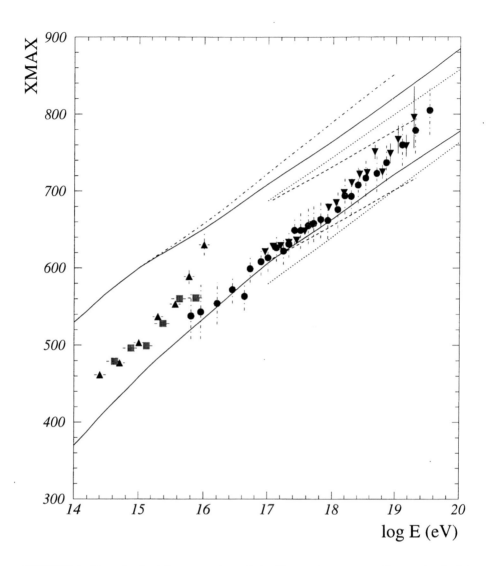

FIGURE 1. Mean depth of shower maximum (g/cm^2) vs. primary energy. Data: triangles below 10^{16} eV—DICE [30]; squares—HEGRA [31]; circles—Yakutsk [32]; inverted triangles—stereo Fly's Eye [27]. The upper set of lines shows simulations of proton showers, while the lower set shows showers initiated by nuclei of iron. Solid lines—SIBYLL [6]; dotted lines—QGSjet [7]; dashed lines—KNP [26] from Ref. [27]; The dash-dot line is from the SIBYLL model but using σ_{p-air}^{inel} calculated from the Donnachie & Landshoff parameterization of σ_{pp}^{tot}.

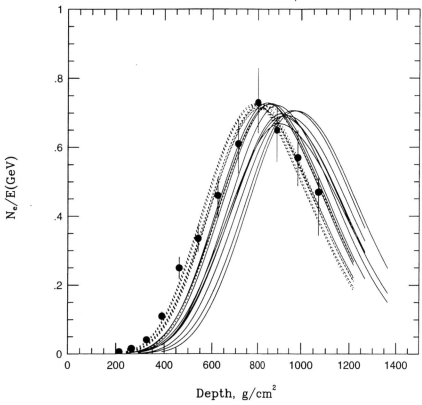

FIGURE 2. Shower size vs. atmospheric depth for the 3×10^{20} eV event of Fly's Eye [34]. Solid lines show 10 random simulations of proton-initiated showers and the dashed lines, 10 showers from incident nuclei of iron (©1998, Cambridge University Press. Reprinted with the permission of Cambridge University Press) [33].

Principles of Wide Angle, Large Aperture Optical Systems

David J. Lamb, Russell A. Chipman[†], Lloyd W. Hillman[†], Yoshiyuki Takahashi, and John O. Dimmock[†]

Department of Physics and [†]Center for Applied Optics
The University of Alabama in Huntsville, Huntsville, Alabama 35899

Abstract. The general principles of designing wide angle, large aperture optical systems are addressed. In particular, the advantages and disadvantages of reflective and refractive optical systems are discussed. It is shown that reflective optics have an inherently smaller field of view than do refractive systems. Conservation of étendue is discussed and directly applied to the determination of focal plane size of all optical systems as well as the central obscuration of systems with reflective primaries.

INTRODUCTION

Many interesting astronomical phenomena occur at unknown or seemingly random locations in the sky. Appearance of new comets and supernova are examples in astronomy. Air showers caused by extremely energetic cosmic rays, Cherenkov lights caused by highly energetic gamma rays, and possible optical counterparts to gamma ray bursts are all further examples of such phenomena. The nature of the origin of all of these different kinds of events is not readily understood, and much data is needed to provide physical insight and justification to existing astronomical theories. A major goal of the experimental astrophysics community is to observe as many events as possible to help unlock the mysteries of the universe.

Optical systems that view these type of events, in general, require a large aperture to collect a detectable signal. Furthermore, because they search the sky for phenomena that occur in unknown locations, such systems must have a large field of view. This will ensure that many events will be observed over a reasonable period of time. Fundamental conservation laws in optics require that the focal plane of large aperture, wide field systems be relatively large. The only way to minimize the size of the focal plane is to make the system as "fast" as possible (i.e., have as large a numerical aperture or as small an f/# as possible).

CP433, *Workshop on Observing Giant Air Showers from Space*
edited by J.F. Krizmanic et al.

In general, fast, large aperture, wide field systems that provide diffraction limited image quality are difficult to design. This task becomes even more difficult when designing systems that must be simple enough to deploy and maintain in space. Fortunately, optical systems that are used to detect events such as air showers do not require diffraction limited image quality. In fact, the resolution requirements for a proposed space based air shower detector (OWL: Orbiting, Wide-angle, Light-collector) are up to four orders of magnitude *above* the diffraction limit.

From an optical design point of view, there are two basic approaches to solving this problem of a large aperture, wide angle system. The first involves the use of many narrow field, large aperture systems in combination to cover the entire field of view. The second involves the use of a single large aperture system with a sufficiently wide field of view. The second option is most attractive for space-based applications in which mission cost increases drastically with the number of apertures that must be deployed. The first option does have its advantages, however. If one of the small field systems fails for some reason, then data can still be taken by the other systems. Also, a narrow field system can produce the same image quality as a large field system in a much simpler configuration (though more individual systems are required to cover the entire field of view). These advantages must be balanced and considered with a risk assessment when deciding which approach to take.

REFLECTIVE VERSUS REFRACTIVE OPTICS

The type (reflective, refractive, diffractive, catadioptric, etc.) of optical system that is used will further determine the approach (a single wide field system versus the combination of many small field systems) that is taken in creating a system. For air shower observatories and most other large aperture systems, diffractive optics prove to be difficult to manufacture on the scale of meters. The properties of diffractive optics tend to vary greatly with angle of incidence which reduces the effectiveness of such optics in large field applications. Reflective optics (mirrors) and refractive optics (lenses and prisms) are the typical components of choice for developing large aperture systems. Each type of optic, however, has its own advantages and disadvantages. A catadioptric optical system utilizes both reflective and refractive optics together to balance these pluses and minuses.

When designing and optimizing optical systems of this nature, it is desired to provide image quality that meets the specified resolution requirements over the entire field of view. The on-axis image is typically the best. The quality of this image can then be sacrificed to provide better image quality for the off-axis fields until all fields meet the resolution specification. The degree to which the field of view can be opened in this manner is highly dependent upon the type of optical components that are used. Reflective optics tend to have a much narrower field of view than do refractive optics. This can be seen through a very simple aberration analysis.

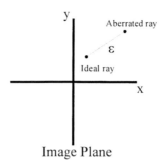

FIGURE 1. Locations of an ideal ray and a real ray in the image plane of an optical system.

Non-perfect imaging is the result of aberrations that are present in optical systems. In the image plane, the distance between a real ray and its ideal location (as determined by perfect imaging) is denoted by the quantity ε. This quantity can be expanded in a Taylor series, and the effects of aberrations can be broken down into different orders. First order, or paraxial, optical analysis yields perfect imaging and is usually used to quantify the general properties of an optical system. When the aberrations are large (as in the case of OWL), the third order correction to the paraxial terms will be at least as large, if not much larger, than any higher order terms (1). The surface contributions of four of the five primary third order aberrations are proportional to a term that involves the difference in index of refraction on either side of the surface (2).

$$\varepsilon \propto \frac{n}{n'}(n - n') \tag{1}$$

The quantities n and n' in equation (1) represent the index of refraction of the medium before and after the surface, respectively. The index of refraction of a typical glass element is approximately 1.5, and the difference in index of refraction between that and air is approximately equal to 0.5. For a reflective surface in air, however, this difference is equal to 2 (the mathematical convention in using these equations with a reflective surface is to set n' = -n). The magnitude of the aberration given by equation (1) is, therefore, different for reflective and refractive surfaces.

$$\varepsilon_{reflective} \propto 2$$
$$\varepsilon_{refractive} \propto 0.75 \; or \; 0.33 \tag{2}$$

With all else being equal, then, the third order surface contributions to the total aberration will be greater for a reflective surface than a refractive surface. Even though a refractive element has two such surfaces, the algebraic sign of the two surface contributions may be opposite such that they actually correct one another. Since the aberrations are smaller for a refractive element at a given angle of incidence, its field of view can be increased until the aberrations are equal to that of a mirror. It is this phenomena that enables lenses to have a larger possible field of view than mirrors.

Mirrors, for this reason, are typically used in small field applications that require excellent image quality (mirrors do not suffer from chromatic aberration which is caused by dispersion). By placing the stop at the center of curvature of a spherical mirror, many aberrations can be completely eliminated due to symmetry about the center of curvature (3). Residual aberrations can be reduced by making the mirror an asphere, but that breaks the symmetry for off axis fields. This is why reflective optics that are used in this manner can provide excellent image quality for only a very small field of view. All attempts to increase the field of view of reflective systems have relied on the use of refracting corrector elements (4).

CONSERVATION OF ÉTENDUE AND DETERMINATION OF FOCAL PLANE SIZE AND CENTRAL OBSCURATION

Étendue is a purely geometrical quantity that defines the flux gathering and transmitting capabilities of an optical system (5). Equation (3) gives an expression for the étendue of an optical system, and figure 2 illustrates the geometrical parameters involved. The index of refraction associated with the surface medium is given by n, and a differential area element on the surface is represented by dA. The differential projected solid angle of the flux striking the optical surface is represented by $d\Omega \cos(\theta)$. Integrating over the entire optical surface and over all angles of flux incidence, the étendue of a particular surface can be calculated.

$$E = n^2 \iint dA \cos(\theta) d\Omega \qquad (3)$$

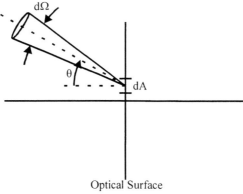

FIGURE 2. Geometrical quantities involved in the calculation of étendue. Total étendue is calculated by integrating over the entire area of the surface and over all angles of incident flux.

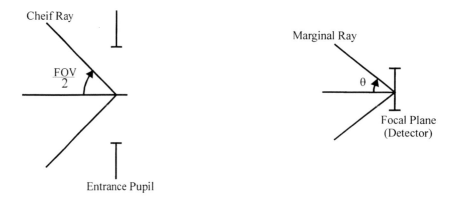

FIGURE 3. Cones of light used to calculate the étendue of the entrance pupil and detector planes.

If the optical system is lossless (i.e. light is not absorbed, vignetted by any limiting apertures, nor lost due to ray splittings at interfaces), then this quantity is invariant throughout the entire system. Because it is a constant, the étendue of any two surfaces in the same optical system can be equated. This enables certain geometrical quantities to be determined for one surface based upon those of another. In particular, the area required for the focal plane of a system can be determined (to first order) by knowing the étendue at the entrance pupil and the f/# of the optical system.

The expression for the étendue in equation (3) is easily evaluated at the entrance pupil and detector planes. Figure 3 depicts the areas and the cones of illumination that are used to perform the calculations. The field of view is denoted by FOV in figure 3. If A denotes the total area and Ω represents the solid angle associated with a given surface, then equation (3) reduces to equations (4) and (5) for the pupil and detector, respectively.

$$E_{EP} = n_{EP}^2 A_{EP} \Omega_{EP} = n_{EP}^2 A_{EP} 4\pi \sin^2\left(\frac{1}{2}\frac{FOV}{2}\right) \tag{4}$$

$$E_D = n_D^2 A_D \Omega_D = n_D^2 A_D 4\pi \sin^2\left(\frac{1}{2}\theta\right) \tag{5}$$

The quantity, θ, in equation (5) and figure 3 is simply related to the inverse sine of the numerical aperture of the optical system. This angle is, further, related to the f/# of the system by equation (6). This is valid for any system in which the object is located very far away from the system compared to the focal length of the system, as in a telescope.

$$n_D \tan\theta = \frac{1}{2f/\#} \Rightarrow \theta = \tan^{-1}\left(\frac{n_D}{2f/\#}\right) \tag{6}$$

Assuming the system is lossless, the étendue at the focal plane can be equated to the étendue at the entrance pupil. Equations (4), (5), and (6) can then be used to develop an expression for the focal plane size of an optical system based upon its field of view, f/#, and entrance pupil radius, r_{EP}. In most cases the optical system is immersed in air or vacuum, and the indices of refraction of the entrance pupil and detector are equal to unity.

$$n_{EP}^2 A_{EP} \Omega_{EP} = n_D^2 A_D \Omega_D \tag{7}$$

$$A_D = A_{EP} \left(\frac{\Omega_{EP}}{\Omega_D} \right) \left(\frac{n_{EP}}{n_D} \right)^2 \tag{8}$$

$$A_D = \pi \cdot r_{EP}^2 \left[\frac{\sin^2 \left(\frac{1}{2} \frac{FOV}{2} \right)}{\sin^2 \left[\frac{1}{2} \tan^{-1} \left(\frac{n_D}{2 f /\#} \right) \right]} \right] \left(\frac{n_{EP}}{n_D} \right)^2 \tag{9}$$

For the case of OWL, the entrance pupil radius will have to vary as the square of the orbit height in order to collect a detectable signal. As the orbit height increases, however, the field of view that is necessary to view a constant area on the earth's surface becomes smaller and smaller. Equation (9) is plotted in figure 4(a) as a function of f/# for various OWL orbit heights assuming that the image and object space indices of refraction are unity. In each case, it is also assumed that a circle of radius 500 km is visible on the ground. This area is to be divided into 1 km² pixels which implies that the resolution of higher orbit systems must be better than that of lower orbit systems. So, the field of view, resolution, and entrance pupil size of the various optical systems at different orbits vary, but the total number of pixels viewed remains constant in the calculation.

In addition to providing information regarding the size of the focal plane required in a particular optical task, étendue analysis can be used in a similar fashion to determine the extent to which a reflective optical system is obscured. Assuming, once again, that the image and object space indices of refraction are unity, the obscuration ratio of the detector plane to the entrance pupil is simply given by the ratio of the respective solid angles (as indicated by equation (7)).

$$Obscuration\ Ratio = \frac{A_D}{A_{EP}} = \frac{\Omega_{EP}}{\Omega_D} = \frac{\sin^2 \left[\frac{1}{2} \frac{FOV}{2} \right]}{\sin^2 \left[\frac{1}{2} \tan^{-1} \left(\frac{1}{2 f /\#} \right) \right]} \tag{10}$$

This equation is valid for telescopes with reflective primaries as well catadioptric

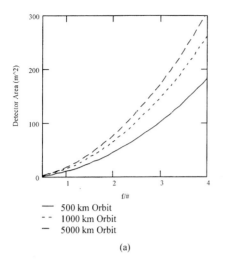
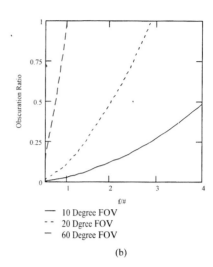

— 500 km Orbit	— 10 Degree FOV
- - 1000 km Orbit	- - 20 Dgree FOV
— 5000 km Orbit	— 60 Degree FOV
(a)	(b)

FIGURE 4. Detector area for a proposed OWL system as a function of system f/# (a). The central obscuration of a reflective system is shown as a function of f/# for various fields of view (b).

systems in which the refractive correctors are placed near the aperture stop and have relatively weak optical power (as is the case in a Schmidt camera). Given a required field of view and an obscuration ratio target value, this equation can be used to determine the necessary speed of the optical system. Very simple analysis of this equation demonstrates that slow, wide field reflective or catadioptric systems are inefficient light collectors. Figure 4(b) shows how the central obscuration varies with f/# for systems with varying fields of view. Equation (10) actually gives a worst case scenario in that the focal plane of an optical system can have curvature. A detector surface with curvature will require the same area as calculated by the above étendue analysis, but the extent to which it obscures the pupil will be diminished due to its reduced projected area.

CONCLUSIONS

Wide angle, large aperture optical systems certainly have vast potential for use in experimental astrophysics. Their development presents many challenges from both optical engineering and fundamental physics perspectives, and various configurations must be investigated for each particular application. The configuration that is used to provide wide field, large aperture imaging is dependent upon the type of primary optics that are used. A system that consists of a single aperture will probably have to be refractive in nature due to the inherently large field of view of refractive optics compared to reflective optics. Many small field, large aperture systems could be used in conjunction to cover an extended field of view. Reflective systems could possibly be used in this manner, although the added complexity may be prohibitive for space applications.

Refractive optics do have the distinct disadvantage of being massive and bulky when scaled to diameters of the order of meters. They also suffer from chromatic aberration induced by dispersion, and they generally tend to have some absorption in the waveband of interest. Much research is currently being performed in the area of Fresnel lenses to minimize these adverse effects (6) (7). Fresnel lenses can be made very thin and, hence, reduce effects of absorption and component mass.

It has been shown that fundamental conservation laws in optics (conservation of étendue) places restrictions on the overall size of the focal plane of a particular system. Due to the amount of information that must be transmitted through large aperture, large field systems, they will necessarily have a large focal plane which could dominate the overall system cost. To minimize the required focal plane size, it is necessary to develop optics that have a low f/# ("fast" optics). Research is also being performed in the area of focal plane minimization for large aperture, wide field systems through the use of re-imaging optics (8).

REFERENCES

1. Hopkins, R.E. and Hanau, R., "Aberration Analysis and Third Order Theory," in *Military Standardization Handbook: Optical Design, MIL-HDBK* 141, U.S. Defense Supply Agency, Washington, D.C., 1962.
2. Hillman, L.W., *Geometrical Optics Notes*, Huntsville, 1994, ch. 16, pp. 1-6.
3. Malacara, D. and Malacara, Z., *Handbook of Lens Design*, New York: Marcel Dekker, Inc., 1994, ch. 15, pp. 468-469.
4. Smith, W.J., *Modern Optical Engineering*, New York: McGraw Hill, 1990, ch. 13, pp. 446-452.
5. Boyd, R.W., *Radiometry and the Detection of Optical Radiation*, New York: John Wiley and Sons, 1983, ch. 5, pp. 89-91.
6. Lamb, D.J., et al, "Wide Angle Refractive Optics for Astrophysics Applications," presented at the Workshop on Observing the Highest Energy Particles ($>10^{20}$ eV) from Space, College Park, MD, November 1997.
7. Lamb, D.J., et al, "Computer Modeling of Optical Systems Containing Fresnel Surfaces," presented at the Workshop on Observing the Highest Energy Particles ($>10^{20}$ eV) from Space, College Park, MD, November 1997.
8. Lamb, D.J., et al, "Focal Plane Reduction of Large Aperture Optical Systems," presented at the Workshop on Observing the Highest Energy Particles ($>10^{20}$ eV) from Space, College Park, MD, November 1997.

Auger: What, Why and How?

Clem Pryke for the Auger Collaboration

Enrico Fermi Institute, University of Chicago, Chicago, IL 60637, USA

Abstract. The Pierre Auger Observatories for the highest energy cosmic rays will be hybrid systems, employing water Čerenkov ground arrays over-looked by atmospheric fluorescence detectors. This will allow accurate reconstruction of primary cosmic ray energy and arrival direction in a cost effective manner, and without dependence on uncertain models. In this paper the design of the proposed instrument is reviewed, together with the simulations on which it is based, and the resulting performance predictions.

I WHAT?

The proposed Auger Observatories are of a hybrid design, employing fluorescence detectors overlooking ground arrays. During clear, moon-less nights all events will be observed both by the nitrogen scintillation light given off in the atmosphere, and also by particle detectors at ground level (this is 10% of the time). The remaining 90% of the time the ground arrays will work alone, aided by the data collected in the hybrid mode, collecting a large sample of events at energies above 10^{19} eV. Figure 1 shows the layout of an Auger observatory. Each ground array consists of 1600 water Čerenkov tanks arranged in a hexagonal grid with spacing of 1.5 km; three fluorescence eyes overlook each array. There will be two identical installations, one each in the Northern and Southern hemispheres, to obtain coverage of the complete celestial sphere.

Since this meeting has focused on the nitrogen fluorescence technique it is perhaps worthwhile to review the ground array method. Whereas a fluorescence detector measures the longitudinal profile of a shower as it grows and then diminishes while passing through the atmosphere, a surface array measures it in a single plane as the shower particles strike ground level.

When considering fluorescence detectors it is usually acceptable to treat the shower as a point source of light moving along the projection of the primary particle's trajectory at velocity c. Since the fall off of particle density at increasing distance from the axis is very steep, the numerical bulk of the particles are within ≈ 50 m of the core. However, there is a measurable density of particles at distances up to several km. Figure 2 shows the longitudinal and lateral distributions of a

CP433, *Workshop on Observing Giant Air Showers from Space*
edited by J.F. Krizmanic et al.
© 1998 The American Institute of Physics 1-56396-766-9/98/$15.00

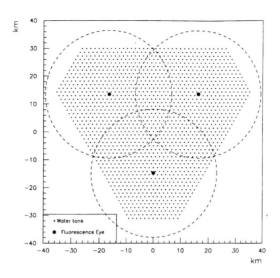

FIGURE 1. Layout of an Auger Observatory. The large circles indicate the approximate viewing range of the fluorescence detectors at 10^{19} eV.

single simulated shower.

It was using ground arrays that air showers were first measured, and it was at the Volcano Ranch Array in 1962 that the first event was observed which was claimed to have a primary energy of $\geq 10^{20}$ eV [1]. As they sweep down through the atmosphere, the shower particles form a curved disc normal to the primary cosmic ray's trajectory. The particle detectors which form a ground array take a number of "samples" of the shower front as it intersects the ground. The relative arrival time of the particles allow the shower direction to be inferred. The amount of signal recorded at each unit allows the core position and shower size to be reconstructed.

A fluorescence detector observes cosmic rays in much the same way as a video camera might observe meteor trails; the difference is that the air shower event is much fainter, and has a much greater angular velocity. A set of photomultiplier tubes view each small patch on the sky, and a series of hits allow the shower direction across the sky to be determined. The operation of hybrid air shower detector is illustrated in figure 3.

II   WHY?

The reason for building a cosmic ray detector with time-averaged aperture on the order of 10^4 km^2 sr is clear; there are events at energies $\gg 10^{20}$ eV and such an aperture will be required to observe a statistically meaningful number of them. The previous and current experiments, Fly's Eye and AGASA, both had apertures of the order of 10^2 km^2 sr, and each observed one event well in excess of 10^{20} eV [2,3]

313

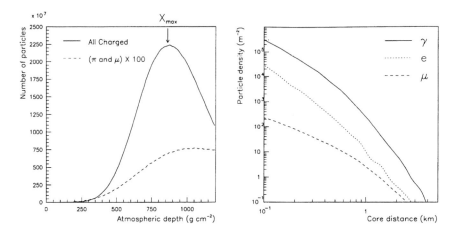

FIGURE 2. Simulated longitudinal and lateral profiles of a giant air shower. At left the number of charged particles is plotted against atmospheric depth. At right the lateral distribution is plotted for particles reaching ground level. The simulation is for a 3×10^{19} eV proton incident at 45° to the zenith, with ground level at 1450 m asl.

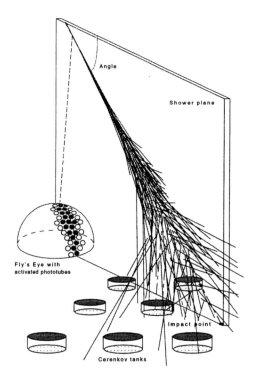

FIGURE 3. Illustration of the operation of a Hybrid air shower detector.

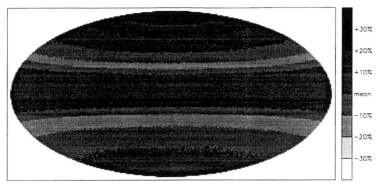

FIGURE 4. Relative all sky exposure of the Auger Observatories.

during a period of approximately 5 years (AGASA is still running). The next generation HiRes detector will achieve 10^3 km^2 sr, but this will still not be enough to settle the formidable questions posed by the existence of such events.

It is difficult enough to construct plausible theories of acceleration to $\gg 10^{20}$ eV even when the most extreme astrophysical objects are invoked (such as AGN's). However, due to interactions with the all-pervading CMB radiation, the search volume must be limited to on the order of 100 Mpc, and there are few, if any, suitable candidates. To compound the mystery, the extra-galactic magnetic fields do not seem strong enough to deflect a proton propagating such a distance by more than a few degrees. Therefore one expects the events to point back towards their sources; there are no obvious candidates so far [4].

The Auger project aims to provide the necessary event statistics, to do it in a cost effective way, and furthermore to be a conservative detector which makes all-important experimental cross checks and inter-calibrations.

The need for sites in the Northern and Southern hemispheres is clear; since the phenomenon under study is so poorly understood it is necessary to survey the complete celestial sphere. A water Čerenkov ground array can reconstruct events at up to 60 degrees from the zenith, while a fluorescence detector can go nearly to the horizon. Even in the latter case it is not possible to observe the entire sky from a single site with acceptable uniformity. Figure 4 shows the relative all sky exposure for the Auger water Čerenkov arrays; variation is on the order of ±20% over the bulk of the celestial sphere.

The combination of a surface array and fluorescence detector characterizes showers to a greater degree than either technique alone. The decision to use the two techniques together is based upon a set of related considerations which can be grouped into the four categories below.

- **Energy Spectrum Measurement:** The two methods have different problems when measuring the intensity of cosmic rays. To do this one must know both the number of events at a given energy, and the effective detection aperture at that energy. A fluorescence detector has an (in principle) direct energy calibration, but the aperture is uncertain. A ground array has a known aperture, but a more indirect energy calibration.

The energy calibration of a ground array comes (in most cases) from calculations of shower development, either analytic or Monte Carlo. Since the first few cascade generations are above the energy range which has been studied with accelerators, large systematic errors could result [1] —this is the Achilles heel of ground array detectors.

The aperture of a ground based fluorescence detector grows with energy as brighter events can be seen from further away. The detection aperture also varies as a function of night sky background, and hence from night to night, and can only be determined by a detector Monte Carlo which includes the details of the trigger electronics.

By combining the two techniques it is possible to have the best of both worlds. The 10% sub-sample of events collected in hybrid mode will allow the energy calibration constants for the ground array to be extracted. The ground array data can then be analyzed without model dependence, and with a well known aperture.

- **Direction measurement:** The direction resolution of a stereo fluorescence detector is somewhat better than that of a ground array. However, optical detection is only possible on clear, moon-less nights (10% duty cycle). Hence the aperture waxes and wanes as a function of both lunar and seasonal cycles. A ground array runs 24 hours a day, 365 days a year, giving a constant aperture. Ground array data is much more straightforward to interpret when investigating the arrival direction distribution.

- **Composition Sensitivity:** A fluorescence detector measures directly the depth of shower maximum (X_{max}). This is a powerful composition parameter, but it is important to note that its interpretation is model dependent. A ground array measures semi-independent composition parameters. A hybrid detector therefore offers the best chance of being able to both select a reasonable shower model, and deduce the primary composition, simultaneously and independently.

- **Practical Considerations:** We believe our design to be highly competitive in terms of dollars spent per shower detected above 10^{20} eV. However, cost considerations do not control the ratio of the apertures of the fluorescence and ground array components. The fact that the bulk of the data is recorded in ground-only mode is simply because the optical detectors can operate only 10% of the time (clear, moon-less nights).

---

[1] In fact on closer examination, it turns out that even quite radical changes to the high energy interaction model have small effects when trying to determine shower energy [5]. Room for maneuver is limited when one considers the existing accelerator data [6].

III HOW?

The proposed Auger fluorescence detectors are similar in design to the Fly's Eye / Hi-Res [7]. The surface array is made up of water Čerenkov tanks, and resembles the array successfully operated by the Haverah Park group for more than twenty years [8] (although on a much larger scale). The tanks are to be cylindrical with 10 m² top area, and 1.2 m deep. They will be lined with a highly reflective material, and filled with purified water. Large photomultiplier tubes (pmts) will view the water volume vertically downwards from above; when an event trigger is generated the pmt waveforms will be recorded by a flash-ADC system. Figure 5 shows an artist's impression of a ground array station. The detectors will be semi-autonomous, relying on battery-backed solar power, radio data communications, and obtaining time synchronization from the Global Positioning Satellite (GPS) system.

There will be 3 fluorescence eyes per site, each viewing an elevation range of 30 degrees, with a complete circle in azimuth. Pixel size will be 1.5 degrees, with 48 mirror units per eye, each unit covering a 15 degree by 15 degree region of sky. A sketch of a possible fluorescence detector unit is shown in figure 6; there would be 12 such units per eye. Sites have been chosen for the Observatories in Utah, USA, and Mendoza Province, Argentina. A large international collaboration has been formed (more than 250 people), and funding is being organized.

Extensive simulations have been carried out to predict the performance of the proposed instrument. The first step is the generation of simulated air showers. These are then passed through simulations of the ground array and fluorescence detectors. The fake data thus generated is reconstructed, and the results compared to the input shower parameters to determine the experimental resolutions. In this work the Mocca shower simulation code [9] is used, with the Sibyll high energy hadronic interaction generator [10]. Individual events from the shower simulation are input to the detector simulation stage. Hence shower-to-shower fluctuation effects are fully included, and if anything, are over estimated.

The fluorescence and ground array detectors are simulated in great detail so that measurement fluctuation effects should be fully included. Figure 7 shows some example Flash-ADC waveforms for tanks taking part in a ground array event.

The hybrid data is reconstructed using a similar method to that described in [11]. The arrival time of the shower particles in the ground array detectors is used in combination with the track orientation on the sky as seen by the fluorescence eye(s). The result is geometrical reconstruction accuracy which is equal to that of stereo fluorescence. Angular error is $\sim$ 0.3 degrees, and energy error $\sim$ 10% for primary energies above 10^{19} eV (68% bounds). A caveat here is that the atmospheric attenuation is assumed to be perfectly known, and hence perfectly correctable.

Note that the hybrid reconstruction does not use the signal size information from the ground array detectors, only the trigger times. This is vitally important as it allows us to overcome a major problem of stand-alone giant arrays. To reconstruct ground array data it is necessary to know the form of the lateral distribution

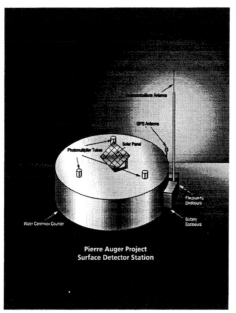

FIGURE 5. An Auger ground array station.

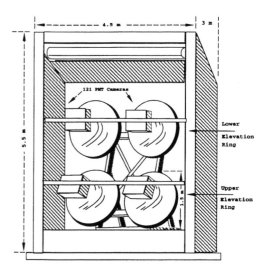

FIGURE 6. Possible fluorescence detector enclosure.

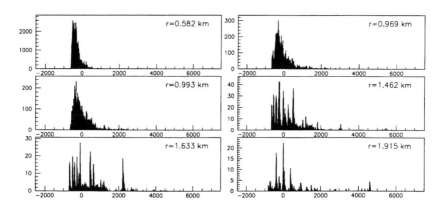

FIGURE 7. Simulated pmt output for six tanks taking part in a ground array event. Note the very different vertical scales; a single muon has an amplitude of approximately 20. The horizontal scale is time in nanoseconds, and distance from the shower axis is noted on each trace.

318

function (ldf); how the particle density varies with axial distance to the shower core. For a sparse array it is difficult to determine the ldf and the shower core positions simultaneously; there is insufficient data. An assumed shape is required, either extracted from lower energy data, or determined by analytic or Monte Carlo calculations.

Hybrid reconstruction will determine the core position to within 50 m. The lateral distribution function can then be accurately measured event-by-event using the independent surface detector density information. This measured ldf will then be used in the reconstruction of the 90% of the data which is ground only. This resolves the first major problem with conventional ground arrays.

The second major problem of ground arrays is how to infer the primary energy from ground level observations. In this work $\rho(1\ km)$ is used as the energy parameter; the interpolated signal density in a water Čerenkov detector at 1 km from the shower core. According to shower simulations this parameter has small fluctuations, and corresponds almost linearly with primary energy. The shower size on axis reaches maximum somewhat above ground level at the Auger sites (since the ldf is approximately a power law, the total number of particles is dominated by those close to the axis). However, for a vertical event, $\rho(1\ km)$ is still growing at ground level; it is maximal for events which arrive at approximately 40 degrees zenith angle. Within the zenith angle range 0 to 60 degrees fluctuations in $\rho(1\ km)$ are minimal since the gradient of this quantity is small.

The important thing to note is that it does not matter if the Mocca-Sibyll predictions of X_{max} etc. are exactly right; the freedom allowed by existing data from the Fly's Eye is not enough to qualitatively alter the situation. The exact form of the relationship between $\rho(1\ km)$ and primary energy will be determined in the final experiment by exploiting the pseudo-calorimetric energy measurements available from the fluorescence detectors for the 10% of the data where hybrid observations are made. This process has been fully implemented in the simulation — the hybrid reconstruction results are used to extract the (2) parameters of a simple model relating primary energy to $\rho(1\ km)$ and shower zenith angle. In all cases *reconstructed* parameters are used to make the process realistic.

Angular error is ~ 1 degree for vertical events, falling with increasing zenith angle. Energy error is $\sim 25\%$ for a worst case scenario where the primary particles are an equal mix of protons and heavy nuclei, and no correction is made. Using hybrid cross calibration the mean energy error is always zero, so if the primary beam is not mixed this figure reduces to 15%. The energy error distribution does not have a high side tail — see figure 8. These resolution figures are entirely appropriate to the problem at hand. The normal assumption is that the primaries are charged particles; if so they will be subject to random magnetic deflections greater than the angular resolution. Likewise the energy resolution will not significantly degrade spectral measurements. What is needed to make progress is event statistics with a well known aperture; high resolution is not required.

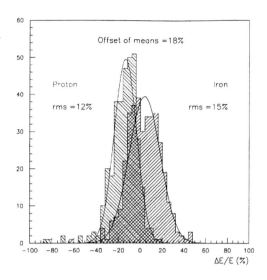

FIGURE 8. Energy error distribution of ground-only events.

IV   CONCLUSIONS

The proposed Auger Observatories of cosmic rays above 10^{19} eV have been described. The background to the project has been outlined, and the reasoning behind the hybrid experimental design given. Finally event simulation and reconstruction has been discussed, including the way in which hybrid data will be used to free ground array data from the problems to which it is normally subject. We believe this experiment to be the logical next step in the exciting field of ultra high energy cosmic rays.

REFERENCES

1. Linsley, J., *Phys. Rev. Lett.* **10**, 146 (1963).
2. Bird, D., et al., *Ap. J.* **441**, 144 (1995).
3. Hayashida, N., et al., *Phys. Rev. Lett.* **73**, 3491 (1994).
4. Elbert, J., and Sommers, P., *Ap. J.* **441**, 151 (1995).
5. Dai, H., et al., *J. Phys. G* **14**, 793 (1988).
6. Auger FAQ 10, http://www-td-auger.fnal.gov:82/Questions/Questions.html
7. Baltrusaitis, R., et al., *Nucl. Instr. and Meth. in Phys. Res. A* **240**, 410 (1985).
8. Lawrence M., Reid, R., and Watson, A., *J. Phys. G* **17**, 733 (1991).
9. Hillas, A., "The MOCCA program: MOnte Carlo CAscades", in *Proceedings of the 24th International Cosmic Ray Conference, Rome*, **1**, 270 (1995).
10. Fletcher, R., et al., *Phys. Rev. D* **50**, 5710 (1994).
11. Dawson, B., et al., *Astropart. Phys.* **5**, 239 (1996).

Workshop Summary

David N. Schramm[1]

Department of Astronomy and Astrophysics
The University of Chicago
Chicago, IL 60637-1433

Due to his untimely death, Prof. David Schramm was unable to complete his workshop summary. We have decided to publish his transparencies, which echo his excitement regarding the physics potential of measuring the characteristics and determining the source of the ultra-high-energy component of the cosmic radiation. David's talk began with a synopsis of the current state of affairs - cosmic rays with energy greater than 10^{20} eV have been measured, and the GZK cutoff implies that if these particles are nuclei, they must originate locally in the universe. However, the absence of identifiable astrophysical sources, such as Active Galactic Nuclei, in this nearby volume leads one to hypothesize that 'unconventional' physics mechanisms may be required for their generation. These include 'bottom-up' acceleration mechanisms such as gamma ray burst fireballs[2] or 'top-down' decays of super-massive ($M_x \sim 10^{24}$ eV) particles resulting from remnant topological defects . Alternatively, the observed ultra-high energy airshowers may have been initiated by super-symmetric particles which would be immune from the effects of the GZK mechanism and thus could travel cosmological distances with minimum energy loss[4]. Yet another mechanism is offered by the interactions of ultra-high energy, massive neutrinos with the 2 K neutrino background, which results in a burst of particles from the decay of a Z gauge boson[5].

David's and our enthusiasm for the this subject can be summarized by his conclusion. The 'conservative' explanation of acceleration by an AGN powered by a black hole (!!) has observational problems. The 'top-down' scenario involving topological defects formed at the time of Grand Unification symmetry breaking (10^{-35} s after the Big Bang) is a viable alternative. The bottom line is that more data is needed in order to identify the source and the underlying physics which generate these ultra-high-energy particles. This task will lead particle astrophysics into the 21st century and could provide a stepping stone to a unified description of the forces of nature, which David referred to as the "Theory of Everything."

---

1) deceased
2) Waxman et al., PRL 75 (1995), astro-ph/9612061 (1998)
3) P. Bhattacharjee, in these proceedings
4) G. Farrar, in these proceedings
5) T. Weiler, in these proceedings

Fig. 1B

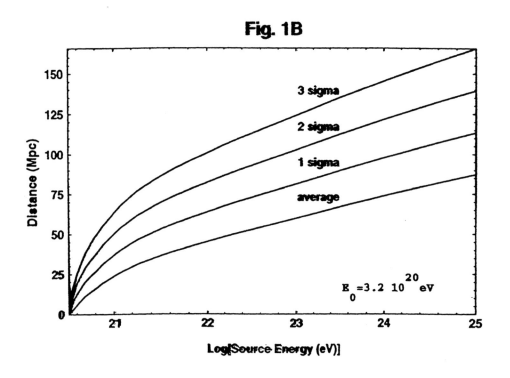

3 sigma

2 sigma

1 sigma

average

$E_0 = 3.2 \ 10^{20} \ eV$

Distance (Mpc)

Log[Source Energy (eV)]

322

Particle-Astrophysics

Into The

21st Century

ULTRA
HIGH
ENERGY

$(E > 10^{20} \text{ eV})$

COSMIC
RAYS
A MYSTERY

BACKGROUNDS

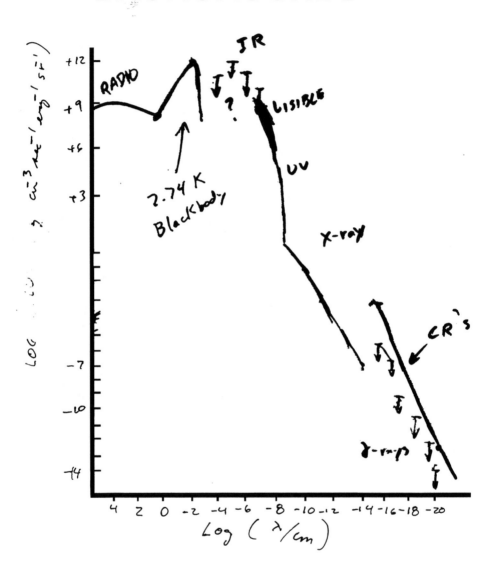

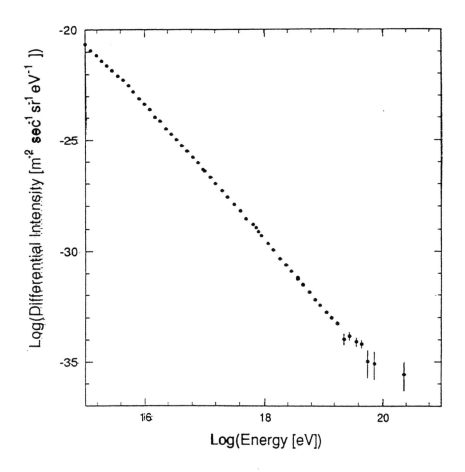

Figure 13: The energy spectrum of primary cosmic rays between $10^{14.5}$eV and $10^{20.5}$eV obtained at Akeno. Open circles are the present results by the AGASA by fitting the observed points to the Eq.(11) in $10^{18.5} \sim 10^{19.0}$eV energy region.

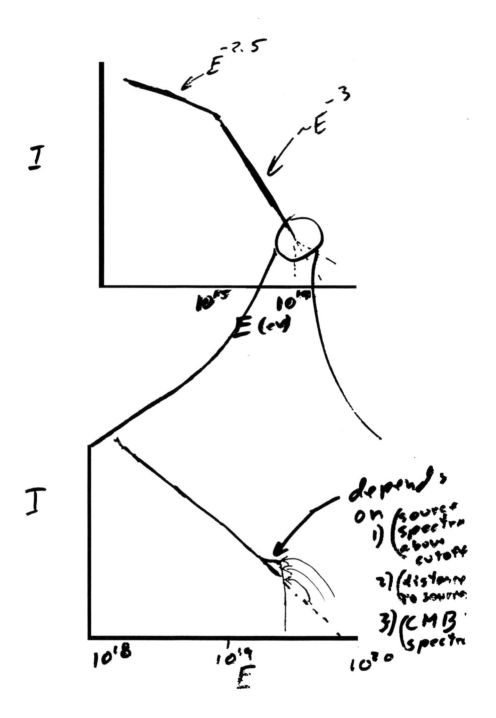

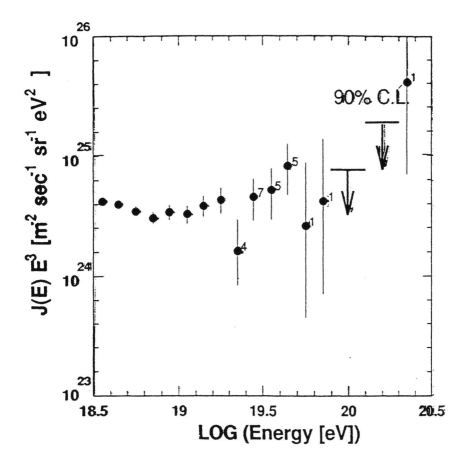

Figure 9: The energy spectrum determined by the present experiment. Numbers attached to the closed circles show the number of events in each bin.

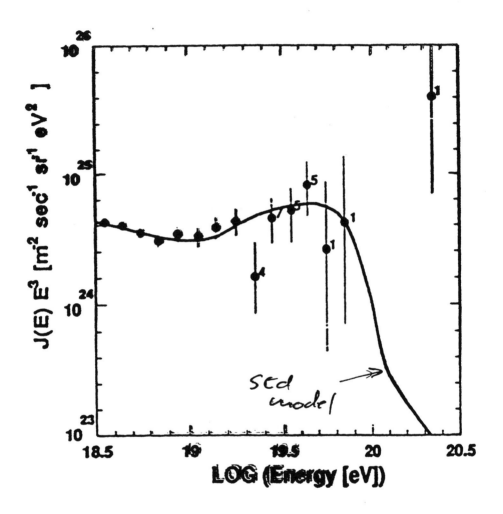

The Cutoff

<u>Cosmic Microwave Background & Cosmic Rays:</u> Griesen;Zatsepin;Kuzmin; Hill & DNS

$$p + \gamma_{3k} = p + \pi + \dots$$
$$\downarrow$$
$$\nu\text{'s}$$

Cutoff for Extragalactic Cosmic Rays for:

$$E > 4 \times 10^{19} \text{ eV}$$
$$\lambda = 1/n\sigma \sim 6 \text{ Mpc}$$

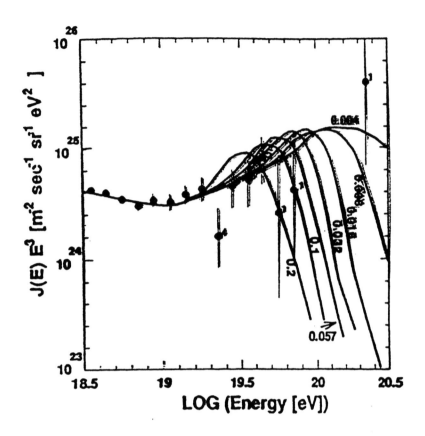

Regardless of path length

Bottom-up has problems
with observed Flux
at $E > 10^{20}$ eV

A LOOP HOLE

in

G–Z Cutoff:

SUSY PARTICLES

with

Long Mean Free

Paths from AGN's

Ferrar & Kolb

but constraints: Okun; Berezinsky et al.

A semi - Loop hole :

⌈ relies on multiple product of low ⌉
└ but finite probability events ┘

Ultra high Energy $\nu \bar{\nu}$ annihilation
off HDM cluster halos

$$\nu \bar{\nu} \rightarrow Z \rightarrow \gamma's , P, \bar{P} \ldots$$

Weiler
Fargion etal

Requires ultra high E primary p's ($\geq 10^{23}$ e
finite mass ν's $\Rightarrow$ HDM halos (strud
fount

For E > 10^{20} eV Protons POINT to Sources

If IGM B < 10^{-9} G
for R < 10^2 Mpc

[Or for B ~ 3 x 10^{-6} G
with R < 20 kpc in
Galaxy]

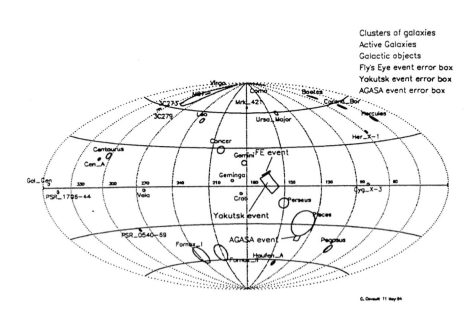

Clusters of galaxies
Active Galaxies
Galactic objects
Fly's Eye event error box
Yokutsk event error box
AGASA event error box

C. Dawson 11 May 94

BOTTOM-UP

Acceleration Mechanisms

Blandford;Hilles;Berezinsky;etc

E < 10-1000 TeV
(10^{13-15} eV)
Supernova Shock
Acceleration in ISM

E < 10^{19} eV AGN's

E > 10^{20} eV ?????

Special case: γ-ray bursts - waxman
but rates require differences
it GRB cosmology
~ ιO Acceleration Mechanism- - - - - - - sig/

GUT SCALE DEFECTS

Hill & D∿S; Sigl, Bhattatarjee & D∿S

$E > 10^{15}$ GeV (10^{24} eV) DECAYING
TOPOLOGICAL
DEFECTS

STRINGS, MONOPOLES, ...

CASCADE DOWN WITH DIP AT
PHOTO-PION THRESHOLD

SUPERHEAVY HALO DECAYS

Berezinsky, Vilenkin & Korchelynss

Small admixture sub Gut Scale DM in CDM halo decays on long Timescale via worm-holes —

338

Defect Decay

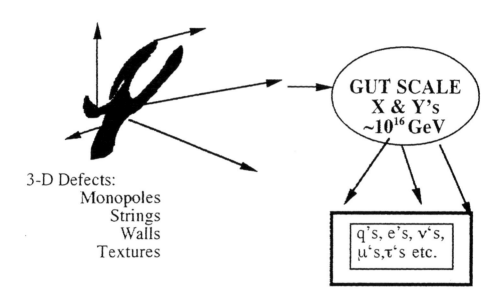

GUT SCALE
X & Y's
~10^{16} GeV

3-D Defects:
Monopoles
Strings
Walls
Textures

q's, e's, ν's,
μ's,τ's etc.

Generic Aspect: Top Down Cascade

Vacuum   Energy
&
Broken   Symmetries
$\Rightarrow$Potential   for   Topological   Defects

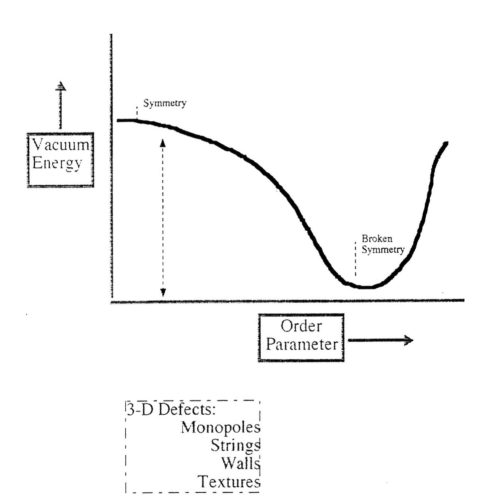

DOMAIN WALL

STRING

MONOPOLE

TEXTURES

(WINDING NUMBER VARIATIONS)
IN 3D DOMAIN VOLUMES

Pencil Network

IF SPRINGS ARE OPTIMAL
THEN CORRELATIONS
OF NEIGHBORING PENCILS
IN THEIR "LOW T" STATE

Figure A-12: Photograph of the defect tangle in a thin film of freely suspended nematic liquid crystal after a temperature quench. The dark, sharp lines in the picture are type-$\frac{1}{2}$ strings. In the top-center of the picture, is a diffuse but visible type-1 string with three monopoles, which appear as black spots on the string. Below that is a type-$\frac{1}{2}$ string attached in two places to a type-1 string which is also supporting a monopole. Various other features in the photograph include boojums, which are defects which are attached to the surface of the film and appear as lines which terminate in dark blobs, and many instances of type-1 strings cutting across horseshoe shaped type $\frac{1}{2}$ strings. The picture is about 790 μm wide.

NOTE:

For "Top Down"

Distribution of

Sub G-2 cutoff

events Is

IRRELEVENT

A GAP?

Sigl, Lee & DNS

IF TRUE $\Rightarrow$

NO "BOTTOM-UP"
Scenario works

Then Would
Need
"TOP-DOWN"

TD model: $m_x = 10^{23}\,eV$ $\dfrac{dN_h}{dx} = \dfrac{15}{16}\,x^{-1.5}(1-x)^2$

$\beta = 10^{-9}\,G$

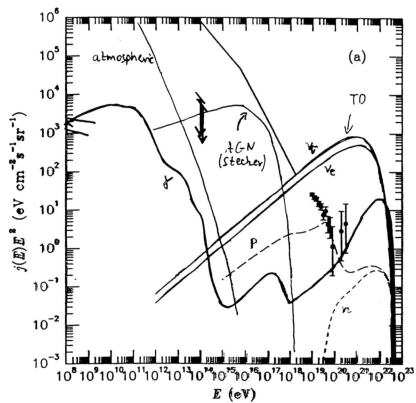

Sigl, Lee, Coppi + DNS.

———— neutrino fluxes

———— γ-ray fluxes (and limits)

———— nucleon fluxes (and data)

346

TD model : $m_X = 10^{23} eV$ $\frac{dN_y}{dx} = \frac{15}{16} x^{-1.5} (1-x)^2$

$B = 0$

ν's The Resolution To The Mystery

(a)

atmospheric

AGN (Stecher)

TD

ν_μ

ν_e

γ

p

n

$j(E)E^2$ (eV cm^{-2} s^{-1} sr^{-1})

E (eV)

Sigl, Lee, Coppi + ONS

——— neutrino fluxes

——— γ-ray fluxes (and limits)

——— nucleon fluxes (and data)

347

Paired Events?

(AGASA etc.)

Single, cos, Os Events.

Probability of Accidental pairing < 2%

Most Interesting Case:

AGASA pair: 2×10^{20} eV
5×10^{19} eV

Within~1-deg.⇒ Can't be a High Z Nucleus since Galactic magnetic Field would yield a larger Splitting

Muon/Hadron ~ 10% for each
⇒Unlikely to be Gamma's

Point Source! But can it be a single burst spread out by intergalactic magnetic fields? Depends on strength & coherence of Intergalactic Field.

Distribution of Arrival
Energies & Times
from a Single Burst

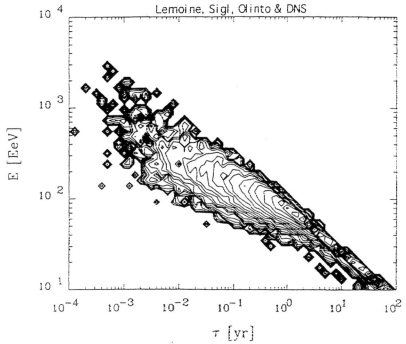

Contour plot for a distribution of events originating from a bursting source, as projected in the time-energy plane. The corresponding parameters are $D = 30$ Mpc, $B_{rms} = 10^{-11}$ G, $n_B = 0$, $l_c = 1$ Mpc; the differential index of the energy injection spectrum is $\gamma = 2.0$. Deviations from the mean correlation $\tau_E \propto E^{-2}$, for $E \geq 70$ EeV, reflect pion production on the CMB. In order to show the statistics at high energies, 40 contours with logarithmic interspacing decrements of 0.15 dex, are shown.

SOURCES

Sigl, DNS & Bhattacharjee; Gaisser; Blandford;
Hilles;Berezinsky; etc

BOTTOM-UP

Acceleration

Mechanisms – ???

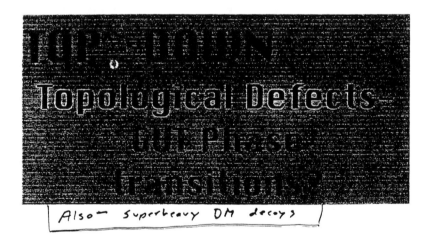

TOP-DOWN

Topological Defects –

GUT Phase

transitions?

Also — Superheavy DM decays

350

Current Situation:

The "Conservative" AGN
(Black Hole!) Bottom-Up
scenario has observational
Problems that are difficult
to dismiss.

The Top-Down (Big Bang
Relics) is also constrained
but is still flexible enough to
be quite viable.

NEED MORE DATA!!

Note: Top-Down(Big Bang Relics)
$\Longrightarrow$ TOE (New Fundamental Physics)

The Future of:
<u>Ultra High Energy</u>
<u>Particle-Astrophysics</u>

Particle Detectors:

Hi-Res Fly's Eye (Utah)
AUGER (Utah-Argentina)
OWL (Orbit)

Gamma Rays:

MILAGRO (New Mexico)
VERITAS (Arizona)
GLAST (Orbit)

Neutrinos:

AMANDA (South Pole)
NESTOR (Mediterranean)
KM^3 (??)

 # RESOLUTION TO THE MYSTERY

AIRWATCH: the Fast Detector

Alippi E., Lenti A.

LABEN S.p.A., Strada Padana Superiore 290, 20090 Vimodrome (Mi), Italy

Attiná P.

ALENIA Aeropazio, Corso Marche 41, 10146 Torino, Italy

Gregorio A., Stalio R., Trampus P.

CARSO, Area Science Park, Padriciano 99, 34100 Trieste, Italy

Bosisio L., Giannini G., Vacchi A.

Università and INFN – Trieste, Area Science Park, Padriciano 99, 34100 Trieste, Italy

Gracco V., Petrolini A., Piana G.

Università and INFN – Genova, via Dodecanneso 33, 16146 Genova, Italy

Catalano O., Giarrusso S.

Istituto di Fisica Cosmica ed Applicazioni dell'Informatica – CNR, Via Ugo La Malfa 153, 90146 Palermo, Italy

Bonanno G.

Osservatorio Astrofisico di Catania, Via Doria 6, 95125 Catania, Italy

Abstract. We propose an instrument to look for Extensive Air Showers (EAS) produced by Extremely High Energy Cosmic Rays (EHECR, $E > 10^{20}$ eV) from space (Airwatch concept) by observing nitrogen fluorescence radiation.

CP433, *Workshop on Observing Giant Air Showers from Space*
edited by J.F. Krizmanic et al.
© 1998 The American Institute of Physics 1-56396-766-9/98/$15.00

EXPERIMENTAL KEY PARAMETERS

The experimental parameters should be adjusted accordingly to the specifications derived from the final optical design. The values in this presentation are just an example; they are reported in Tab. 1.

TABLE 1. Detector parameters. (*): at $E=1\times10^{20}$ eV, hitting a single pixel.

Projected area	circle of 400 km diameter on Earth
Total number of pixels	about 130,000
Background events	0.2 photons/3.3 μsec/pixel
Signal events	40 photons/3.3 μsec(*)
Single photo-electron sensitivity	yes
Average track image	20–40 pixels
Time resolution	20–100 nsec
Focal plane segmentation	yes, 1 pixel $\leftrightarrow$ 1 $\times$ 1 km^2

DETECTOR CONCEPT

The focal plane detector has two geometrical requirements:

- large sensitive area, of the order of square meters;

- concave surface to match the optical design.

These two requirements can be achieved by using a mosaic of detectors such as the Hybrid Photo Diode Tube detector (HPDT) [1] or the Silicon Intensified Micro–Strip detector (SIMS) [2].

Both have single photoelectron sensitivity, provide a very fast response (of the order of 20–100 nsec) and have a low noise contamination with high gain. The HPDT detector has furthermore an image demagnification capability.

The two detectors are briefly described in the following sections.

HPDT Detector

A schematic diagram of a HPDT detector is given in Fig. 1.

In a HPDT tube, photoelectrons generated on a photocathode are accelerated and focused by an electrostatic field onto a silicon pixel detector.

The sensitive area consists of a two dimensional array of 100 tubes covering the focal plane. Each tube has a 10 cm diameter circular shape to match the shape of electrodes which generate the focusing (demagnifying) field. A ratio sensitive area to total area of about 70% is achieved.

The gain of a HPDT tube is linear with applied voltage and reaches about 5000 at 20 kV. A demagnification factor of about five is feasible with no loss in spatial information; we are considering a value of four. The read–out system has approximately 0.6 mm pixel size and is bump–bonded to the chip. The vacuum inside the tube is 10^{-9} mbar.

The effective photoelectron detection efficiency depends on the high–voltage and the discriminator threshold of the front–end electronics. A value of 0.75 is likely.

The charge collection time is of the order of a few nsec. Read–out speed is thus limited by front–end electronics.

Pixel layout and sizes are optimized by taking into account the size of the focal plane, the demagnification factor, the noise level and the read–out system characteristics.

SIMS Detector

A schematic diagram of a SIMS detector is given in Fig. 2.

In a SIMS detector, photons hit a mosaic of Micro Channel Plates (MCP) with opaque photocathode. The produced electron cloud is proximity focused onto a silicon microstrip detector. Ionization is produced within the silicon wafer, opposite charges (electrons and holes) are revealed and the position information of the cloud is obtained.

The sensitive area consists of a two dimensional array of 100 MCPs covering the focal plane. Each MCP has a square shape of 10×10 cm^2. A ratio sensitive area to total area of about 85% can be achieved.

The focal plane has 196 7×7 cm^2 double sided silicon microstrip detectors with approximately 2.5×2.5 mm^2 pixel size which is bigger than the typical size of MCP pores. We are also considering the possibility of "pixelizing" the detector with pixel size of the order of the strip width.

The SIMS detector has similar performances in charge collection time, read–out speed and noise level as HPDT. The signal identification technique is based on pulse height discrimination from background light. Algorithms based on signal persistency on strips and signal evolution along a linear track provide alternative positional information.

The performance comparison between the two detectors is given in Tab. 2

FRONT END ELECTRONICS

Front end electronics interfaces the two silicon detectors and is implemented with fully custom-made analogue ASIC located close to the detector to minimise distur-

TABLE 2. Comparison between HPDT and SIMS. (*): new algorithms for image reconstruction look promising.

Unambigous 2-dim image reconstruction	yes	no(*)
noise, radiation hard, data acquisition speed	good	good
front-end electronics	complex	simple
development status	in progress	advanced
packaging characteristics	bulky	thin

bances. It performs the following tasks: amplification, shaping and discrimination of the detector signal.

Very high speed, low power consumption and small dimensions are fundamental requirements for such circuit.

High speed, low power electronics for HPDT and SIMS is already available and tested.

CONCLUSIONS

A detector meeting the airwatch requirements does not exist yet even if there are strong lines of research in that direction.

The critical point is the ultra–fast and low noise electronics. At this time the read–out system is a factor 3–5 slower than required.

Possible back–up solutions to these detectors, such as the multianode photomultiplier or the Silicon Drift chamber (SD2000) [3], still deserve a study even if at first sight they don't look promising.

REFERENCES

1. HPDT: *Nucl. Instrum. Methods* **A355**, 386 (1995), *Nucl. Instrum. Methods* **A365**, 76 (1997), *Nucl. Instrum. Methods* **A397**, 92 (1997), *Nucl. Instrum. Methods* **A397**, 167 (1997).
2. SIMS: *Nucl. Instrum. Methods* **A315**, 121 (1992), *Nucl. Instrum. Methods* **A326**, 183 (1993), *Nucl. Instrum. Methods* **A379**, 101 (1996).
3. SD2000: *Optical Engineering* **36**, 2143 (1997).

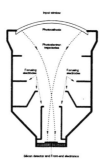

FIGURE 1. The HPDT detector

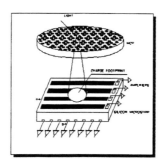

FIGURE 2. The SIMS detector

Background Measurement with UVSTAR

Gregorio A., Stalio R., Trampus P.

CARSO – Area Science Park, Padriciano 99, 34100 Trieste, Italy

Scarsi L.

Istituto di Fisica Cosmica ed Applicazioni dell'Informatica – CNR, Via Ugo La Malfa 153, 90146 Palermo, Italy

Abstract. Within the AIRWATCH collaboration, we propose to measure the Earth night side background with high accuracy using the UVSTAR instrument from the Shuttle.

INTRODUCTION

The AIRWATCH purpose is to look for Extensive Air Showers (EAS) produced by Extremely High Energy Cosmic Rays (EHECR, $E > 10^{20}$ eV) from space by observing nitrogen fluorescence radiation. The great uncertainties on the evaluation of the expected background from the Earth night push us to study a new experiment to measure this kind of background. This can be achieved quite easily and with high accuracy with UVSTAR [1–3], a Ultra–Violet telescope for the Shuttle. In this paper we will try to explain how this purpose could be achieved.

UVSTAR CHARACTERISTICS

UVSTAR (Ultra Violet Space Telescope For Astronomical Research) is a joint collaboration between the University of Trieste and Arizona (Tucson). The experiment is intended as a facility devoted to solar system and astronomy studies. It covers the wavelength range from 50 to 125 nm with sufficient resolution to separate emission lines and to form spectrally resolved images of extended plasma sources including the Io plasma torus of Jupiter, hot stars, planetary nebulae, supernova remnants and bright galaxies. Minor but still important UVSTAR target is the Earth airglow.

CP433, *Workshop on Observing Giant Air Showers from Space*
edited by J.F. Krizmanic et al.

Supported by the Italian Space Agency and NASA, UVSTAR has been designed and built by CARSO and the University of Arizona. A picture of the instrument is given in Fig. 1.

The instrument consists of a movable platform and an optical system.

After a rough pointing, the platform provides the fine pointing (controlled by CARSO tracking system) within $\pm 3^o$ from the nominal view direction perpendicular to the long axis of the Shuttle and in the plane of the wings.

The optical system has two channels, each one consists of a telescope and a Rowland concave–grating spectrograph with intensified CCD detectors. The first channel, FUV (Far–Ultra Violet), operates in the 80–125 nm spectral range, the second one, EUV (Extreme–Ultra Violet), covers the 50–90 nm region.

UVSTAR uses two 30 cm diameter off–axis parabolic mirrors having focal length of 1.5 m. They have selectable spectral resolution of 0.1, 0.4 and 1.2 nm. The experiment has capability for long slit spectral imaging of extended cosmic sources.

UVSTAR has recently flown as a Hitchhiker–M payload on the STS–69 mission (September 7–18, 1995) of the Shuttle Endeavour, and on the STS–85 mission (August 7–19, 1997) of the Shuttle Discovery. A third flight is scheduled for October 1998.

THE MISSION

Since NASA has allocated a total of five flights for UVSTAR we propose for the fourth and fifth flight to modify the EUV channel in order to measure the Earth night side background. This could be done by exchanging the EUV spectrograph assembly with a CCD camera operating around 330 nm. The instrument will be oriented to Earth during the measurements.

In order to further improve our knowledge of the Earth N_2 fluorescence phenomenum, we are also considering the possibility to have a signal to background evaluation by exciting the fluorescence during UVSTAR observations. This could be done either with a dedicated laser positioned on the Shuttle and aligned with the instrument (a better option) or with a laser operating from the Earth. The two options still deserve a deep analysis for a better understanding of all conditions.

THE CAMERA CHARACTERISTICS

The main characteristics of the camera are given in Tab. 1.

Using these parameters we can evaluate the expected event rate from the Earth night side background impinging on the detector. The very spread measurements existing nowadays give an average value of the background intensity of 100 photons/m^2/nsec/sr. Assuming this value, for an integration time of one second we get $22 \cdot 10^6$ photons on the detector, that means one photon per pixel (considering a total efficiency of 0.05).

TABLE 1. The UVSTAR camera parameters.

Mirror diameter	0.3 m
Focal length	1.5 m
Intensified CCD detector	22 x 22 mm (1000 x 1000 pixel)
FOV	0.0008 sr

With these values we can conclude that UVSTAR should be able to provide a very precise measurement of this kind of background.

We haven't yet made any evaluation on the signal to natural noise ratio in the case of N_2 exitation by laser.

CONCLUSIONS

The uncertainties on the expected background intensities from the Earth night side lie in the range from 30 to 300 photons/m^2/nsec/sr. The background could be measured with high accuracy by UVSTAR even if the new options still need a deep study.

REFERENCES

1. Broadfoot A.L., Sandel B.R., Stalio R., *Optical Engineering* **32**, 3009 (1993).
2. De Carlo F., Stalio R., Trampus P., Broadfoot A.L., Sandel R.B., *Optical Engineering* **33**, (1994).
3. De Carlo F., Stalio R., Trampus P., Broadfoot A.L., Sandel R.B., Sicuranza G., *1993 Shuttle Small Payloads Symposium – NASA Conference Symposium* **3233** 153 (1993).

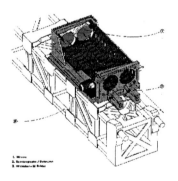

FIGURE 1. The UVSTAR instrument

Air Fluorescence efficiency measurements for AIRWATCH based mission: Experimental set–up

Biondo B., Catalano O., Celi F., Fazio G., Giarrusso S., La Rosa G., Mangano A.

Istituto di Fisica Cosmica ed Applicazioni dell'Informatica – CNR, Via Ugo La Malfa 153, 90146 Palermo, Italy

Bonanno G., Cosentino R., Di Benedetto R., Scuderi S.

Osservatorio Astrofisico di Catania, Via Doria 6, 95125 Catania, Italy

Richiusa G.

DEAF, Università di Palermo, Palazzo Steri Piazza Marina 61, 90133 Palermo, Italy

Gregorio A.

CARSO – Area Science Park, Padriciano 99, 34100 Trieste, Italy

Abstract. In the framework of the AIRWATCH project we present an experimental set–up to measure the efficiency of the UV fluorescence production of the air using hard X–ray stimulus. The measures will be carried out at different pressure and temperature to emulate the same condition of the upper layers of the atmosphere where X–ray and gamma ray photons of Gamma Ray Bursts are absorbed.

THE ATMOSPHERE AS TARGET AND SIGNAL GENERATOR FOR EXTRATERRESTRIAL RADIATION.

A very interesting technique to study the Extreme Energy Cosmic Rays (EECR) with E> 10^{19} eV (including primary Gamma Rays and Neutrinos) is the use of the atmosphere as a giant absorber scintillator. The method suggested several years

CP433, *Workshop on Observing Giant Air Showers from Space*
edited by J.F. Krizmanic et al.

ago by Linsley [1] and others consists in observing from space (let's say from 300–500 km) the physical phenomenon of the emission of the UltraViolet fluorescence radiation (or UV scintillation light) induced in the atmosphere during the passage of the EECR particles.

EECR particles in the collision with air nuclei produce hadrons that in turn collide with air nuclei giving rise to a propagating cascade of particle (Shower). In the complex hadron–electromagnetic cascade the more numerous particles are electrons. The number of electrons (size Ne of the Shower) at the cascade maximum is proportional to the primary energy $Eo(eV)$: $Ne_{max} \sim Eo/1.3 \cdot 10^9$.

The basis of the observation is the UV fluorescent light emission from atmosphere (Nitrogen) excited by the Air Shower electrons. Shower electrons (and other charged particles) moving through atmosphere are ionising the air atoms and also exciting the methastable electron levels in the Nitrogen atoms and molecules. In a short relaxation time, the excited atoms (molecules) decay to the ground level emitting the characteristic UV fluorescence light with peaks at wavelengths from 330 nm up to 400 nm (narrow lines at 337, 357, 391 nm).

The emitted light is isotropic and proportional to the shower size at the given depth in atmosphere. An estimate of the UV fluorescence signal in a shower track can be done following the relation between primary energy and the light flux in the wavelengths 330–400 nm from the shower maximum: $Q \sim (Eo/1.3 \cdot 10^9) \cdot F \cdot \Delta L/4\pi R^2$ (photons m^{-2}). ΔL is the observed track length in meters, F is the fluorescence luminosity of the track in photons/meters and R is the distance in meters between the detector and the track. For electrons $F \sim 4$ photons/meter constant at different depth in the atmosphere.

GAMMA RAY BURST DETECTION BY UV ATMOSPHERIC FLUORESCENCE

The detection of the UV atmospheric fluorescence radiation induced by X and Gamma rays is important to catch the Gamma Ray Bursts (GRB) that represent one of the most intriguing phenomena in Astrophysics. Search for coincidence between GRB and Extreme Energy Cosmic Rays and Neutrinos is, moreover, the future challenge to advance in the GRB phenomenon knowledge as well as in the EECR origin. In fact, if the suggested association of the GRB with cosmological (extragalactic) origin is confirmed, the high energy release involved ($10^{51} - 10^{52}$ ergs) could involve the probable contemporary emission of high energy neutrinos or EECR, associated with the low energy gamma burst normally observed. The GRB are characterised by flat spectra from few keV up to several hundreds keV (up to the GeV, in some cases of particularly flat and intense GRB). The entire energy of an intense GRB ($10^{19} - 10^{21}$ eV) impacting with the atmosphere is absorbed by the upper atmospheric layers (30–50 km heights). A fraction of the absorbed energy is released in an UV flash of the same GRB duration (from few msec up to tens seconds) and of intensity proportional to the total GRB energy.

UV AIR FLUORESCENCE EFFICIENCY

As mentioned above an EECR particle (including neutrinos) or a Gamma ray photon produce in air a cascade of secondary particles (mainly electrons) as well as photons of moderately high energies. The interaction of these secondary through air results in the ionisation and excitation of its molecules. The electronic excitation energy of the molecules is either dissipated non–radiatively or emitted as visible or ultraviolet photons. It has been shown experimentally that, down to very low pressure, the fluorescence radiation from air results almost entirely from electronic transitions in the N_2 molecule and N_2^+ molecular ion. The radiative emission from these electronic transitions is detected as very narrow lines in the near UV band (main lines at 337, 357 and 391 nm). Due to the highly competitive non–radiative process, it is crucial to estimate the quality of air as a scintillator finding the efficiency of the fluorescence, that is, the fraction of the energy lost by ionisation and excitation that goes into UV fluorescence photons. Some experimental measurements and theoretical studies have been carried out in the past to determine the efficiency of the fluorescence process. In all experiments, included the recent measurements made by a Japanese team of the ICRR [2], the fluorescence was induced by high energy particles (electrons) with energies above 1 MeV up to few GeV. In the framework of the AIRWATCH project, in which the detection of GRBs in coincidence with EECRs or neutrinos is an important task, it is of interest to determine the air fluorescence yield at low energies by using electrons or directly photons of energies below 100 keV. Measurements of fluorescence efficiency as a function of pressure, temperature and chemical composition of the air, to reproduce the same condition in the upper layers of the atmosphere (where Gamma Ray photons are absorbed), are crucial.

THE EXPERIMENTAL SET–UP

The measurement of the fluorescence efficiency induced by low energy photons (X–ray) is in charge of the experimental team of AIRWATCH collaboration and it will carried out by IFCAI in Palermo. As starting point we are reusing a device designed, in the past, for the measurement of the fluorescence yield of pure Xenon excited by X–ray photons of 22 keV coming from a collimated radioactive source of Cd_{109}.

This apparatus, shown in Figure 1, is very simple and it consists of a cylindrical gas cell of ceramic closed at top and bottom by quartz windows. Two photomultipliers (pmt's) looking through the quartz windows detect the UV light produced inside the cell when a X–ray is absorbed (see Figure 2).

The coincidence technique eliminates the uncorrelated noise of the pmt's permitting to detect the very few UV photons produced. Random coincidence as well as coincidence induced by cosmic rays are cancelled by background subtraction. If the efficiency of the UV production is very low (as in our case where the expected value for air at the pressure of tens of mbar is of the order of 0.1%) the rate of coincidence due to UV photons is a key parameter to determine the efficiency. We have planned to use the 22 keV X–ray source because this is a trade–off with to have as much UV photons as possible and low X–ray Compton diffusion. The measurements are in course. Due to the very low photoelectric cross section and UV yield of air a high flux of X–ray photons is mandatory. This implies the use of an intense X–ray source. For this purpose, in Palermo, is installed an X–ray beam of adequate flux available for this experiment: the LAX [3] (LAboratory for X–ray experiment). This facility managed in collaboration between IFCAI–CNR and DEAF (University of Palermo), consists of a beam 12 m long with available lines at 5.4, 6.4, 8, 22 keV. The maximum flux of photons at the end of the beam line on axis is of the order of $10^4 - 10^5$ ph mm^{-2} s^{-1}. The vacuum inside the chamber is in the range of $(2–4)\cdot10^{-6}$ mbar.

FUTURE PLANS

We are also planning to repeat the measurements of UV air efficiency at higher energies both with photons and high energy particles with an appropriate redesign of the shown above detector. These measurements will be carried out at a Synchrotron Facility and at a Particle Accelerator.

REFERENCES

1. Linsley, J., *Proc. Workshop on Very High Energy Cosmic Ray Interactions, Univ. Of Pennsilvania*, 476–491 (1982)
2. Kakimoto, F., Loh, E. C., et al., *ICRR Report 346-95-12, Univ. Of Tokyo*, (1995)
3. Celi, F., et. al., *Vakuum in Forschung und Praxis* **1**, 21–24, (1996)

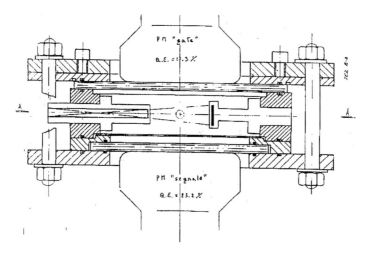

FIGURE 1. Schematic of gas cell.

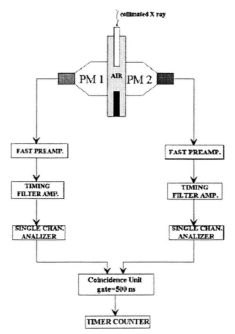

FIGURE 2. The coincidence electronic chain.

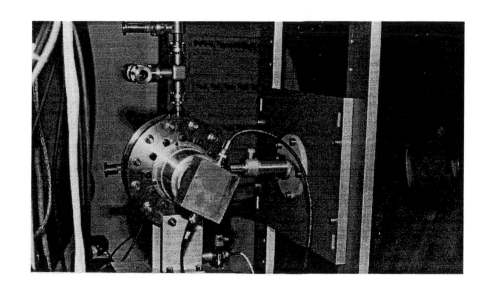

FIGURE 3. A picture of the apparatus.

Balloon Borne Detector for Cosmic Ray Energy Spectrum Measurements and its Possibilities in Connection with AIRWATCH Experiment

R. A. Antonov, D. V. Chernov, A. N. Fedorov, E. E. Korosteleva, M. I. Panasyuk, E. A. Petrova

Institute of Nuclear Physics, Moscow State University, Moscow, Russia

Abstract. The polar balloon–borne experiment to detect atmospheric fluorescence and Cerenkov light of Extensive Air Showers (EAS) may serve as a useful step of preparation of orbital experiment. A small prototype of spacecraft–borne detector is proposed to be used in balloon–borne experiment.

There are some reasons for such a balloon-borne experiment:

1. first of all, it will allow to tie together the data and method of orbital experiment with results of lower energe ground–based experimental arrays;

2. two different methods of primary particle energy measurement, EAS fluorescence and Cherenkov light reflected from the snow detection, can be used in the experiment (see Figure 1);

3. the technique of data preliminary analysis by on-board computer may be optimized in the experiment;

4. real light background value at the atmosphere boundary will be determined experimentally;

5. the protection of light detectors from night–day light conditions changes and local light background sources may be tested in such an experiment;

6. detector for balloon–borne experiment will by some orders cheaper than that for the spacecraft.

Estimated threshold energy of EAS fluorescent track detection for the spacecraft ($H = 500-800$ km) detector is $E_{thr} \sim 10^{19}$ eV for mirror area $S \sim 1-3$ m^2 and

CP433, *Workshop on Observing Giant Air Showers from Space*
edited by J.F. Krizmanic et al.
© 1998 The American Institute of Physics 1-56396-766-9/98/$15.00

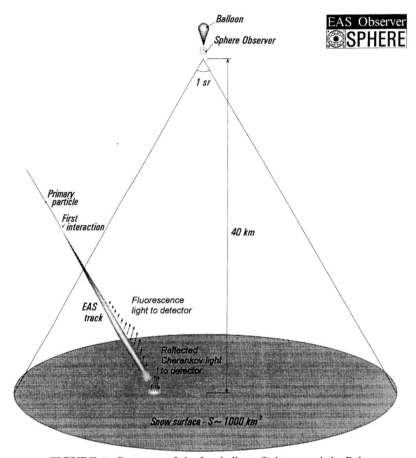

FIGURE 1. Geometry of the free balloon flight around the Pole.

angular resolution of light detector pixel $\omega \sim 10$ mrad. According to

$$E_{thr} \sim H^2 \cdot \sqrt{\frac{\omega^2}{S}}.$$

the same mirror area and angular resolution $\omega \sim 100$ mrad in balloon–borne experiment ($H \sim 40$ km) may ensure threshold energy $E_{thr} \sim 10^{18}$ eV. Threshold energy for Cherenkov detection in balloon experiment should be $E_{Chr} \sim 10^{17}$ eV due to large reflection coefficient of the snow surface (~ 0.9).

The integral flux of fluorescent light as well as the integral flux of Cherenkov light is a good measure of primary particle energy. But due to the anisotropy of Cherenkov radiation the Cherenkov image of EAS is easier to detect than the fluorescent one. Simultaneous using of these two methods might increase the methodical accuracy of primary particle energy determination.

About 150 events of energy $> 10^{19}$ eV may be detected during 20-day polar night balloon flight around Pole at 40 km altitude above the surface.

One needs an array with large number of pixels in the focal plane for a space experiment. To decrease the data flux intelligent data acquisition system is required which selects events of a certain type. Analysing system may include the trigger electronics as well as on–board computer procedure. Its efficiency might be tested and optimized in balloon experiment with small pixel number by transmission of data without selection as well as selected data.

Pixel number might be by 2−3 order smaller in balloon–borne configuration to provide the same spatial resolution as in spaceship–borne configuration because the balloon altitude is lower by $10 - 20$ times than spaceship altitude. Simple optics might be used for small light detector number. A spherical mirror of 1−2 m diameter with corrective diaphragm with the field of view ~ 1 sr ensures 100 mrad angular resolution that implies 60−100 detectors on the focal plane. Angular resolution of $\omega = 100$ mrad at 30−40 km altitude corresponds to the 3−4 km spatial resolution.

EAS track resolution of balloon detector as well as that of spacecraft detector may be improved by each pixel pulse shape detection. It is equal to increasing of pixel number.

Array becomes simple and cheap enough if the number of light detector pixels in mosaic is small $N \sim 10^2$. Electronics may be identical to that of the space detector.

The same array elevated to 1–3 km makes it possible to study primary cosmic ray energy spectrum structure in the region $10^{15} - 10^{17}$ eV and sensitive to cosmic ray composition shape of Cherenkov light lateral distribution function by detecting Cherenkov light, reflected from the snow.

To sum up, suggested balloon–borne detector is characterized by following properties:

1. detector provides the same spatial resolution as in spaceship–borne configuration;

2. detector consists of small pixel number $\sim 10^2$ and simple optics;

3. electronics is identical to the space detector electronics;

4. each pixel pulse shape detection improves the resolution of detector.

SPHERE detector may serve as a prototype of the detector for the balloon-borne AIRWATCH test flight.

SPHERE detector array

SPHERE detector array was elaborated for balloon–borne experiment [1–5]. This work is based on the Prof. A.E. Chudakov's [6] suggestion to detect the Cherenkov

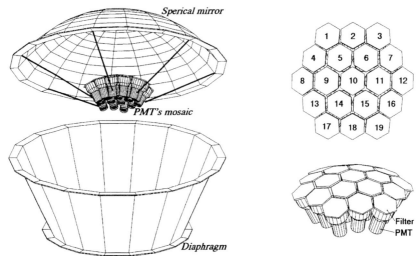

FIGURE 2. 3-D scheme of the optical part of the SPHERE detector.

light reflected from the snow surface.

Figure 2 shows the scheme of this array. The light spots are detected by 19 photomultipliers (FEU–110) situated on the focal surface of the spherical mirror of 1.2 m diameter. Dark violet filters and shifters are used with photomultipliers to decrease the influence of the starlight background. The angular aperture of detector is about $50^o \times 50^o$. Detector lifted to the altitude H make it possible to have a sensitive area $\sim H^2$.

The first measurements were carried out in the Thien–Shan mountains in winter 1993 (Figure 3). SPHERE detector was situated on the 160 m high mountain ledge nearby the B.Alma–Ata lake (2500 m above sea level) to detect Cherenkov light reflected from the snow surface of the lake.

The area of the lake is about 0.7 km^2.

The average inclination angle of the detector optical axis to the horizon was 10°. Cherenkov light is reflected from the snow surface according to the Lambert

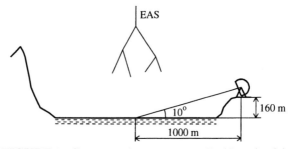

FIGURE 3. Geometry of experiment on B. Alma–Ata lake.

370

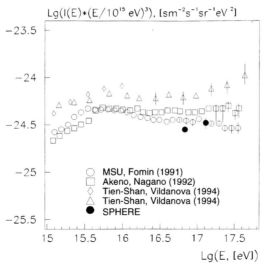

FIGURE 4. The differential energy spectrum.

reflection law:

$$I(\Theta) = I_0 \cdot \cos\Theta,$$

where $I(\Theta)$ is flux reflected at angle Θ, I_0 — normally reflected flux. So only about $\cos 80^o \sim 0.17$ of normally reflected light was detected in this experiment.

Primary cosmic ray flux at the energy 10^{17} eV obtained in this experiment is in agreement with other experimental data (Figure 4).

Energy threshold in this experiment was due to the large dead time of the device, not by starlight background.

During 1994–96 the detector SPHERE was improved significantly.

The amplitude measurements are completed by the time analysis of PMT pulses. It will allow us to analyse the detected events more completely.

The trigger rate ability is increased up to 50 Hz by using fast electronics and microcomputer. In balloon–borne experiment the reflected light intensity increases by 6 times according to the Lambert reflection law. This two reasons allow us to decrease energy threshold to $\sim 10^{15}$ eV under condition of starlight background only. The size of detector storage is sufficient to store $\simeq 3.6 \cdot 10^6$ events.

The detector electronics measures the integral of light pulse in PMT, pulse duration and intervals between pulses. The 30-ns discreteness allows to reject events simulated by charged particles in the PMT tubes and filter glass reliably . It makes possible to determine the arrival direction of EAS too.

The electronics for each pixel pulse shape detection is elaborated.

The first methodical lifting of SPHERE detector to 0.9 km altitude by fastened balloon was carried out in winter 1996–1997.

We plan to carry out measurements at altitudes 1–3 km above snow surface using

TABLE 1. Estimation of EAS with $E > E_o$ number to be detected by SPHERE detector

E_o, eV	$I(> E_o)$, $(\mathrm{m}^2\cdot \mathrm{hour}\cdot \mathrm{sr})^{-1}$		fastened balloon		4 flights of balloon around the South Pole
		H, km	1	3	40
		S, m^2	$\simeq 10^6$	$\simeq 10^7$	$\simeq 1.6\cdot 10^9$
		E_{thr},eV	$\simeq 10^{15}$	$\simeq 5\cdot 10^{15}$	$\simeq 5\cdot 10^{17}$
		t, hour	$\simeq 100$	$\simeq 100$	$\simeq 500$ (20 days)
$1\cdot 10^{15}$	$5.0\cdot 10^{-3}$		$1.5\cdot 10^6$	—	—
$1\cdot 10^{16}$	$6.5\cdot 10^{-5}$		$2.0\cdot 10^4$	$2.0\cdot 10^5$	—
$1\cdot 10^{17}$	$6.5\cdot 10^{-7}$		$2.0\cdot 10^2$	$2.0\cdot 10^3$	$1.6\cdot 10^6$
$1\cdot 10^{18}$	$6.5\cdot 10^{-9}$		2.0	20	$1.6\cdot 10^4$
$1\cdot 10^{19}$	$6.5\cdot 10^{-11}$		—	—	$1.6\cdot 10^2$
$3\cdot 10^{19}$	$6.5\cdot 10^{-12}$		—	—	16

the fastened balloon in winter 1997–98 under condition of small light background. In the future it is desirable to perform the large–scale measurements in the Arctic or Antarctic to detect EAS with energy up to $\sim 10^{20}$ eV. One such session will be enough to get the amount of data on EAS with $E \geq 10^{19}$ eV comparable with that of Yakutsk array.

Table 1 shows the estimated event number to be detected by SPHERE detector for given flight height H and exposure time t.

1−3 km high balloon–borne experiment is supported by Russian Foundation for Basic Research (grants 95–02–04325a and 96–02–31005k).

REFERENCES

1. Antonov, R.A., et al. *Proc. 14th ICRC*, **9**, 3360 (1975).
2. Antonov, R.A., Ivanenko, I.P. and Kuzmin V.A., *Izvestiya Academii Nauk*, (in Russian), **50**, pp. 2217-2220, (1986).
3. Antonov, R.A., Petrova, E.A., Fedorov A.N., *Vestnik MGU*, ser.3, (in Russian), **36**, 4, pp. 102-105, (1995).
4. Antonov, R.A., Chernov, D.V., Fedorov, A.N., et al. *Nuclear Physics B (Proc. Suppl.)* **52B**, pp. 182-184, (1997).
5. Antonov, R.A., Chernov, D.V., Fedorov, A.N., et al. *Proc. 25 ICRC*, Durban, OG 6.2.11, **4**, 149 (1997).
6. Chudakov, A.E, in *Trudy conf. po cosm. lutcham* (in Russian), pp. 69, Yakutsk, (1972).
7. Fomin Yu.A., et al. *Proc. 22nd ICRC*, **2**, 87 (1991).
8. Nagano M., et al., *J.Phys. G: Nucl.Phys*, **18**, 423, (1992).
9. Vildanova L.I., et al. *Izvestiya Academii Nauk*, (in Russian), **58**, p. 79, (1994).

LOW FREQUENCY RADIO RADIATION FROM CLUSTERS OF GALAXIES

V.S. Berezinsky[a], P. Blasi[b]

[a] *INFN Laboratori Nazionali del Gran Sasso*
67010 Assergi (AQ) - ITALY

[b] *Department of Astronomy & Astrophysics, and*
Enrico Fermi Institute, The University of Chicago,
5640 S. Ellis Av. Chicago, IL 60637

Abstract. Clusters of galaxies might be the main sources of extragalactic diffuse radio flux at the lowest frequencies $\nu < 1\ MHz$. This radition is produced as synchrotron radiation of low-energy electrons in intracluster space, where magnetic field is known to be of order of $B \sim 1\mu G$. Two sources of low-energy electrons are considered: the secondary electrons produced by the proton component of cosmic rays in the intracluster space and the primary electrons leaking from the galaxies. The low-frequency synchrotron radiation from intracluster space suffers very small absorption due to the low density of the gas and its high temperature. The diffuse radio flux at low frequencies is calculated. It gives the main cotribution to the absorption of ultra-high energy gamma-radiation.

CP433, *Workshop on Observing Giant Air Showers from Space*
edited by J.F. Krizmanic et al.

I INTRODUCTION

The propagation of high energy gamma rays in the universe for energies above $\sim 100 \ GeV - 1 \ TeV$ is mainly determined by the pair production energy losses in photon-photon scatterings on some radiation field. In particular, TeV photons are affected by infrared-optical (IR-O) backgrounds, while starting from $\sim 10^6 GeV$ the collisions on the photons of the microwave background become important. Ultra-high energy gamma rays (UHEGR), with $E \geq 10^{10} GeV$, produce $e^+ e^-$ pairs when they scatter on the low frequency radio radiation. In this paper we shall be interested in radio background radiation responsible for absorption of UHEGR.

The direct measurements of the isotropic radio background and the extraction from it of the extragalactic component were performed since early 60s (for a review see ref. [1]. A milestone was the direct satellite RAE-1 measurement of the radio background spectrum below 1 MHz [2]. In all cases the extraction of the extragalactic component from observational data needed additional assumptions, bacause of the absorption of low-frequency radiation in the galactic disc and quasi-isotropic radiation from the galactic halo. In particular, one should use cautiously the interpretation of ref. [3], where the absence of galactic radio halo is assumed.

Apart from the extraction of the extragalactic component from observational data, the low-frequency cutoff of extragalactic radiation remains a problem. An observed cutoff can be due to absorption in the disc or in the galactic halo, while the cutoff frequency of extragalactic component could be lower.

In ref. [1] the data of 60s on the extragalactic radio background were summarized and the absorption length for UHEGR was calculated. Some scepticism to this calculation was often expressed starting from beginning of 80s, when the radio emission from the halo of our galaxy was firmly established. Many people suspected that most part of the diffuse radio flux attributed to extragalactic radiation was in fact the halo component. A clear answer to this criticism was given only recently in ref. [4], where the catalogue data of individual extragalactic radio sources were summed up. The obtained flux coincides well at $\nu > 100 \ MHz$ with extragalactic flux as obtained in 60s.

The detailed calculations of the extragalactic radio flux were made recently by Protheroe and Biermann [5]. In this paper the authors used the data of catalogue sources, together with data from some particular populations of radio sources (most notably normal galaxies). The absorption lengths of UHEGR were also calculated.

However, the position of low-frequency cutoff remains a theoretical problem. The cutoff is provided by absorption in the sources and cannot be directly observed because of absorption in the disc and halo of our galaxy.

In this paper we argue that clusters of galaxies most probably give the dominant contribution to the low-frequency part of the extragalactic flux due to weak absorption of radio radiation in these sources. We shall consider here only two scenarios in which electrons are produced, that is a bright phase in the galaxy evolution,

374

discussed in section II, and secondary electrons resulting from pp collisions in the intracluster medium, discussed in section III.

II   THE BRIGHT PHASE

There are several types of cosmic ray (CR) sources in the clusters of galaxies: normal galaxies, radiogalaxies, and accretion shocks. An interesting source is the burst production of CR in the past, related to a short time of enhanced luminosity of protogalaxies. This period is usually called 'bright phase'. Since the confinement time of CR in a cluster exceeds the age of Universe, the CR protons accelerated at early epochs in the clusters are still present there, producing the secondary particles due to the interaction with the intracluster gas.

The idea of a bright phase was first put forward by Partridge and Peebles [8]. During this stage massive stars were produced and gave birth to SNeII. The duration of this phase can be estimated as $10^7 - 10^8$ yrs from the lifetime of the massive protostars.

Partridge and Peebles [8] estimated the total energy released during this stage as $\sim 3 \times 10^{61}$ erg per galaxy, and similar numbers were found by Schwartz et al. [9], who considered the bright phase at red shift $z \sim 2 - 10$. Volk et al. [6] assumed a bright phase at moderate red shift $z \gtrsim 1$ with a relatively long duration $\tau \sim 10^9$ yrs.

During this phase, the intergalactic medium was enriched of heavy elements produced in the supernovae explosions, and this could explain the large abundance of iron in the intracluster medium, which cannot be accounted for by the rate of supernovae observed in normal galaxies at present (see [6] and references therein).

Further indications of a bright phase come from the measurements of the abundances of light elements (B and Be) in population II stars in the halo of our galaxy (e.g. [10], [11], [12]): if these elements have a spallogenic origin, then a period of enhanced cosmic ray (CR) production is needed in the past of our galaxy in order to explain the observed abundances. This enhanced rate of CR production is related to the increased rate of supernovae explosions.

In the present calculations we shall assume that a bright phase occurred at moderate red shift, as in ref. [6]. The total number of SNe II in a cluster (N_{SN}) can be estimated from the iron abundance in the intracluster gas as $N_{SN} \approx \epsilon_{Fe} M_{gas}/M_{Fe}^{SN}$, where $M_{gas} \sim 10^{14} M_\odot$ is the average gas mass in a cluster, $\epsilon_{Fe} \simeq 7 \times 10^{-4}$ is the iron mass fraction in clusters and $M_{Fe}^{SN} \simeq 0.1 \ M_\odot$ is the average mass in iron ejected per SN II. The total energy produced in CR by these SNe II in each galaxy in the cluster is $W_{CR} \approx N_{SN} E_{CR}^{SN}/N_{gal}$, where $E_{CR}^{SN} \sim 3 \times 10^{49} erg$ is the energy in CR per SN II and $N_{gal} \simeq 100$ is the average number of galaxies in a cluster.

The radio radiation from this stage is produced in two ways: (i) The direct acceleration of electrons at the bright phase and their escape into intracluster space and (ii) acceleration of protons (nuclei) with their accumulation in the intracluster space and production of secondary electrons. In this section we shall concentrate on the process (i).

It is usually assumed that in a SN event a fraction $\sim 1\%$ of the total CR energy is in the form of electrons: $W_e^{SN} \approx 0.01\ W_{CR}^{SN}$. The rate of injection of electrons in the interstellar medium is easily estimated as W_e^{SN}/τ, where τ is the duration of the bright phase.

We calculated the radio flux produced by these electrons assuming that parent galaxies are ellipticals. This assumption is motivated by some observations which shows the dominance of the elliptical galaxies in the clusters. The arguments in favor of a bright phase in elliptical galaxies were recently considered by Zepf and Silk [13].

The most essential part for the estimate of the radio background is the calculation of the electron spectrum. The equation for the number density of electrons per unit energy $n(E, \vec{r})$ can be written in the general form as [14]:

$$-div(D\nabla n) + \frac{d}{dE}[b(E)n] = q(\vec{r}, E) \tag{1}$$

where $D(\vec{r}, E)$ is the diffusion coefficient, in general dependent on the position, $b(E) = dE/dt$ is the rate of energy losses for electrons with energy E, and $q(\vec{r}, E)$ is the number density of electrons produced at position $\vec{r}$ with energy E per unit energy interval. This last function depends on the type of galaxies we are interested in.

For boundary conditions at infinity the solution of eq. (1) can be written in the form given by Syrovatskii [15]:

$$n(E, \vec{r}) = \int d^3\vec{r_0} \int_E^\infty dE_0 Q(E_0, \vec{r_0}) \frac{1}{|b(E)|(4\pi\lambda)^{3/2}} e^{-(\vec{r}-\vec{r_0})^2/4\lambda}, \tag{2}$$

provided that the diffusion coefficient and the energy losses do not depend on the spatial coordinates. Here we put

$$\lambda(E, E_0) = \int_E^{E_0} dE' \frac{D(E')}{b(E')}. \tag{3}$$

In this section we shall use the Syrovatskii solution for elliptical galaxies, under the assumption of spherically symmetric space ditribution of the sources (SNe):

$$q(E, r) = \frac{KE^{-\gamma}}{\pi^{3/2}R_g^3} exp\left[-\frac{r^2}{R_g^2}\right], \tag{4}$$

where r is the radial distance from the center and R_g is the radius of the galaxy.

The constant K is fixed here according with the criteria mentioned above, based on the estimate of the total electron energy in the form of electrons. The number density of electrons is easily obtained from integration in eq. (2) over volume, which gives

$$n(E, r) = \int_E^\infty dE_0 K E_0^{-\gamma} exp\left[-\frac{r^2}{4\lambda + R_g^2}\right] \frac{1}{\pi^{3/2}|b(E)|(4\lambda + R_g^2)^{3/2}}. \qquad (5)$$

As mentioned above, the Syrovatskii solution can be strictly applied only if the energy losses and the diffusion coefficient in eq. (1) are distance-independent; therefore the averaged values for some parameters will be assumed here. The more accurate numerical solution of Eq. (1) will be given in another work [16]. In the present calculations we use the following parameters: $N_{gas} = 10^{-3}$ cm^{-3} and $T = 10^7 K$ for the gas density and temperature inside the galaxy and $B = 1$ μG for the magnetic field. These parameters have approximately the same values in the intracluster gas [17] and in the elliptical galaxies, so that the assumption about space homogenity is justified in this case and the Syrovatskii solution can be applied.

The energy dependence of a diffusion coefficient $D(E)$ is quite model dependent; in this estimate we shall assume for $D(E)$ the same form as in Ref. [5]: $D(E) = 2.5 \times 10^{28} cm^2 s^{-1}$ for $E < 3$ GeV and $D(E) = 2.5 \times 10^{28}(E/3 \ GeV)^{1/3}$ $cm^2 s^{-1}$ for $E \geq 3$ GeV.

We considered bremsstrahlung, synchrotron and ICS as the relevant energy losses of electrons, taking the correspondent expressions for $b(E)$ from ref. [18].

For the known equilibrium electron spectrum the calculation of the synchrotron radio emission $J(\nu, \vec{r})$ is straightforward and we will not enter here into the details. The important feature of these calculations is the absorption of radio radiation, which results in the low-frequency cutoff of the radiation from a source. The resultant intensity I_ν is described by a transfer equation. The total absorption coefficient, χ_ν, is given as the sum of the free-free and the synchrotron absorption coefficients: $\chi_\nu = \chi_\nu^{ff} + \chi_\nu^{syn}$, where

$$\chi_\nu^{ff} = 10^{-2}\frac{N_{gas}^2}{T^{3/2}\nu^2}\left[17.7 + \ln\frac{T^{3/2}}{\nu}\right] cm^{-1} \qquad (6)$$

and

$$\chi_\nu^{syn} = -\frac{c^2}{8\pi\nu^2}\int dE_e E_e^2 \frac{\partial}{\partial E_e}\left[\frac{n(E_e, s)}{E_e^2}\right] p(\nu, E_e). \qquad (7)$$

Here N_{gas} is the gas density, T is the temperature of the electron gas and $p(\nu, E_e)$ is the emissivity of an electron with energy E_e in the form of photons of frequency ν.

The details of this calculation will be given somewhere else [19]. The results are shown in fig. 1 by the dash-dot-dot-dot curve as compared with the contribution of spiral normal galaxies (not in clusters), obtained following the same steps as before, for two different values of the average gas density, $N_{gas} = 0.1$ cm^{-3} (dash-dot lower curve) and $N_{gas} = 0.01$ cm^{-3} (dash-dot upper curve). The density of spiral galaxies in the universe was taken from Ref. [20], while the density of clusters at $z = 0$ is taken from [21].

The effect of a bright phase at $z \sim 1$, though very model dependent at the present stage of the calculations, is an overall increase of the radio background. The cut off frequency is $\sim 10^5 \, Hz$, so that the effect of the low energy electrons escaping from the galaxies is not very evident. This is due to the huge electron injection, which is responsable not only for a strong radio emission but also for an effective synchrotron self absorption at moderately high frequency.

III SECONDARY ELECTRONS

As was mentiond above the proton component of CR in the clusters can be produced by accretion shocks, normal galaxies, radiogalaxies and SN explosions during the bright phase. The confinement time of high energy protons at interest is larger than the age of Universe and thus the secondary electrons responsable for radio emission are produced now, even if CR protons are accelerated in the past. In this section we shall consider two extreme cases for CR production: a) acceleration in normal galaxies, which gives the lower limit, and b) acceleration in active galaxies, which can provide the CR luminosity $L_{CR} \sim 10^{44} \, erg/s$ per cluster. In the case of normal galaxies, we take the CR luminosity $L_{CR} \approx 3 \times 10^{40} \, erg/s$ per galaxy with $N_{gal} \sim 100$ the number of galaxies per cluster. The generation spectrum is taken as $Q_p(E) \propto E^{-\gamma_g}$, normalized to the total CR luminosity and with two different values of the slope, $\gamma_g = 2.1$ and $\gamma_g = 2.4$.

The production of secondary electrons in pp-collisions occurs through pion production. At low energy we used for the pion production the formalism of ref. [22], recently reviewed in [23]; at energies higher than 10 GeV the scaling approximation of ref. [24] was used. The equilibrium spectrum of electrons was calculated using the energy losses due to bremsstrahlung and synchrotron energy losses in the intracluster medium with density $N_{gas} \approx 10^{-3} \, cm^{-3}$ and magnetic field $B \approx 1 \, \mu G$. In the low frequency region in which we are interested in, the main contribution to the radio background comes from low energy electrons ($E \lesssim 1 \, GeV$), which lose a very small fraction of their energy in the intracluster medium, due to the small gas density existing there ($N_{gas} \approx 10^{-3} cm^{-3}$). In this regime, if $q_e(E_e)$ is the rate of production of electrons in pp collisions per unit volume and per unit energy, then the equilibrium electron density is $n_e(E_e) \propto q_e(E_e)\tau_e(E_e)$ where $\tau_e(E_e) = E_e/b(E_e)$ is the typical time for the losses ($b(E_e)$ here is the rate of energy losses).

With the known electron spectrum it is easy to calculate the synchrotron spectrum of radio emission $J(\nu)$. The absorption is taken into account with the help of absorption coefficients given in the previous section. The transport equation can be analytically solved for the case of homogeneous distribution of electrons valid for this case. It results in

$$I(\nu) = \frac{J(\nu)}{4\pi\chi_\nu} \left(1 - exp(\chi_\nu L)\right),$$ (8)

where $I(\nu)$ is the radio intensity in units of energy per unit time per unit frequency range per unit area per unit solid angle, and $L \sim 1 \ Mpc$ is the average size of the cluster.

As discussed in the previous section two main absorption processes have been considered here, free-free absorption and synchrotron self absorption. It is easy to see from Eq. (6) that for $N_e = 10^{-3} cm^{-3}$ and $T = 10^8 K$ free-free absorption is relevant only at very small frequency, $\nu \lesssim 1 \ kHz$. Thus the cut off at low frequency is mainly determined by synchrotron self absorption.

The calculated diffuse radio fluxes from secondary electrons are shown in fig. 1 (dashed and solid curves). The solid curves are referred to $L_p = 10^{44} \ erg/s$ for $\gamma_g = 2.4$ (upper curve) and $\gamma_g = 2.1$ (lower curve). The dashed curves correspond to luminosity $L_p = 3 \times 10^{42} \ erg/s$ for $\gamma_g = 2.4$ (upper curve) and $\gamma_g = 2.1$ (lower curve).

In all cases the cut off frequency is $\sim 6 \times 10^3 \ Hz$, much smaller than that obtained both for normal galaxies and bright phase models.

IV DISCUSSION

We estimated the extragalactic diffuse radio flux due to radiation of electrons in clusters of galaxies. The clusters of galaxies are very efficient for production of low-frequency radio radiation, because the electrons escaping from the parent galaxy continue to radiate in the intracluster space, where magnetic field is high. The absorption of radio emission in this region is weaker than in the galaxies.

We considered here two cases: the radiation of electrons accelerated at the bright phase of galaxy evolution and the radiation of the secondary electrons produced by primary protons in the intracluster gas.

In case of the bright phase, considered at relatively low red shift $z \sim 1$, the low frequency cutoff corresponds to relatively high frequency $\nu \sim 0.1 \ MHz$. It occurs due to synchrotron self-absorption and provided by large density of emitting electrons in the intracluster space. The predicted flux in this case is rather model-dependent.

In case of the secondary electrons the cut off frequency is about $6 \times 10^3 \ Hz$, i.e. much smaller than in the case of individual spiral galaxies and the bright phase. At these frequencies the diffuse flux from clusters of galaxies dominate over that produced by normal galaxies.

The low cut off frequency $\nu_{min} \sim 6 \times 10^3 \ Hz$ corresponds to absorption of very high energy gamma-rays, $E_\gamma \approx 10^{13} \ GeV$. It can be easily estimated that the absorption length of gamma rays at this energy varies between $\sim 40 \ Mpc$ for the lower dashed curve in Fig. 1, and $\sim 400 \ kpc$ for the upper solid curve.

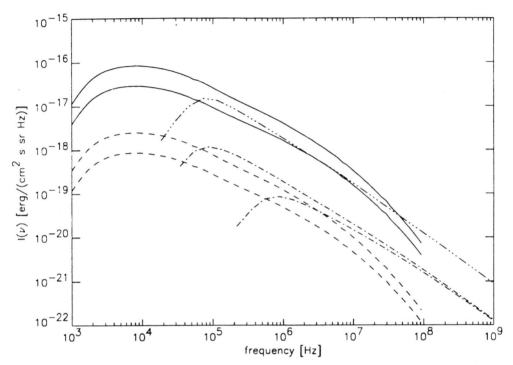

FIGURE 1. Diffuse radio flux from the clusters of galaxies. The solid curves refer to radio radiation from secondary electrons with $L_p = 10^{44} erg/s$ and $\gamma_g = 2.4$ (upper curve) or $\gamma_g = 2.1$ (lower curve). The same is for dashed curves but with $L_p = 3 \times 10^{42} erg/s$. Dash-dot curves describe the radio emission from normal galaxies for $N_{gas} = 0.1 \ cm^{-3}$ (lower curve) and $N_{gas} = 0.01 \ cm^{-3}$ (upper curve). Radio flux due to bright phase is plotted as dash-dot-dot-dot curve.

REFERENCES

1. Berezinsky V.S., *Soviet Journal of Nucl. Physics* **11**, 222 (1970).
2. Alexander J.K. et al. *Astron. Astroph.* **6**, 476 (1970).
3. Clark T.A., Brown L.W. and Alexander J.K. *Nature* **228** 847 (1970).
4. Berezinsky V. and Rigitano R. *in preparation.*
5. Protheroe R.J., and Biermann P.L. *Astrop. Physics* **6** 45 (1996); erratum-ibid **7** 181 (1997).
6. Völk H.J., Aharonian F.A., and Breitschwerdt D., *Space Sci. Rev.* **75**, 279 (1996).
7. Berezinsky V.S., Blasi P., and Ptuskin V.S., *Ap. J.* **487**, 529 (1997).
8. Partridge, R.B., and Peebles P.J.E., *Ap. J.* **147**, 868 (1967).
9. Schwartz J., Ostriker J.P., and Yahil A., *Ap. J.* **202**, 1 (1975).
10. Walker T.P., Steigman G., Schramm D.N., Olive K.A., and Fields B., *Ap. J.* **413**, 562 (1993).
11. Fields B.D., Schramm D.N., and Truran J.W., *Ap. J.* **405**, 559 (1993).
12. Fields B.D., Olive K.A., and Schramm D.N., *Ap. J.* **435**, 185 (1994).
13. Zepf S.E., and Silk J., *Ap. J.* **466**, 114 (1996).
14. Ginzburg V.L., and Syrovatskii S.I., *The Origin of Cosmic Rays*, Pergamon Press, Oxford (1964).
15. Syrovatskii S.I., *Sov. Astron.* **3**, 22 (1959).
16. Berezinsky V.S., Blasi P., and Sigl G., *in preparation.*
17. Kronberg P.P., *Phys. Rep.* **57**, 325 (1994).
18. Longair M.S., *High Energy Astrophysics*, Cambridge University Press, Cambridge (1994).
19. Berezinsky V.S., and Blasi P., *in preparation.*
20. Tinsley B.M., *Ap. J.* **220**, 816 (1978).
21. Bahcall N.A., and Cen R.Y., *Ap. J.* **407**, L49 (1993).
22. Dermer C.D., *Astron. Astrophys.* **157**, 223 (1986).
23. Moskalenko I.V., and Strong A.W., *preprint astro-ph/9710124.*
24. Berezinsky V.S., Blasi P., and Hnatyk I.B., *Ap. J.* **469**, 311 (1996).
25. Colafrancesco S., and Blasi P., *To be published in Astrop. Physics.*
26. Blasi P., and Colafrancesco S., *Submitted to Ap. J. Lett.*

The OWL Detector: Aperture and Resolution

H.Y. Dai, E.C. Loh, P. Sokolsky

High Energy Astrophysics Institute
Physics Department, University of Utah
Salt Lake City, UT 84112

Abstract. The OWL trigger aperture as a function of the detector spacing and threshold are presented. The detector resolution of the primary energy and X_{max} is also studied by Monte Carlo.

DETECTOR CONFIGURATION

The orbital height of the detectors is assumed to be one tenth of the earth radius (638 km), and the light collection area is $4.9m^2$ (2.5 meter diameter mirror or lens). A typical UV filter (80% transmission at UV) and a bi-alkali PMT (0.28 maximum quantum efficiency) are used in the simulation. The pixel size is chosen to be 0.05 degree and the full field of view of 60°. The number of detector units in this simulation is 2 with a variable distance between these two units. We always require that more than 5 pixels within one detector unit to form a trigger. Events are recorded when both units trigger.

APERTURE AS A FUNCTION OF DETECTOR SPACING

For this study, we set the threshold at 3 photo-electrons/pixel and vary the distance between two detectors from 500 to 2000 km in a step of 500 km. For each detector separation, the viewing direction of the detector is optimized to maximize the stereo aperture.

Figure 1 summarizes the trigger aperture as a function of the detector separation. This figure shows that the high energy trigger aperture increases with the detector separation, as so does the threshold energy. The detector fully turns on at $10^{19.25}$ eV when the detector separation is 500 km. At large detector separation, the aperture varies gently with energy .

CP433, *Workshop on Observing Giant Air Showers from Space*
edited by J.F. Krizmanic et al.
© 1998 The American Institute of Physics 1-56396-766-9/98/$15.00

If we assume the neutrino cross section is 3×10^{-32} cm^2 at 10^{20} eV, the neutrino trigger aperture is 20.7 km^2Sr with detector separation of 2000 km.

APERTURE AS A FUNCTION OF DETECTOR THRESHOLD

The detector separation is fixed at 2000 km for this set of simulations. The detector threshold depends on the background light level as viewed from space. We are designing an experiment to measure this important number. We vary the threshold from 3 photo-electrons per pixel to 10 photo-electrons per pixel in our simulation. The sky background measured by the HiRes detector on the ground corresponds to a threshold 4 to 5 pe/pixel.

Figure 2 shows the aperture as a function of primary energy for different threshold values. At 10^{21} eV, the aperture only changes by 15% when the threshold changes from 3 pe to 10 pe/pixel. However at 10^{20} eV, the aperture changes by factor of 5 over the same range of threshold.

DETECTOR RESOLUTION

The detector resolution is studied by reconstructing the Monte Carlo generated events using the method described in our last report (UUHEP 97-5). The pixel signals are Poisson fluctuated. The time resolution is assumed to be 1 μs, the shower-detector plane resolution is 0.1 degree, and the resolution of the centroid is assumed to be one tenth of the pixel size (0.005 degree). No atmospheric uncertainty is included.

Figures 3, 4, 5, and 6 show the energy and X_{max} resolution at four different energies. The systematic shift in energy is an artificial effect (inconsistent assumptions in the Monte Carlo and the reconstruction about the maximum shower size and energy) and should be ignored. The cuts applied to the data are: the first visible depth (the depth corresponding to the earliest pixel in an event) must be in the range of 250 g/cm^2 to 750 g/cm^2 and the shower width expressed in X_{max}-X_o must be less than $1000 g/cm^2$. More than 90% of the events survived the cuts (2000 events generated for each data set and more than 1800 events passed the cuts).

As we can see from figures 3 through 6, both energy and X_{max} resolution improves with energy. The energy resolution due to the geometrical resolution and signal fluctuation is typically at 15%. The X_{max} resolution is about 65 g/cm^2 at threshold and decreases to about 44 g/cm^2 at 10^{21} eV.

Acknowledgment

All calculations are carried out on the IBM SPs of CHPC, University of Utah. Dr. Stuart Taylor is acknowledged for reading this draft.

OWL APERTURE vs DETECTOR SEPARATION

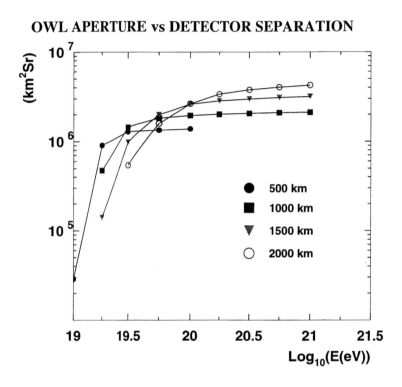

FIGURE 1. Trigger aperture $(km^2 Sr)$ as a function of the detector separation.

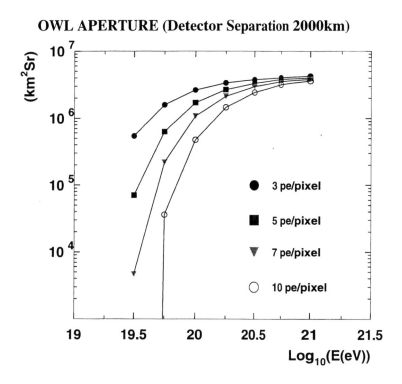

FIGURE 2. Trigger aperture $(km^2 Sr)$ as a function of the detector threshold.

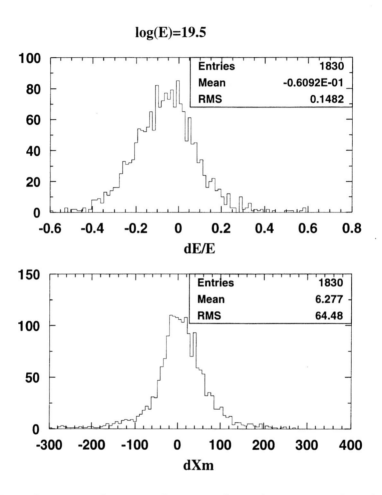

FIGURE 3. Detector resolution on the energy (relative) and X_{max} (in g/cm^2) for $E = 3.16 \times 10^{19}$ eV.

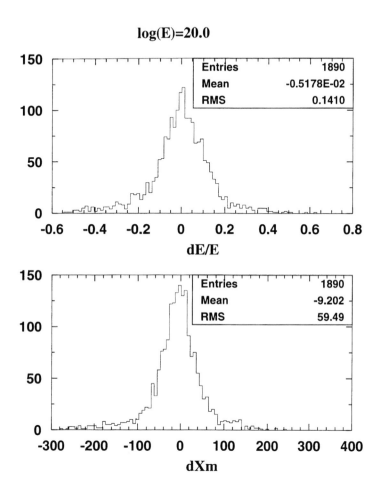

FIGURE 4. Detector resolution on the energy (relative) and X_{max} (in g/cm^2) for $E = 10^{20}$ eV.

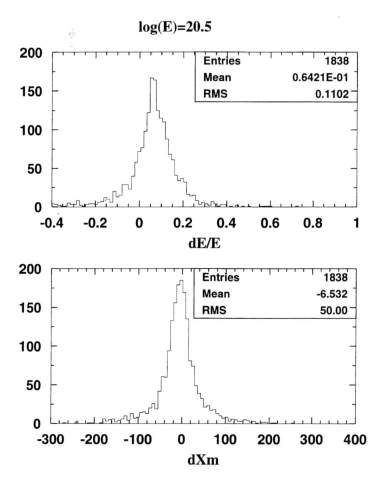

FIGURE 5. Detector resolution on the energy (relative) and X_{max} (in g/cm^2) for $E = 3.16 \times 10^{20}$ eV.

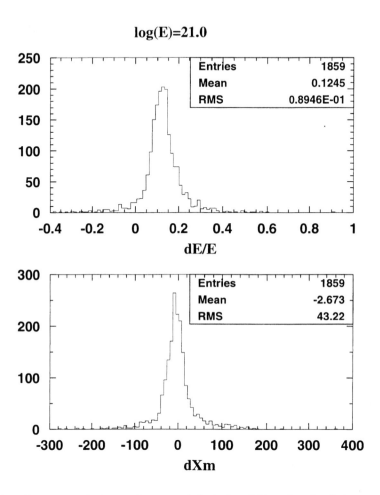

FIGURE 6. Detector resolution on the energy (relative) and X_{max} (in g/cm^2) for $E = 10^{21}$ eV.

Observation of UHE Neutrino Interactions from Outer Space

G. Domokos and S. Kovesi-Domokos

Department of Physics and Astronomy
The Johns Hopkins University
Baltimore, MD 21218[1]

Abstract. The interaction of UHE neutrinos can be observed from outer space. The advantage of the proposed method is that the Earth can be used as an energy filter as well as a target. We sketch the potentials of these observations in searching for particle physics beyond the Standard Model of elementary particle interactions as well as observations of astronomical objects such as active galactic nuclei.

I INTRODUCTION

It has been recognized that the observation of UHE neutrino interactions can serve a dual purpose:

- The observation of objects of high activity and high column density in the sky, (a typical example is an AGN) which otherwise are inaccessible to observation using electromagnetic waves at almost any wavelength;

- a search for particle physics not described by the current "Standard Model" of elementary particle interactions.

The first entry on our list has been widely explored in a number of articles and it has been discussed at a variety of conferences; the first meeting with the purpose of exploring all the astrophysical uses of neutrino telescopy has been the Hawaii conference [1], followed by a series of meetings on neutrino astronomy in Venice [2] and in Pylos [3].

The particle physics aspect of neutrino telescopy is somewhat less known; however, in our opinion, it is equally important. A reasonably comprehensive discussion of the particle physics uses of neutrino telescopy is contained in an article published in one of the Venice proceedings [3].

In this paper we report the results of a first exploration of the particle physics uses of a novel type of neutrino detectors: the use of orbiting detectors, such as the

---

[1] E-mail: skd@haar.pha.jhu.edu

CP433, *Workshop on Observing Giant Air Showers from Space*
edited by J.F. Krizmanic et al.

Orbiting Wide angle Light collector (OWL) as well as a detector flown on a space station.

This paper is organized as follows. In the next section we outline the potential advantages of an orbiting detector used as a neutrino telescope. Section 3 contains a brief description of our first results. The final section contains a discussion and outlook.

Most of the results reported here are preliminary in nature: they will be followed up by more detailed investigations. Nevertheless, the ones reported here look exciting and, for that reason, worth discussing.

II "EYE IN THE SKY"

Conventional (?) neutrino telescopes, such as AMANDA, BAIKAL, SUPER-KAMIOKANDE and the future NESTOR have the advantage that they are based on technologies which, by now, are reasonably well explored. Their basic feature is that they are capable of a good directional resolution (currently of the order of 1° or so. (This is about the resolution of Galileo's telescopes: neutrino telescopy is at the age optical telescopy was in Galileo's time...) These telescopes therefore provide a valuable tool for exploring the neutrino sky for both point like (AGN?) and diffuse sources, such as atmospheric neutrinos and neutrinos emitted by baryonic dark matter interacting with the primary cosmic radiation [4]. Likewise, on the particle physics side, such detectors are able to explore such questions whether the neutrino-nucleon cross section shows any anomaly as compared to the Standard Model [3].

What is missing is that a typical underground or underwater neutrino telescope cannot be easily made to be sensitive to the primary neutrino energy or to explore the development of a shower resulting from the interaction of the primary neutrino. To be sure, there *are* some solutions available to bridge the gap [5]. Most of those are, however, based on software development. It would be also desirable to have detectors which are sensitive to the primary neutrino energy directly. (The importance of this feature has been emphasized in ref. ([3]).)

The basic idea we are here proposing is a very simple one. UHE neutrinos originating either from a point source or from a diffuse background will interact in the Earth. If the interaction takes place sufficiently close to the surface of the Earth, at least a substantial part of the shower originating from the primary interaction will develop in the atmosphere and it can be viewed by OWL or, possibly, by a future detector mounted on the space station. Due to the fact that the interaction cross section of neutrinos is energy dependent, there is a one to one correspondence between the interaction mean free path (mfp) of the neutrino and its impact parameter with respect to the center of the Earth. (Equivalently, by an elementary exercise in geometry, the impact parameter dependence can be translated into a nadir angle dependence.)

The Earth acts as a filter: given a nadir angle and the corresponding mfp, more energetic neutrinos will interact deep in the Earth and the shower will die out before reaching the surface. Conversely, less energetic neutrinos either do not interact at all or they interact high in the atmosphere and there will be insufficient target thickness for the shower to develop. It is to be emphasized, of course that this are statements which hold for the average shower. Due to fluctuations in the multiplicity of the first interaction and in the shower development, one may not be able to determine shower energies on an event by event basis. (Nevertheless, simulations carried out for the Fly's Eye detector indicate that even an event by event energy determination may be feasible with a tolerable error.)

III   AN EXAMPLE

In order to determine the shower development under these assumptions, our preliminary study concentrated upon the feasibility of observing a neutrino induced shower. The question one has to decide is whether the upward going shower develops over a substantial length in the atmosphere so that it can be observed. (One has to recall that an upward going shower starts in the dense part of the atmosphere, near the Earth.) For this purpose, we calculated the longitudinal development of showers in approximation A, assuming Feynman scaling. It is known that Feynman scaling is violated in hadronic interactions, due to QCD loop effects. The violation is, however, logarithmical. We took the neutrino mfp from the work of Gandhi et. al. , ref. [6]. An average hadronic cross section of 60 mb was used, even though we find that the development of the electromagnetic component does not depend very sensitively on variations of the hadronic cross section within a reasonable range. The results were obtained (as usual) in target depth, measured in g/cm^2. Conversion to distances was obtained assuming an exponential atmosphere, of a scale height of 7.8 km.

In the following Figure, we show the result of a calculation for a neutrino incident at a nadir angle of 80°. According to the neutrino mfp given in ref. [6] and the specifications given above, this corresponds to a neutrino energy of about 10^9 GeV. In calculating the electromagnetic component, we assumed an initial hadron multiplicity of ≈ 50 and an inelasticity of ≈ 0.5. The Figure contains the longitudinal profile of the leptonic component induced by one hadron. The reason for this is that one may contemplate other multiplicities and inelasticities. In that case, the result is obtained by a simple rescaling of the ordinate. (This is a property of the diffusion approximation used here; for the present purposes, it is adequate. However, in the case of more sophisticated models, this scaling property is only approximately valid.)

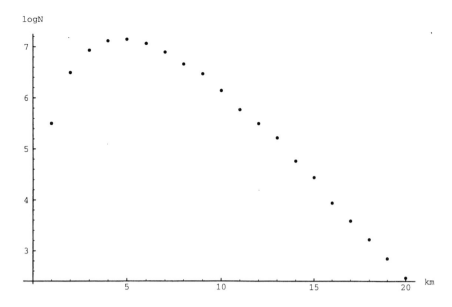

FIGURE 1. Longitudinal profile of a shower described in the text.

IV DISCUSSION

The example shown here suggests that one may be able to use orbiting detectors in conjunction with the Earth as a target for the detection of UHE neutrino interactions. At present, the properties of the detectors are not yet sufficiently well known and thus, no quantitative conclusions can be drawn. A more detailed calculation is needed in order to explore particle physics capabilities of the orbiting neutrino telescopes. Work along these lines is in progress.

We wish to thank the organizers of the Giant Air Shower meeting for a very stimulating conference and Bianca Monteleoni for several useful discussions on neutrino detection.

REFERENCES

1. High Energy Neutrino Astrophysics, editors V.J. Stenger *et. al.* World Scientific, Singapore (1992)
2. Proceedings of the Workshops on Neutrino Telescopes, edited by M. Baldo-Ceolin. Published by the Academy of Arts, Sciences and Letters, Venice.
3. G. Domokos, B. Elliott, S. Kovesi-Domokos and S. Mrenna, in Proc. Third Workshop on Neutrino Telescopes. (*cf.* previous reference.)
4. G. Domokos, B. Elliott and S. Kovesi-Domokos, Jour. Phys. **G 19**, 899 (1993).
5. B. Monteleoni, private communication.
6. R. Gandhi *et. al.* Astropart. Phys. **5**, 81 (1996)

FOCAL PLANE DETECTORS
Possible Detector Technologies for OWL/AIRWATCH

Esso Flyckt

Photonis SAS[1]
BP250, F-18106 Brive, France[2]

Abstract. New satellite-born projects **OWL** and **AIRWATCH** will need single-photon focal-plane detectors of a million pixels in a design which is optimized to the focusing optics and electronics at acceptable cost. We discuss different phototube possibilities and their pros and cons with crude cost estimates. We conclude that a multichannel-photomultiplier solution is safe. A better compromise may be to adapt a 6 or 9 inch X-ray image intensifier tube or develop a 12 inch image intensifier for detecting individual photons, and adapt the optics to have many mirror modules. The possibility of developing super-large-area phototubes is also discussed.

GENERAL CONSIDERATIONS

The **OWL** baseline design discussed in [1] is used as an example. The proposed optics leads to at least 2.5×10^5 pixels or about $36,000$ pixels of 4 mm × 4 mm dimensions in the focal plane of each of seven mirrors/lenses. Recent discussion [2] indicates that the pixels may be as large as 10 mm × 10 mm and as many as one million, considerably increasing the challenge and costs.

[1] formerly Philips Photonics
[2] Email: so.flyckt@photonis.com, Fax:+33-555-863773

CP433, *Workshop on Observing Giant Air Showers from Space*
edited by J.F. Krizmanic et al.

The focal-plane detector should be able to handle the following scenario:

- few-photon signals $\rightarrow$ single photoelectron resolution
- 337, 357 and 391 nm fluorescence $\rightarrow$ window transparent at 337 nm
- < 100ns time resolution $\rightarrow$ rules out traditional CCD arrays
- $< 4 - 10$ mm $\times$ $4 - 10$ mm pixel size $\rightarrow$ larger pixels simplifies optics
- Many pixels per detector element $\rightarrow$ e.g., $\sim 36,000$ per focal plane array
- "Launchable" $\rightarrow$ ruggedized
- Low-power consumption $\rightarrow$ no power-hungry voltage dividers
- Low noise $\rightarrow$ < 1 count per s per pixel
- Few readout channels $\rightarrow$ smart pixel-grouping electronics
- Minimal dead space $\rightarrow$ tubes with large photocathode area
- Acceptable Cost $\rightarrow$ 3×10^5 pixels ~ 10 million dollars

The fast-timing, single-photoelectron, and noise demands eliminate cheap semi-conductor pixel detectors like CCDs, and require further development on Si photo-diodes and APDs. These requirements point to the usage of clean-gain technology or *phototube technology*. However, it may be wise to evaluate the possibility of fast Si diode/CCD shift-register readout technology behind a phototube to decrease the number of readout channels, favoring an image intensifier tube solution. Several phototube options are discussed below.

PHOTOMULTIPLIERS

Discrete photomultipliers

The Hamamatsu R5600 TO-8 metal-dynode tube is the only discrete tube, small enough to be a candidate. Positive parameters are speed and the possibility of single photoelectron resolution. Negative parameters are low photocathode sensitivity, useful-to-total photocathode area ratio and price. We therefore exclude the idea of handling $> 500,000$ discrete PMTs.

Current multichannel photomultipliers

Such tubes already exist from two companies. Philips Photonics offers 10-stage *foil-multiplier* MC-PMTs as catalogue items with 64 or 96 pixels of dimensions 2.54 mm $\times$ 2.54 mm [3,4]. Such tubes offer single-photoelectron resolution, a com fortable gain of $\sim 10^6$ and, if enough funding is made available, they can be adapted to $4 - 10$ mm $\times$ $4 - 10$ mm readout pixels. For 4 mm $\times$ 4 mm pixel dimensions, the pixels can be arranged in a 12×12 array by moderate redesign. Larger-size arrays can be envisaged with some more efforts/costs. Parameters to optimize are:

- *Useful-to-total photocathode area*
  For a 12×12 pixel design in a square-envelope tube this ratio will be only

$50 - 60\%$, as some area will be needed for supporting a (ruggedized) multiplier structure. Moreover, the tube envelope needs to have a certain (ruggedized) thickness. This will demand some light piping and corresponding photon loss; or simpler, one accepts the loss of $\sim 50\%$ of the available photons.

- *Gain spread between pixels*
  In the worst case, the gain spread may be as large as 5:1 for a 12×12-pixel tube. Gain spread of adjacent pixels may be no more than 2:1.

- *Photocathode sensitivity down to 337 nm*
  The demand for high transmission at this wavelength excludes fiber-optic windows. However, the large pixels will be very forgiving as to optical crosstalk contribution in, e.g., a borosilicate glass window. Optical crosstalk will also depend on how well the photons are focused in the photocathode plane illustrating the interplay between the optics and the phototube.

In the standard method of producing a MC-PMT, the bialkali, SbKCs, photocathode is processed by pre-depositing a Sb layer, resulting in lower sensitivity than when the Sb layer is deposited in-situ. Consequently, the photocathode sensitivity is typically 60 mA/W at 337 nm (quantum efficiency $\sim 22\%$). Higher cathode sensitivity, up to ~ 80 mA/W at 337 nm (QE $\sim 30\%$) can be made in a so-called transfer-technology process at a price premium of $30 - 40\%$.

- *Ruggedized*
  The MC-PMTs are at least semi-ruggedized. Work and tests have to be done to guarantee their surviving the acceleration and vibrations at launch.

- *Price per pixel*
  In the chosen example, 250 square tubes (12×12 pixels of 4 mm $\times$ 4 mm) would be packed in an "octagonal" array $\sim 36,000$ pixels for each of the seven lenses/mirrors. Philips Photonics estimates that the price can come down to $35 - 40$ for a 144-pixel tube with 4 mm $\times$ 4 mm pixels (non-transfer technology tubes). Therefore, the total cost of ~ 1750 MC-PMTs can be estimated to be $9 - 10$M.

Hamamatsu offers a "copy" technology referred to as *metal-dynodes* in a 4×4 pixel arrangement, 24 mm $\times$ 24 mm envelope [5]. The useful area is thus $4 \times 4 \times 16/24 \times 24 \sim 50\%$ when tubes are packed into an array, similar to what Philips estimates.

Parameters for the Philips design that need to be optimized also apply for this design. So far, prices indicate a similar cost structure per pixel. Designing the focal-plane detector with such MC-PMTs would also be in the order of $9 - 10$ million dollars but the design will need 2250 tubes per mirror or a total of $15,750$ tubes.

It should be underlined that both concepts are single-photoelectron devices with a comfortable gain of $\sim 10^6$ at moderate high voltages of $1000 - 1500$ V. Special voltage dividers must be designed to limit the power consumption.

PMT life-time

5,000 Philips PMTs, set a gain of 10^5, used in HiRes (2-meter diameter mirror with each PMT covering an angular aperture of $1° \times 1°$ have been in operation for more than three years without measurable degradation [2]. If space-borne PMTs are operated with a similar gain, observing a similar level of light, PMTs can be expected to last for many years.

HYBRID PHOTOMULTIPLIERS

Such single-channel devices are slowly available in *sample* quantities from some suppliers including Hamamatsu and DEP [6]. The same companies are developing techniques for producing multi-channel hybrid PMTs.

Common properties for these devices are:

- Semi-transparent photocathode

- $7 - 10$ kV accelerating the photoelectrons toward

- A silicon photodiode, a silicon APD or pixelized arrays thereof.

Due to the 3.2 eV band gap of intrinsic silicon, thousands of electron-hole pairs are created and collected. The feature of such a tube is a higher dynamic range than traditional PMTs capable of resolving the peaks in the spectrum of one, two and several photoelectrons. However, the intrinsic gain is only some thousands, demanding additional fast, low-noise preamplifiers. The time resolution of such hybrid tubes is more than sufficient for **OWL** application, so some time degradation can be accepted in adding the preamplifiers. However, if the photodiode area is large, 4 mm $\times$ 4 mm or larger, the diodes will have a considerable capacitance contributing to additional noise.

This will limit the so-called proximity design. An unanswered question is how many diodes one can have in one device and still have a good production yield - or tolerate one or two dead pixels per tube. Large diodes are expensive, a broad guess being in the order of $10 per diode chip/pixel. Such a hybrid tube could be similar in size to the 144 pixel MC-PMTs. However, because the tube has to withstand the > 10 kV acceleration voltage, its design is more complicated than that of the MC-PMT. Assuming the same yield and same number of pixels, we estimate that the total cost per pixel will certainly be in the range of $50 - 60$. The focal plane detector cost will then be in the order of $12 - 15$M. The uncertainty in our cost estimate is due mainly to the amplifier design and production.

Another type of design should also be considered where the photoelectrons are focused onto a small Si-diode array (or more futuristic - onto a thinned and back-bombarded, *fast* CCD chip). This technology is a direct spin-off from so-called first generation image intensifiers (with no MCP).

IMAGE INTENSIFIERS

Non-MCP image intensifiers

As things stand today, it may still be safer to use a standard off-the-shelf image intensifier and place a Si-photodiode (or fast CCD) array outside the tube to read out the phosphor screen. Such tubes have resolutions of > 20 line-pairs per millimeter with a good point spread function (contrast), much better than our needs. The limitation will be how many 4mm × 4 mm pixels we can get onto the front-end of one tube.

A typical such tube with ~ 50 mm useful photocathode could take about 100 pixels, so one would need about 360 tubes per mirror. In quantity, such tubes could be fairly cheap, $< \$4000$ each with special fast phosphors (P46/P47). Adding another $\sim \$1000$ for the small pixel photodiodes should bring the pixel cost to $\sim \$50$ or a total of $\sim \$13$M. Such round tubes in hexagonal packing will pack poorly and the effective light sensitive area would be less than 50%. To redesign for square (or hexagonal) tubes in this technology will cost too much to be worth the trouble. Winston light cones could be used to improve the light collection efficiency. Moreover low-noise, high-gain preamplifiers would be needed as the tube gain is only about $15 - 20$. This approach is, therefore, regarded as a weak candidate. However, because tubes of this type contain neither dynode multipliers nor MCPs, tubes would have an excellent lifetime and consume low power.

A larger version of this concept is the 80 mm diameter de-magnifying image intensifier (without MCP) from DEP, Photek or Hamamatsu (used in the CHORUS experiment at CERN). With 270 pixels per tube, it could take 135 tubes per mirror or a total of 950 tubes for the experiment. Prices in larger quantities could probably be in the order of $\$12,000$ each, including the Si array and fast, high-gain preamplifiers behind the phosphor. The total project cost would then become $\$11 - 12$M. The packing ratio will be $< 50\%$ and the power consumption will be moderate.

X-ray image intensifier concepts

A larger version of the last concept is based on modifications of very large image intensifiers as used in medical X-ray imaging. Such 6 inch tubes (Philips) could be modified to take a bialkali photocathode instead of the X-ray screen, and de-magnify the 6 inch area to about 20 mm × 20 mm where it could be covered by a suitable amount of small-pixel photodiode arrays. Such a 150 mm tube could take > 1000 of the 4 mm $\times$ 4 mm pixels. This means that **OWL** needs approximately 250 of these tubes. Packing such round tubes will again have a poor useful photon coverage of $< 50\%$.

The initial R&D of defining such 6-inch tubes will be expensive. A broad guess is that the tubes could sell for $\sim \$20,000$ each, leading to a favorable total cost of $\$5-6$M. Maybe some money could be saved in the optics design of 250 smaller mirrors. The gain of such tubes would only be about $30 - 50$. The power consumption will be moderate.

NOTE: The development money given by a member of the **OWL** collaboration to Hamamatsu for modifying the 20" SUPER-KAMIOKANDE *PMT* into a 100×100 pixel MC-PMT probably will lead to a tube (with very strong crosstalk) producible below $\$70,000$. The gain will be comfortably $> 10^6$ but it will have lifetime problems and high power consumption like the compact MC-PMTs. Ruggedizing this design will be a special problem. Total cost for 3×10^5 pixels will be of the order of $2M.

The Philips DUMAND "smart" PMT [7] was originally developed from a 9" X-ray image intensifier into a hemispherical tube by giving up the image quality for two-π angular response. This 9" Philips tube has in the meantime received a changed design with a Ti input window but could be reverse-engineered to its original glass envelope. Such a tube would take ~ 2700 pixels and we would need only ~ 100 tubes, each with its own mirror/lens. Including the reverse-engineering needed, such tubes could be made for $\sim \$50,000$ each (note that the quantity is relatively small) meaning a total cost of $5M plus the Si-diode arrays and amplifiers to read out 2700 pixels per tube. The estimated cost for **OWL** would still be $\sim \$5$M. If development cost can be made reasonable, we may be able to develop a 12" version greatly reducing the number of tubes with a slight increase in cost ($\sim \$7 - 9$M).

MCP image intensifiers

If the gain in the first generation image intensifiers is not enough, or the cost and complications of the preamplifiers is a hurdle, a MCP (microchannel plate) could be added just in front of the phosphor screen into a second generation inverter tube. It is of the utmost importance to conclude early how much signal per pixel the MCP has to process - including all raw background signals - as the typical life of a phototube incorporating an MCP is in the order of $0.1 - 0.5$ coulomb/cm^2 of output charge. The second-generation tubes are more expensive than the first-generation tubes, but the difference is probably in the error bar of the rough estimates above. These tubes are not suitable because of their short lifetime.

It would also be possible to further explore 40 mm image intensifier tubes, but it is not very clear if this will bring any real advantages over the multitude of tube versions discussed above.

LARGE SENSITIVE CATHODE SURFACE

We emphasize again that **OWL** optics requires the use of large photocathode photon detectors. Large cathode phototubes would provide the best coverage. It is possible to develop large cathode surfaces (~ 1 m$\times \sim 1$ m) with uniform sensitivity, low noise, and transparent to 337 nm light. This surface could be followed by appropriate electron multipliers or Si arrays. To ensure UV transparency, such windows may have to be made reasonably thin. Cathodes need to either be encased by a strong cover to be removed in space or to be supported internally in the tube. The development cost of such surfaces is estimated to be well above $1M with a considerable capital equipment cost of several million dollars resulting in $7 - 10$ tubes for the **OWL** project at a total cost of $4 - 7M.

RECOMMENDED BASELINE DESIGN

The above survey gives us a choice between the following technologies:

1. MC-PMT (two versions) $9 $- 10M straightforward

2. Hybrid PMT $12 - 15M R&D

3. Small image intensifier tube + diodes $12 - 14M straightforward

4. 80 mm image intensifier tube + diodes $11 - 12M development

5. 6" X-ray image intensifier tube + diodes $5 - 6M development

6. 9" X-ray image intensifier tube + diodes $5 - 6M development

7. MC-PMT 20" low price, low spec $2 - 3M development

8. 12" X-ray image intensifier tube + diodes $7 - 9M R&D

9. Large cathode surface technology $4 - 7M R&D + capital investment

Three things should be repeated:

- Phototube technology is needed in one form or another.

- The costs of $\sim$ $10 million for the focal plane detector is deemed to be achievable.

- R&D money would be required.

We conclude that three baseline photon detectors should be investigated in any proposed **OWL** optics design:

- MC-PMT - a safe technology

- X-ray technology possibilities (6" or 9")

- New 12" image intensifier design

400

WILD CARDS

Many wild-card ideas are around, with one large photocathode taking all $36,000$ pixels into one tube per mirror. If the pixel size is 10 mm × 10 mm, it may lead to a round 1.5 m diameter single tube containing $36,000$ large Si diodes or avalanche diodes. For the initial **OWL** baseline design [1] where seven light collectors are used, one such tube is required for each light collector. As attractive as this idea is, the development of such a phototube is faced with many challenges. To offset the risks involved, one must invest considerable resources to develop large cathode technology and invest in sophisticated processing equipment.

ACKNOWLEDGEMENT

The author thanks Eugene C. Loh of the University of Utah for providing valuable **OWL** information and for reading of the manuscript. His help in making a poster paper for the Workshop from a faxed text is greatly appreciated.

REFERENCES

1. Orbiting array of Wide-angle Light collectors (OWL): A Pair of Earth Orbiting "Eyes" to Study Air Showers Initiated by $> 10^{20}$ eV Quanta by Jonathan F. Ormes et al.
2. Eugene C. Loh, private communication, October 1997
3. Philips Photonics catalogue XP1700 multi-channel PMTs
4. E. Flyckt: Low-crosstalk multi-channel PMTs from UV to IR; Proceedings of SCINT95, Delft, The Netherlands
5. Hamamatsu data sheet on R5900 tube family
6. DEP hybrid PMT catalogue
7. van Aller et al. Proceedings of IEEE Nuclear Science Symposium,1983

Camera For Detection Of Cosmic Rays Of Energy More Than 10 Eev On The ISS Orbit

Garipov G.K.[i], Khrenov B.A.[i], Nikitsky V.P.[ii], Panasyuk M.I.[i], Saprykin O.A.[ii], Sholokhov A.V.[ii] and Syromyatnikov V.S.[ii]

[i] *D.V. Skobeltsyn Institute of Nuclear Physics of the Moscow State University, Vorobjevy Gory, Moscow, 119899, Russia*
[ii] *RSC " Energia", Korolev, Moscow region, 141070, Russia*

Abstract. Concept of the EHE CR observation from the ISS orbit is discussed. A design of the camera at the Russian segment of the ISS comprising a large area (60 m^2) parabolic mirror with a photo multiplier pixel retina in its focal plane is described.

Introduction

We gave thought to a space camera based on the concept of the HiRes optical module of the fluorescence detector, i.e. the wide open angle camera with the PMT photo sensor retina placed at the focal distance f≅D equal to the diameter of the mirror. When applied to the ISS project this concept leads to the retina pixel of the same size as in the surface HiRes detector (1) or in the Auger optical detector (2). This equality of the photo sensor gives a good chance for collaboration in the construction of surface and space optical detectors[1]. The number of pixels in the photo sensor retina in the space camera is less than in the surface one site detector and the data acquisition and data analysis is simpler in the space camera. The real problem in the space camera is construction of a large area good quality mirror at ISS. This paper presents an approach to a design of the camera for the Russian segment of ISS.

Camera For Detection of the EHE Fluorescence EAS Tracks From the ISS

In Fig. 1 the camera concept is presented. Parabolic mirror of the diameter D=9-10 m concentrates the light on the pixel mosaic placed in the focal plane of the mirror, focal distance f=10 m. Every pixel has a square window of size 5x5 cm so that the angular pixel size is 5 mrad. The pixel is a photo multiplier tube (PMT) with a light guide of

---

[1] One of the authors (B.K.) being a member of the Auger collaboration, presented this approach to his Russian colleagues.

CP433, *Workshop on Observing Giant Air Showers from Space*
edited by J.F. Krizmanic et al.

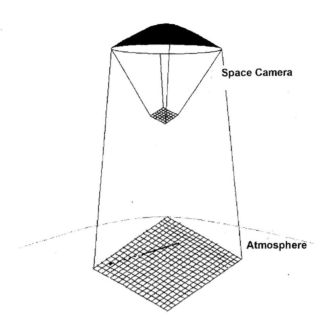

Figure 1. Camera on the ISS observing the track of EHE CR .

Winston cone type. The square cone window is covered with the UV filter transparent to light with the wavelengths 300-400 nm. PMT parameters are discussed below.

The camera field of view 15x15 degrees is filled with the mosaic of 2500 pixels. From ISS orbit this camera observes the atmosphere area 100x100 km , the "pixel" area is 2x2 km. The shadow of the photo sensor construction is 10% of the mirror area and the effective mirror area is $S=63$ m^2.

For this camera the fluorescence EAS track signals in pixels were calculated for various primary energies E_o and various zenith angles θ of entering tracks. The electron size along the track was calculated by Kalmykov et al (3) in the frame of QGSJET model (4). The fluorescence photon yield in the wavelengths 320-400 nm was taken as in the Auger Design Report (2) equal to 5 photons per meter of the electron track (independent of height in atmosphere). The pixel PM tube quantum efficiency was taken as 20%, the mirror reflection coefficient as 0.83. The atmosphere light extinction was presented as exp(-X/ X_o (λ)) where X_o (λ) was taken due to the data on the standard atmosphere extinction (5).

In Table 1 the parameters of the tracks in the camera for various E_o and θ are presented:

E_o EeV	$\theta°$	n, p.e.	ΔL,km	H_m, km	$L_{1/2}$, km	$T_{1/2}$, μs	A_{cl}, p.e.	t_{cl}, μs	A_{gr}, p.e.	t_{gr}, μs
10	30	700	4	4.5	7	43			$3\ 10^2$	37
10	60	550	2.3	9	13	65	$2.5\ 10^3$	22	$2\ 10^2$	90
10	70	480	2	11.5	19	85	$2.5\ 10^3$	35		
100	30	7000	4	4	6.5	40			$3\ 10^3$	30
100	60	5500	2.3	8.6	12	62	$1.6\ 10^4$	20	$2\ 10^3$	80
100	70	4800	2	11	18	78	$1.6\ 10^4$	31		

TABLE 1. Parameters of the EAS track in atmosphere (see text) for various primary energies and zenith angles.

n is the number of photo electrons (p.e.) in the pixel containing the EAS maximum, ΔL is the length of the EAS track in this pixel, H_m is the EAS maximum height in atmosphere (for primary protons), $L_{1/2}$ is the track length at the half maximum, $T_{1/2}$ is the width of the signal at the half maximum. Also presented are signals from the Cherenkov light reflected either from the clouds (average height 6 km, albedo coefficient 0.7) or from the ground surface (albedo coefficient 0.1). The Cherenkov signal amplitude and the delay to the time of the maximum track signal are presented. As it is seen the tracks with zenith angles more than $60°$ are longer than $L_{1/2}= 10$ km, their height of maximum is higher than 8 km and their duration $T_{1/2}$ is longer than 50 μs. The track width is approximately equal to the Molierre diameter of the shower at the given height and at all zenith angles is much less than the track length. The track width is also small to compare with a reasonable camera angular resolution (5 mrad for example). The camera pixels divide the track length in intervals of the atmosphere depth. At the zenith angles of about $80°$ the pixel size 5 mrad is corresponding to the depth interval 40 g/cm^2 . The pixel depth interval increases when the zenith angle decreases. But due to the measurement of the temporal profiles of showers the resolution in cascade curve could be high even at small zenith angles. For example, the cascade curve of the EAS with zenith angle $0-30°$ is divided in bins of 20 g/cm^2 if the time measurements are made with time intervals $\cong 1$ μs.

The presence of the reflected Cherenkov signal (RChS) makes possible to measure the absolute track depth (depth of shower maximum X_{max}) in atmosphere. For this purpose after every selected EAS event the height of the reflective surface should be determined by sending a light flash from the camera (see below). This method of X_{max} determination is relevant as in geometrical analysis of the "hit" pixels so in the analysis of the temporal profile of the track signals (in measurement of the RChS pulse delay

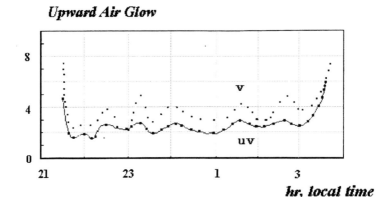

FIGURE 2. COSMOS 45 data on the atmosphere night glow. Solid Line - ultraviolet range, light intensity units 10^7 cm^{-2} s^{-1} sr^{-1}. Dotted Line - visual range, light intensity units 10^8 cm^{-2} s^{-1} sr^{-1}.

relative to the fluorescence signal at shower maximum, the expected average delays t_{cl} and t_{gr} are presented in Table 1).

Noise In The Camera Pixels At The ISS Orbit

Our knowledge of the noise expected in camera pixels at the night side of the Earth at the ISS orbit is based on the previous measurements of the light from the atmosphere. In measurements of the light in the wavelength band of the UV filter (λ=260-400 nm) at the Cosmos 45 satellite (R=200 km) it was shown (6) that the average intensity of light is in the range of 200-600 photons/ m^2 ns sr. In Fig. 2 one of the Cosmos 45 measurements on the night (moonless) side in the region of oceans is shown. There the light intensity measured in the visual band λ=400-600 nm is also presented. The variation of the intensity is correlated with the cloud cover. The intensity in visual band is approximately 10 times higher than in the UV band. The authors (6) noted that in the UV band correlation with clouds is less than in visual light. The lowest intensity of about 200 photons/ m^2 ns sr relates to the clearest (cloudless) regions and should be compared with the surface "downward" sky light intensity. In a simple estimate of the "upward-downward"(u/d) intensity ratio due to molecular scattering (only) one should expect the u/d ratio in the UV band (320-400 nm) of $\cong$0.7 and the upward intensity 200 should be converted to the downward intensity 300 which agrees with the experimental figures at sites where the surface EHE CR optical detectors are used.

At the Cosmos 900 satellite (R=500 km) only a narrow band of the 391.4 ±5 nm line was monitored (7). In those measurements the flashes of several minutes duration were observed not only in polar regions (aurora lights) but also at low latitudes. The lights from the cities were excluded as the observations were made above the oceans. It was noted that the longitudes of those flashes are correlated with the Brazilian magnetic anomaly. Authors of (7) were inclined to explain the flashes in this region as fluorescence initiated by the "anomalous" low energy electrons accelerated in the Earth plasma turbulence. The flash intensity is hundreds times higher than the average light intensity but as the flashes are short they will not take much of the time in measurements of the EHE CR.

The observations of the average upward light intensity with the proposed ISS camera for EHE CR study can reveal much more detailed picture of the UV light flashes: first of all it will give a map of the flash light intensity. It was noted (7,8) that the UV light intensity could be correlated with the Earth seismic activity and the data on the atmosphere glow might be used for prediction of the earthquakes. In this line the ISS camera will have a valuable practical application.

In quiet regions the light noise of the order of v=200-600 photons/ m^2ns sr will produce noise in camera pixels (in number of photoelectrons) equal to

$$\sigma=\sqrt{(vpS\omega\Delta t)}=\sqrt{(v \times 0.2 \times 63 \times 2.5 \ 10^{-5} \times 10^4)}=25\div44 \ \text{p.e.} \qquad (1)$$

where Δt is the time interval of the signal observation, p is the pixel quantum efficiency, S is the mirror area and ω is the pixel solid angle.

This noise level is 10 times lower than the fluorescence signals at the maximum of the EAS tracks at primary energy of E_o= 10 EeV (Table 1) and the observation of the tracks starting with the threshold energy of about 10 EeV is reliable. The gain of the PM tube should be of about 10^4 if the pixel front end electronics measures the signal charge in units of 10^{-13} C and the tube average anode current will be of about 10^{-7} A.

On full moon nights when the expected upward light intensity in the UV band is $\cong$8 times higher (the pixel noise is $\cong$3 times higher) the energy threshold is of about 30 EeV and is still much lower than the most interesting energy range of $E_o > 100$ EeV. On moon nights the average current in the PM tubes is higher (of about 10^{-6} A) but almost all types of tubes operate with that current.

When the local orbit time comes closer to the day time or the camera meets the region with the extra light from a flash (or human activity) the average current could exceed the limit level for the chosen tube gain. By decreasing the gain and so increasing the energy threshold in the EHE CR measurements we can continue to operate the tubes. For a given energy threshold it is important to choose a type of the PM tube capable of operating in the presence of high light noise. The choice of the PM tube for operation in the presence of high light noise was discussed in paper (9). The conclusion is that the tube with a multialcali cathode and a venetian blind dynode system is preferable to compare with the bialcali cathodes and other dynode systems. The tube with a properly designed electric field in the gap between the cathode and the first dynode can operate in linear range at average cathode currents up to 0.1 mA and anode current up to several mA. This kind of the tube was tested on aging in the presence of day light (with no voltage on the electrodes). It was shown that the tube gain is constant (with 10% errors) after being exposed to the equivalent ISS camera day light conditions during 5 years. That allows to consider the design of the camera pixel system without the

407

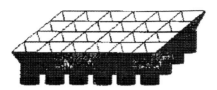

Figure 3. Segment of the photo sensor retina in the mirror focal plane.

shutters closing the system on the day side of the orbit (the direct sun light supposed to be prevented by the camera blend).

The internal noise current in the tube with a multialcali cathode is higher than in tubes with a bialcali cathode but even with a multialcali cathode the internal current is always much lower than the current caused by external light noise. It was shown (9) that the known effect of the increase of the tube internal noise current after the tube exposure to the day light is not important when the tube performs in the optical detector. For example, the 75 mm diameter multialcali cathode after exposure to day light has a 1 p.e. noise rate of about 300 kHz (10 times higher than the "normal" dark rate) but the 1 p.e. rate expected due to the "night light" current is more than 10 MHz. It should be noted also that the relaxation time for the after exposure current is short: in about 5 minutes the tube current comes down almost to the normal level.

Nevertheless the pixel tubes should be protected against sudden light overflow by using a system automatically decreasing voltage on the dynodes if current is higher than a critical value.

The Camera Retina Design

The design of the pixel retina might follow the solar panel design that is segmented and packed for transportation to the orbit. In the case of the ISS camera for the EHE CR observations the full size retina could be segmented in 2 pieces (1250 pixels a piece). In Fig. 3 a segment of the retina is shown. The important part of the camera pixel is a light guide and the UV filter. The light guide is chosen with a square window to form an orthogonal system of pixels that is better for the time-amplitude (charge) analysis of the track data. The UV filter should be transparent to the air fluorescence wavelengths band and the PM tube window should be made of the UV transparent glass. By the time of the workshop the "candidate" pixel was tested with the help of the air fluorescence flash tube. The flash tube produces the spectrum shown in Fig. 4. The tube is the FEU-115 type (20 mm in diameter) of the Moscow plant with a UV glass window. The cathode is multialcali with the quantum efficiency shown in Fig. 4, the dynode system is of linear focused type. The UV filter transparency is 90% in the

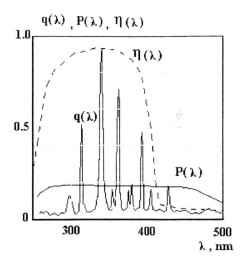

Figure 4. Air fluorescence spectrum q(l) (arbitrary units), the pixel quantum efficiency p(l) and UV filter transparency h(l).

wavelength band 290-370 nm (Fig. 4). The efficiency of light collection over the square window of the light guide is in the range of 0.85-0.95. The resulting efficiency of the pixel to the air fluorescence light was found equal to 0.15.

The Mirror Design

As the first suggestion for the mirror construction the parabolic mirror of diameter 9.5 m designed in the RSC "Energia" for the other purpose was considered. In this design the mirror has to be launched to the orbit in a compact cylinder and there a special mechanism opens the mirror to its full size.

Parameters of the mirror are: mass - about 600 Kg, focal distance- 3.7 m, geometrical formula for curvature- $y^2=14800$ x (x,y in mm), opening angle -130.8°, full area in operation-65.7 m^2.

The mirror comprises 36 hard "petals" packed for transportation into the cylinder with the diameter 2.2 m and the length 4.2 m.

Technological errors in the surface curvature (errors in the normal angle to the surface) are expected to be not more than 8', errors related to the petal opening and fixing mechanics- not more 7'.

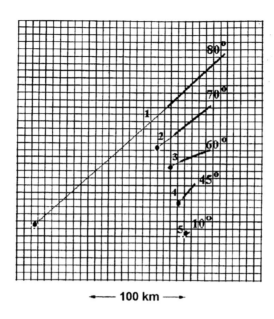

Figure 5. EHE CR tracks examples (map of the ISS camera pixels)

Reflection coefficient at the light wavelengths 330-500 nm of the aluminum surface (covered with a special coating) is 0.83 and is expected to be stable to 1% during one year of operation.

Analysis of the operation of this mirror for our purposes showed that with the focal distance 3.7 m it is impossible to have the field of view 15°x15° due to coma much wider than 5 mrad at the edges of the pixel retina. The coma size decreases with the focal distance for a given mirror diameter . The coma size was considered small enough if 50% of the coma light is in vicinity of the 5 mrad pixel at the edge of the field of view. With this criteria the mirror focal distance should be of about 9-10 m. The constructors of the mirror are ready to make changes to the mirror design keeping the technological angular errors of the mirror at the level of 10'. Not solved yet is the problem of stability of the mirror parameters during the flight due to temperature effect (deformations of the mirror surface could of the order of 1°). For stabilization of the mirror surface in flight an expensive additional construction should be applied.

The other disadvantages of this design are: the need of large size space ship for transportation of big mirror "petals" to the station, the need of heavy equipment for the final construction of the mirror at the station and an unique character of the whole design that could not be developed to bigger scale needed for the study of the higher energy cosmic rays.

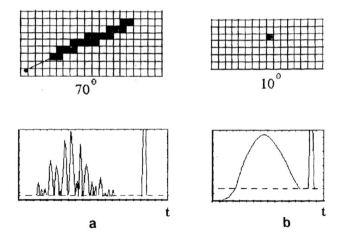

Figure 6. Pixel map and time profiles of the pixel signals for two EHE CR tracks.

The other suggestion is based on the idea also developed in the RCS "Energia": the parabolic mirror designed as the Fresnel lens by making circular grooves in the plane plate reflecting light due to formula of the parabolic mirror. The mirror plane could be done from the light material, the reflecting profile of the groove should be aluminized.

The advantages of that plane mirror are: an easy way for making a compact pack of the plane mirror pieces (for transportation) and a simple mechanism for the opening it to full mirror size- solar panels are a prototype of such construction, the possibility to develop much larger mirrors for the next generation of the space camera (or solar energy concentrators).

Both mirror designs for the ISS camera need to be developed in detail and tested on stability in the ISS temperature conditions.

Control Measurements Of The Surface Height In Atmosphere that Reflects Cherenkov Light

The measurements of the height of the surface which reflected the Cherenkov light shower can be made with the help of the same mirror immediately after the reception of the reflected Cherenkov signal (RChS) while the position of the space station has not changed yet. In order to do this in the vicinity of every pixel in the mirror focal surface the flash lamp should be installed (for example at the cross point of 4 pixels). On signal from the camera trigger selected the EHE CR event the lamp near the pixel registered

411

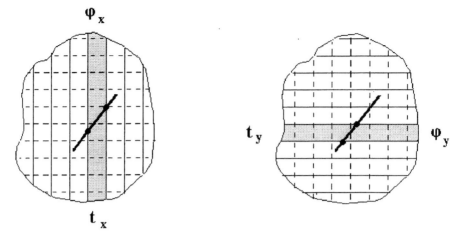

Figure 7. Illustration to the method of the time-angular coordinate method of the directional angles determination. Pixel signals in shaded areas are summed up and used as signals from two orthogonal strips.

RChS. In time needed for light to pass the distance R=400 km and back (10^{-3} sec) the station is shifted only by 10 m- the distance much smaller than the size of the RChS. It is convenient to trigger the lamp by the clock signal next after the trigger pulse (delay time to the real time of the RChS less than $2 \cdot 10^{-4}$ sec, see below). Moderate accuracy in time measurement (like 200 ns in the FADC of the pixel electronics, see the next section) gives a high accuracy in the measurement of the distance between camera and the reflective surface: the error of about 30 m.

As the mirror area is big the reflected signal (for reflective coefficient k≥0.1) would be high (about $2 \cdot 10^4$ p.e. in the pixel) even from a comparatively low intensity flash: 10^{15} photon with $\lambda \cong 380$ nm . The lamp of that luminosity is miniature and might be placed in the vicinity of the pixel without shadowing it. Measuring the reflected signal for a known lamp luminosity it is possible to measure the combined factor kη where η is the light absorption coefficient in atmosphere. With a known factor kη the signal of the RChS could be used as an additional estimator of the primary energy (along with the measure of fluorescent signal in the shower maximum).

Examples Of Expected Events. Data Analysis. Expected Errors In Energy And Direction Of Primary Particle

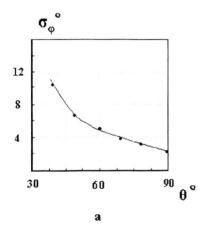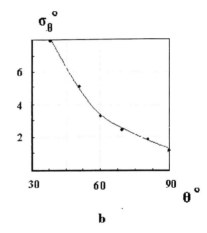

Figure 8. Error in azimuth (a) and zenith (b) angles as a function of zenith angle.

In Fig. 5 and 6 several examples of track registrations at the energy $E_o=10^{19}$ eV are presented for various zenith angles. In Fig. 5 tracks are presented as lines going through the camera pixels. The thick line represents the fluorescent "half track" with the length $L_{1/2}$. The thin line represents a mixed fluorescent and Cherenkov scattered light track. The point at the end of the track is RChS. Presentation of the track as a line is relevant as the width of the track is much less than the size of the pixel (2x2 km). In Fig. 6 along with the map of "hit" pixels the temporal profiles in the pixels are shown (the RChS pulse is the last of the pulses there). On the pulse profiles the noise level is shown for the time intervals of 200 nsec.

The track data allow to determine the main parameters of the primary particle: energy E_o , directional angles θ (zenith) and ϕ (azimuth) , position of shower maximum X_{max}.

The amplitude (PMT charge) in every pixel or the amplitude in a given time interval (when one pixel is considered) directly corresponds to the size of the shower at a given depth in atmosphere measured from the position of the shower maximum. Primary energy is in proportion either to the shower size in the maximum or to the integral shower size over the whole track (calorimetric measurement of the primary energy).

As it is seen from the above examples of tracks of the 10^{19} eV showers where the amplitudes near the shower maximum are much higher than the noise level, there are many pixels (or time intervals in one pixel) with significant amplitudes and so the energy determination in showers above energy 10^{19} eV is precise (statistical errors less than 10%).

The determination of directional angles θ and ϕ could be done in analysis of timing of the track at pixel borders, i.e. at the given angular coordinates of the track. In this method the square pixel shape is preferable. Summing the data of the "hit" neighbor square pixels in both orthogonal directions one can determine the time of the track crossing every border of pixels (Fig. 7).

Azimuth angle φ is determined as a weighted (the signal amplitude is a weight) average of the ratio of the pixel pulse duration $\Delta t_{i,j}$ in two orthogonal pixel directions:

$$\tan\varphi = \Sigma \; A_i \; \Delta t_i \; / \Sigma A_j \; \Delta t_j \qquad (2)$$

In the same way $\tan\varphi$ could be determined as the ratio of the whole track duration in two orthogonal directions – in this case the RChS could be added to the analysis and the accuracy in azimuth angle is higher.

With the decrease of the zenith angle the number of the "hit" pixels is decreasing and the angle φ error is increasing. In Fig. 8a the estimate of angle φ error as a function of the zenith angle is presented ($E_o = 10^{19}$ eV). Zenith angle θ is determined in analysis of the angular velocity of the track:

$$\Delta\Psi/\Delta t = c \; \sin\theta / R \; (1+\cos\theta) \qquad (3)$$

where $\Delta\Psi$ is the angular size of the track in the pixel plane and Δt is the duration of the track in this angular size. It is evident that the error in zenith angle θ is increasing when θ is decreasing (the number of "hit" pixels is decreasing). As in case of the azimuth angle the RChS is helpful for more accurate determination of the zenith angle. In Fig. 8b the zenith angle θ error is presented as a function of θ.

The important parameter of the shower is the position of shower maximum X_{max}. As was mentioned above the reflected Cherenkov spot (RChS) from the surface with a known height in atmosphere is the way to have a reference height (depth) in atmosphere for the whole track. The reference height might be measured with a high accuracy (30 m) and the accuracy in position of X_{max} is limited due to error in the

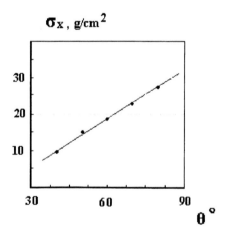

Figure 9. Error in position of the EAS maximum as a function of zenith

relative position of X_{max} in the track. For inclined tracks ($\theta > 60°$) the position of maximum would be determined with the error $\sigma_{x\,max}$ much less than the pixel size $\Delta X \cong$ 40 g/cm^2 ($\sigma_{x\,max} \cong 10$ g/cm^2). For near vertical tracks X_{max} is determined from the temporal profile of the track with accuracy dependent on the accuracy of timing. If the profile is measured with time intervals 200 nsec then the relative position of X_{max} could be determined with the error of about 4 g/cm^2. The resulting absolute position of X_{max} depends on the zenith angle. Interesting that the big errors in zenith angle determination in near vertical showers do not affect the absolute value of X_{max} (as $\cos\theta \cong 1$). For near horizontal showers the error in X_{max} increases with error in zenith angle. In Fig. 9 the dependence of error $\sigma_{X\,max}$ on the zenith angle is presented for the suggested error 10 g/cm^2 in relative position of the shower maximum.

Data Acquisition System

In Fig. 10 the block diagram of the camera data acquisition system is presented. The signal from every PM tube is going to two channels of FADC (signals are collected from the anode and from the dynode with gain 50 times less than at the anode- for making dynamic range wide). The FADC is the ADC with 10 bit dynamic range operating with the time interval equal to 200 ns (it is an economic value, better time resolution could be achieved with less time intervals, but introducing less than 10 ns time intervals is useless due to the limited time resolution of the available PM tubes). The PMT data acquisition in FADC is permanent. Time is measured relative to the camera clock pulses of frequency 10 kHz. The data flux is recorded as 100 μs frames in a cycle memory (CM). The CM volume is equal to 4 frames of 10 bit information in every time interval 200 nsec, i.e. the memory volume is equal to $\cong 3$ kbyte.
The block of Preliminary Event Selection (PES) considers the data in 3 frames previous to the last one and selects the candidates to EHE CR events. If the data in those 3 frames contains a candidate, the data from CM are sent to the operative memory (OM) and wait there for the final decision from the central camera selection system (CSS). On signal from CSS the data from OM are sent to the camera computer. The selection of the useful events is a very important part of the camera operation. It should be developed in the process of the camera final design. We present here an example of the selection of the EHE CR event.
At stage 1 (preliminary selection in one pixel) the candidate for the useful event is selected either as the expected fluorescent track with duration from 6 μs (near horizontal track) to 40 μs (near vertical track) or the expected RChS with duration from 0.2 μs to 6 μs. The output charge (amplitude) of the PM tube should be higher than some noise level-let it be 3σ. Following the estimates in section 3 the candidate event in one pixel should have either the number of p.e. more than 90 p.e. in serial of 30 time intervals of 200 ns (fluorescence track candidate) or the number of p.e. more than 45 p.e. in serial of 2 intervals of 200 ns (RChS candidate). With this selection criteria the energy threshold for near horizontal track is about 4 10^{18} eV and for near

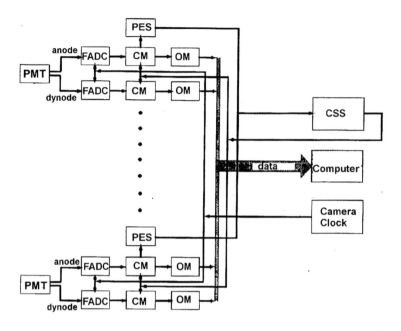

Figure 10. Block scheme of the camera electronics.

vertical track it is about 10^{19} eV. The RChS in those showers will be registered even if the reflective surface has poor reflectivity coefficient k=0.1. At stage 2 the candidate events from any pixel of the camera mosaic are considered in CSS. Inclined tracks $\theta > 45°$ are expected to be registered in more than 4-5 pixels so the selection criteria for those tracks should include the presence of more than three first stage selected neighbor pixels. On signal from CSS all pre- selected "candidate" pixel data are sent to the computer. Those data include the pre-selected RChS data.

The EHE CR events could be separated from other phenomena of light flashes in atmosphere as events with extremely short flashes (to compare for example with lightning that has a duration of 0.001-.01 seconds or meteor –with duration of 0.1-1 sec). To use this feature of the EHE CR events the constraint on the signal duration should be installed in the first stage of event selection: the signals with amplitudes higher than 3σ of noise repeated in more than 500 time intervals of 200 ns should be excluded from "candidates".

The acquisition of useful event data to the computer should not break the operation of the pixels or operation of the selection system. The large volume of memory in CM and introduction of the additional OM memory helps to resolve this problem. The data

416

exchange between CM and OM (on the signal from CSS) is fast and does not interrupt the data collection in CM. The data exchange between OM and the computer might be slow as the rate of useful events is low. The real EHE CR events in the camera with the estimated geometric factor of about $3\ 10^4\ \text{km}^2$ sr with the energy threshold $4\ 10^{18}$ eV is about 30 events per hour. Events at the energy threshold contain about 100 Kbyte of data and so the average useful information flux is very low: 1Kbyte per sec.

In final design of the camera electronics the flux of other possible data (selected by criteria different from EHE CR events) should be taken into account as well as the flux of "noise" events selected under criteria of useful events.

ACKNOWLEDGMENTS

Authors are grateful to J. Linsley, L. Scarsi and O. Catalano for many discussions of the ISS camera project. The work on the ISS camera for the EHE CR observations should be international and all contributions to the project are welcome. One of the authors (B.A.K.) is grateful to support from RFFI, grant number 96-15-96783.

REFERENCES

1. Bird D.J. et al, 23-d ICRC (Calgary) (1993), @, 450.
2. The Pierre Auger Project, Design Report, (1996) (2-nd edition).
3. Kalmykov N.N. and Pavlov A. I. , private communication (1997).
4. Kalmykov N.N. and Ostapchenko S.S.., Yadernaya Fisika (Russ Nucl Phys), (1993), 56, 105.
5. Allen C.W., Astrophysical Quantities, University of London,The Athlone Press, (1955).
6. Lebedinsky A.I. et al, Space Research (in Russian), (1965), Nauka, Moscow, p. 77-88.
7. Tverskaya L.V. and Tulupov V.I. , Geomagnetism and Aeronomia, Ac Sci, of USSR, (1984), №4, 695.
8. Biryukov A.S. Grigoryan O.R., Kuznetsov S.N., Oraevsky v.N.,Panasyuk M.I., Pulinets S.A., Chmyrev V.N., Engineering Ecology, (1996), 6, 92-115.
9. Bezrukov L.B. et al, paper prepared for the "Instruments and Technique of Experiment", (1997).

Observing Air Showers from Cosmic Superluminal Particles

Luis Gonzalez-Mestres[*,†]

*Laboratoire de Physique Corpusculaire, Collège de France, 75231 Paris Cedex 05, France
†L.A.P.P., CNRS-IN2P3, B.P. 110, 74941 Annecy-le-Vieux Cedex, France

Abstract. The Poincaré relativity principle has been tested at low energy with great accuracy, but its extrapolation to very high-energy phenomena is much less well established. Lorentz symmetry can be broken at Planck scale due to the renormalization of gravity or to some deeper structure of matter: we expect such a breaking to be a very high energy and very short distance phenomenon. If textbook special relativity is only an approximate property of the equations describing a sector of matter above some critical distance scale, an absolute local frame (the "vacuum rest frame", VRF) can possibly be found and superluminal sectors of matter may exist related to new degrees of freedom not yet discovered experimentally. The new superluminal particles (**"superbradyons"**, i.e. bradyons with superluminal critical speed) would have positive mass and energy, and behave kinematically like "ordinary" particles (those with critical speed in vacuum equal to c, the speed of light) apart from the difference in critical speed (we expect $c_i \gg c$, where c_i is the critical speed of a superluminal sector). They may be the ultimate building blocks of matter. At speed $v > c$, they are expected to release "Cherenkov" radiation ("ordinary" particles) in vacuum. Superluminal particles could provide most of the cosmic (dark) matter and produce very high-energy cosmic rays. We discuss: a) the possible relevance of superluminal matter to the composition, sources and spectra of high-energy cosmic rays; b) signatures and experiments allowing to possibly explore such effects. Very large volume and unprecedented background rejection ability are crucial requirements for any detector devoted to the search for cosmic superbradyons. Future cosmic-ray experiments using air-shower detectors (especially from space) naturally fulfil both requirements.

"The impossibility to disclose experimentally the absolute motion of the earth seems to be a general law of Nature"

H. Poincaré

"The interpretation of geometry advocated here cannot be directly applied to submolecular spaces... it might turn out that such an extrapolation is just as incorrect as an extension of the concept of temperature to particles of a solid of molecular dimensions"

A. Einstein

CP433, *Workshop on Observing Giant Air Showers from Space*
edited by J.F. Krizmanic et al.

RELATIVITY, MATTER AND CRITICAL SPEEDS

If Lorentz symmetry is viewed as a dynamical property of the motion equations, no reference to absolute properties of space and time is required [1]. In a two-dimensional galilean space-time, the equation:

$$\alpha \, \partial^2\phi/\partial t^2 \; - \; \partial^2\phi/\partial x^2 = F(\phi) \tag{1}$$

with $\alpha = 1/c_o^2$ and c_o = critical speed, remains unchanged under "Lorentz" transformations leaving invariant the squared interval:

$$ds^2 = dx^2 - c_o^2 dt^2 \tag{2}$$

so that matter made with solutions of equation (1) would feel a relativistic space-time even if the real space-time is actually galilean and if an absolute rest frame exists in the underlying dynamics beyond the wave equation. A well-known example is provided by the solitons of the sine-Gordon equation, obtained taking in (1):

$$F(\phi) \quad = \quad - \; (\omega/c_o)^2 \; sin \; \phi \tag{3}$$

where ω is a characteristic frequency of the dynamical system. A two-dimensional universe made of sine-Gordon solitons plunged in a galilean world would behave like a two-dimensional minkowskian world with the laws of special relativity. Information on any absolute rest frame would be lost by the solitons, as if the Poincaré relativity principle (see [2] to [5] for the genesis and evolution of this deep concept) were indeed a law of Nature, even if actually the basic equation derives from a galilean world with an absolute rest frame (a system built "on top of a table", with $c_o \ll c$). The actual structure of space and time can only be found by going beyond the wave equation to deeper levels of resolution, similar to the way high-energy accelerator experiments explore the inner structure of "elementary" particles (but cosmic rays have the highest attainable energies).

At this stage, two crucial questions arise: a) is c (the speed of light) the only critical speed in vacuum, are there particles with a critical speed different from that of light? [1,6]; b) can the ultimate building blocks of matter be superluminal? [7,8]. These questions make sense, as: a) in a perfectly transparent crystal it is possible to identify at least two critical speeds, those of light and sound, and light can interact with phonons; b) the potential approach to lattice dynamics in solid-sate physics is precisely the form of electromagnetism in the limit $c_s \, c^{-1} \to 0$, where c_s is the speed of sound. Superluminal sectors of matter can be consistently generated [1,9], with the conservative choice of leaving the Planck constant unchanged, replacing in the Klein-Gordon equation the speed of light by a new critical speed $c_i \gg c$ (the subscript i stands for the i-th superluminal sector). All standard kinematical concepts and formulas [10] remain correct, leading to particles with positive mass and energy which are not tachyons. We shall call them **superbradyons** as, according to standard vocabulary [11], they are bradyons with superluminal critical speed in vacuum. The energy E and momentum p of a superluminal particle of mass m and critical speed c_i will be given by the generalized relativistic equations:

$$p = m\,v\,(1 - v^2 c_i^{-2})^{-1/2} \qquad (4)$$
$$E = m\,c_i^2\,(1 - v^2 c_i^{-2})^{-1/2} \qquad (5)$$
$$E_{rest} = m\,c_i^2 \qquad (6)$$

where v is the speed and E_{rest} the rest energy. Energy and momentum conservation will in principle not be spoiled by the existence of several critical speeds in vacuum: conservation laws will as usual hold for phenomena leaving the vacuum unchanged. Each superluminal sector will have its own Lorentz invariance with c_i defining the metric, and is expected to generate a sectorial "gravity". Interactions between two different sectors will break both Lorentz invariances. Lorentz invariance for all sectors simultaneously will at best be explicit (i.e. exhibiting the diagonal sectorial Lorentz metric) in a single inertial frame (**the vacuum rest frame**, VRF, i.e. the "absolute" rest frame). In our approach, the Michelson-Morley result is not incompatible with the existence of some "ether" as suggested by recent results in particle physics: if the vacuum is a material medium where fields and order parameters can condense, it may well have a local rest frame whose identification would be prevented by the sectorial Lorentz symmetries in the low-momentum limit (where different sectors do not mix and the sectorial Lorentz symmetries become exact laws, so that each sector feels a "Poincaré relativity principle").

If superluminal particles couple weakly to ordinary matter, their effect on the ordinary sector will occur at very high energy and short distance [12], far from the domain of successful conventional tests of Lorentz invariance [13,14]. In particular, superbradyons naturally escape the constraints on the critical speed derived in some specific models [15,16] based on the $TH\epsilon\mu$ approach [17] , as their mixing with the ordinary sector is expected to be strongly energy-dependent [8,18]. High-energy experiments can therefore open new windows in this field. Finding some track of a superluminal sector (e.g. through violations of Lorentz invariance in the ordinary sector or by direct detection of a superluminal particle) may be the only way to experimentally discover the VRF. Superluminal particles lead to consistent cosmological models [6,7,9], where they may well provide most of the cosmic (dark) matter [19]. Although recent criticism to this suggestion has been emitted in a specific model on the grounds of gravitation theory [20], the framework used is crucially different from the multi-graviton approach suggested in our papers where we propose (e.g. [1,9]) that each dynamical (ordinary or superluminal) sector generates its own gravitation associated to the sectorial Lorentz symmetry and couplings between different "gravitons" are expected to be weak. Superbradyons can be the ultimate building blocks from which superstrings would be made and a "pre-Big Bang" cosmology would emerge. Nonlocality at Planck scale would then be an approximation to this dynamics in the limit $c\,c_i^{-1} \rightarrow 0$, where superluminal signals undergo apparent "instantaneous" propagation similar to electromagnetic interactions described by a potential model of lattice dynamics in solid state physics.

IMPLICATIONS FOR HIGH-ENERGY COSMIC RAYS

Accelerator experiments at future machines (LHC, VLHC...) can be a way to search for superluminal particles [12,18]. However, this approach is limited by the attainable energies, luminosities, signatures and low-background levels. Although the investigation at accelerators provides unique chances and must be carried on, it will only cover a small domain of the allowed parameters for superluminal sectors of matter. Cosmic-ray experiments are not limited in energy and naturally provide very low background levels: they therefore allow for a more general and, on dynamical grounds, better adapted exploration. It must also be realized that, if the Poincaré relativity principle is violated, a 1 TeV particle cannot be turned into a 10^{20} eV particle of the same kind by a Lorentz transformation, and collider events cannot be made equivalent to cosmic-ray events.

The highest observed cosmic-ray energies (up to 3.10^{20} eV) are closer to Planck scale ($\approx$ 10^{28} eV) than to electroweak scale ($\approx$ 10^{11} eV): therefore, if Lorentz symmetry is violated, the study of the highest-energy cosmic rays provides a unique microscope directly focused on Planck scale [7,8,21]. The search for very rare events due to superluminal particles in AUGER, AMANDA, OWL, AIRWATCH FROM SPACE... can be a crucial ingredient of this unprecedented investigation [12,18,22,23]. In what follows we assume that the earth is not moving at relativistic speed with respect to the local vacuum rest frame.

Superluminal kinematics

The kinematical properties and Lorentz transformations of high-energy superluminal particles have been discussed elsewhere [12]. If an absolute rest frame exists, Lorentz contraction is a real physical phenomenon and is governed by the factor γ_i^{-1} = $(1 - v^2 c_i^{-2})^{1/2}$ for the i-th superluminal sector, so that there is no Lorentz singularity when a superluminal particle crosses the speed value v = c in a frame measured by ordinary matter. Similarly, if superbradyons have any coupling to the electromagnetic field (adding in the standard way the electromagnetic four-potential to the superluminal four-momentum to build the covariant derivative in the VRF), we expect the magnetic force to be proportional to v c_i^{-1} instead of v c^{-1} . Contrary to tachyons, superbradyons can emit "Cherenkov" radiation (i.e. particles with lower critical speed) in vacuum. If c_i $\gg$ 10^3 c , and if the VRF is close to that defined requiring isotropy of cosmic microwave background radiation, high-energy superluminal particles will be seen on earth as traveling mainly at speed v $\approx$ 10^3 c , as can be seen from the following analysis. Since we expect to measure the energy of superluminal particles through interactions with detectors made of "ordinary" particles, we can define, in the rest frame of an "ordinary" particle moving at speed $\vec{V}$ with respect to the VRF, the energy and momentum of a superluminal particle to be the Lorentz-tranformed of its VRF energy and momentum taking c as the critical speed parameter for the Lorentz transformation. Then, the mass of the superluminal particle will depend on the inertial frame. The energy E_i' and momentum $\vec{p}_i'$ of the superluminal particle i (belonging to the i-th super-luminal sector and with energy E_i and momentum $\vec{p}_i$ in the VRF) in the new rest frame, as measured by ordinary matter from energy and momentum conservation (e.g. in decays of superluminal particles into ordinary ones), will be:

$$E_i' = (E_i - \vec{V}.\vec{p}_i)(1 - V^2 c^{-2})^{-1/2} \tag{7}$$

$$\vec{p}_i' = \vec{p}_{i,L}' + \vec{p}_{i,\perp}' \tag{8}$$

$$\vec{p}_{i,L}' = (\vec{p}_{i,L} - E_i c^{-2} \vec{V})(1 - V^2 c^{-2})^{-1/2} \tag{9}$$

$$\vec{p}_{i,\perp}' = \vec{p}_{i,\perp} \tag{10}$$

where $\vec{p}_{i,L} = V^{-2}(\vec{V}.\vec{p}_i)\vec{V}$, $\vec{p}_{i,\perp} = \vec{p}_i - \vec{p}_{i,L}$ and similarly for the longitudinal and transverse components of $\vec{p}_i'$. We are thus led to consider the effective squared mass:

$$M_{i,c}^2 = c^{-4}(E_i^2 - c^2 p_i^2) = m_i^2 c^{-4} c_i^4 + c^{-2}(c^{-2} c_i^2 - 1) p_i^2 \tag{11}$$

which depends on the VRF momentum of the particle. m_i is the invariant mass of particle i , as seen by matter from the i-th superluminal sector (i.e. with critical speed in vacuum $= c_i$). While "ordinary" transformation laws of energy and momentum are not singular, even for a superluminal particle, the situation is different for the transformation of a superluminal speed, as will be seen below. Furthermore, if the superluminal particle has velocity $\vec{v}_i = \vec{V}$ in the VRF, so that it is at rest in the new inertial frame, we would naively expect a vanishing momentum, $\vec{p}_i' = 0$. Instead, we get:

$$\vec{p}_i' = -\vec{p}_i(c^{-2} c_i^2 - 1)(1 - V^2 c^{-2})^{-1/2} \tag{12}$$

and $p_i' \gg p_i$, although $p_i' c \ll E_i'$ if $V \ll c$. This reflects the non-covariant character of the 4-momentum of particle i under "ordinary" Lorentz transformations. Thus, even if the directional effect is small in realistic situations (f.i. on earth), the decay of a superluminal particle at rest into ordinary particles will not lead to an exactly vanishing total momentum if the inertial frame is different from the VRF.

In the rest frame of an "ordinary" particle moving with speed $\vec{V}$ with respect to the VRF, we can estimate the speed $\vec{v}_i'$ of the previous particle i writing:

$$\vec{v}_i = \vec{v}_{i,L} + \vec{v}_{i,\perp} \tag{13}$$

where $\vec{v}_{i,L} = V^{-2}(\vec{V}.\vec{v}_i)\vec{V}$, and similarly for the longitudinal and transverse components of $\vec{v}_i'$. Then, the transformation law is:

$$\vec{v}_{i,L}' = (\vec{v}_{i,L} - \vec{V})(1 - \vec{v}_i.\vec{V} c^{-2})^{-1} \tag{14}$$

$$\vec{v}_{i,\perp}' = \vec{v}_{i,\perp}(1 - V^2 c^{-2})^{1/2}(1 - \vec{v}_i.\vec{V} c^{-2})^{-1} \tag{15}$$

leading to singularities at $\vec{v}_i = c^2$ which correspond to a change in the arrow of time (due to the distorsion generated by the Lorentz transformation of space-time) as seen by ordinary matter traveling at speed $\vec{V}$ with respect to the VRF.

Experimental implications

At $v_{i,L} > c^2 V^{-1}$, a superluminal particle moving forward in time in the VRF will appear as moving backward in time to an observer made of ordinary matter and moving at speed $\vec{V}$ in the same frame. On earth, taking $V \approx 10^{-3} c$ (if the VRF is close to that suggested by cosmic background radiation, e.g. [24]), the apparent reversal of the time arrow will occur mainly at $v_i \approx 10^3 c$. If $c_i \gg 10^3 c$, phenomena related to propagation backward in time of produced superluminal particles may be observable in future accelerator experiments slightly above the production threshold. In a typical event where a pair of superluminal particles would be produced, we expect in most cases that one of the superluminal particles propagates forward in time and the other one propagates backward. As previously stressed, the infinite velocity (value of v_i') associated to the point of time reversal does not, according to (7) and (9), correspond to infinite values of energy and momentum. The backward propagation in time, as observed by devices which are not at rest in the VRF, is not really physical (the arrow of time is well defined in the VRF for all physical processes) and does not correspond to any real violation of causality. The apparent reversal of the time arrow for superluminal particles at $\vec{v}_i.\vec{V} > c^2$ would be a consequence of the bias of the laboratory time measurement due to our motion with respect to the absolute rest frame. The distribution and properties of such superluminal events, in an accelerator experiment or in a large-volume cosmic-ray detector, would obviously be in correlation with the direction and speed of the laboratory's motion with respect to the VRF. It would provide fundamental cosmological information, complementary to informations on "ordinary" matter provided by measurements of the cosmic microwave background.

From (14) and (15), we also notice that, for $V \ll c$ and $\vec{v}_i.\vec{V} \gg c^2$, the speed $\vec{v}_i'$ tends to the limit $\vec{v}_i^\infty$, where:

$$\vec{v}_i^\infty (\vec{v}_i) = - \vec{v}_i c^2 (\vec{v}_i.\vec{V})^{-1} \tag{16}$$

which sets a universal high-energy limit, independent of c_i , to the speed of superluminal particles as measured by ordinary matter in an inertial rest frame other than the VRF. This limit is not isotropic, and depends on the angle between the speeds $\vec{v}_i$ and $\vec{V}$. A typical order of magnitude for $\vec{v}_i^\infty$ on earth is $\vec{v}_i^\infty \approx 10^3 c$ if the VRF is close to that suggested by cosmic background radiation. If C is the highest critical speed in vacuum, infinite speed and reversal of the arrow of time occur only in frames moving with respect to the VRF at speed $V \geq c^2 C^{-1}$. Finite critical speeds of superluminal sectors, as measured by ordinary matter in frames moving at $V \neq 0$, are anisotropic. Therefore, directional detection of superluminal particles would allow to directly identify the VRF and even to check whether it can be defined consistently, simultaneously for all dynamical sectors. If a universal, local VRF cannot be defined, translational and rotational modes may appear between different kinds of matter generating significant cosmological effects (e.g. a cosmic rotation axis for "ordinary" matter).

A superbradyon moving with velocity $\vec{v}_i$ with respect to the VRF, and emitted by an astrophysical object, can reach an observer moving with laboratory speed $\vec{V}$ in the VRF at a time, as measured by the observer, previous to the emission time. This remarkable astronomical phenomenon will happen if $\vec{v}_i \cdot \vec{V} > c^2$, and the emitted particle will be seen to evolve backward in time (but it evolves forward in time in the VRF, so that again the reversal of the time arrow is not really a physical phenomenon). If they interact several times with the detector, superbradyons can be a directional probe preceding the detailed observation of astrophysical phenomena, such as explosions releasing simultaneously neutrinos, photons and superluminal particles (although causality is preserved in the VRF). For a high-speed superluminal cosmic ray with critical speed $c_i \gg c$, the momentum, as measured in the laboratory, does not provide directional information on the source, but on the VRF. Velocity provides directional information on the source, but can be measured only if the particle interacts several times with the detector, which is far from guaranteed, or if the superluminal particle is associated to a collective phenomenon involving several sectors of matter and emitting also photons or neutrinos simultaneously. In the most favourable case, directional detection of high-speed superluminal particles in a very large detector would allow to trigger a dedicated astrophysical observation in the direction of the sky determined by the velocity of the superluminal particle(s). If d is the distance between the observer and the astrophysical object, and Δt the time delay between the detection of the superluminal particle(s) and that of photons and neutrinos, we have: $d \simeq c \, \Delta t$.

Annihilation of pairs of superluminal particles into ordinary ones can release very large kinetic energies and provide a new source of high-energy cosmic rays. Decays of superluminal particles may play a similar role. Collisions (especially, inelastic with very large energy transfer) of high-energy superluminal particles with extraterrestrial ordinary matter may also yield high-energy ordinary cosmic rays. Pairs of slow superluminal particles can also annihilate into particles of another superluminal sector with lower c_i, converting most of the rest energies into a large amount of kinetic energy. Superluminal particles moving at $v_i > c$ can release anywhere "Cherenkov" radiation in vacuum, i.e. spontaneous emission of particles of a lower critical speed c_j (for $v_i > c_j$) including ordinary ones, providing a new source of (superluminal or ordinary) high-energy cosmic rays. High-energy superluminal particles can directly reach the earth and undergo collisions inside the atmosphere, producing many secondaries like ordinary cosmic rays. They can also interact with the rock or with water near some underground or underwater detector, coming from the atmosphere or after having crossed the earth, and producing clear signatures. Contrary to neutrinos, whose flux is strongly attenuated by the earth at energies above $10^6 \; GeV$, superluminal particles will in principle not be stopped by earth at these energies. In inelastic collisions, high-energy superluminal primaries can transfer most of their energy to ordinary particles. Even with a very weak interaction probability, and assuming that the superluminal primary does not produce ionization, the rate for superluminal cosmic ray events can be observable if we are surrounded by important concentrations of superluminal matter, which is possible in suitable cosmologies [7]. Atypical ionization properties would further enhance background rejection, but ionization can be in contradiction with the requirement of very weak coupling to ordinary matter unless the coupling is energy-dependent.

The possibility that superluminal matter exists, and that it plays nowadays an important role in our Universe, should be kept in mind when addressing the two basic questions raised by the analysis of any cosmic-ray event: a) the nature and properties of the cosmic-ray primary; b) the identification (nature and position) of the source of the cosmic ray. If the primary is a superluminal particle, it will escape conventional criteria for particle identification and most likely produce a specific signature (e.g. in inelastic collisions) different from those of ordinary primaries. Like neutrino events, in the absence of ionization we may expect the event to start anywhere inside the detector. Unlike very high-energy neutrino events, events created by superluminal primaries can originate from a particle having crossed the earth. An incoming, relativistic superluminal particle with momentum p and energy $E_{in} \simeq p_i\, c_i$ in the VRF, hitting an ordinary particle at rest, can, for instance, release most of its energy into two ordinary particles with momenta (in the VRF) close to $p_{max} = 1/2\ p_i\, c_i\, c^{-1}$ and oriented back to back in such a way that the two momenta almost cancel. Then, an energy $E_R \simeq E_{in}$ would be transferred to ordinary secondaries. More generally, we can expect several jets in a configuration with very small total momentum as compared to c^{-1} times the total energy, or a basically isotropic event. Corrections due to the earth motion must be applied (see previous Section) before defining the expected event configuration in laboratory or air-shower experiments, but the basic trends just described remain. At very high energy, such events would be easy to identify in large volume detectors, even at very small rate. If the source is superluminal, it can be located anywhere (and even be a free particle in the case of "Cherenkov" emission) and will not necessarily be at the same place as conventional sources of ordinary cosmic rays. High-energy cosmic-ray events originating form superluminal sources will provide hints on the location of such sources and be possibly the only way to observe them. The energy dependence of the events should be taken into account.

At very high energies, the Greisen-Zatsepin-Kuzmin (GZK) cutoff [25,26] does not in principle hold for cosmic-ray events originating from superluminal matter: this is obvious if the primaries are superluminal particles that we expect to interact very weakly with the cosmic microwave background, but applies also in practice to ordinary primaries as we do not expect them to be produced at the locations of ordinary sources and there is no upper bound to their energy around 100 EeV. Besides "Cherenkov" deceleration, a superluminal cosmic background radiation may exist and generate its own GZK cutoffs for the superluminal sectors. However, if there are large amounts of superluminal matter around us, they can be the main superluminal source of cosmic rays reaching the earth. To date, there is no well-established interpretation of the highest-energy cosmic-ray events. Primaries (ordinary or superluminal) originating from superluminal particles are acceptable candidates and can possibly escape several problems (event configuration, source location, energy dependence...) faced by cosmic rays produced at ordinary sources.

POTENTIALITIES OF AIR-SHOWER DETECTORS

Since the discovery of superluminal matter would be an unprecedented event in the history of Physics, and we do not know at what energy scale it would manifest itself, direct detection of cosmic superluminal particles (CSL) deserves special consideration having in mind the exceptional potentialities of future cosmic-ray detectors. As we expect a very weak coupling between superluminal and "ordinary" matter, except possibly at Planck scale, it is crucial to be able to cover an unusually large target volume. If the coupling increases with energy, it can compensate the possible fall with energy of the CSL flux and make the highest-energy experiments especially adapted to the search for CSL. Future air-shower detectors devoted to the highest-energy cosmic rays will observe the largest target volumes ever reached in a particle physics experiment (especially in the case of satellite-based programs such as OWL or AIRWATCH FROM SPACE). Due to the energies they are able to cover, and considering the possibility that Lorentz symmetry be violated at Planck scale, such experiments are as sensitive to phenomena generated by Planck-scale physics as any possible particle physics experiment can be.

To possibly observe CSL, background rejection must be unprecedentedly powerful. This would be the case for ultra-high energy events generated by CSL. As previously stressed, the ratio $E_{in} \simeq p_i \, c_i$ (in the VRF) provides a unique event profile: since the total momentum of the produced ordinary particles is very small as compared to the total available energy (using c as the conversion factor), the event cannot have the usual, sharply forward-peaked shape of showers produced by "ordinary" cosmic rays. Instead, it can be made of two or more (broad) jets, or be basically isotropic. No "ordinary" ultra-high energy particle can produce such an event shape. The discussion remains valid in any reference frame moving at low speed with respect to the VRF, with the corrections discussed previously. Therefore, air-shower detectors should basically look for events originating at any depth in the atmosphere (like neutrino-induced events) but which, unlike neutrino events where a single elementary particle gets part of the incoming neutrino momentum and subsequently produces a conventional shower profile, do not present a single privileged direction for the produced particles and have instead a tendency to be isotropic. Furthermore, if the earth moves at a speed $\approx 10^{-3} \, c$ with respect to the VRF and $c \, c_i^{-1} \ll 10^{-3}$, the total momentum of the produced particles must in most events cancel with $\approx 10^{-3}$ precision as compared to the total energy, up to fluctuations due to unobserved neutrals and to measurement uncertainties.

REFERENCES

1. Gonzalez-Mestres, L., Proceedings of the Moriond Workshop on "Dark Matter in Cosmology, Clocks and Tests of Fundamental Laws", Villars January 1995 , Ed. Frontières, p. 645, paper astro-ph/9505117 of LANL (Los Alamos) electronic archive.
2. Poincaré, H., "A propos de la théorie de M. Larmor", *L'Eclairage électrique*, Vol. **5**, p. 5 (1895).

3. Poincaré, H., "Electricité et Optique: La lumière et les théories électriques", Ed. Gauthier-Villars, Paris 1901.

4. Poincaré, H., Speech at the St. Louis International Exposition of 1904 , *The Monist* **15** , 1 (1905).

5. Poincaré, H., "Sur la dynamique de l'électron", *Comptes Rendus de l'Académie des Sciences*, Vol. **140** , p. 1504, June 5, 1905.

6. Gonzalez-Mestres, L., Gonzalez-Mestres, L., Proceedings of the IV International Workshop on Theoretical and Phenomenological Aspects of Underground Physics (TAUP95), Toledo September 1995, Ed. Nuclear Physics Proceedings, p. 131.

7. Gonzalez-Mestres, L., paper physics/9704017 of LANL electronic archive.

8. Gonzalez-Mestres, L., talk given at the International Conference on Relativistic Physics and some of its Applications, Athens June 1997, paper physics/9709006 of LANL archive.

9. Gonzalez-Mestres, L., contribution to the 28th International Conference on High Energy Physics (ICHEP 96), Warsaw July 1996, paper hep-ph/9610474.

10. Schweber, S.S., An Introduction to Relativistic Quantum Field Theory", Row, Peterson and Co., Evanston and Elmsford (1961).

11. Recami, E., in "Tachyons, Monopoles and Related Topics", Ed. E. Recami, North-Holland, Amsterdam (1978).

12. Gonzalez-Mestres, L., papers physics/9702026 and physics/9703020.

13. Lamoreaux, S.K., Jacobs, J.P., Heckel, B.R., Raab, F.J. and Forston, E.N., *Phys. Rev. Lett.* **57** , 3125 (1986).

14. Hills, D. and Hall, J.L., *Phys. Rev. Lett.* **64** , 1697 (1990).

15. Coleman, S. and Glashow, S.L., *Phys. Lett.* B **405**, 249 (1997).

16. Glashow, S., Halprin, A., Krastev, P.I., Leung, C.N. and Pantaleone, J., *Phys. Rev.* D **56** , 2433 (1997).

17. See, for instance, Will, C. "Theory and Experiment in Gravitational Physics", Cambridge University Press (1993).

18. Gonzalez-Mestres, L., contribution to the Europhysics International Conference on High-Energy Physics (HEP 97), Jerusalem August 1997, paper physics/9708028.

19. Gonzalez-Mestres, L., Proceedings of the First International Workshop on the Identification of Dark Matter, Sheffield September 1996, Ed. World Scientific, p. 93.

20. Konstantinov, M.Yu., "Comments on the Hypothesis about Possible Class of Particles Able to Travel faster than Light: Some Geometrical Models", paper physics/9705019.

21. Gonzalez-Mestres, talk given at the International Workshop on Topics on Astroparticle and Underground Physics (TAUP 97), Gran Sasso September 1997, paper physics/9712005.

22. Gonzalez-Mestres, L., Proceedings of the 25th International Cosmic Ray Conference, Durban July-August 1997 (ICRC 97), Vol. **6** , p. 109 (1997).

23. Gonzalez-Mestres, L., talk given at the Pre-Conference "Pierre Auger" Workshop of ICRC 97, paper physics/9706032.

24. Peebles, P.J.E., "Principal of Physical Cosmology", Princeton University Press 1993.

25. Greisen, K., *Phys. Rev. Lett.* **16** , 748 (1966).

26. Zatsepin, G.T. and Kuzmin, V.A., *Pisma Zh. Eksp. Teor. Fiz.* **4** , 114 (1966).

Wide Angle Refractive Optics for Astrophysics Applications

David J. Lamb, Russell A. Chipman[†], Lloyd W. Hillman[†], Yoshiyuki
Takahashi, and John O. Dimmock[†]

Department of Physics and [†]Center for Applied Optics
The University of Alabama in Huntsville, Huntsville, Alabama 35899

Abstract. Optical systems consisting of several Fresnel lenses are demonstrated to provide large aperture, wide field imaging for systems with forgiving imaging requirements. Fresnel lenses are shown to be made very thin which makes them ideal for space applications in which system mass and absorption losses are critical. Optics for a proposed space-based air shower detector (Orbiting Wide-angle Light-collector, OWL) are displayed which are the result of a feasibility study.

INTRODUCTION

A space-based air shower observatory has been proposed that detects and reconstructs the trajectories of highly energetic cosmic rays (above 10^{20} eV). The optical system required for such an observatory has a very unusual combination of specifications. Air showers occur at unknown locations in the sky which necessitates a wide field of view in order to maximize the detection probability of events [1]. Furthermore, the amount of light generated from an air shower is relatively small, and the aperture of the system must be correspondingly large in order to collect enough signal. Large-field, large-aperture optical systems are generally extremely complex and bulky (not well suited for space applications), but this application has forgiving imaging requirements. In order to accurately reconstruct the trajectory of a cosmic ray, an angular resolution of approximately $0.1°$ is required for the proposed system. This corresponds to imaging at approximately four orders of magnitude *above* the diffraction limit. Such low resolution requirements enable the consideration of novel and unusual optical systems.

The primary collecting optic of most telescopes is a reflecting mirror. Such telescopes are very well suited for observing known locations in the sky at very high resolution, but they are not well suited to searching the sky for events that take place in unknown locations. This is because reflective optics have an inherently small field of view [1]. Past attempts to increase the field of view of such telescopes have relied on the use of

CP433, *Workshop on Observing Giant Air Showers from Space*
edited by J.F. Krizmanic et al.

arrays of several small field optical systems to provide large field coverage. This technique is not well suited for space applications in which system cost increases rapidly with the number of apertures.

Refractive optics (lenses), on the other hand, have an inherently larger field of view (1). When scaled to very large apertures, however, lenses become prohibitively thick and massive. By converting traditional lenses into Fresnel lenses, the element thickness and mass can be reduced while maintaining the desirable large field of view inherent to refractive optics.

FRESNEL LENSES

A Fresnel lens is a lens in which the surface curvature of one or both of its surfaces have been collapsed into annular zones to form a thin plate (2). If both surfaces of the lens are Fresnel surfaces, then the lens is said to be a "double sided Fresnel lens." The thin plate that consists of the Fresnel lens may be allowed to take on a curvature as in figure 1. This curvature, which could in general be aspherical, is referred to as the "base curvature" of the Fresnel lens. This base curvature acts as an additional degree of freedom in the design and can be used to further enhance the system's performance.

In practice, the number of zones in a Fresnel lens is usually quite large such that the width of each individual facet is very small. Because the facets can be made so small, the curved profile of each facet face can be replaced by a straight profile for ease of fabrication. The facet size must be kept small enough such that the error due to the omission of the curved profile is negligible. The size of the facets is further limited by the substrate thickness of the lens and mechanical considerations.

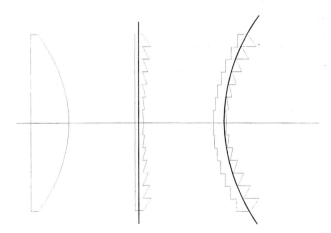

FIGURE 1. Two Fresnel lenses and the standard lens from which they are derived. One is a single sided Fresnel lens with a flat base curvature, while the other is a double sided Fresnel lens with a spherical base curvature.

The reduced thickness benefit realized by using Fresnel lenses comes at a cost. Each Fresnel zone has a back-cut region associated with it that does not contribute to the imaging ability of the lens. Any light that strikes a back-cut will be scattered from its intended image to a (possibly) undesirable location. Ghost images formed in this manner can contribute to the noise associated with the system. If too much light is lost due to this vignetting, then the aperture of the system must be increased to, once again, collect a detectable signal.

AIR SHOWER DETECTOR SYSTEM DESIGN

The vignetting and ghost imaging aspects of Fresnel lenses were temporarily disregarded, and systems were designed and optimized based solely on their imaging ability. Traditional lens design software was used to perform the optimizations. This software idealizes the faceted structure of the Fresnel lens by modeling each Fresnel surface as a substrate who's surface normal at every location has been redefined by some prescription surface (3). This corresponds to an infinitesimal zone spacing. While such a surface could never be manufactured, this does indicate how the surface will focus light for relatively small zone spacings. Once systems were designed in this fashion, complete vignetting and image quality analysis could be performed using more complex optical system models (4).

The optical specifications of the air shower detector design were based on flux collection, resolution, and total field of view requirements. The proposed system was to have a two meter entrance pupil diameter and a 60° full field of view. The system was to resolve a 1 km² area on the ground from a 500 km orbit. This corresponds to approximately 0.1° angular resolution. The system was to observe fluorescence of three N_2 lines of 337, 357 and 391 nm. One of the goals of the design was to make the system as simple as possible while achieving the specified resolution. Also important was to minimize the size of the detector surface. This corresponded to making the focal length of the system as short as possible.

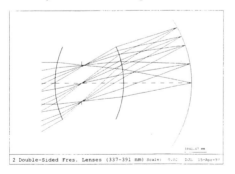 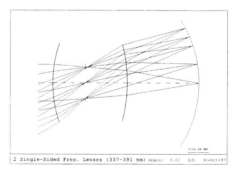

FIGURE 2. Profiles of two Fresnel lens system air shower detectors. The system on the left contains four Fresnel surfaces while the system on the right contains only two Fresnel surfaces.

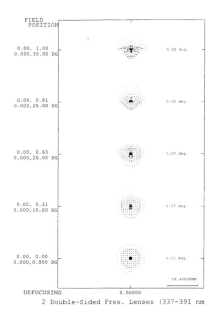

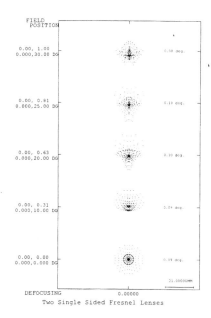

FIGURE 3. Spot diagrams for the two air shower detectors. Numbers to the right indicate the angular resolution associated with that spot.

Figure 2 shows profiles of two of the systems designed for this application. Each design consists of two Fresnel lenses with spherical base curvatures. In the first design, both lenses are double sided for a total of four Fresnel surfaces. The lenses of the second design are both single sided Fresnel lenses for a total of two Fresnel surfaces. The second design was created in the hope that a reduction in the number of Fresnel surfaces would minimize the negative effects associated with Fresnel lenses. The four Fresnel surface system is f/2.6 while the imaging requirements could only be met for the two Fresnel surface system at f/3. This corresponds to focal surfaces with areas of approximately 22 and 45 m² respectively. All lenses are 10 mm thick and made of fused silica.

Figure 3 shows spot diagrams for the two systems. To the right of each spot is the corresponding angular resolution associated with that particular field. Because the focal length of the second system is larger than that of the first system, the spot sizes of the second system can be larger and still yield acceptable resolution.

The results indicate that each system is capable of providing the specified resolution over the entire field of view. Detailed analysis indicates that chromatic aberration is the dominant factor that limits the possible resolution. This is, in part, due to the fact that the dispersion increases dramatically in the ultraviolet region of the spectrum. Any future improvement in the chromatic behavior of these systems will dramatically increase the resolution of these systems. The chromatic behavior could be improved, for example, by using different materials for the different lenses (different indices of refraction and dispersions), thus creating an achromat.

CONCLUSIONS

The two systems shown here demonstrate that Fresnel lenses can be used to provide acceptable images for the purpose of air shower detection. Systems of Fresnel lenses have also been considered for other applications such as ground based gamma ray detectors. Fresnel lenses have the wide field capabilities that are necessary for applications in which the sky must be searched for events that occur in unknown locations. They have the added advantage of being very thin which facilitates their use in space applications.

Some work has been done in the past using a single Fresnel lens to observe giant air showers. Research groups at Cornell and, later, Tokyo used single Fresnel lenses to observe the sky to fields of view of up to a maximum of 26 degrees (5) (6). The novel approach developed by the UAH (University of Alabama in Huntsville) group of combining several Fresnel lenses in a single system enables the field of view of observatories to be opened to more than 60 degrees. This is a significant achievement that can open new possibilities to the experimental astrophysics community.

With the promising results shown here, the research of Fresnel lens systems must continue for the purpose of answering several pertinent questions. If too severe, vignetting of off axis light could have rendered these systems useless as wide angle light collectors. Vignetting, scattering, and ghost image analysis has been performed using more complex optical design software. This software has also been used to evaluate the image quality of systems containing Fresnel surfaces that can actually be manufactured (4). The results of this work are, once again, very promising with vignetting losses being on an acceptable scale.

Also a potential problem is the large focal surface associated with these relatively large f/# optical systems. The cost and complexity of the detectors and associated electronics will scale dramatically with focal surface size. Depending on the type of detector used, it may be necessary to reduce the f/# of these systems even further. By using re-imaging optics, the system f/# (and, hence, necessary detector area) can be reduced dramatically while maintaining the specified resolution (7).

Finally, questions of manufacturing and materials must also be answered before a workable system can be designed. The designs shown here employ the use of curved Fresnel lenses that have grooves cut in either one or both sides of the lens. Difficulties in manufacturing large lenses (more than 3 meters in diameter) of this nature are abundant. Furthermore, a material must be found that will survive in space for several years and retain its transmission characteristics. Funding has been secured to carry out materials research as well as manufacture scaled down prototype optics beginning in spring of 1998.

Despite all of the questions that must be answered and technological hurdles that must be overcome, the notion of using Fresnel lenses as large field observatories is both interesting and promising. By moving away from reflective optics, problems of pupil obscuration and narrow field of view can be eliminated. Fresnel lenses have been shown to posses the potential of providing large field imaging in simple configurations for applications with modest resolution requirements.

REFERENCES

1. Hillman, L.W., et al, "Principles of Wide Angle, Large Aperture Optical Systems," presented at the Workshop on Observing the Highest Energy Particles (>10^{20} eV) from Space, College Park, MD, November 1997.
2. Smith, W. J., *Modern Optical Engineering*, New York: McGraw Hill, 1990, ch. 9, pp. 257-258.
3. Optical Research Associates, *Code V Reference Manual*, Pasadena, 1996, ch. 2A, pp. 415-416.
4. Lamb, D.J., et al, "Computer Modeling of Optical Systems Containing Fresnel Surfaces," presented at the Workshop on Observing the Highest Energy Particles (>10^{20} eV) from Space, College Park, MD, November 1997.
5. Greisen, K., *Annual Report*, Cornell University, 1965; Bunner, A., "Opening Remarks," presented at the Workshop on Observing the Highest Energy Particles (>10^{20} eV) from Space, College Park, MD, November 1997.
6. Tanahashi, G., "Early Air Fluorescence Work: Cornell and Japan," presented at the Workshop on Observing the Highest Energy Particles (>10^{20} eV) from Space, College Park, MD, November 1997.
7. Lamb, D.J., et al, "Focal Plane Reduction of Large Aperture Optical Systems," presented at the Workshop on Observing the Highest Energy Particles (>10^{20} eV) from Space, College Park, MD, November 1997.

Computer Modeling of Optical Systems Containing Fresnel Lenses

David J. Lamb, Russell A. Chipman[†], Lloyd W. Hillman[†], Yoshiyuki
Takahashi, and John O. Dimmock[†]

Department of Physics and [†]Center for Applied Optics
The University of Alabama in Huntsville, Huntsville, Alabama 35899

Abstract. Computer modeling and analysis of optical systems containing Fresnel lenses is performed
for various systems that have been designed for a proposed air shower detector. In particular, the
complex faceted structure of Fresnel lenses is modeled, and the image quality is compared to that of
models that idealize the Fresnel surface. Stray light analysis is also performed, and effects of
vignetting and scattering are quantitatively related to light loss over the entire field of view.

INTRODUCTION

Optical systems containing one or more Fresnel surfaces are currently being considered
for applications that require large fields of view and have modest imaging requirements
(1). Several such systems have been designed and optimized for image quality using
traditional lens design software (2). This software idealizes the structure of the Fresnel
surface. It is incapable of analyzing the adverse effects associated with Fresnel lenses
such as the vignetting of off axis flux that strikes the back-cuts of the Fresnel facets.
Furthermore, the Fresnel surface generated by traditional lens design software is a
mathematical construct and could never be manufactured.

To address the issues of stray light analysis and image quality of a surface that can
actually be manufactured, more complex optical design software must be utilized. Several
such commercially available software packages currently exist that enable Fresnel surfaces
to be modeled and analyzed accurately. Such programs use nonsequential ray tracing and
are capable of modeling many types of optical surfaces, optical coatings, and scattering
surfaces (3, 4).

The primary goal of this research was to determine if the imaging predictions of the lens
design software could be obtained from realistic Fresnel surfaces. Imaging results of the
two different models were, therefore, compared to one another to determine the feasibility
of imaging to the specified resolution with Fresnel lenses. A second goal was to

CP433, *Workshop on Observing Giant Air Showers from Space*
edited by J.F. Krizmanic et al.

determine the amount of light lost due to vignetting in each of two systems incorporating Fresnel surfaces.

Two optical systems that were designed using traditional lens design software (2) were analyzed using complex optical design software. Both contain two Fresnel lenses that have spherical base curvature. In one design, both surfaces of both lenses are Fresnel surfaces, while in the other design only one surface of each lens is a Fresnel surface. These optical systems were modeled for a feasibility study of a space based air shower detector (OWL: Orbiting Wide-angle Light-collector).

SYSTEM MODELING AND EXPERIMENTS

Figure 1 illustrates the manner in which a real Fresnel lens is modeled using advanced optical design software. The systems that were designed and optimized for image quality using traditional lens design software were translated into the complex modeling software. In these models, each Fresnel surface is defined as a series of faceted zones with a finite extent denoted by δy. The facet faces are modeled with linear segments rather than curved. The complex model, then, approaches the idealized model in the limit that δy goes to zero, and it is expected that the images formed by the real Fresnel lens will degrade as δy increases.

To determine the effects of vignetting, rays were traced through the various systems for several different field angles. The rays exiting the system were then collected at the known image location. Any rays that did not reach the image were lost due to either vignetting or surface reflections. Knowing the amount of energy in the incident field and how much was collected at the image, it could be determined how much energy was lost due to vignetting and surface reflections. The results of the vignetting study are depicted

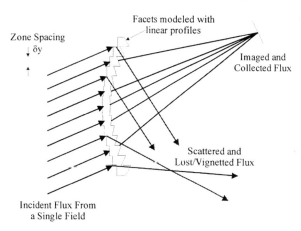

FIGURE 1. A Fresnel lens as modeled in complex optical design software. The facets of each zone have a finite extent, and effects of scattering an stray light can be readily analyzed.

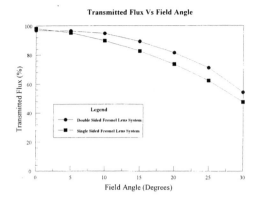

FIGURE 2. Plot of transmitted and imaged flux versus field angle for the two optical systems under investigation. Losses are attributed to light that strikes the back-cuts of the real Fresnel surfaces and is scattered away from its intended image location.

in figure 2 which shows the amount of light that reaches the intended image plane for various field angles when surface reflections are ignored.

The vignetting study indicates that losses due to the faceted structure of Fresnel lenses are not prohibitive. Surprisingly at first, the vignetting of the two Fresnel surface system is slightly higher than that of the four Fresnel surface system. The initial assumption that vignetting is dependent solely upon the number of Fresnel surfaces in a system was not entirely correct. More important than the number of Fresnel surfaces in a system is the total back-cut area of all of the Fresnel facets combined. This back-cut area is approximately the same for both systems considered here with the four Fresnel surface system having a slightly larger area. This explains the similarity in vignetting results and indicates that other parameters are further determining the amount of light that is lost. Furthermore, the amount of light that is lost is relatively independent of the zone spacing, δy. This is because whether there are many small zones or very few large zones, the total back-cut area necessary to create the lens will remain approximately the same.

In addition to vignetting information, it was also necessary to compare the image quality of these systems as modeled in both traditional lens design software and complex optical design software. This was important to verify that the resolution achieved with the idealized models could indeed be achieved with lenses that could actually be manufactured. Towards this end, the spot distributions of the two different models were compared. Since the complex model approaches the ideal model as the zone spacing, δy, goes to zero, the image quality was expected to diminish as the zone spacing increased. The spot distribution results are shown in figures 3 and 4 for the various models and zone spacings.

436

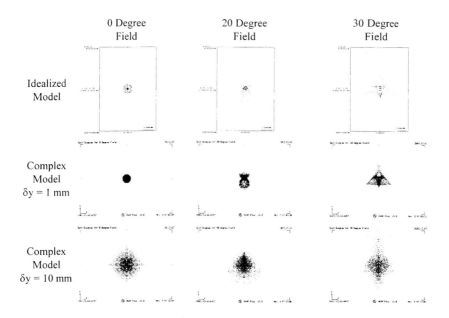

FIGURE 3. Spot Diagrams for the four Fresnel surface design for various models.

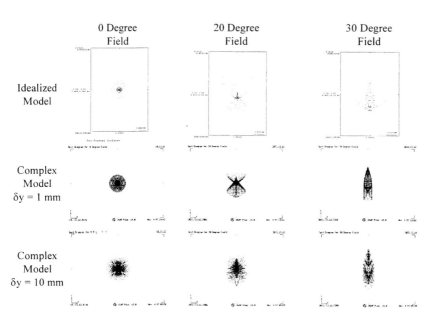

FIGURE 4. Spot Diagrams for the Two Fresnel surface design for various models.

The image quality results indicate that the imaging requirements can be met for both systems if a sufficiently fine zone spacing is used. The spot diagrams and RMS spot diameters imply that the four Fresnel surface system is far more sensitive to zone spacing than is the two Fresnel surface system. Since the number of zones is inversely proportional to the size of each zone, it may be easier to fabricate the two Fresnel surface system with larger zones without sacrificing image quality. The most important result obtained is that systems of Fresnel lenses (that can actually be manufactured) are capable of providing sufficient resolution for the application at hand.

CONCLUSIONS

This research has provided crucial information regarding the feasibility of using Fresnel lenses in imaging applications. First, the systems designed and optimized using the traditional lens design software can be translated into realistic systems that can be manufactured while still providing the specified image quality for this particular application. Furthermore, the vignetting of off axis flux is not prohibitively high. The entrance pupil of the system can be increased to compensate the vignetting losses, and there may exist more efficient schemes for defining the faceted structures of the various lenses.

There are additional features of these lenses that can be further analyzed using this optical design software. Adding curvature to the facets may enable the image quality specifications to be met while using a larger zone spacing (which could make the fabrication process simpler). Furthermore, all of the results here have been obtained using monochromatic illumination. While the results of this research should hold true for all wavelengths, studies must be performed that include wavelengths over the entire spectrum of interest. Finally, this software can be used to analyze systems that contain Fresnel surfaces which have aspherical base curvatures. By making the base curvatures of the lenses aspheric, it is believed that the total back-cut area can be reduced which could further reduce the amount of vignetting.

REFERENCES

1. Hillman, L.W., et al, "Principles of Wide Angle, Large Aperture Optical Systems," presented at the Workshop on Observing the Highest Energy Particles ($>10^{20}$ eV) from Space, College Park, MD, November 1997.
2. Lamb, D.J., et al, "Wide Angle Refractive Optics for Astrophysics Applications," presented at the Workshop on Observing the Highest Energy Particles ($>10^{20}$ eV) from Space, College Park, MD, November 1997.
3. Breault Research Organization, Inc., *ASAP Reference Manual*, Tucson, 1996.
4. Optical Research Associates, *LightTools User's Guide*, Pasadena, 1997.

Focal Plane Reduction of Large Aperture Optical Systems

David J. Lamb, Russell A. Chipman[†], Lloyd W. Hillman[†], Yoshiyuki
Takahashi, and John O. Dimmock[†]

Department of Physics and [†]Center for Applied Optics ·
The University of Alabama in Huntsville, Huntsville, Alabama 35899

Abstract. The size of the focal plane of an optical system is shown to be reduced through the use of re-imaging optics. The general principles of image segmentation and relay optics are discussed as well as trade-offs that must be performed to develop a practical system. A system of re-imaging optics is designed that reduces the necessary detector area of a proposed air shower detector by almost an order of magnitude.

INTRODUCTION

In any optical system, the cost of detectors and associated electronics plays a significant (if not the most important) role in determining total system cost and feasibility. Due to the conservation of étendue (or throughput), optics that are both large aperture and operate over a wide field of view will necessarily have a large focal plane (1). Since the theoretical minimum image size is already quite large for systems of this nature, focal plane minimization is crucial in reducing the cost and complexity of the focal plane.

A space based air shower detector (Orbiting Wide-angle Light-collector, OWL) has been proposed for study that is to have a 60 degree full field of view and a 2 meter entrance pupil diameter. The theoretically smallest focal plane for such a system (operating at f/0.5) would have an area of over 1 m². One optical system that has been designed for this detector consists of two double sided Fresnel lenses that form an f/2.6 image onto a 22 m² spherical focal surface (2). The goal of this research was to determine the feasibility of reducing the size of this image through the use of re-imaging optics while maintaining the proper image quality and resolution. It was also important to form flat rather than spherical images to more readily match available detectors.

In order to accomplish re-imaging for this Fresnel lens system in an efficient manner, the intermediate image formed by the Fresnel lens system must be segmented. This will enable the optical elements of the re-imaging system to be small and relatively simple.

CP433, *Workshop on Observing Giant Air Showers from Space*
edited by J.F. Krizmanic et al.

The principles involved in segmenting and re-imaging the intermediate image formed by the Fresnel lens system will be presented, and the concept will be demonstrated through the use of specific example re-imaging systems.

RE-IMAGING OPTICS AND INTERMEDIATE IMAGE SEGMENTATION

The term "re-imaging" refers to the process of forming an additional image of an image that has been formed by a previous optical system. Figure 1 illustrates the concept. Rays of light form an intermediate image at the dashed surface. Rather than collecting this image on a focal surface, the light is allowed to propagate to another optical system. This re-imaging optical system, then, demagnifies the image and forms a final image on a smaller focal surface. The entire purpose of this process is to decrease the total system f/# and obtain as compact an image as possible.

The re-imaging system shown in figure 1 may consist of any combination of optical components (i.e. lenses, mirrors, diffraction gratings, etc.) that are necessary to demagnify the image while maintaining image quality and acceptable resolution. If the intermediate image is very large, then the re-imaging system components will have to be comparably large. By segmenting the intermediate image and re-imaging small segments individually, the optical elements can be kept small and manageable.

Figure 2 shows a scheme by which an intermediate image is segmented and relayed to several different re-imaging systems. Losses due to both intermediate image segmentation and cross talk between the various image segments and re-imaging systems are abundant unless field lenses are used. By placing field lenses very close to the intermediate image, the light rays from the various intermediate image segments can be bent back towards the desired re-imaging system. Losses due to cross-talk can, hence, be eliminated. Field lenses can, in principle, be placed very near one another such that segmentation losses are also minimized. The amount of light and information that is lost will vary with the degree to which the intermediate image is segmented. The size and complexity of the optical components will also vary with intermediate image is segmentation. All of these parameters must be considered and traded off as the entire system is designed.

Figure 3 shows a "black box" schematic of the entire re-imaging concept. A large

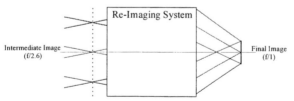

FIGURE 1. A "black box" re-imaging system. An optical system (not shown) forms an intermediate image at a relatively large f/#. The re-imaging system then forms an image of the intermediate image at a smaller f/#.

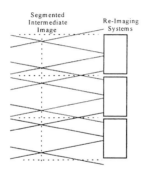 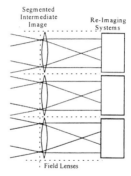

FIGURE 2. Intermediate image segmentation with and without field lenses. Notice that cross-talk losses are abundant in the absence of field lenses. Segmentation losses can be minimized by placing the field lenses as close to the intermediate image as possible in addition to using as small an edge thickness as possible.

objective optical system forms a large intermediate image onto a spherical focal surface. Field lenses are placed at the intermediate image where they act to efficiently segment the image while minimizing losses. The intermediate image is then re-imaged by the various re-imaging optical systems onto smaller focal planes. Since the total system f/# has been reduced from that of the objective alone, the total area of all of the detectors combined is smaller than that of the intermediate image surface. This enables the detectors and associated electronics to be packaged in conveniently sized and easily managed modules, the size of which may be used as a system design parameter.

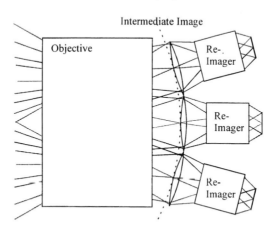

FIGURE 3. "Black box" schematic of a complete re-imaging system with a reduced focal plane. The objective optical system forms an intermediate image that is segmented and relayed by field lenses to a series of re-imaging optical systems. The total area of the final image is reduced from that of the intermediate image.

441

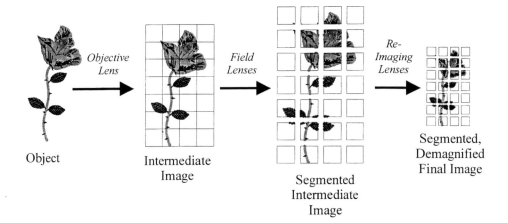

FIGURE 4. The images of an object are shown at the various stages. The intermediate image that is formed by the objective lens is broken up into smaller and more manageable sections by the field lenses. The re-imaging lenses act to demagnify each individual segment.

Figure 4 shows how the various optical components act on a hypothetical object. The objective lens forms an intermediate image of this object, and the intermediate image is then segmented by an array of field lenses. Note that the field lenses act only to separate the image sections and do not contribute to the optical power of the system or to its image forming abilities. Also note that each section consists of many pixels which, for the sake of simplicity, are not depicted. Each individual section is then demagnified by the re-imaging optical systems, and the total image area has thus been reduced.

RE-IMAGING SYSTEM DESIGN AND APPLICATION

This concept of focal plane reduction through the use of re-imaging optics was applied to an existing Fresnel lens design for the OWL project (2). Two different system designs were obtained for various degrees of image segmentation. The purpose of these different configurations was to demonstrate some of the trade offs associated with designing systems of this nature.

During the design process, the Fresnel lens system did not change, and two re-imaging optical systems were developed for different intermediate image segmentation. In both cases, re-imaging systems were designed for the on axis, next to on axis, and full field-of-view portions of the intermediate image. In one instance, the full 60 degree field of view of the Fresnel lens system was segmented into 2.5 degree increments and re-imaged. In the other design, the full field of view was segmented into 5 degree increments. The re-imagers were designed and optimized to provide the required $0.1°$ angular resolution at as small an f/# as possible.

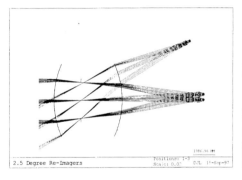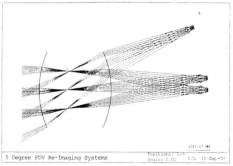

| 2.5 Degree Re-Imagers | Positions: 1-3 Scale: 0.0? DJL 1?-Sep-97 |
| | 1086.96 MM |

| 5 Degree FOV Re-Imaging Systems | Positions: 1-3 Scale: 0.02 DJL 15-Sep-97 |
| | 1041.67 MM |

FIGURE 5. Double Fresnel lens objective with two different re-imaging configurations. The 2.5 degree field of view systems are smaller and simpler, but many more are needed to cover the entire field of view. Fewer of the 5 degree field of view systems are needed, but each system is larger and more complex.

Figure 5 shows all three re-imaging segments for both systems. Figure 6 shows a close up of the on axis and next to on axis re-imagers for both systems. Note how the field lenses act to separate the adjacent intermediate images which are 0.1° apart from one another. In the absence of the field lenses (which do not contribute to the optical power of the system), light would propagate in a straight line from the intermediate image to the re-imaging segments and result in cross talk losses. Also important in Figure 6 is the angular extent of the image cones before and after the re-imaging segments. The increased cone angle at the final surface indicates a smaller system f/# and, hence, a smaller final image surface.

Table 1 compares the attributes of the two re-imaging schemes in an attempt to demonstrate the trade-offs between system complexity and image segmentation. The 2.5

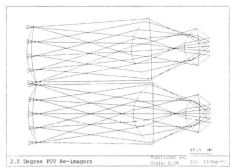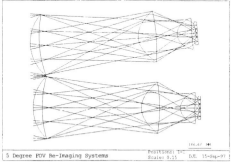

| 2.5 Degree FOV Re-imagers | Positions: 1-? Scale: 0.28 DJL 13-Sep-97 |
| | 89.?? MM |

| 5 Degree FOV Re-Imaging Systems | Positions: 1-? Scale: 0.15 DJL 15-Sep-97 |
| | 166.?? MM |

FIGURE 6. Close up view of the on axis and next to on axis re-imaging systems for the various designs. Note that the scales are not the same. The 2.5 degree field of view systems are smaller and require one fewer element than the 5 degree systems. Notice how the field lenses act to seperate the adjacent fields, and note the reduced size of the final image.

443

TABLE 1. Comparison of Various Re-Imaging Systems

Sub-System FOV (Degrees)	Total System f/#	Individual Detector Radius (cm)	Total Detector Area (m²)	Individual Segment Mass (kg)	Axial Glass Thickness (m)	Elements per Segment	Total # Segments
5	1.00	8.7	3.1	36.1	0.315	4	141
2.5	0.90	3.9	2.5	6.9	0.182	3	563

degree field of view re-imaging systems bring the total system down to f/0.9 with three elements, while four elements are required to bring the 5 degree field of view systems down to f/1. Nonetheless, many more total segments are necessary to cover the entire field of view with the 2.5 degree systems. It should be emphasized that these systems were optimized for image quality at the smallest possible f/# and not for other parameters such as axial glass thickness or total system mass. Other configurations exist in which total system mass is reduced.

CONCLUSIONS

It has been demonstrated that the 22 m² image area of the f/2.6 Fresnel lens system can be reduced significantly. The degree to which this intermediate image is reduced and the overall complexity of the system are dependent upon many factors. Two examples have been shown that segment the intermediate image to a different extent. From these examples it is clear that trade-offs exist in the design, and many iterations must be accomplished before an optimum system can be developed.

While field lenses can be used to minimize the amount of light lost due to cross talk and image segmentation, they cannot completely eliminate it. Real lenses and re-imaging systems will have to be mounted in structures that will have finite thicknesses. Segmentation losses are, therefore, inevitable in a re-imaging scheme such as this. This, however, is not only a problem of the segmented and re-imaged optical system discussed here. If a large spherical 22 m² detector were to be constructed, it would necessarily be made up many smaller detectors with boundaries at the edges. Depending upon the type of detector device used, this could cause as much, or even more, light loss as in the re-imaging scheme. Both scenarios must be considered as the trade-offs between optical and detector complexity are analyzed.

From an overall system perspective, the re-imagers add a significant amount of complexity to the optics (compared to the double Fresnel lens objective alone). What they add in optical complexity they more than compensate for in reduced detector area and configuration. The electronics associated with, and the power supplies necessary to operate a monolithic 22 m² are phenomenal. So, while the optics may look a bit more complex, the detectors have been simplified tremendously.

Re-imaging optics provide many additional degrees of freedom with which optical designers can work. Chromatic aberration is what currently limits the resolution of the Fresnel lens objective (2). By using more than one type of glass in the re-imaging segment, it is believed that the chromatic aberration of the Fresnel lens system can be reduced. This would lead to better resolution and make these types of systems useful in many applications that require a large aperture and a wide field of view.

Also important in this configuration is the use of a single large aperture to cover a wide field of view. The alternative is to use many large aperture, narrow field of view optical systems to cover the entire field of view. By using a single aperture, manufacturing, deployment, and operation costs can be minimized. Furthermore, by utilizing the re-imaging concept, the system f/# can be brought down to a level that is comparable to many (reflecting), narrow field systems. This leads to a minimized detector area in a relatively simple configuration.

REFERENCES

1. Hillman, L.W., et al, "Principles of Wide Angle, Large Aperture Optical Systems," presented at the Workshop on Observing the Highest Energy Particles ($>10^{20}$ eV) from Space, College Park, MD, November 1997.
2. Lamb, D.J., et al, "Wide Angle Refractive Optics for Astrophysics Applications," presented at the Workshop on Observing the Highest Energy Particles ($>10^{20}$ eV) from Space, College Park, MD, November 1997.

Connection between the Statistical Parameters of Hadronization and the QCD Coupling Constant .

L.Popova[1], Institute of Nuclear Research and Nuclear Energy, Sofia, Bulgaria
G. Kamberov, Washington University at St Louis, USA

ABSTRACT

We propose a statistical parameterization of hadron production which is verified up to the highest EAS energies. This parameterization leads to Stephan's type equation of state relating the 'temperature' of the relativistic hadron gas with the collision energy if a power multiplicity law is assumed. A correspondence between the lattice theory expansion on the coupling constant and the statistical sum of temperature expansion in statistical mechanics reveals a simple relation between the multiplicity power law index and the coupling constant. Cosmic ray data yield estimates of the running values of the coupling constant beyond the range of collider experiments. These estimates indicate the asymptotic behavior of the running values of the coupling constant and provide a hint that color confinement and asymptotic freedom of quarks co exist in an unified phase of QCD.

1   Statistical Parameterization of Hadron Production

This parameterization is based on the empirical spectra of hadrons

$$\frac{2E}{\sqrt{s_0}\sigma_{in}dx} = Ae^{-Bx}, \qquad x = \frac{2p_l}{\sqrt{s}} \tag{1}$$

obtained at low energy FNAL experiments [1]. Hadron spectra are extrapolated in the range of EAS assuming power law dependence of multiplicity

$$\bar{n} \sim \sqrt{s^{\alpha}} \tag{2}$$

and energy invariance in respect to statistical scaling variable [2]

$$x_s = \frac{E}{\bar{\bar{E}}} \simeq x(\frac{s}{s_0})^{\alpha/2} \tag{3}$$

In the range of EAS the average energy $\bar{E}$ is expressed, by virtue of (2) The $\frac{4(m^2 + p_t^2)}{s}$ term in the expression for E is neglected.

---

[1]Supported in part by the National Science Foundation of Bulgaria under Grant 498.

CP433, *Workshop on Observing Giant Air Showers from Space*
edited by J.F. Krizmanic et al.

Changing variables in (1) we parameterize the spectra of hadrons at energy $\sqrt{s} > \sqrt{s_0}$ as follows:

$$\frac{2E}{\sqrt{s}\sigma_{in}dx} = A\left(\frac{s}{s_0}\right)^{\alpha/2} e^{-B\left(\frac{s}{s_0}\right)^{\alpha/2}x} \tag{4}$$

The parameters A and B are fitted to data at a basic energy $\sqrt{s_0} = 19.7\text{GeV}$ In a detailed semi-inclusive model version (SISM) they are parameterized by parabolic functions on the relative multiplicity $z = n/\overline{n}$ in several intervals of p_t.(Fig. 1 shows the approximation curves for π^+ and π^-). A thermodynamic distribution $g(p_t)$ is proportional to e^{-p_t} was used for the transverse momentum. It has been shown that equation (4) fits experimental data from strong, week and electromagnetic interactions up to the highest accelerator energies [3]. In Fig.2 and Fig.3 statistical scaling distributions are compared with proton-antiproton [2],[4] and electron-positron [5],[6] collision data. The calculated results are obtained assuming $\alpha = 0.26$.

2 Thermodynamical presentation

Replacing x in (4) by E (using $E \simeq x\sqrt{s}/2$ at high energies) one can obtain the energy spectrum of relativistic hadrons

$$\text{probability} = \frac{d^3p}{E}g(p_t)\exp\left(-\frac{E}{(s_0^{\alpha/2}/(2B))\sqrt{s}^{1-\alpha}}\right) \tag{5}$$

We assign

$$\overline{E} = \frac{s_0^{\alpha/2}\sqrt{s}^{1-\alpha}}{2B}, \tag{6}$$

and put it in the denominator of the Boltzmann term of (5) to obtain the energy spectrum of hadrons in a thermodynamical form:

$$\text{probability} = \frac{d^3p}{E}g(p_t)(s/s_0)^{\alpha/2}\exp\left(-\frac{E}{\overline{E}}\right) \tag{7}$$

The hadron spectrum differs from the Boltzmann distribution of free gas by the Bloch-Nordsieck term d^3p/E appearing in the QCD collisions of quark- antiquark pair production and gluon bremsstrahlung.

A similar to (7) spectrum of hadrons

$$\text{probability} = \frac{d^3p}{E}g(p_t)(s/s_0)^{\alpha/2}\exp\left(-\frac{E}{kT}\right) \tag{8}$$

was derived by Chou and Yang [7] for fixed interaction energy applying the method of Darwin and Fowler [8] for the steepest descent in statistical mechanics. Although for the gas model the temperature T is an equilibrium concept, for the high energy collision problem, T is just a mathematical parameter that governs the partition of energy and neither requires nor implies equilibrium. The partition energy T and the equation (8) have a physical meaning for a gas in three dimensional region of x, y, x space, where

$$z = \frac{y}{(m^2 + x^2 + y^2)^{1/2}} g(y), \qquad \text{and } y = \frac{1}{2} \log \frac{E + p_l}{E - P_l}.$$

3 Equation of state

From the similarity of (7) and (8) we can assume that

$$\overline{E} = kT \tag{9}$$

and insert it in (6) to obtain the equation of state

$$T = \text{const} s^{(1-\alpha)/2}, \qquad \text{const} = \frac{s_0^{\alpha/2}}{2Bk}. \tag{10}$$

This equation can be considered as a generalization of Stephan's law (for free gas $\alpha = 0.5$). Fermi's *ideal gas* approach to relativistic hadrons, [9], was not confirmed by the early accelerator data at low energies. The thermodynamical (Cocconi-Koester-Perkins) energy spectrum was replaced by the empirical equation (1). Obviously a relativistic hadron gas should obey different thermodynamical distribution, e.g. the statistical mechanics equation of Chou and Yang (7). The latter distribution was postulated for a fixed interaction energy and the problem was to find out the theoretical dependence of T on the interaction energy. The assumption for statistical scaling allows us to derive the equation of state (10) and to use our equivalent equation (7) to parameterize low energy accelerator distributions in the range of EAS energies by the help of only one statistical parameter α.

For the relativistic hadron gas we assign $\alpha = 0.26$. This value was found by fitting to existing data for p-p and $e^+ - e^-$ data at low energies [3]. Recent collider data on p-p collisions up to the highest energies 1.8TeV confirmed that α is at least 0.26 ([10]). Experimental data from $e^+ - e^-$ annihilation processes [11] at high energies are best fitted by QCD [12] approximation leading to a stronger rise of multiplicity $0.26 < \alpha < 0.5$ (see Fig.4). In this figure we just-impose the asymptotic behavior of multiplicity according to various laws. They are compared with accelerator data from p-p and $e^+ - e^-$ collisions. The number of hadrons in $e^+ - e^-$ annihilation is larger than in p-p interactions due to their different intelasticity.

We present graphically in Fig.5 the equation of state (10) for $\alpha = 0.26$ and that of Stephan-Boltzmann gas ($\alpha = 0.5$). Feynman's upper boundary of scaling, $\alpha = 0$, is also included in the figure (the parameter T does not represent a true temperature). The lowest Hagedorn limit [13] is also included. The curve of the Stephan-Boltzmann gas (S-B) has been adjusted to the critical temperature T_{cr} according to the GUT and the Big Bang theories.

The S-B curve intercepts the Hagedorn limit and BECAT line. The latter has been derived [14] in the limited interval of SPEAR and PETRA experiments postulating for high energy electron-positron annihilation the existing of hadron gas in thermodynamical equilibrium before the hadrons themselves decouple (freeing out) and decay giving rise to observable particles. Boltzmann limit of Fermi and Bose statistics is used in Becattini calculations for the mass distribution of different species.

Becattini's estimations for T differ from those of Chou and Yang [15]. These estimates were obtained by a fit to the momentum spectra of hadrons in the same $e^+ - e^-$ collisions and in the same energy interval of SPEAR and PETRA experiments. These authors extrapolate their hadron spectrum (8) assuming Feynman scaling.

We include also a curve in Fig 5 for p-p interactions according to DPM and QGSM which is similar to that of Chou and Yang for $e^+ - e^-$ annihilation processes. This curve tends to the statistical scaling curve at low energies. It could be assumed a smooth transition from color (resonance) states to some termalization of hadron gas. In that figure we mark symbolically a low and a high energy phase transition.

4   Correspondence between statistical parameters and the QCD coupling constant

In order to examine the asymptotic behavior of the statistical parameter α one could use the correspondence [16] between the structure of the expansion on the coupling constant $\alpha_s(Q)$ in Lattice theory and the high temperature expansion (on $1/kT$) in statistical mechanics. In Fig. 6 we extrapolate the Lattice calculation curve for $\alpha_s(Q)$ beyond accelerator energy range. The Lattice calculations are fitted to ALEPH, DELPHI, L3, OPAL, deep inelastic scattering, $e^+ - e^-$ annihilation, hadron collisions and heavy quarkonia data up to $Q = 161 GeV$. On the same figure the right-up plot presents the correspondent energy dependence of the statistical index n that has been derived using the equation of state (10) and equation (2).

Next we present our analysis of EAS data in this aspect.

5   Analysis of EAS data

As a first step we compare in Fig.7 the data of shower maximum with different model calculations. The thick lines present the results obtained on the basis of SISM applied in MOCCA algorithm [17]. They are confirmed by our calculations with SISM (using our full Monte Carlo computer generator. In these calculations a constant value (0.26) was assumed for the statistical parameter α. The dash lines present the results from the MOCCA generator implementing the Hillas algorithm for Feynman scaling ($\alpha = 0$) [17]. The thin and the point-dash curves correspond to the new versions of DPM [18] and QGSM [19] models with additional production of mini jets. The multiplicity increment in the two latter models corresponds more or less to $\alpha = 0.26$ but these models predict different momentum spectrum of secondary hadrons e.g. strong rise of differential cross section in vicinity of $x \simeq 0$ and Feynman scaling in the fragmentation region. We use also the results of Heck and Knapp [20] obtained with the other model versions at low energies (100-1000 TeV).

The ellipse in Fig.7 serves as a reference point. It was obtained with SISM assuming mixed composition of primary cosmic rays (14% p, 25% He,26% C-O-N,13% Ne-Mg-Si and 24% Fe) from the balloon experiments [21] performed to 470 TeV/nucleon.

In the range of high energies we present the calculated results for two extreme cases of pure proton and pure iron primary flux. They are compared with the depth of maximum obtained in recent Akeno [22] and Yakutsk [23] experiments. Obviously there is no essential difference between the statistical model predictions and those from the new versions of DPM and QGSM. (We neglect the contradiction between the slopes of the curves at low and high energies obtained by different authors with the new versions of DPM and QGSM.). Excluding MOCCA calculations with Feynman scaling all those results seem to conform to the trend of shower data without any essential change of the primary mass composition. A stronger rise of multiplicity ($\alpha > 0.26$) would also fit the data if an increasing contribution of light nuclei is assumed in the range of giant showers.(Our early calculations [24] of shower absorption length fitted the data in the range of giant showers with $\alpha \geq 0.5$ assuming pure proton composition.

Figure 8 suggests a similar conclusion for the increase of the multiplicity power law index if the contribution of light nuclei in the primary mass composition increases in the range of giant showers. In this figure the calculated results are compared with Akeno [22] and Yakutsk [23,25] data. QGSM with a strong rise of mini jet cross sections fail to fit the data from small showers. A drastic conclusion for heavier than iron primary mass composition is needed to get agreement with the data. It could be explained by the special character of hadron spectra in that model version: Feynman scaling in the fragmentation region and very big differential cross sections for x approaching 0. In these conditions a great deal of low energy hadrons give rise to muons that decay above the observation level. Figure 9 shows that the mini jet production has a negligible effect on the number of muons in the range of small showers. .

6 Conclusions

This analysis suggests a new aspect of the study of primary mass composition. Giant shower data could be used to examine the asymptotic behavior of the QCD coupling constant, provided that there exists a correspondence between the high-temperature expansion in statistical mechanics and the expansion on the coupling constant in Lattice theory. Our preliminary estimations give a hint for a continuing phase transition. It seems that asymptotic freedom and color confinement exist as an unified phase of QCD. More detailed calculations are in run to verify such conclusions.

References

1. T. Kafka et al, Phys.Rev.D 16 (1977) 5, 1261
2. J. Wdowczyk and A. Wolfendale, N.Cim.A 54 (1979) 433
3. G. Kamberov and L. Popova, Proc. 33 CERN Workshop (1977) World Scientific N.Y.
4. P. Capiluppi et al, Nucl.Phys.B 70 (1974) 16
5. M. Basile et al, N.Cim.A 79 (1984), 1, 1
6. M. Althoff et al, Z.Phys.C 22 (1984) 307
7. T. T. Chou and C. N. Yang, Phys.Rev.D 32 (1985) 7, 1692
8. C. G. Darwin and R. H. Fowler, Proc. Cambridge Phys.Soc. 21 (1923) 730
9. E. Fermi, Progr. Theor. Phys. 5 (1950) 4, 570
10. F. Abe et al, Phys.Rev.D 50 (1994) 9,1550
11. A. W. Zheng et al, Phys.Rev. 42 (1990) 3, 737
12. B. Bassetto, Phys.Lett.B 83 (1997) 2, 207
13. R. Hagedorn, La Riv. N.Cim.10(1983) 1
14. F. Becattini, Proc. 33 CERN Workshop (1997) World Scientific N.Y.
15. T. T. Chou and C. N. Yang, Proc. CERN Workshop (1984) 1013
16. T-P Chang and L-F Li "Gauge Theory of Elementary particle Physics" (1984) Clarendon Press - Oxford
17. A. M. Hillas Nucl.Phys.B(suppl) 52 (1997) 29
18. T. Gaisser et al, Proc. 25 ICRC 6 (1997) 281
19. N. N. Kalmykov et al, Nucl.Phys.B(suppl) 52 (1997) 17
20. D. Heck and J. Knapp, Nucl.Phys.B(suppl)52 (1997) 139
21. S. Dake et al, Proc. ISVHECR, Tokyo (1994) 513
22. N. Hayashida et al, Proc. 25 ICRC 25 (1997) 241
23. B. N. Afanasiev et al, Poster ISVHCRI (1996) Karlsruhe
24. L. Popova and J. Wdowczyk, Proc.15 ICRC 8 (1975) 150
25. N. Dyakonov et al, Proc. 23 ICRC 4 (1993) 303

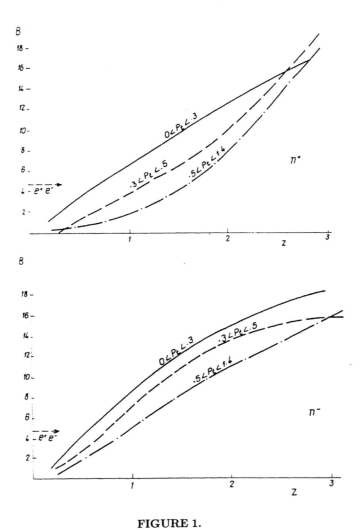

FIGURE 1.

Fig.1 Dependence of the slope parameter on the relative multiplicity in pion spectra used in SISM

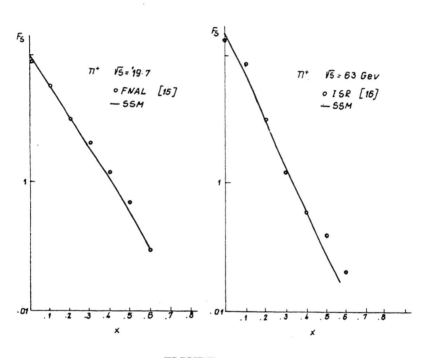

FIGURE 2.

Fig.2 Comparison of SSM spectra for positive pions with FNAL [2] and ISR [4] data

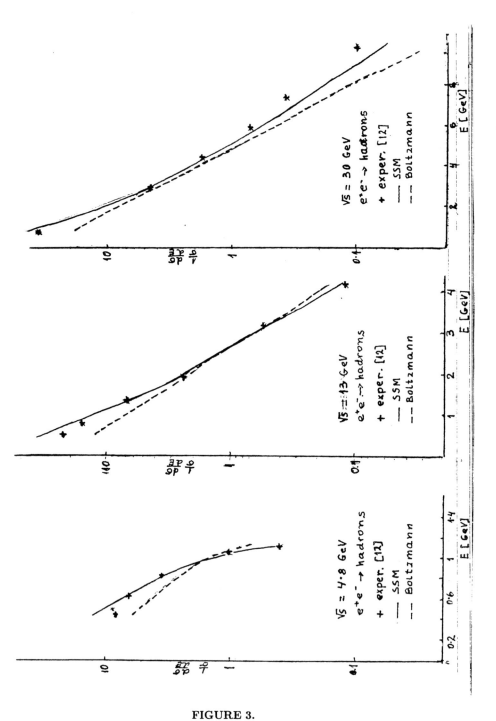

FIGURE 3.

Fig.3 Comparison of hadron energy spectra from electron-positron annihilation
[5] with SSM and Boltzmann distributions

454

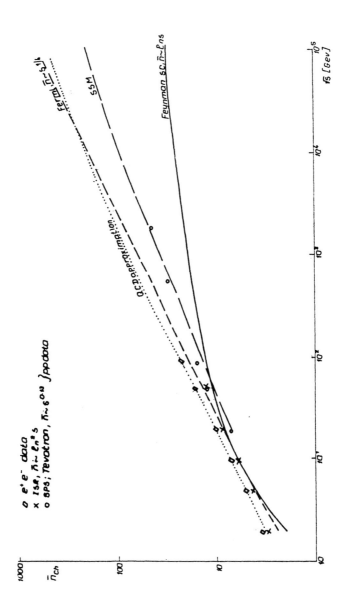

FIGURE 4.

Fig.4 Comparison of accelerator data ($p - p$ and $e^+ - e^-$ collisions) for the energy dependence of charged particle multiplicity with different model predictions

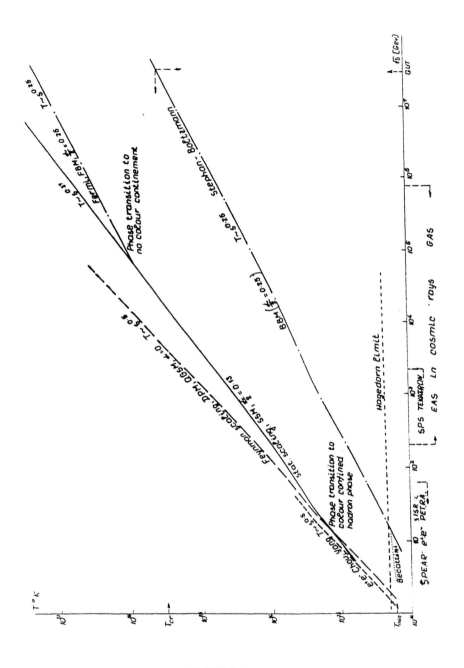

FIGURE 5.

Fig.5 Graphical comparison of equation of state with different values of the statistical parameter α

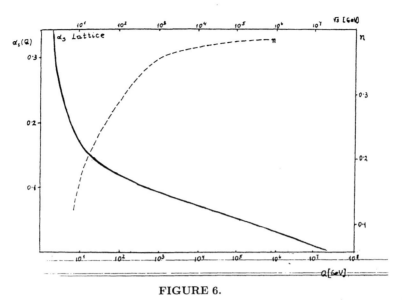

FIGURE 6.

Fig.6 Extrapolation of Lattice calculations for $\alpha_s(Q)$ (down-left plot) Corresponding values of multiplicity power index n (up-right plot)

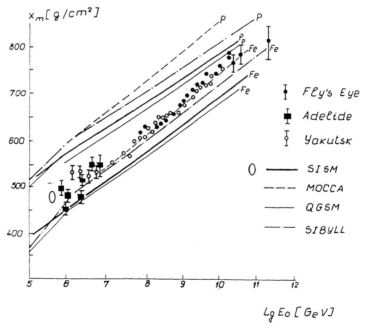

FIGURE 7.

Fig.7 Energy dependence of shower depth maximum

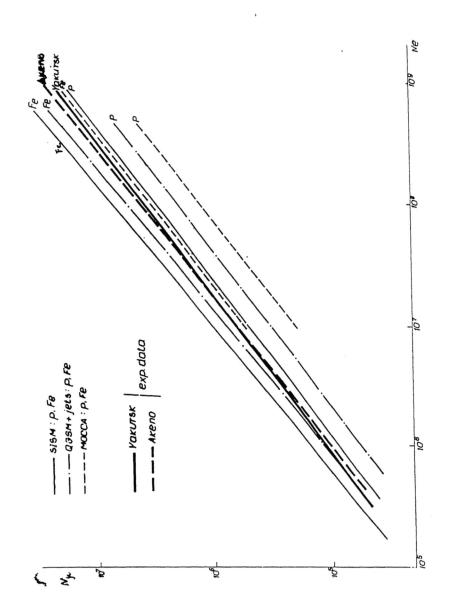

FIGURE 8.

Fig.8 Number of muons (above 1 GeV) in EAS with different sizes

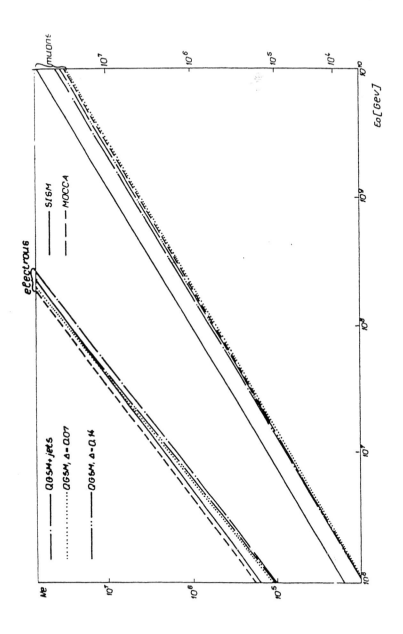

FIGURE 9.

Fig.9 Energy dependence of muon number calculated with different models

Computer Aided Optimal Design of Space Reflectors and Radiation Concentrators

Oleg A. Saprykin*, Yuriy K. Spirochkin†, Vladimir G. Kinelev‡, Valeriy D. Sulimov‡

*RSC Energiya, Design Department, 4a Lenina St., Korolev, Moscow Region, 141070 Russia, †Dynamika Co., 16-58 Kosmonavtov Ave., Korolev, Moscow Region 141070 Russia, ‡Scientific Institute of Applied Mathematics and Mechanics, BMSTU, 5 Baumanskaya 2ⁿᵈ St., Moscow, 107005 Russia

Abstract. The goal of space radiation receiver design is achievement of its maximal reflecting properties under some technological and financial restrictions. Optimal design problems of this type are characterized by nonconvex nondifferentiable objective functions. A numerical technique for optimal design of the structures and an applied software **REFLEX** under development are proposed.

INTRODUCTION

A quality and an area of a reflecting surface are the most important functional characteristics of space radiation receivers. There are different structural solutions for reflectors and concentrators, among them:
- rigid non-transformable structures (e.g., the Hubble space telescope);
- rigid transformable structures (e.g., space satellite IAE);
- frame-film transformable structures (Fig. 1);
- frameless film transformable structures (Fig. 2).
Most of present-day designs of reflectors and concentrators are based on a rigid non-transformable structure concept. This makes possible maximizing of the reflecting surface quality. However technical performance of space vehicles imposes stringent restrictions on overall dimensions of concentrators of this type. At present the maximal dimension of such concentrators does not exceed 4.5 m (a diameter of the Space Shuttle cargo compartment). This does not suffice to potential needs of spacecraft energetics or scientific observations.
The other types of both reflector and concentrator designs are transformable structures. Thanks to up-to-date level of space technology, their overall dimensions run into 400 - 600 m depending of specific aims. But reflecting surface qualities of these structures are inferior to the qualities of non-transformable ones. An optical efficiency of transformable type structures depends essentially on structural features,

CP433, *Workshop on Observing Giant Air Showers from Space*
edited by J.F. Krizmanic et al.

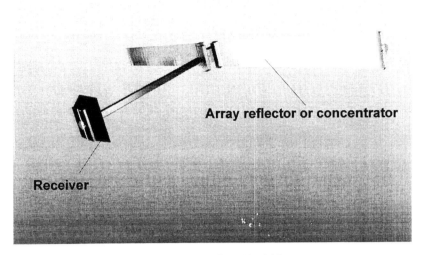

FIGURE 1. Transformable rigid structure.

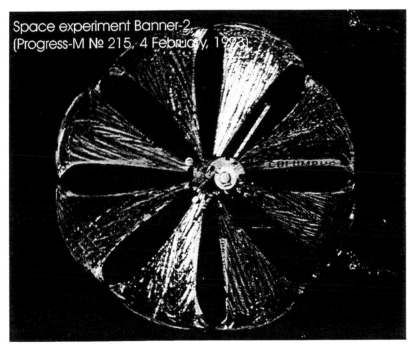

FIGURE 2. Frameless film transformable structure (space experiment Znamya-2). With permission of Consortium Space Regatta.

461

unfolding techniques when staying in orbit, the quality and the number of assembling elements. So the actual purposes are as follows:
- to evaluate the functional suitability of the structure at the initial design stage;
- to choose the most significant design variables defining a reflection capacity;
- to find optimal values of these variables with due regard for financial, technological, and other restrictions.

An application software **REFLEX** is proposed to solve the above problems. This software is intended for numerical analysis and optimization of optical, strength and dynamic characteristics of space reflectors and radiation concentrators by means of computer simulation. Such design environment as AutoCAD or other 3D CAD-programs can be used for preparation of data on a structure under design. The purpose of this paper is to briefly present information about the problem formulation, solution methods and on the software development state.

1. OPTIMAL DESIGN PROBLEM FORMULATION

Spinning frameless film concentrators and reflectors of solar radiation are subjects of inquiry. The essential feature required of spacecraft of the type is to provide maximal quantity of the reflecting light flux. This flux depends of a cloth area and film reflecting properties under conditions of a fixed sun orientation of a reflector. The optical quality of the reflecting surface can be characterized by a value of a *mirror reflection coefficient* (MRC) representing a ratio between values of reflecting (Φ_R) and falling (Φ) fluxes correspondingly in a divergence angle $32'$:

$$K = \frac{\Phi_R}{\Phi} < 1. \tag{1}$$

The objective of design problem is to maximize the MRC value taking proper account of technological, energy and other restrictions. Generally speaking, there are following three groups of optimal design criteria: functional, technological and financial. The first group consists of:
- maximal MRC value;
- maximal reflecting area;
- maximal period of active stay in orbit.

The most essential criteria (restrictions) in the second group:
- useful load limitation;
- overall dimension limitation (in transporting configuration);
- limitations on the density of useful load configuration;
- minimal time of the structure unfolding operation;
- maximal dynamic stability of the unfolded structure;
- restrictions on the stress level in all the structural elements.

The following criteria are worth consideration in the third group:
- minimal costs of launch into orbit;

- minimal costs of the unfolding operation of the structure;
- minimal operational costs.

In order to find the best design it is necessary to solve the constrained optimization problem, that is to find optimal value of the objective function subjected to constraints on design variables.

2. STRUCTURAL OPTIMIZATION TECHNIQUES

When formulating the optimal design problem the following main assumptions are essential:
- all the criteria are continuous functions of a bounded set of structural design variables;
- all the design restrictions (permissible stress levels in structural elements, maximal area of the reflecting surface, maximal operational costs and others) are given, that is to say the ideal design is given;
- during the optimization process all the current designs must stay permissible.

On the k-th step of the optimization procedure the current design is characterized by a vector x^k of design variables. Depending of the problem under consideration this vector can contain all the design variables or some of them: structural characteristics, angular velocity et cetera. The quality of the current design can be evaluated on base of given criteria: the best design is one whose parameters are nearest to corresponding parameters of the ideal design p^a. Let the current design be characterized by a set of parameters $p^k(x^k)$. In general case it is possible to build the bounded set of mismatch criteria as follows

$$f_i(x^k) = \left| p_i^k(x^k) - p_i^a \right|, \; x^k \in X \subset R^n, \; i \in I^0, \tag{2}$$

where X is the feasible domain, R^n is the set of real numbers, I^0 is the index set, $I^0 = \{1, 2, \ldots, N\}$, N is the number of design parameters under consideration. The problem is to find such vector of design variables that gives minimal differences between the ideal and current designs. So it is necessary to minimize simultaneously all the mismatch criteria (2).

Below the optimization problem statement is considered in the following scalar form: find the vector of design variables $x \in X$ such that minimizes the maximal value of mismatch criteria (2) which is equivalent to the Discrete Minimax Problem

$$\min_{x \in X \subset R^n} \; \max_{i \in I^1} \; \{f_i(x)\}. \tag{3}$$

The solution of the problem (3) is the feasible vector $x^* = \{x_1^*, \ldots, x_n^*\}^T$, which minimizes the objective function $f(x) = \max\{f_1(x), \ldots, f_N(x)\}$. It should be pointed that $f(x^*) = 0$ is the case when the current and the ideal designs are coincident. In practice it is sufficient to solve the problem (3) approximately provided that the design requirements are satisfied with the given accuracy.

Theoretically it is well known that the Discrete Minimax Problem (3) is a non-differential optimization problem [1]. Moreover, in general case a convexity of the objective function is not guaranteed too. That is why traditional methods are not successful when solving such a problem [2]. So the special deterministic method has been developed to solve the nonconvex nondifferential optimization problem in its general statement. In time of solution the original objective function is substituted by a virtual smoothed function. The global minimization technique contains the following main steps:
- the objective function formulation;
- smoothing approximation of the objective function;
- search for a current local minimum;
- tunneling transformation and search for a new local minimum at which the objective function takes its lower value.

Benchmark results demonstrate numerical efficiency of this technique in comparison with such methods as fast simulated annealing and stochastic approximation combined with convolution smoothing.

The objective function in the optimal design problem is generated as follows:
- determine parameters of the stress-deformable state of structural elements for the current vector of design variables;
- evaluate the MRC using the analysis of stress and deformation data: $f_1(x) = K$;
- evaluate other current criteria, e.g.

$$f_2(x) = S \text{ (the cloth area);}$$
$$f_3(x) = F \text{ (required costs), et cetera;}$$

- formulate the objective function in a vector or in a scalar form:

$$f(x) = w_1 f_1(x) + w_2 f_2(x) + w_3 f_3(x) + \ldots,$$

where w_i is the weight of the i-th criterion.

Stresses and deformations are computed using the finite-element (FE) analysis program *NewTone* [3]. Sometimes the MRC of the frameless film reflector can be evaluated approximately on base of data on the main stresses field (Fig. 3). In this case it is possible to use empirical relationships between film reflecting properties and stress values. Other criteria of the current design quality can be evaluated by means of corresponding database approximations. The subject of designer

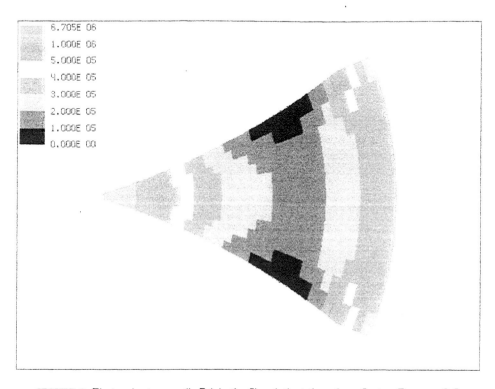

FIGURE 3. First main stresses (in Pa) in the film cloth ot the solar reflector *Znamya-2.5*

competence is determining the criterial set and evaluating the corresponding relative weights w_i.

3. SOFTWARE PRODUCT *REFLEX* FOR OPTIMAL DESIGN OF TRANSFORMABLE REFLECTORS AND CONCENTRATORS

There are some popular high-performance program products for the FE-analysis of structures, among them *NASTRAN*, *ANSYS*, *MARC*. These products give user powerful tools to solve optimization problems in a traditional form. Owing to well-structured interfaces they can be included practically into any integrated program system. But possibilities of the products are restricted by class of problems with convex differentiable functions. The interfaces use symbolic or binary files and do not allow fast sensitivity analysis processing when calls for solver are multiple. Besides these products have no any tool for evaluating of optical characteristics and other specifically features of reflectors or concentrators under design. That is why developing of the problem-oriented software product *REFLEX* is quite actual.

The main components of the *REFLEX* software product being developed for Pentium under Windows 95/NT are following (Fig. 4):

■ *optimization module* that builds desirable objective function and find its global minimum taking into account restrictions defined by the user;

■ *calculation module* that evaluates the objective function by components under given values of design parameters (structure and load characteristics, etc.) and other simulated factors;

■ *interfaces* with CAD-systems which provide translating data on initial design;

■ *user shell* that realizes user control at problem solution, connection with operation system, and data exchange between the modules.

At present the optimization module is developed in the form of subroutine group, which can be called by any other program. E.g., this module is built into an application program system *OPS* designed for the KED firm (Germany) in order to make an optimal setting of pipeline supports.

The structural analysis program *NewTone* completed with blocks for optical analysis is the main part of the calculation module. This program have been used in design calculations of the solar reflectors *Znamya-2*, *Znamya-2.5*, *Znamya-3*. *NewTone* contains build-in blocks for 3D structure modeling and visualization of results. It is compatible with pre-, postprocessors and other FE-programs in NASTRAN-format.

These features enable to investigate just the same models by different analysis instruments supplied with such the interface, and also to carry out the investigations by different teams of specialists connected through INTERNET.

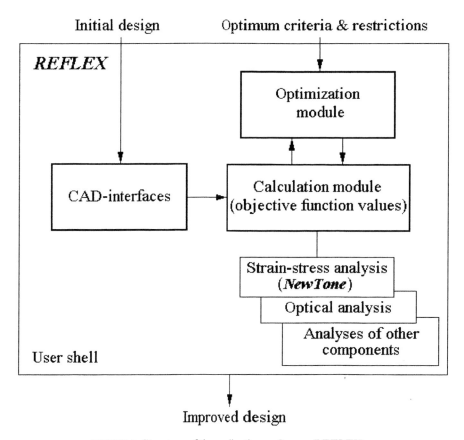

Initial design   **Optimum criteria & restrictions**

REFLEX

Optimization module

CAD-interfaces

Calculation module (objective function values)

Strain-stress analysis (*NewTone*)

Optical analysis

Analyses of other components

User shell

Improved design

FIGURE 4. Structure of the application software *REFLEX*

The designers of the **REFLEX** software started with writing of user shell and CAD-interfaces which are called to support convenient graphical working and monitoring through the whole of optimal design process.

REFERENCES

1. Fletcher, R., *Practical methods of optimization*, 2nd ed., Chichester et.al.: Wiley, 1987.
2. Pinter, J.D., *Global optimization in action. Continuous and Lipschitz optimization: Algorithms, implementations and applications*, Dodrecht et al.: Kluwer Academic Publishers, 1996.
3. Spirochkin , Yu., *NewTone/386. Finite-Elemente-Programmsystem f r die Berechnung nichlinearer Statik und Dynamik von Strukturen*, Kaliningrad: Dynamika Co. - Berlin: Dr. Krause Software GmbH - Rodenbach: KED, 1994.

A Study of the Correlation of EHE Cosmic Rays with Gamma Ray Bursts

Yoshiyuki Takahashi

Department of Physics, University of Alabama in Huntsville, Huntsville, AL35899, USA;
and LINAC Laboratory, RIKEN, 2-1 Hirosawa, Wako-shi, Saitama 351-01, Japan

Abstract. A study of space angles and temporal spacing was made for extremely-high energy (EHE) cosmic ray events to see if there are any correlation with Gamma Ray Bursts (GRB's) recorded by the BATSE experiment on the Compton Gamma Ray Observatory. The results on the most generic correlation using all the recorded GRB's and EHECR's show no significant correlation. Nevertheless, the highest-energy cosmic ray "pair" events observed by the AGASA experiments appear to be correlated with the very high fluence GRB's.

Some basis to form a GRB and a fireball is discussed. Empirical analysis of the GRB events strongly implied that the photonic field energy density in the source region should have exceeded the electric energy density of Schwinger field. A possible generation of an initial GRB, its fireball and relativistic shocks therein, is considered in terms of Schwinger field generated by radiation pressure of transient, high luminosity photons provided by collective nuclear collisions of neutron matter. Acceleration of electrons, and some protons, may be possible in the radial electrostatic Schwinger field. Ultra-relativistic shocks might also accelerate particles to certain high energies ($\gamma \leq 10^{12 \cdot 15}$). Neutral secondaries, including gamma rays, neutrinos, "strangelets," and Farrar's SUSY S_0 particles, are discussed as plausible EHECR pair candidates from GRB fireballs. The OWL/AIRWATCH may be able to explore them from 4×10^{19} eV to well beyond 10^{21} eV.

INTRODUCTION

Highest energy particle acceleration in the universe must account for the recent observation of air showers at energies greater than 2×10^{20} [1, 2]. Classical shock acceleration is so far limited by the maximum energy less than about 10^{19} eV. New models have emerged in the last decade to go beyond the propagation limit of 5×10^{19} eV, considering the ultra-relativistic shock [3] in fireballs [4] at the source of Gamma Ray Bursts, and decays of massive particles such as the Topological Defects [5]. Extremely high energy cosmic rays (EHECR) reported by AGASA group [6] indicate some

CP433, *Workshop on Observing Giant Air Showers from Space*
edited by J.F. Krizmanic et al.
© 1998 The American Institute of Physics 1-56396-766-9/98/$15.00

tantalizing mysteries of spatial and temporal correlation among themselves and possibly, with the Gamma Ray Bursts.

Gamma Ray Bursts [7] were finally observed at optical, x-ray and radio wavelengths as well, thanks to Beppo-Sax's fine-spatial resolution [8] to precisely locate the GRB's for optical search [9]. Interestingly, the optical and x-ray light curves of the counterpart indicated $t^{-(1.1 \sim 1.5)}$ decay, being consistent with Sedov phase fireballs, as expected by Meszaros and Rees [4] several years ago. This new evidence for fireball-like behavior of the GRB's [10] indeed supports an idea of Waxman model [11] that the GRB could be a very powerful acceleration site for cosmic ray particles. Milgrom and Usov [12] already reported in 1995 that some of the highest energy cosmic ray events had possible correlation with the high fluence GRB's. More recently, Vietri published rigorous arguments on generation of high energy neutrinos from GRB's [13]. The ultra-relativistic shock as introduced by Quenby and Lieu [3] seems promising in this context. Notably, the GRB's (and also AGN's) are believed to have shocks with Lorentz factor exceeding 10^{3-4}. Although the initial "bang" of the GRB's and creation of relativistic shocks are yet to be accounted for, optical counterparts seem to provide sufficient observational grounds for the existence of electron-dominant fireballs, *at least after the initial GRB event.*

Two more features of optical observation of GRB's are critically important to understand the GRB's and fireballs. KECK telescope data [14] provided the redshift of $0.835 < z < 2.3$ for GRB970508 event, which clearly indicated the cosmological distance (> 3.6 Gpc), and consequently, the extreme energy of the GRB's in the range of 10^{51-53} erg. The optical data by HST [10], X-ray data by Beppo-Sax, ASCA, ROSAT and R-XTE and radio data [15] gave the time profile over 7 months for GRB970228 and others. They showed an interesting time dilation of about several days to 60 days before reaching a peak.

A highly relativistic "fireball" was introduced [4] for GRB to solve difficulties to extract high energy gamma rays from an optically very thick environment at the presumed source region. Such a fireball concept has a long history in the Cosmic Ray Physics and multi-particle dynamics, since its introduction by Heisenberg and Wataghin in 1930-40's, and detailed developments by Fermi, Pomeranchuk and Landau in early 1950's.

Generally speaking, fireballs have been believed to be the GRB itself [4]. However, long delay of the optical peak and a subsequent, smooth time dependence of the light curve for a long time suggest to the present author that *a relativistic fireball may not necessarily be the direct cause of the GRB, but possibly, a consequence of GRB or other grand mechanisms.* The delayed optical peak was followed by a clear power-like ($t^{-1.1-1.5}$) decay-phase [10, 15], supporting the existence of the Sedov-phase fireballs [16] which have much less baryonic matter than electrons.

Motivated by the AGASA mysteries and recent developments of optical observation of GRB's, a physics concept study was performed for an potential opportunity to observe EHECR's and GRB's with the satellites, the Orbiting-array of Wide-angle Light-collector (OWL) [35, 36, 37] and the AIRWATCH [8, 37].

This paper reports two relevant studies regarding the EHECR and GRB's:

Firstly, a manifest of extremely high energy density of photons in GRB is described. Then, its possible relevance to the QED runaway by Schwinger field is suggested. A scenario for the onset of a GRB and subsequent generation of a Sedov fireball will be discussed in the following sequence.

(1) Schwinger field [17] as a possible cause of relativistic shocks, gamma rays, and fireballs in which protons can be accelerated to extremely high energies,

(2) Quark-Gluon Plasma (QGP) phase transition emits non-thermal gamma and meson radiation at the tangent of the rotating and merging neutron stars, which can produce a direct emission of GRB. Other particles hitting the neutron stars thermalize the merging matter, igniting the Schwinger field that can produce a fireball and relativistic shocks.

(3) A possible role of neutral secondaries for EHECR's: They can be produced by proton interactions with the circumstellar matter after high energy acceleration of protons.

Next, the temporal and spatial correlation of the GRB and EHECR data is examined. A generic analysis using all the GRB events of any luminosity turned out to have no evidence for correlation. High luminosity GRB's and the highest energy EHECR-pair events appeared to have a correlation, as Milgrom and Usov already pointed out [12].

VERY HIGH DENSITY PHOTONS IN GRB'S - A CRUDE SCENARIO

All the higher order vacuum polarization loops of e^+e^- will become equally probable and unstable (explosive) in the Schwinger field which can be reached by either a strong magnetic field over 5×10^{13} Gauss, or a high electric field generated by the high intensity of real photons. A magnetic field of the static kind alone is less likely to induce an instability of quantum vacuum, and will require a much higher magnetic field than the critical field strength to cause an explosion. The electric field, on the other hand, is very unstable for fluctuations, and could lead to an explosive instability even at a field lower than the critical value; in particular, when an independent high magnetic field is applied.

This QED catastrophic field was considered in the past for the high magnetic field beyond that of typical neutron stars. The merging neutron stars may increase the magnetic field towards (but lower than) the critical value for the QED breakdown. Some GINGA data on a few GRB's [18] showed Landau cyclotron lines, indicating some (but not very high) static magnetic field. BATSE data, however, didn't clearly confirm them. If fireballs (1) emerge only after the GRB emission or (2) do not disturb or mask the interior, cyclotron lines may be expected. The latter case (2) contradicts with rapid motion of matter in fireballs. The present BATSE data [19] favor an understanding that the majority of GRB's except some do not bear cyclotron lines. If these lines could be convincingly observed, it could indicate the former case, the static magnetic source of Landau lines *during the GRB, preceding an onset of a fireball.*

High electric field can create the Schwinger limit. The low energy heavy-ion collision experiments at GSI [20] and other laboratories have been trying to create a microscopic, transient, high charge state of a nucleus ($Z > 137$) for this limit, but a coalescence/fusion of two colliding nuclei barely succeeded to reach the electric Schwinger field except a

471

regular Coulomb effect on passing nuclei. Electromagnetic accelerators of terrestrial laboratories [21, 22, 23] recently began to see different acceleration frontiers of high intensity laser wakefield [24]. Wakefield and Snowplow acceleration indicate the highest acceleration rate per unit spatial displacement of particles by extremely high electric field generated from high intensity laser (~ 100 TeV/m at 10^{21} W/cm^2). This field may not be sustained in a large scale such as in the astrophysical settings, so long as we are concerned in the laser wakefield. Nevertheless, the high electric field does not have to be created by laser, but by incoherent photons, so long as there is high-enough intensity.

Considering the very high photon intensity at the GRB site, one might possibly relate them to a catastrophic explosion of QED vacuum. If ignited, it could cause a gamma ray burst, relativistic shocks, and an electron-dominant fireball (Sedov). Understanding of an extremely dense photonic field created in the laboratories is now approaching towards the Schwinger field, albeit the laboratory realization is ~ 10^5 times too low at present. Nevertheless, physics of high intensity plasma can suggest a possible catastrophe for making a GRB and a fireball, once a GRB's high photon energy density data is assumed.

Schwinger Field - Catastrophic Explosion of QED Vacuum

Quantum vacuum is characterized by the time constant (Δt_S), for which vacuum fluctuation, $\Delta \varepsilon \Delta t \sim \hbar$, takes the rest mass energy of electron (mc^2): $\Delta t_S = \hbar / mc^2 = 1.3$ x 10^{-21} sec . In terms of acceleration (a), the Schwinger field ($\varepsilon_S \sim 10^{26}$ erg/cm^3) realizes this process as follows with a vacuum polarization rate of 10^{48} ($\varepsilon/\varepsilon_S$)^2 pairs/(cm^3 sec),

$$(a_S) = c/\Delta t_S = mc^3/\hbar = 2.3 \text{ x } 10^{31} \text{ cm}^2/s \ (= 2.3 \text{ x}10^{29} \text{ g}), \qquad (1)$$

GRB's observed on earth have gamma ray fluence (L_{ob}) of 10^{-7} - 10^{-4} erg/cm^2. The average gamma ray energy ($<E_\gamma>$) is of an order of 800 keV, and the number density on earth is I = 1 - 1000/cm^2. We use GRB970508 as a reference, because it is typical (L_{ob} = 3 x 10^{-6} erg/cm^2) and have a known minimum distance. If we push this photon intensity back to the "source" of a sphere with a radius of ~1 light-milisec (300km), assuming the measured source distance (D ~ 1.49 x 10^{28} cm. [14]), we get the photon's column density ($n_\gamma^0 = L_{ob}/<E_\gamma>$ x D^2/R^2) and the energy column density ($\varepsilon' = n_\gamma^0 <E_\gamma>$):

$$n_\gamma^0 \geq 3 \text{ x } 10^{-6} \text{ erg/cm}^2 /1.6 \text{ x } 10^{-7} \text{ erg x } (3 \text{ x } 10^7 \text{ cm}/1.49 \text{ x } 10^{28} \text{ cm})^2$$
$$\approx 7.6 \text{ x}10^{43} \text{ } \gamma\text{-rays / cm}^2,$$
$$\text{or, } \varepsilon' \geq 1.2 \text{ x}10^{37} \text{ erg/cm}^2. \qquad (2)$$

The rise time of ~1 msec is what we are interested in with regard to the size of the region (R) where a GRB begins. Let's take R =1 msec x c (= 3 x 10^7 cm) for simplicity and estimate the photon density during this period: $n_\gamma^{msec} > n_\gamma^0$ x (1 msec/ 1 sec = R/c) ~ 7.6 x 10^{40} γ-rays/cm^2. If we take a radial spread of the source as R in unit of 300 km, the photon density (n^0) would reach $n^0 > 2.5$ x 10^{33} (300 km/R)^3 photons/cm^3, and the energy density(ε) would become $\varepsilon > 3$ x 10^{27} (300 km/R)^3 erg/cm^3, much larger than ε_S.

In the Peta-Watt laser plasma formation, the acceleration of the wakefield (a_w) reaches 10^{27} cm/sec^2 at the intensity of 10^{22} W/cm^2, where the number density [n] is ~ 10^{18}/cm^3. The acceleration (a) is proportional to the square root of the number density [n]. The photonic density in the GRB source is 10^{15} times larger than the Peta-Watt laser case. Therefore, the acceleration in the GRB source is inductively shown as higher than a_s:

$$a > 10^{34.5} \text{ cm/sec}^2 > a_s. \tag{3}$$

The radiation pressure, i.e., Thomson-Compton scattering of electrons by high density photons, is essentially responsible for a generation of high electric field, as in the Plasma Wakefield. Unlike Eddington limited luminosity of a star surface plasma, transient high density plasma field does not require Eddington condition for photon luminosity. The Eddington limiting luminosity is essentially due to a condition to maintain the luminous matter in a steady state, keeping the Debye length and equating the accelerating force of radiation with the decelerating force of gravity of a star. The *transient process* does not have to sustain materials. The transient luminosity is only bounded by the photon source luminosity, and is free from Eddington limit and Debye length at low temperature. This is why we can allow extremely high column density of photons in observed GRB's, exceeding the Eddington limit by orders of magnitude.

We already estimated the *GRB photon density at source* **empirically from the observed GRB fluence**, indicating that the GRB's photon density alone should have realized the Schwinger field at some GRB source region. In a GRB case, the electric Schwinger field could have been achieved in a *macroscopic scale* and the vacuum depolarization would cause an enormous explosion with extremely relativistic shocks. In this argument we ignored three important elements: existence of strong magnetic field and baryons, and e+e- opaqueness. The magnetic field might not help igniting an early explosion with photon density less than that for the critical Schwinger electric field. The effect of the bayons is unclear. However, we remark here that high intensity photons can be generated by synchrotron radiation from quarks, if Quark-Gluon Plasma were formed in a high magnetic field at source. About e+e- production with many photons, relative momentum difference of two photons > 1 MeV/c would still remain as a severe caveat.

Quark-Gluon Plasma Would Pump Photons and Hadrons

Schwinger field of macroscopic volume can cause ultra-relativistic shocks and an electron-dominant fireball. Once they are created, we can probably apply Quenby-Lieu's ultra- relativistic shock scheme [3] for particle acceleration in the fireball, as Waxman noted [11] and Vietri and others adopted [13]. However, questions remain on the first cause: how are the high density photons created? Answering this question with the current knowledge is by no means possible, and would perhaps be very much reckless. Nevertheless, there is a reason to consider it, when we accept arguments that there are significantly high probabilities for a merge of neutron stars in early galaxies, which, if integrated over the entire universe, can be as frequent as GRB's (> 1 merge per day).

Merging neutron stars have been studied by many people [25], particularly taking into account its hydrodynamical flow in strong gravitational tidal field during the merging process. The bulk of two neutron stars are shown to merge into a high mass state. From this picture, another gravitational collapse (supernovae or hypernovae [26]) is surmised. However, this scenario is rather simplistic, ignoring the internal disturbance of neutron matter, and plagued by an inability to produce a burst of predominantly gamma rays.

The interior super-fluid matter of a neutron star at merge do not co-rotate as a rigid body. Although it is v ~ 0.21c for a rigid, 1 msec neutron star with 1.4 solar mass, the tangential, rotational velocity at the merging edges is faster than 0.4c, due to the gravitational energy provided by the second neutron star. Collisions with the facing neutron matter beams are *massive and collective* as illustrated in **FIGURE 1**. This process finally converts the gravitational energy given by the second neutron star into the thermal energy of matter. This temperature of neutron matter (~ 210 MeV) is high

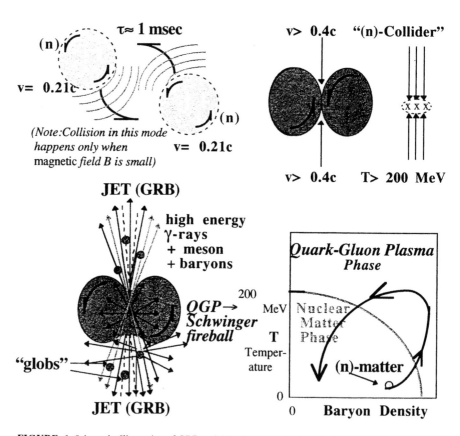

FIGURE 1 Schematic Illustration of GRB and QGP formation in the spinning merge of neutron stars.

enough to cause a phase transition into high density Quark-Gluon Plasma (QGP). More than 10% of its thermal energy will be converted into photons in its final state.

When two neutron stars come closer, gravity near the merging edges cancel out, making the outer crusts flying away like sprouting gases. In the inner crusts of neutron matter, superfluidity may be destroyed by the extremely aspheric and oscillatory (~ 1000 Hz) gravitational potential. Rapid oscillation of aspheric gravity (U_G^r) before merge could cause non-superfluid matter to convert about (> 1/3 ~ 1/2) of the pre-merge, radial gravitational energy (U_G^r ~ 180 MeV per neutron) into thermal energy. Superfluid matter would be tangentially accelerated in about 10^{-4} sec to > 0.4c, or kinetic energy (KE) ~ 210 MeV, before colliding with the merging matter beam. Due to a loss of effective radial gravity at the merging sides the virial theorem (KE < U/2) cannot hold and the neutron stars would partially break up. Merging, superfluid neutron matter would be in the "collider" mode at the CMS energy of (1.21 GeV/n x 1.21 GeV/n) if B-field is null.

Nuclear matter without having baryons is predicted by Quantum Chromo-dynamics (QCD) to make a phase transition into QGP at about 200 MeV [27]. Neutron matter with degenerate neutrons should turn into QGP phase at temperature lower than 200 MeV.

Although a pre-merge, steady-state neutron star has a rotational KE of only 40 MeV per neutron, the head-on collisions forced by this internal energy, or accelerated energy (~ 210 MeV), or high collective pressure is sufficient to turn neutron matter into a huge QGP at high baryon density, whose free expansion proceeds into the tangential plane.

The QGP droplets in strong magnetic field would be polarized both electrically and color-wise and make *plasmon* oscillations, emitting soft photons which go out of the surfaces. They would rapidly expand and cool down, going back to nuclear matter by emitting hadrons, leptons and photons for which one can even apply relativistic thermal equilibrium. *Neutron matter and QGP arerather transparent to low frequency photons.* Most of the thermal energy in dense neutron matter would be quickly lost by vv-bar emission and cooling until kT becomes less than ~500 keV. Low momentum photons may be able to carry thermal energy in an ideal neutron matter at below that temperature.

A neutron star having about 1.4 solar mass might keep at merge most of its 1.4 solar mass, in which ~ 2.8 x 10^{57} baryons are contained. Besides the colliding part, other merging "matter" are bombarded by the continuing collisions and by secondary particles at the merging common side and burn fast into QGP, which propagates into the rest of the matter. They receive a high pressure from the QGP volume and the increasing gravitational force, giving rise to high densities and instability in neutron stars. Assuming ~ 1 second for this process, or ~1,000 "scraping" spins, some photons could run away from the surface. (High orbital momentum is assumed for two neutron stars). The photons' remaining rate in the neutron stars would be larger than R/c = 10 km/(3 x 10^{10} cm/sec) = 3 x 10^{-5}/sec. If we assume in average only one photon emission per baryon from 1% of the neutron star during this fast process, the photonic density in the volume of (4π/3) x 10^{18} cm^3 would be ~ 10^{32}/cm^3. It gives the photonic acceleration field of the order of

$$a \sim 10^{34} \text{ cm/sec}^2 > a_s \ (=10^{32} \text{ cm/sec}^2),$$ (3)

which is also far larger than the required Schwinger field. Hadrons and electrons are completely ignored in this rough estimate of the photon density at merge. Here we also note that we neglected the extremely high magnetic field of $> 10^{12}$ Gauss. It is already close to the critical static field of 4.4×10^{13} Gauss, but would require a high electric field for depolarization of QED vacuum, if we consider the effective density $(E^2 - H^2)$. The electric acceleration rate (a) in eq. (3), or $\varepsilon > 10^{26}$ erg/cc, is high enough to create the critical Schwinger field and to ignite a catastrophic runaway.

We discussed, in the above paragraph, a very primitive picture of the QGP formation of bulk neutron matter in the tangential plane of rapidly-rotating and merging neutron stars. Multiple collisions and secondary hadrons of the central rapidity region, as well as very high pressure would also cause a fast phase transition of bulk neutron matter from each side of the neutron stars. However, all of them would be consumed in the thermalization inside the neutron stars, and most of which escape in terms of neutrinos. It can bring out only some low energy thermal photons but no high energy gamma rays.

GRB's and EHECR Acceleration at a Merging Site

High energy particles in tangential plane would be directly emitted into free space. The gravitational forces from each neutron stars momentarily cancel in this direction, and the "sawdusts" (or scraped droplets) or bulk flow would also escape. These blobs emit gamma rays very efficiently in their cooling process, as they have a large surface/volume ratio for an emission of gamma rays into space. Because this first generation of the secondary photons is most likely those of π° decays, direct thermal photons from QGP, and nuclear level transitions of leading baryons, their energies are of the order of 100 keV - 100 MeV. This part will be created in seconds by rotations of "sawing" edges.

If the line of sight partially meets the tangent plane of the merge, we would see GRB's with a harder energy spectrum. Otherwise, visible GRB's should mainly consist of lower energy gamma rays from the QGP blobs and others; and the emission angle of gamma rays from them should be nearly isotropic. A variety of fluctuations in energy spectrum and spiky temporal profile are naturally expected for GRB's in this scenario of a spinning merger that certainly bangs and wobbles rapidly by reactions of the collision.

This crude QGP scenario has very many caveats, and may not be correct. It is not very certain in what way high photonic density would be affected by the presence of strong magnetic field, high density quarks and gluons, and tremendous gravitational forces in the region.

There are two possible acceleration mechanisms in the present scenario for extremely high energy particles. One is ultra-relativistic shocks in a fireball, as were generally pointed out by many authors [3, 9, 11, 13]. Another, but remote, mechanism is a direct electrostatic acceleration by Schwinger field before an explosion. Its electric field is E ~ 10^{16} eV/cm. The Schwinger field hardly seems to extend to the size of the source size (R km) for a single proton acceleration. If ever it extends by unknown reasons, E_{max} can be

$$E_{max} = 2 \times 10^{22} \ (R/300 \text{ km}) \text{ eV}. \qquad (3)$$

NEUTRAL SECONDARIES FROM GRB'S FOR EHECR'S

GRB's can hardly be justified for the EHECR's without having any compelling experimental facts. Before examining the correlation of EHECR and GRB events for that purpose, we should first check if there are any reasonable candidate particles that are consistent with the observed EHECR's at above 4 x 10^{19} eV.

The AGASA data indicated that the highest energy events of these EHECR's were in pairs in a time span of 1 - 3 years. The chance coincidence rate to be observed as pairs among EHECR's were studied by AGASA simulations, which concluded its likelihood to be less than 2.9% for AGASA events, and 0.1% for the World-Wide data [5].

Protons would scatter in arrival time over 1,000 - 10,000 years during their propagation within a passage of ~ 30 Mpc, mainly due to local fluctuations of the intergalactic magnetic field of the order of about $10^{-(7 \sim 9)}$ Gauss. This time-spread for protons cast doubts on the nature of the particle species of the EHECR's, if "pairing" among EHECR's are seriously taken. Above all, it is inconsistent with the GZK cut-off beyond 30 Mpc [29] if we assume that protons reach us from a GRB, while the typical GRB distance is of the order of Gpc. These points would eliminate any possibility of GRB's for the origin of EHECR protons in normal sense. Furthermore, it should be noted that there are no experimental evidences that the EHECR's are indeed protons.

If they are long-lived neutral particles, e.g., gamma rays, neutrinos, strangelets, and SUSY S_0 particles [30], the EHECR's would not be cut off by the GZK process. Also, effects due to magnetic fields of intergalactic space are null for neutral charges. Thus, the observable EHECR's from GRB's must necessarily be the secondary neutral particles or those having very low charge/mass ratios ($z/m < 0.1$ e/GeV/c^2) like strangelets [31].

Gamma rays and neutrinos should arrive at earth without any measurable time delay, if they were indeed produced at the same instance as the GRB occurred. However, their parent protons would be produced in a fireball which could last as a powerful accelerator for a year. Because of this time-span, it is not right to assume that EHECR gamma rays and neutrinos should come to earth at the same time as the GRB arrives. It is, therefore, highly meaningful to investigate the temporal and spatial correlation of EHECR's with the observed GRB events within one year of delay.

The baryon number of strangelets could be as high as 100 - 1000, containing a lot of strange quarks (- 1/3e). The GZK cut-off threshold energy for them would be greater than 10^{21} eV. The SUSY S_0 (a bound state of uds + gluino) does not interact with photons. Both S_0's [30] and "strangelets" [31] are free from GZK cut-off and deflection for propagation through a distance of 1 Gpc [30]. Their hadronic mean free path is about 100 g/cm^2 and are capable to create air showers upon arrival at the earth's atmosphere.

POSSIBLE CORRELATION OF EHECR'S WITH GRB'S ?

Thus, these "neutral secondaries" from GRB fireballs could reach earth within ~1 year after the arrival of the GRB itself. In the following, we examine the correlation of the EHECR and GRB data for the observed period from April, 1992 to October, 1995.

Using the AGASA data on 27 events, we searched for GRB events from the Second [32] and the Fourth BATSE Catalog [33]. The time window for the search was within 1 year prior to, (and posterior to), each observed EHECR event. For evaluating the random background rate, the search was divided into four equal time segments of $0 \sim 6$ months, $6 \sim 12$ months, $-(0 \sim 6)$ months, and $-(6 \sim 12)$ months. In this search we considered three different sizes of the error regions of measured GRB events relative to the EHECR angles of Right Ascension and Declination. They are $5°$, $10°$, and $15°$. The chance coincidence rate was estimated by using the actual GRB data, taking into account the BATSE's orbital correlation of viewing the GRB's and isotropic nature of the spatial distribution of events. No fluence cut was made for BATSE GRB's for generic search.

We subsequently performed a similar search by limiting the burst size to the top 100 events within this period. We consider that the generic search would be strongly overshadowed by high background coincidence due to a very high population of weak and distant bursts that could barely deliver high flux of EHECR to earth. GRB's with high luminosity (that might be close to our galaxy) would have higher signal-to-noise ratio for the EHECR-GRB correlation search. Just from these primitive probabilistic considerations on the EHECR flux, the high luminosity GRB search was made.

Statistical Summary of Generic Correlation

The results of the generic search are listed in **TABLE 1**. As it clearly shows, there was no significant correlation of GRB's with the EHECR events. The probability of random coincidence to reproduce the number of GRB's found within the given space-time window of search can be summarized as 84%, consistent with zero correlation.

A remark of caution is due, however. Random coincidence rate for a generic search is very high, 24.6% ($< 5°$) per event, and this search is very insensitive to signals, even if there are some signals. To increase the sensitivity of generic correlation test to a sufficiently high level, it is crucial to improve the experimental angular resolutions for both EHECR's and GRB's to better than 1 degree.

Teshima et al. just recently tried to find a generic correlation of GRB's with the entire AGASA events at lower energies (10^{17} eV - 4×10^{19} eV). Their preliminary results were said [34] to have no correlation with GRB's, as expected by the arguments in our generic correlation analysis.

TABLE 1 Statistical summary of the search for generic correlation

A, B, C denote the degree of the angular similarity of GRB candidates for the EHECR events.
Time spacings are $\Delta t < 0 - 6$ Months for A,B, and C; $6 < \Delta t < 12$ Months for D, E, and F.

GRB's	A < 5°	0° < B < 10°	10° < C < 15°	D < 5°	5° < E < 10°	10° < F < 15°
before EHECR	8	16	32	7	18	31
after EHECR	8	26	24	4	17	40
Expected	6.7	20	33.3	6.7	20	33.3

TABLE 2 EHECR pairs and GRB's 0 - 12 months prior to EHECR first event. A,B, C denote the class of the angular similarity, as in Table 1. GRB events underlined are high-luminosity events.

(I) Pair-1

PAIR 1	log $_{10}$ E (eV)	G-Longitude	G-Latitude
AGASA 910819	19.79	165.64	10.39
FE 911015	20.477	163.4	9.4

Δangle (A,B,C), Δt (< t months)	BATSE Trigger No.	BATSE Catalog No.	G-Longitude (deg)	G-Latitude (deg)
B; (3, 5)	*143*	*GRB910503*	*172.030*	*5.79*
B; (-1/4, 2)	727	GRB910826	156.370	0.81
A; (-1.3, 0.5)	840	GRB910930	165.420	13.47

(II) Pair-2

PAIR 1	log $_{10}$ E (eV)	G-Longitude	G-Latitude
AGASA 920801	19.74	143.45	56.92
AGASA 950126	19.89	145.75	55.32

Δangle (A,B,C), Δt (< t months)	BATSE Trigger No.	BATSE Catalog No.	G-Longitude (deg)	G-Latitude (deg)
C; (0, 30)	*1733*	*GRB920801*	*139.460*	*43.46*
C; (4, 34)	1556	GRB920418	141.800	44.62
C; (6, 36)	*1425*	*GRB920221*	*138.220*	*67.60*

(III) Pair-3

PAIR 1	log $_{10}$ E (eV)	G-Longitude	G-Latitude
AGASA 931203	20.33	131.24	-41.08
AGASA 950126	19.89	145.75	55.32

Δangle (A,B,C), Δt (< t months)	BATSE Trigger No.	BATSE Catalog No.	G-Longitude (deg)	G-Latitude (deg)
A; (4, 27)	*2450*	*GRB930720*	*131.950*	*-36.980*
C; (6, 29)	2391	GRB930612D	146.420	-27.680
C; (8, 31)	2304	GRB930415	144.100	-42.670
C; (8, 31)	2283	GRB930404	146.610	-43.510
B; (10, 33)	2141	GRB930121	139.450	-49.92
A; (13, 36)	2110	GRB921230	131.980	-45.19

479

Possible Correlation of EHE-Pair-Events with High Fluence GRB Events

The generic correlation search is in principle hopeless with the current large experimental uncertainties on angles. Hence, we consider a further study, limiting the GRB's to the highest fluence events (100 among ~1,000) in correlation with the striking EHE-pair events. **TABLE 2** lists 3 AGASA pair events and GRB candidates for them without applying a fluence cut. The high fluence GRB events are underlined with *italic-bold* types.

The result is interesting in the sense that these three pairs have high fluence GRB's, and their association cannot be easily accounted for by random coincidence. The overall probability that the random coincidence can reproduce these correlations is less than 1.4%. If we combine this probability with the fact that these "pairings" of EHECR's within themselves had a probability less than 2.9%, the observed EHECR-GRB phenomena would turn out to be of a very small likelihood of 0.04% for chance coincidence.

SUMMARY

Some interesting correlation of GRB's are indicated only for the highest energy AGASA pair events of EHECR at above the Greisen-Zatsepin-Kuzmin cut-off energy, 4×10^{19} eV. Generic correlation survey indicated no correlations, but it might be largely due to the high background rate caused by poor angular resolutions for the required sensitivity.

A primitive, qualitative scenario of the cause of the Gamma Ray Bursts is considered with a hope of accounting for high energy particle acceleration in a fireball after the onset of a GRB but within about one year. Quantum Chromo-dynamics phase transition of neutron matter into Quark-Gluon Plasma is considered for a rapidly-rotating and merging pair of mili-second neutron stars. It was shown that QGP could generate a part of GRB at impact and could also realize some Schwinger field from the bulk neutron matter. It could then produce a GRB, ultra-relativistic shocks and an explosive electron fireball.

By comparing this to an analogous acceleration of electrons in the high intensity plasma field, a possible collapse of QED is discussed in terms of Schwinger field for very high photon densities when provided by QGP phase transition. It was discussed that the macroscopic, large-scale Schwinger field could cause an instantaneous onset of ultra-relativistic shocks and a macroscopic Sedov fireball from QED vacuum. Acceleration of extremely high energy protons could be achieved in a fireball that has ultra-relativistic shocks. Neutral secondaries, e.g., gamma rays, neutrinos, neutral (or low-charge) strangelets and SUSY S_0's, produced by proton interactions near the acceleration site, are positively discussed as *prominent and observable* EHECR's beyond the GZK cut-off energy for the required propagation range of > 100 Mpc, allowing the source distance even exceeding 1 Gpc. These neutral particles are also most promising for EHECR pairing and their correlation with GRB's within 1 year.

Future investigations are needed with high statistics of EHECR's and appropriate angular resolution for GRB's. The OWL/AIRWATCH can observe _both_ EHECR's and GRB's by the same apparatus [35, 36, 37]. An emphasis of high angular resolution in the design studies is recommended for the OWL/AIRWATCH experiments that could collect thousands of EHECR's beyond 10^{20} eV and GRB's to as low as 10^{-8} erg/cm^2.

ACKNOWLEDGMENTS

This work is supported by NASA (NAS8-38609 D.O. 161, NCC8-65 Mod. 6, and NAG5-3905).

The author is indebted to a number of colleagues for helpful and stimulating discussions and cooperation. All the errors and inappropriate statements that might be found in this article are, however, solely author's responsibility.

The author would like to thank Dr. T. Tajima of the University of Texas, particularly, for discussions and suggestions of extremely high intensity plasma. He is indebted to Dr. C. Meegan and Dr. G. Fishman of NASA/MSFC for their independent survey of the GRB-EHECR correlation, as well as for their cautious advice. Dr. Nagano of the ICRR, the University of Tokyo, kindly provided the AGASA data for this study. Dr. M. Teshima of the ICRR kindly informed the author of their preliminary analysis performed after the present work, which were also negative results on the generic correlation of GRB's with the low energy (10^{17} - 4 x 10^{19} eV), unpublished AGASA events. The author is grateful to Dr. G. Farrar for her stimulating discussions and explanations of the S_o candidacy.

Thanks for discussions are also due to Dr. L. Scarsi of the University of Palermo, Dr. John Linsley of the University of New Mexico, Dr. R. Lieu and J. van Paradijs of the University of Alabama in Huntsville, Dr. T. Kajino of National Astronomical Observatory, Dr. T. Ebisuzaki of RIKEN, and Dr. T. Kifune of University of Tokyo.

REFERENCES

1. N. Hayashida et al., Phys. Rev. Lett., 7 3, 3491 (1994); Phys. Rev. Lett., 7 7, 1000 (1996); S. Yoshida et al., Astropart Phys., 3, 105 (1995); M. Nagano, this volume (1998).
2. D.J. Bird et al., ApJ 4 2 4, 491 (1994); P. Sokolsky, this volume (1998).
3. J.J. Quenby and R. Lieu, Nature 3 4 2, 654 (1989); R. Lieu and J.J. Quenby, ApJ 3 5 0, 692 (1990).
4. M. Rees and P. Meszaros, Mon. Not. Roy. Astron. Soc. 2 5 8, 41P (1992); ApJ 476, 232 (1997).
5. C. T. Hill, Nucl. Phys. B 2 2 4, 469 (1983); C.T. Hill, D. Schramm and T.P. Walker, Phys. Rev. D 3 6, 1007 (1987); G. Sigl et al., Science, 2 7 0, 1977 (1995); G. Sigl., Space Sci. Rev. 7 5, 375 (1996); P. Bhattacharjee, G. Sigl, D. Schramm, each in this volume (1998).
6. N. Hayashida et al., Phys. Rev. Lett. 7 7, 1000 (1996); Y. Uchihori et al., Proc. Int. Symp. on EHECR, Tokyo, ed. M. Nagano, 50 (1996).
7. G.J. Fishman and C.A. Meegan, Ann. Rev. Astron. Astrophys. 3 3, 415 (1995).
8. L. Scarsi, this volume (1998); Beppo-Sax team, E. Costa et al., Nature 3 8 7, 783 (1997).

9. J.van Paradijs et al., Nature, **386**, 686 (1997).

10. T. Galama et al., Nature 387, 479 (1997).

11. E. Waxman, ApJ **452**, 1 (1995); E. Waxman and P. Coppi, ApJ **464**, L75 (1996); E. Waxman and J. Bahcall, Phys. Rev. Lett, **78**, 2293 (1997).

12. M. Milgrom and V. Usov, ApJ **449**, L37 (1995).

13. M. Vietri, Phys. Rev. Lett., submitted, 1997; ApJ, **453**, 883 (1995).

14. M.R. Metzger et al., Nature **387**, 878 (1997); S.G. Djorgovski et al., Nature **387**, 876 (1997).

15. E. Costa et al., Nature **387**, 783 (1997); D.A. Frah et al., Nature, **389**, 261 (1997); T. Murakami, private communication (1997).

16. L.I. Sedov, "Similarity and Dimensional Methods in Mechanics," Academic Press, New York, 1959.

17. J. Schwinger, Phys. Rev. **82**, 664 (1951); Proc. Nat. Acad. Sci. **40**, 132 (1954).

18. T. Murakami et al., Nature **290**, 378 (1990).

19. D.M. Palmer et al., Proc. Gamma Ray Bursts, AIP **307**, 247 (1994); M. Briggs, Proc. 4th Huntsville GRB Symposium, AIP (in press), 1997.

20. P. Kienle, Nucl. Phys. **A478**, 297 (1988); M. Clemente et al., Phys. Lett. **137B**, 41 (1984).

21. A. Modena et al., Nature **377**, 606 (1995); M. Everett et al., Nature **368**, 527 (1994).

22. K. Nakajima et al., Phys. Rev. Lett. **74**, 4428 (1995); KEK preprint, (1997).

23. "Laser-Plasma Acceleration of Electrons using the Peta-Watt Laser," ,T. Cowan et al., LLNL preprint (1998); S. Wilks et al., UCRL-JC-126964 (1997).

24. T. Tajima and J.M. Dawson, Phys. Rev. Lett. **43**, 267 (1979); T. Tajima, Laser & Particle Beams, **3**, 351 (1985).

25. D. Eichler, M. Livio, T. Piran, D.N. Schramm, Nature **340**, 126 (1989); B. Paczynski, Acta Astronomica, **41**, 257 (1991); R. Narayan et al., ApJ **395**, L83 (1992)

26. B. Paczynski, Proc. of 4-th Huntsville Gamma-Ray Burst Symposium, 1997, AIP (in press).

27. T. Celik, J. Engels and H. Satz, Phys. Lett. 133B, 427 (1983); **125B**, 411 (1983); J. Kogut et al., Phys. Rev. Lett. **50**, 353 (1984).

28. R. Hagedorn, Supple. Nuovo Cimento **6**, 311 (1968); Astron. Astrophys. **5**, 184 (1970); K. Huang and S. Weinberg, Phys. Rev. Lett. **25**, 895 (1970); J. Wheeler, ApJ **169**, 105 (1971).

29. K. Greisen, Phys. Rev. Lett., **21**, 1016 (1966); G.T. Zatsepin and V.A. Kuzmin, JETP Lett. **4**, 78 (1966).

30. D. Chung, G. Farrar and E. Kolb, Astron-ph/9707036, Phys. Rev. D (submitted 1997); G. Farrar, this volume (1998). .

31. E. Witten, Phys. Rev. D**30**, 272 (1984);E. Fahri and R.L. Jaffe, Phys. Rev. Lett. **34**, 1353 (1984).

32. "Compton Observatory BATSE Second Burst Catalog," The Compton Observatory Science Support Center and the BATSE Instrument Team, NASA, 1995.

33. "Compton Observatory BATSE Fourth Catalog," The Compton Observatory Science Support Center and the BATSE Instrument Team, NASA, 1997.

34. M. Teshima et al., private communication.

35. Y. Takahashi, Proc. 24th International Cosmic Ray Conference, Rome, **3**, 595 (1995); *MASS/AIRWATCH Huntsville Workshop Report*, 1 - 16, Univ. of Alabama in Huntsville (1995); Y. Takahashi et al., SPIE **2806**, 102 (1996); Y. Takahashi et al., Proc. Int. Symp. on EHECR , ed. Nagano, 310 (1997); Y. Takahashi, "OWL efforts in Japan," this volume (1998).

36. B. Sacco, Proc. of First AIRWATCH Symposium, Palermo, 1997; L. Scarsi, this volume (1998).

37. R. Streitmatter et al., "OWL,"; C.N. De Marzo et al., "AIRWATCH," this volume (1998).

On the Origin
of Ultra High Energy
Cosmic Ray Particles

K.O. Thielheim
University of Kiel
Kiel, Germany[1]

Abstract. Understanding physical mechanisms by which Nature succeeds in bestowing 'macroscopic' amounts of energy on single 'microscopic' particles remains one of the greatest challenges for both, astrophysicists seeking to understand the functioning of cosmic objects, and accelerator designers inventing new conjectures for more powerful machines.

The first and introductory chapter of this contribution is a short retrospective on our early research work, in the 'Mathematical Physics Division' at Kiel, on the origin of Ultra High Energy Cosmic Ray Particles (UHECRP), beginning in the sixties and seventies with studies on Atmospheric Particle Propagation, proceeding in the seventies and eighties to investigations on Galactic Particle Transfer, and leading in the eigthies and nineties to our present work on mechanisms for particle acceleration in pulsar magnetospheres. Cosmic rays, more than many fields of research, are found to be closely related to other branches of physics[2].

When looking on powerful cosmic accelerators from the point of view of fundamental physics, one has to study charged particle dynamics in extremely strong electromagnetic fields, of a kind that is expected near the surface of rapidly rotating, strongly magnetized neutron stars. Conventional Maxwell Theory (MT), for clearly discernable reasons, does not provide an adequate theoretical means of description in this case. Selfconsistent Electrodynamics (SCED), distinguished from MT through some of its essential premises offers itself, alternatively, as an appropriate language. Therefore, in the second chapter of my talk, I shall outline some features of SCED and reproduce the equation of particle motion on these grounds.

In the third and concluding chapter, I shall discuss a mechanism to create very narrow bundles of energetic particles in the polar regions of aligned rotators[3].

---

[1] Postal address: Arbeitsgruppe Mathematische Physik, Universität Kiel, PO-BOX 51 51, 24 063 Kiel, Germany, Email: thielheim@email.uni-kiel.de .

[2] This retrospective is presented here with reference to the fourtieth anniversary of the former Institut für Reine und Angewandte Kernphysik, Kiel, Germany, on May, 1^{rst}, 1997, which -in the four decades of its existence- and with its Divisions for 'Hochenergie-Physik', 'Mathematische Physik', and 'Extraterrestrische Physik' in various ways has contributed to cosmic ray research.

[3] Report financially supported under Proj. Ref. Numb. INTAS-RFBR 95-301.

CP433, *Workshop on Observing Giant Air Showers from Space*
edited by J.F. Krizmanic et al.

LOOKING BACK:
FROM THE SURFACE OF THE EARTH
TO THE SURFACE OF A NEUTRON STAR

Atmospheric and Galactic Particle Propagation

Simultaneously with the contruction of an Extensive Air Shower (EAS) experiment by the 'Hochenergie' group at Kiel in the early sixties[4], we have developped three-dimensional Monte-Carlo codes from corresponding techniques of neutron transport theory for simulations of the hadronic component of EAS[5] investigating, thereby, possibilities to estimate the mass composition of primaries from observable parameters at later stages of cascade development, for example, from the multi-core structure of EAS registered at sea level in the Kiel Neon hodoscope[6]. We then reached the conclusion that very light primaries, e.g. Hydrogen-induced showers, should *in principle* be distinguishable from very heavy primaries, e.g. Iron-induced showers, if more precise data would become available from accelerator experiments on high energy hadron collisions[7].

In that context, R. Zöllner and myself also have developed numerical methods for the integration of Electromagnetic Cascade Equations from analogous procedures of radiative transfer theory[8]. Results for longitudinal cascade development, computed with the help of QED cross-sections, then were used to estimate the range of applicability and accuracy of earlier analytical approaches obtained from Mellin-Laplace transformations of cascade equations, in what was then called approximations A and B, respectively.

---

[4] E. Bagge, E. Böhm, R. Fritze, U. Roose, M. Samorski, C. Schnier, R. Staubert, K.O. Thielheim, J. Trümper, L. Wiedecke, and W. Wolter, Proc. Int. Conf. Cosmic Rays, London, England **2**, 738 (1965).

[5] K.O. Thielheim and S. Karius, Proc. Int. Conf. Cosmic Rays, London, England **2**, 779 (1965); K.O. Thielheim and R. Beiersdorf, Journ. Phys. A **3**, 341 (1969). When we reported on that work first at the 1965 ICRC in London, we learned that H. Bradt and his colleagues had chosen a similar computational approach, in association with the Bolivian Air Shower Joint Experiment: [H. Bradt, M. La Pointe and S. Rappaport, Proc. Int. Conf. Cosmic Rays, London, England, **2**, 651 (1965)].

[6] K.O. Thielheim, Nucl. Instr. Meth., **31**, 341 (1964).

[7] The author wishes to apologize for that it will not be possible, within given limits of this lecture, to discuss or even just to mention all the numerous and relevant results that have been published by other authors on this topic, as well as on further problems considered here.

[8] K.O. Thielheim and R. Zöllner, Journ. Phys. **5**, 1054 (1972). Preceeding work on radiative transfer, aiming at a more stringent deduction of Diffusion Theory from Transport Theory and at an extension of the range of applicability of the former, later also was found to be useful in the context of *diffusive* Cosmic Ray propagation in irregular fields: [K.O. Thielheim, Nucl. Sci. Eng. **20**, 111 (1964)]. First Monte-Carlo codes for *electromagnetic* cascade development were developped even earlier: [R.R. Wilson, Phys. Rev. **86**, 261 (1952)].

Proceeding from atmospheric to galactic particle propagation, we then designed a phenomenological model of the *mean* galactic magnetic field structure, relevant for *deterministic* particle transport to be described by *individual particle trajectories*, from the possible locations of their origin through interstellar space to the position of the solar system[9].

The persistence of a *well-ordered* component of the galactic magnetic field suggested by the above-mentioned model was primarily deduced from concepts about how the magnetic disk-field, originating from earlier stages of galactic evolution and 'frozen' into the interstellar plasma, is continuously regenerated by differential rotation.

At the same time, dissipative effects, births and deaths of stars for example, permanently reproduce the *irregular* field component relevant for *stochastic* particle transport to be described by *particle diffusion* inside the galaxy[10].

Exploring the regime in parameter space where deterministic motion merges with diffusive motion, we have shown that the limits of applicability of the single particle approach can be expressed in terms of the mean energy density and the auto-correlation length of the stochastic field component[11].

When constructing the above-mentioned model of large-scale magnetic disk field topography, it was also necessary to adjust its parameters to observational astrophysical data, e.g. of gamma-ray background[12], synchrotron background[13], Faraday rotation[14], and the scattering of star light by elongated dust particles partially aligned to the galactic magnetic field[15] within the local region, i.e. in a range of distance up to several 100 pc from the location of the Earth.

Since the galactic magnetic field is 'frozen' into the interstellar gas, while the spatial density distribution of the latter is governed by the structure of the gravitational field generated, essentially, by the density distribution of the stellar component to

---

[9] K.O. Thielheim and W. Langhoff, Proc. Phys. Soc. (London), **A**, **1**, 694 (1968). When computing galactic particle trajectories we used a method to fit together continuously pieces of *exact* analytical solutions for magnetic fields, locally homogenous (in space) and constant (in time), rather than to apply more conventional procedures for straight foreward numerical integration of the equations of motion. Analogous methods were applied later, when computing particle trajectories under the influence of radiation reaction forces, in more complicated field topographies of rotating magnetized neutron stars, as will be discussed later in this contribution.

[10] Again, we refer to footnote 8.

[11] K.O. Thielheim, Journ. Phys., **A 7**, 444 (1974).

[12] R. Schlickeiser and K.O. Thielheim, Astronomy and Astrophysics **34**, 167 (1974).

[13] C.E. Jaeckel, K.O. Thielheim and H. Wiese, Astrophysics and Space Science **51**, 329 (1977).

[14] D. Nissen and K.O. Thielheim, Astrophysics and Space Science **33**, 441 (1975).

[15] [K.O.Thielheim, Proc. Int. Symp. on Progress in Cosmology, Oxford, England, 14-18 September, 1981, 305.] These investigations on the reflection of star light by elongated interstellar dust particles related to studies which we performed, by that time, on the scattering of electromagnetic waves from rough surfaces and irregularly shaped particles: [R. Schiffer and K.O. Thielheim, Journ. Appl. Phys. **50**, 4 (1979); K.O. Thielheim, Fundamentals of Cosmic Physics, **12**, 75 (1987)], with possible applications to the calculation of radar/lidar cross-sections, and to studies on the properties of artificially bifringent optical material: [K.O. Thielheim, DE 36 42 897 C2; D. Klusch and K.O. Thielheim, Waves in Random Media, **5**, 329 (1995).]

which it responds, we also found it useful to study mechanisms responsible for the formation of galactic spiral structure.

Different from what had been widely accepted by that time[16], we then suggested that gravitational interaction among stars inside the galactic disk is of lesser relevance for the generation of spiral arms, and that the formation and persistence of spiral structure is an (essentially linear) response phenomenon stimulated by the (essentially non-linear) evolution of a central, rotating, non-axially symmetric 'oval' mass configuration[17]. The functioning of the underlying symmetry-breaking mechanism, which exhibits some analogy to the symmetry breaking in a driven harmonic oscillator, could be demonstrated by (two-dimensional) N-body simulations[18].

With the help of correspondingly designed models of the well-ordered component of the magnetic disk field, we then evaluated trajectories (in the appropriate regime of rigidity) of *anti-primaries* of EAS propagating from the Earth back to possible locations of UHECRP sources inside our galaxy. For well-known reasons, rotating magnetized neutron stars are possible candidates for UHECRP acceleration [12].

Plasmaborder of Rotating Magnetized Neutron Stars

Even before observational evidence for the existence of pulsars became available [11], certain features of surrounding plasma configurations had been clearly predicted: 'Once the electromagnetic waves are emitted and propagate in the supernova remnant, they will be reflected by the circumstellar gas if the plasma frequency exceeds the radio frequency'[19] [9].

Thus, non-aligned rotating magnets embedded in ambient plasma are predicted to clean surrounding space from all electrically charged particles up the outmost layer surrounding the rotator, at a certain distance r_P from the centre, referred to as the *plasmaborder* (PB), - if the angular velocity of rotation ω is sufficiently high, and if the magnetic dipole moment μ is sufficiently strong, and if the ambient plasma density is sufficiently low.

---

[16] C.C. Lin and F.H. Shu, Proc. Nat. Acad. Sci., **55**, 229 (1966).

[17] K.O. Thielheim, Astrophysics and Space Science **73**, 499 (1980). Accordingly, spiral density waves cannot be understood as 'solitons' of the stellar density distribution propagating inside the galactic disk. The non-axially symmetric 'oval' mass configuration was seen in analogy to a certain type of *triaxial elliptical galaxies*. These investigations on non-linear spiral wave propagation paralleled our studies on 'solitons in distensible tubes' [K.O. Thielheim, Journ. Appl. Phys., **54**, 3036 (1983)] intended to relate observable parameters of systolic blood propagation to physiologically relevant parameters of arterial walls.

[18] K.O. Thielheim and H. Wolff, Ap.J. **276**, 135 (1984). Looking back on that early work, it appears more likely to me now that a small orbital motion of the central ovoid relative to the rest of the galactic disk might have an even stronger influence on spiral density response than an adiabatic change of the quadrupole moment of the central oval mass configuration can exert.

[19] M.J. Rees suggested an analogous mechanism for certain types of galactic nuclei: [Nature,**229**, 312 (1971)].

We have analysed mechanisms underlying the formation of the PB[20] [5–7], calculated its location r_P and internal structure as well as particle acceleration and generation of synchrotron radiation, distinguishing two cases, namely, that the *ambient plasma* is (a) *comparatively thin* or, alternatively, (b) *comparatively dense*. In more detailed investigations, it is necessary to distinguish between two further alternatives, namely, that the outgoing spherical electromagnetic wave field can be either (A) a *vacuum* wave (if most of the outgoing particles *have* decoupled from the e.m. wave field) or else (B) a (possibly non-linear) *plasma* wave (if most of the outgoing particles *have not* decoupled from the e.m. wave field).

Individual particle dynamics was studied in vacuum fields of a rotating, homogeneously magnetized, ideally conducting sphere[21]. *Individual* as well as *collective* (plasma) particle dynamics was investigated within *selfconsistent* plasma fields.

The location r_P of the PB was estimated with the help of certain topological properties of reflected particle orbits which, in the case of *comparatively thin* ambient plasma, are expressed in terms of the *osculation theorem* [5] and, alternatively, in the case of *comparatively dense* ambient plasma, in terms of the *oscillation theorem* [7]. Obviously, in the latter case, particle orbits cooperate in the formation of *plasma oscillations*, as described by an appropriate dispersion relation. Still another way of looking at the formation of a PB, in the case of *comparatively dense* ambient plasma, is that the radiation pressure of the outgoing spherical wave compensates the pressure of the ambient plasma [9]. Recently, we have reproduced the formation of the PB in relativistic N-body plasma simulations[22], which largely confirm preceeding calculations.

---

[20] K.O. Thielheim, ApJ., **409**, 333 (1993); Nucl. Phys. **B 33**, 181 (1993); In: Ilmar Ots and Laur Palgi, (Eds.), Proc. 2nd Tallinn Symp. on Neutrino Astrophysics, Lohusalu, Estonia:, 138 (1993);

[21] This configuration is referred to as *inclined or oblique rotator* if $\chi \neq 0, \pi/2, \pi$, as *orthogonal rotator* if $\chi = \pi/2$ or, respectively, as *parallel (antiparallel) rotator* if $\chi = 0$ (π), as *aligned rotator* if $\chi = 0$ or $\chi = \pi/2$, where χ is the angle of inclination between the vector of angular velocity $\boldsymbol{\omega}$ and the vector of magnetic dipole moment $\boldsymbol{\mu}$. The electromagnetic *vacuum* field of inclined rotators has been calculated by A.J. Deutsch: [Ann. d'Astrophys. **18**, 1 (1955)] and by O. Kaburaki: [Astrophysics and Space Science, **58**, 427 (1978); **67**, 19 (1980).].

The following parameters, all with the dimension of a length, are useful to characterize this configuration: the stellar radius r_N, the radius of light cylinder $r_L = c/\omega$, the typical radius $r_T = \sqrt{e\mu/mc^2}$, the classical particle radius $r_e = c\tau_0$, and the gravitational radius $r_G = GM_S/c^2$ of the rotating neutron star. Numerical values given below, if not stated otherwise, are for the 'standard set of parameters', i.e., for the radius of the neutron star: $r_N = 10^6 cm$, for the light radius: $r_L = 4.8 \cdot 10^8 cm$, for the typical radius: $r_T = 2.4 \cdot 10^{13} cm$ (electrons) and $r_T = 5.6 \cdot 10^{11} cm$ (protons), corresponding to the frequency of rotation $\nu = 10$ sec^{-1} and to the magnetic dipole moment $\mu = 10^{30}$ Gcm^3. For comparison: The magnetic dipolmoment of the Sun is about: $\mu_S \cong 10^{32}$ Gcm^3 The magnetic dipolmoment of a sphere made of neutron matter with all neutrons aligned in saturation would be $\mu^{sat} \cong 3$ M_N μ_N $/m_N = (-)$ $3.5 \cdot 10^{34}$ Gcm^3, where $M_N = 1.98 \cdot 10^{33} g$ is the mass of a Neutron star (which, in the standard set of parameters, is adopted to be equal to the mass of the Sun), $\mu_N = (-)$ $0.98 \cdot 10^{-23}$ Gcm^3 is the magnetic dipolemoment of the neutron, and $m_N = 1.68 \cdot 10^{-24}$ g is the mass of the neutron.

[22] C. Greve and K.O. Thielheim, to be published. Preliminary results are given in [7].

487

The distance r_P of the PB from the centre of the rotating magnet is so large and, consequently, the electromagnetic field strength inside the PB so low that (Lorentz-) acceleration is unable to generate UHECRP. It is necessary, therefore, to consider particle dynamics further inside the magnetosphere. Within that regime, radiation reaction forces play an essential rôle in particle acceleration and, therefore, have to be taken into account in the equations of motion.

FROM MAXWELL'S TO SELF-CONSISTENT ELECTRODYNAMICS

Principle of Selfconsistency

The dynamics of charged particles under the combined influence of extremely strong Lorentz- and radiation-reaction forces mentioned in the preceeding chapter of this talk, is an essential ingredient in virtually *all* theoretical work on the origin of Ultra High Energy Cosmic Ray Particles. Therefore, it is of eminent importance to make sure that the equation of motion commonly in use *is correct*.

As I have argued in detail on earlier occasions[31] [7,8] and will outline shortly here, it is necessary in this context, to rediscuss the appropriate strength of the *principle of equivalence*, as well as the tricky *border between classical and quantum pictures* to prepare, under given extreme conditions, the grounds for a classical equation of *stochastic* particle motion.

MT is resting on a notion of the principle of equivalence, which amounts to saying that electromagnetic radiation is created and radiation reaction is felt as a consequence of charged particle acceleration, *irrespective* of the nature of forces responsible for that acceleration. Consequently, *Larmor's radiation formula*

$$P^{RAD} = m\tau_0 (d\mathbf{v}_{MRS}/dt)^2 = m\tau_0 \|d\underline{u}/d\tau\|^2 \tag{1}$$

is assumed to be governed by the (four-component) vector of *kinematical* acceleration, $d\underline{u}/d\tau$. $d\mathbf{v}_{MRS}/dt$ is the corresponding (three-component) vector in the momentary rest frame (MRS) of that particle[23].

In consistence with that, MT also adopts that in the *equation of motion*

$$du_j/d\tau = \eta_0 F_{jk}^{EXT} u^k + \tau_0 G_{jk} u^k, \tag{2}$$

radiation reaction is represented by the *Abraham-Lorentz tensor*

$$G_{jk} = G_{jk}^{A-L} := ([d^2 u_j/d\tau^2] u_k - u_j [d^2 u_k/d\tau^2])/c^2, \tag{3}$$

---

[23] m is the mass of that particle, $\eta_0 = e/mc$, and $\tau_0 = 2e^2/3mc^3$ is the radiation constant. $\eta_0 = 1.76 \cdot 10^7$ g$^{-1/2}$cm$^{1/2}$ for the electron and $\eta_0 = 9.58 \cdot 10^3$ g$^{-1/2}$cm$^{1/2}$ for the proton. $\tau_0 = 6.27 \cdot 10^{-24}$sec for the electron and $\tau_0 = 3.41 \cdot 10^{-27}$sec for the proton.

governed by (this time the second) *kinematical acceleration*[24]. F_{jk}^{EXT} is the tensor of the external electromagnetic field. The latter is understood as being generated by all other electromagnetically interacting particles around.

If acceleration is due to external *electromagnetic* fields, Larmor's radiation formula (1) specializes to the *radiation formula*

$$P^{RAD} = m\tau_0(c\eta_0 \mathbf{E}_{MRS}^{EXT})^2 = m\tau_0\|\underline{u}^L\|^2, \tag{4}$$

where $u_j^L := \eta_0 F_{jk}^{EXT} u^k$ is used as an abbreviation for the *Lorentz acceleration*. $\mathbf{E}_{MRS}^{EXT}$ is the electric field vector of the external field, in the MRS.

Accordingly, the Abraham-Lorentz tensor (3) specializes to the *radiation reaction tensor*

$$G_{jk} = \eta_0 u^l \partial_l F_{jk}^{EXT} + G_{jk}^T, \tag{5}$$

with the *Thomson tensor*

$$G_{jk}^T := (u_j^{LL} u_k - u_j u_k^{LL})/c^2, \tag{6}$$

and with $u_j^{LL} := \eta_0^2 F_{jk}^{EXT} F^{EXTkl} u_l$ as an abbreviation for the *second Lorentz acceleration*[25].

Equation (2) with (5) and (6), decendent from the A-L equation (2) with (3), can be seen as a specialization (restricted to acceleration through *electromagnetic* forces), and as an iterative approximation (restricted to no higher than the *fourth order* of the interaction constant e).

But there are, at least, two essential discrepancies arising from the A-L equation (2) with (3): (a) no stringent proof has been found, so far, for the A-L equation, neither on the basis of MT[26], nor on the basis of QED[27], and: (b) the A-L equation is known to deliver unphysical solutions (run-away solutions), which have widely been discussed in literature [5]. For these reasons we have suggested that *the A-L equation (2) with (3) is not correct* [7].

As a consequence of that, specializations and approximations deduced from the A-L equation can neither be seen to rest on solid grounds. The fact that the

---

[24] Equation (2) with (3) often is referred to as *Abraham-Lorentz equation* (A-L equation), [H.A. Lorentz, Enzykl. d. math. Wissensch., **V 1**, 188 (1903). M. Abraham, Ann. der Physik, (4) **14**, 236 (1904); Theorie der Elektrizität, Verlag B.G. Teubner, Leipzig, 1. Auflage, Bd. **2**, '15 (1905)], M. Laue, Ann. d. Physik, *28*, 436 (1908) or, *Lorentz-Dirac equation* (L-D equation), with reference to Dirac's early approach towards a more stringent deduction from MT, [P.A.M. Dirac, Proc. Roy. Soc., **A167**, 148 (1938)].

[25] A review is given, for example, in: T. Erber, Fortschritte der Physik, **9**, 343 (1961); J.D. Jackson, Classical Electrodynamics, J. Wiley, N.Y. 2nd ed. (1975); A.D. Yaghjian, Relativistic Dynamics of a Charged Sphere, Springer Verlag, Berlin, Heidelberg, New York (1992).

[26] For example: C. Teitelboim, Phys. Rev. **D1** 1572 (1970).

[27] V.S. Krivitskij and V.N. Tystovich, Usp. Fiz. Nauk **161**, 125 (1991). Quantum effects have also been discussed by E.J. Monitz and D.H. Sharp: [Phys. Rev. **D15**, 2850 (1977).]

equation of motion (2) with (5) and (6) *does not* have run-away solutions cannot, by itself, invalidate this statement.

Thus, we have undertaken to stringently deduce whatever ultimately comes up as the correct equation of motion, from fundamental principles of *Selfconsistent Electrodynamics* resting on a notion of the principle of equivalence, (referred to as *principle of selfconsistency*,) which amounts to saying that (I) *radiation is to be expected and radiation reaction is felt only and insofar as acceleration is due to electromagnetic interaction* [7,8].

Photon Wave Mechanics

Maxwell's Theory (MT) is understood as a *classical field theory* which, through *classical field equations* in the form of partial differential equations[28] governs *classical field amplitudes* represented by the *real* (three-component) vectors of the electric and magnetic field, $\mathbf{E}$ and $\mathbf{H}$, respectively. $\mathbf{j}^{(e)}$ and $\mathbf{j}^{(m)}$ are referred to as the *electric and magnetic current vectors*, and $\varrho^{(e)}$ and $\varrho^{(m)}$ as the *electric and magnetic charge densities*, respectively.

A justification for this kind of a description is often seen in the fact that both field vectors, $\mathbf{E}$ and $\mathbf{H}$, appear to be accessible to 'direct' measurement, for example, through observations of the acceleration of a charged 'test' particle subjet to Lorentz force.

As is well known, *Maxwell's equations* can be written more compactly by means of a *complex* (three-component) vector $\boldsymbol{\psi} := C(\mathbf{E} + i\mathbf{H})$ of which the *real* vectors $\mathbf{E}$ and $\mathbf{H}$, respectively, are the real and imginary parts: $[\nabla, \boldsymbol{\psi}] = i\partial_t \boldsymbol{\psi} + 4\pi i\boldsymbol{\zeta}$, $(\nabla, \boldsymbol{\psi}) = 4\pi\eta$. Likewise, $\boldsymbol{\zeta} = C\{\mathbf{j}^{(e)} + i\mathbf{j}^{(m)}\}$ and $\eta = C\{\varrho^{(e)} + i\varrho^{(m)}\}$ are the 'sources' in complex form. C is a constant.

Of course, the interpretation of *gedankenexperimente* involving charged particles and electromagnetic fields *is not changed* just by introducing a complex notation.

But then, obviously, many people have recognized the isomorphism existing between Maxwell's and *Schrödinger's equations*,

$$i\hbar\partial_t\psi = \mathcal{H}^+\psi - 4\pi i\hbar\boldsymbol{\zeta}. \tag{7}$$

Now, $\boldsymbol{\psi} = C(\mathbf{E}+i\mathbf{H})$ is the *photon wave function*. $\mathbf{s} := ((s_\lambda)_{\mu\nu}) := -i\hbar(\varepsilon_{\lambda\mu\nu})$ is the spin, $\varepsilon_{\lambda\mu\nu}$ are the Levi-Civita symbols, and $\mathbf{p} := (p_\lambda) := -i\hbar(\partial_\lambda)$ is the momentum of the photon, while $\mathcal{H}^+ := (c/\hbar)(\mathbf{s}, \mathbf{p})$ describes the (differential) evolution in time of a photon state $\boldsymbol{\psi}$.

If there are no sources, the photon wave equation (7) reduces to

$$i\hbar\partial_t\psi = \mathcal{H}^+\psi. \tag{8}$$

---

[28] Brackets [,] are for the vector product, (,) for the scalar product. In what follows, Latin indices are running from 0 through 3, Greek indices are running from 1 through 3. Bold letters are used for three-component vectors. Gaussian units will be used. Accordingly, the electric as well as the magnetic field strength is measured in units of 1 G = 300 V/cm.

Alternatively, one might have chosen the complex conjugate wave equation

$$i\hbar\partial_t\psi^* = \mathcal{H}^-\psi^* \tag{9}$$

for $\psi^* := C^*(\mathbf{E} - i\mathbf{H})$ with $\mathcal{H}^- := -(c/\hbar)(\mathbf{s}, \mathbf{p})$.

Only waves corresponding to positive frequency (i.e. positive energy $\mathcal{E}$ of the photon) have a physical counterpart in classical electromagnetic fields and thus need to be selected from the solutions of (8) and (9). Of these, the ones with positive sense of rotation of the electric vector with respect to the direction of propagation (i.e. left-circularly polarized waves in the usual notation, corresponding to positive helicity of the photon) are delivered by (8), while those with negative sense of rotation (i.e. right-circularly polarized waves, corresponding to negative helicity) are described by (9). Thus, one may restrict to states with positive frequency from $i\hbar\partial_t\psi = \mathcal{H}\psi$ with $\mathcal{H} := \chi\mathcal{H}^+$, where χ delivers ± 1 for positive or negative helicity, respectively.

Selfconsistent Electrodynamics is also based on the suggestion that (II) *Maxwell's equations are to be seen as single photon wave equations* [7,8].

Transition from Maxwell's to de Broglie-Schrödinger's picture of the electromagnetic field, as can be expected, *does have* strong bearings on the interpretation of *gedankenexperimente* involving charged particles and electromagnetic fields.

Statistical Interpretation of Electromagnetic Fields

Multiplication of (8) with the adjoint photon wave function $\psi^\dagger = C^*(\mathbf{E}^T - i\mathbf{H}^T)$, where ^{T} stands for transposition, and of (9) with ψ, followed by substraction of one equation from the other, delivers $\partial_\mu w_\mu + \partial_t w = q$, where $w(\mathbf{x}, t) := \psi^\dagger\psi = (\psi^*, \psi)$, $\mathbf{w}(\mathbf{x}, t) = (w_\mu) := -ic[\psi^*, \psi] = ((c/\hbar)\psi_\lambda^* s_{\mu\lambda\nu}\psi_\nu)$ and $q(\mathbf{x}, t) := -8\pi Re(\psi^*, \zeta)$. Obviously, $w \geq 0$. Consider a non-trivial electromagnetic field and a volume V in 3-dimensional position-space with $\int_V w \, d^3\mathbf{x} \neq \infty$, where the following condition holds:

$$\int_V q \, d^3\mathbf{x} - \oint_{O(V)} w_\mu \, d^2o_\mu \equiv 0. \tag{10}$$

Then, normalization is possible so that

$$\int_V w \, d^3\mathbf{x} \equiv 1. \tag{11}$$

Consequently, following Born's interpretation of wave mechanics, w can be seen as the *position probability density* and $\mathbf{w}$ as the *position probability current* of the photon.

The *expectation value of photon energy*, for example, may then be written

$$<\mathcal{E}>= \int_V \psi^\dagger \mathcal{H} \psi d^3 \mathbf{x}. \tag{12}$$

Having translated 4 components of Maxwell's energy-momentum tensor into the language of photon wave mechanics, one may proceed now one step further translating the remaining 6 components: Multiplying (8) with $\psi^*_\varrho \varepsilon_{\kappa\varrho\mu}$ and subtraction of the complex conjugate delivers $\partial_\nu w_{\mu\nu} + \partial_t w_\mu = q_\mu$, where $w_{\mu\nu} := (\psi^*, \psi)\delta_{\mu\nu} - (\psi^*_\mu \psi_\nu + \psi^*_\nu \psi_\mu)$ and $q_\mu := C\{\psi^*_\mu(\mathbf{\nabla}, \psi) + \psi_\mu(\mathbf{\nabla}, \psi^*)\} = 8\pi C Re(\psi^* \eta)$. $Re...$ stands for the real part.

If w is normalized according to (11), $\mathbf{w} = \psi^\dagger \mathbf{c} \psi$ can be seen as the density of $\mathbf{c} := (c/\hbar)\mathbf{s}$ with $c_\mu c_\mu = c^2$. For an interpretation of $\mathbf{c}$, we note that the only possible result of a simultaneous measurement of $\mathbf{c}$ and $\mathbf{p}$, as far as $\mathbf{c}$ is concerned, is $+c$, parallel to $\mathbf{p}$. Thus, $\mathbf{c}$ may be seen as the *velocity* of the photon, $\mathbf{w}$ as the *velocity density* and $w_{\mu\nu}$ as the *velocity current* in the photon wave field. Correspondingly, the *expectation value of photon velocity* is

$$<\mathbf{c}>= \int_V \psi^\dagger \mathbf{c} \psi d^3 \mathbf{x}. \tag{13}$$

Reproducing the Classical Equation of Motion

With these suggestions adopted, conventional nomenclature with two real field vectors $\mathbf{E}$ and $\mathbf{H}$ can be used again, instead of one complex field vector $\psi = \mathbf{E} + i\mathbf{H}$.

If in the MRS, $O(V')$ is an appropriately chosen closed surface, (e.g. a sphere of radius $R > 0$,) surrounding the scattering particle, then the rate of *expected momentum transfer* onto the charged particle is

$$K_{MRS\mu} = - \oint_{O(V')} \sigma_{\mu\nu} d^2 o_\nu \mid_{'ret'}, \tag{14}$$

with *Maxwell's tensor* $\sigma_{\mu\nu} = \epsilon_{MRS}\,\delta_{\mu\nu} - (1/4\pi)(E_{MRS\ \mu}\,E_{MRS\ \nu} + H_{MRS\ \mu}\,H_{MRS\ \nu})$ and $\epsilon_{MRS} = (1/8\pi)(|\mathbf{E}_{MRS}|^2 + |\mathbf{H}_{MRS}|^2)$. 'ret' stands for retardation of field components originating from that charged particle. $\mathbf{E}_{MRS} = \mathbf{E}^{ext}_{MRS} + \mathbf{E}^{Coul}_{MRS} + \mathbf{E}^{rad}_{MRS}$ is the total electric field vector, and $\mathbf{H}_{MRS} = \mathbf{H}^{ext}_{MRS} + \mathbf{H}^{Coul}_{MRS} + \mathbf{H}^{rad}_{MRS}$ is the total magnetic field vector in the MRS.

Two components of Maxwell's tensor contribute to (14):

(A) Contributions from the dyadic product of the external field vector $\mathbf{E}^{ext}_{MRS}$ with the Coulomb field vector $\mathbf{E}^{Coul}_{MRS} = e\mathbf{R}_0/R^2$ deliver the *expected rate of momentum transfer through Coulomb force,*

$$\mathbf{K}^{Lor}_{MRS}(t) = (1/4\pi) \oint_{O(V')} \mathbf{E}^{ext}_{MRS}(t')\,(\mathbf{E}^{Coul}_{MRS}(t'), d^2\mathbf{o}), \tag{15}$$

with $t' = t + \delta t$ and $\delta t = R/c$, resulting in $\mathbf{K}_{MRS}^{Lor}(t) = e\mathbf{E}_{MRS}^{ext}(t)$. R is the distance from the location of the scattering particle to the location where the electromagnetic field is under consideration. $\mathbf{R}_0$ is the corresponding unit vector, and $\boldsymbol{\beta} = \mathbf{v}/c$ is the velocity of that particle in units of the velocity of light, c. Lorentz transformation to an arbitrary inertial frame of reference (IS) leads to the four-compoment vector of *Lorentz force,*

$$K_j^{Lor} = m\eta_0 F_{jk}^{ext} u^k. \tag{16}$$

(B) Contributions from the dyadic product of the radiation field vector $\mathbf{E}_{MRS}^{rad}(t')$, with the Coulomb field vector $\mathbf{E}_{MRS}^{Coul}(t')$ deliver the *rate of expected momentum transfer through radiation reaction force,*

$$\mathbf{K}_{MRS}^{rad}(t) = (1/4\pi) \oint_{O(V')} \mathbf{E}_{MRS}^{rad}(t')(\mathbf{E}_{MRS}^{Coul}(t'), d^2\mathbf{o}). \tag{17}$$

Evaluation of (17) leads to the four-component vector of radiation reaction force

$$K_j^{rad} = m\tau_0 G_{jk} u^k, \tag{18}$$

with the radiation reaction tensor (5) considered earlier in this chapter.

Thus, we have reproduced, *under the premises of Self Consistent Electrodynamics,* the *equation of motion* (2), $du_j/d\tau = \eta_0 F_{jk}^{ext} u^k + \tau_0 G_{jk} u^k$, with the *radiation reaction tensor* (5), $G_{jk} = \eta_0 u^l \partial_l F_{jk}^{EXT} + (u_k^{LL} u_k - u_j u_k^{LL})/c^2$. Within the frames of SCED, therefore, the problem of *run-away solutions* is not existent. In Maxwell Theory, as was mentioned above, (2) with (5) is seen just as a specialization and approximation deriving from (2) with (3), wherein the problem of *run-away solutions* is inherent.

Solutions of (2) with (5) can exhibit rather surprising features, as is illustrated by the following *gedankenexperiment:* Consider an electric field, homogeneous (in position space) and constant (in time), within an arbitrarily given inertial frame of reference (IS). An electrically charged particle is positioned within that field at an arbitrarily chosen point of space, at rest at an arbitrarily chosen instant of time. This particle, by (2) with (5), is predicted to be *accelerated at a rate as if there were no radiation losses* at all! Nevertheless, the same particle also is predicted to *radiate at a rate described by the radiation formula* (4)[29]!

Part of the *radiation reaction force* is provided by the *Thomson force*

$$K_j^T := m\tau_0 G_{jk}^T u_k, \tag{19}$$

the non-covariant form of which, in the MRS, is reducing to

$$\mathbf{K}_{MRS}^T = (\sigma^T/c)\mathbf{S}_{MRS}^{ext}. \tag{20}$$

---

[29] This is *not* a perpetuum mobile! K.O. Thielheim, in: Proc. Particle Accelerator Conference, Washington, D.C., 17-20 May 1993, **1**, 276 (1993).

493

$\sigma^T = 8\pi e^4/3m^2c^4$ is the *Thomson cross-section*. $\mathbf{S}^{ext}_{MRS} = (1/4\pi)[\mathbf{E}^{ext}_{MRS}, \mathbf{H}^{ext}_{MRS}]$ is the *Poynting vector* of the external field. (19) can be deduced from (20) through Lorentz transformation[30].

In the special case of an external *null* field, a plane vacuum wave field, for example, where $|[\mathbf{E}^{ext}_{MRS}, \mathbf{H}^{ext}_{MRS}]| = {\mathbf{E}^{ext}_{MRS}}^2$, Maxwell Theory is able to interpret (a) the radiation field 'created' by Lorentz acceleration according to radiation formula (4) as the Thomson scattered wave field, and (b) the force due to radiation pressure acting on the Thomson cross-section of that particle (20) as (part of) radiation reaction force[31] and, by making use of the quantum picture insofar, as the *knock-on force* due to many-photon Thomson scattering, not unlike *Brownian motion*[32].

In the general case of an external field of *arbitrary topography* different from null fields, where $|[\mathbf{E}^{ext}_{MRS}, \mathbf{H}^{ext}_{MRS}]| \neq (\mathbf{E}^{ext}_{MRS})^2$, this interpretation meets with difficulties within Maxwell Theory. In contrast to SCED, MT is not made to describe momentum transfer between external fields of *arbitrary topography* and charged particles as a scattering phenomenon and hence recurs to the notion of radiation 'creation'.

For illustration, I will mention here still another discrepancy, inherent to Maxwell Theory and widely discussed in cosmic ray physics, which disappears under the aspects of Selfconsistent Electrodynamics, namely, the question of the possible existence of 'magnetic monopoles'. While Maxwell's Equations, for reasons of symmetry, strongly suggest the existence of source terms also for 'magnetic charges',

---

[30] K.O. Thielheim, in: Proc. PAC, Washington, D.C., **1**, 276 (1993); Int. Journ. of Modern Physics D **3.1**, 289 (1993); In: Ilmar Ots and Laur Palgi, (Eds.), Proc. Tallinn Symp. on Neutrino Astrophysics, Lohusalu, Estonia: 2nd, 138 (1993) & 3rd, 138 (1995); Il Nuovo Cimento, **109 B**, 103 (1994); In: R. Cowsik, (ed.), Proc. Int. Conf. on Non-Accelerator Physics, Bangalore, India, World Scientific, 341 (1995); In: Proc. EPAC, London, **1**, 808 (1994); In: Proc. 24th ICRC, Rome, Italy, **3**, 398 (1995).

[31] L.D. Landau and E.M. Lifschitz, The Classical Theory of Fields, Pergamon Press (1959).

[32] (a) In an external *null* field, the *rate of energy* emitted according to Larmor's formula $P^{rad}_{MRS} = (\tau_0/m)(e\mathbf{E})^2$, is found to be in agreement with the rate at which energy is extracted from the incoming wave through Thomson scattering $P^{scatt}_{MRS} = \sigma^T \mathbf{S}^{ext}_{MRS}$, where $\mathbf{S}^{ext}_{MRS} = (c/4\pi)[\mathbf{E}^{ext}_{MRS}, \mathbf{H}^{ext}_{MRS}]$ is the Poynting vector of the external field, in the MRS since, in the MRS, energy *cannot* be transferred to the particle. (Indeed, for an external wave field, the Thomson cross-section can be *defined* through $\sigma^T = (1/|\mathbf{S}^{ext}_{MRS}|) \cdot \oint_{O(V')}(\mathbf{S}^{rad}_{MRS}, d^2\mathbf{o})$.)

Likewise, the *directional energy flow* radiated from a charged particle, $\mathbf{S}^{rad}_{MRS} = (3/8\pi) \cdot P^{rad}_{MRS} sin^2\theta$, within an external null field, is found to match the directional energy flow $\mathbf{S}^T_{MRS}$ resulting from Thomson scattering. For example, in the case of a *linearly polarized* external wave field, $\mathbf{S}^T_{MRS} = \sigma^T(\varphi, \vartheta) \cdot \mathbf{S}^{ext}_{MRS}$, where $\sigma^T(\varphi, \vartheta) = r_e^2(1 - cos^2\varphi \, sin^2\vartheta)$ is the differential Thomson cross-section for linear polarization and $cos\,\theta = cos\,\varphi \, sin\,\vartheta$, where ϑ is the angle between the vector of the incoming wave and the vector of the outgoing wave. φ is the angle between the scattering plane spanned by these two wave vectors and the polarization plane spanned by the vector of the incoming wave and the electric vector of the (linearly polarized) incoming wave. θ is the angle between the electric vector of the incoming wave and the vector of the outgoing wave. (b) Also, due to the forward-backward symmetry of the outgoing wave, in the MRS, the *rate of momentum transferred* to the outgoing wave vanishes. Consequently, the total rate of momentum extracted from the incoming wave through Thomson scattering, is transferred to the particle, leading to (20).

the photon wave equation (7), for reasons of unitarity ('conservation of position probability') and of universality of the electromagnetic interaction constant, suggests just one kind of 'charge' which, according to what has been chosen as the (arbitrary) basis of photon wave functions, can either be given the appearance of 'electric' or 'magnetic', or even some mixture of the two kinds of 'charges'.

A TENTATIVE SCENARIO: UHECRP FROM NEUTRON STARS

Particles and Radiation from the Acceleration Boundary

Equations of motion (2) with (5) are the basis for investigations on particle acceleration inside neutron star magnetospheres.

Immediately after the discovery of pulsars [11], test particle dynamics in vacuum fields ('stage one') was studied analytically [10] as well as numerically[33].

Simultaneously, collective particle dynamics within self-consistent plasma fields ('stage two') was investigated[34], for example, in search for stable rotationally symmetric plasma configurations around aligned rotators [13,18].

To study the possible functioning of rotating, magnetized neutron stars as cosmic accelerators, we have investigated systematically charged particle dynamics around rotating magnets[35] [5–7]. Obviously, for this purpose, it is necessary to understand the structure and dynamics of pulsar magnetospheres. Therefore, we have undertaken to develop a code for N-body-simulations of three-dimensional, relativistic plasma configurations associated with rotating cosmic magnets, taking into account radiation reaction[36].

Inside the magnetosphere of non-aligned rotators though still outside the regime dominated by near-field contributions, the *acceleration boundary* (AB) is defined as the region, where *very* high energy particles can decouple from outgoing spherical wave fields. In spherical *vacuum* wave fields of an orthogonal rotator this occurs at a distance $r_B = (2\pi)^{1/3} r_L (r_T/r_L)^{4/3}$ from the centre of the rotating magnet, and the energy of such particles is of the order of the Gunn-Ostriker energy [10]

---

[33] A. Ferrari and E. Trussoni, Lettere al Nuovo Cimento, **1**, 137 (1971); Astrophysics and Space Science, **24**, 3 (1973); Astronomy and Astrophysics, **36**, 267 (1974); Astrophysics and Space Science, **33**, 111 (1975). V.D. Endean, Mon. Not. Roy. Astron. Soc., **158**, 13 (1072); **204**, 1067 (1983). A.A. da Costa and F.D. Kahn, Mon. Not. Roy. Astron. Soc.,**199**, 211 (1982); **204**, 1125 (1983).
[34] L.G. Kuo-Petravic, M. Petravic and K.V. Roberts, Phys. Rev. Lett., **32**, 1019 (1974); ApJ., **202**, 762 (1975).
[35] K.O. Thielheim, Proc. 125th IAU Symp., Nanjing, China, 555 (1986); Proc. 2nd ESO/CERN Conf., München, Germany. 317 (1986). H. Laue and K.O. Thielheim, ApJ. Suppl., **61**, 465 (1986). R. Leinemann and K.O. Thielheim, ApJ. Suppl., **66**, 19 (1988).
[36] H. Laue and K.O. Thielheim, to be published.

$$\gamma_{GO} = (9/2)^{1/3} \cdot (r_T/r_L)^{4/3}. \tag{21}$$

With the 'standard set of parameters' mentioned before, the corresponding Lorentz factor for protons is $\gamma_{GO} = 2.0 \cdot 10^4$, which is far below the regime of ultra high energies[37]. (Remarkably, in the energy range below γ_{GO}, neutral 'seed' particles invading the region between AB and PB from outside,. after ionization or decay to charged particles with homogeneous spatial source distribution, generate a differential energy spectrum proportional to $E^{-2.5}$, similar to observational findings [5]).

The question arises, whether (possibly non-linear and stable) spherical *plasma* waves can develop in the region between transition zone (TZ) and AB, where the latter would be located, and what energy particles can achieve[21] when decoupling from outgoing spherical *plasma* waves[38].

Physical conditions inside that region are interesting for still another reason: The mechanism responsible for the generation of radio frequency signals which gave name to *pulsars*, is generally adopted to be due to the rotating pattern of a beam born in the polar region of rotating magnetized neutron stars [14], 'similar to radiation in the pattern of a rotating beacon' [12]. As I have mentioned earlier, the existence of non-linear 'seesaw' *plasma* waves could eventually constitute a competing mechanism for the production of highly coherent signals of the same frequency, (covering, though, a wider range of solid angle, with resulting consequences for its probability of being detected.)

Particle and Radiation Beams
from the Polar Zone of Aligned Rotators

So we are left with the possibility that UHECRP originate from locations *very* near to the surface of magnetized neutron stars. There, the magnetic field strength is known to be, for example, of the order of $|\mathbf{H}^{pole}| = 10^{12}G$ with a frequency of rotation of the order of $\nu = 10 \ sec^{-1}$. Then, by induction, an electric field strength of about $|\mathbf{E}^{pole}| = 10^{10}G$ can be expected.

I shall consider here, in a local approximation, a homogeneous static electromagnetic field, the strength of which is in accordance with these values. Under such circumstances an electrically charged particle is expected to move *practically* along a magnetic field line, accelerated by the component of the electric vector $\mathbf{E}^{ext}_{\parallel}$, parallel (or antiparallel) to the magnetic vector $\mathbf{H}^{ext}$. When Lorentz transforming the external fields into the MRS, the parallel component of the electric vector, $\mathbf{E}^{ext}_{\parallel}$, does not change, $\mathbf{E}^{ext}_{\parallel} \rightarrow \mathbf{E}^{ext}_{MRS\parallel} = \mathbf{E}^{ext}_{\parallel}$. But the transverse component of the electric vector, $\mathbf{E}^{ext}_{\perp}$, does change, $\mathbf{E}^{ext}_{\perp} \rightarrow \mathbf{E}^{ext}_{MRS\perp} = \gamma\mathbf{E}^{ext}_{\perp}$. Also, a transverse component of the magnetic vector, $\mathbf{H}^{ext}_{MRS\perp} = -\gamma \ [\boldsymbol{\beta}, \mathbf{E}^{ext}_{\perp}]$, is produced. Inserting

---

[37] K.O. Thielheim, Journ. Phys., Math. Gen. **20**, 203 (1987).

[38] This problem is also dealt with by N-body plasma simulations: T. Narjes and K.O. Thielheim, to be published.

these fields into (2) with (5) delivers an upper limit of particle energy[39]. Making use of a dimensionless notation [6], this upper limit of the Lorentz factor may be written $\gamma \leq (f_\parallel/l_0 f_\perp^2)^{1/2}$, where $f_\parallel$ is the component of the external electric field (in units of mc^2/er_L), tangent to the magnetic field line and $f_\perp$ is its component orthogonal to the latter. $l_0 := c\tau_0/r_L$. The interval of arc length needed to achieve about half the maximum of energy is $l_{max} \cong (l_0 f_\parallel f_\perp^2)^{-1/2}$. $r_L = c/\omega$ is used as a unit of length.

Applying the *energy limiting theorem* to particles within the *vacuum* fields of the polar region of *orthogonal* rotators delivers

$$\gamma^{ortho} \leq 2l_0^{-1/2}(r_N/r_T) \cdot \frac{\sqrt[4]{\cos^2\theta_0(3\sin^2\theta_0 \cdot \cos^2\phi_0 + 1)}}{\sqrt{4 + 5\cos^2\theta_0 + 11\sin^2\theta_0 \cos^2\phi_0 + \sin^4\theta_0 \cos^2\phi_0}}. \qquad (22)$$

Obviously, radiation reaction sets drastic energy limits for particles originating from the surface of *orthogonal* rotators. Given the standard set of parameters, for example, the upper limit in the polar region, $\theta_0 \cong 0$, is $\gamma^{ortho} \leq 1.4 \cdot 10^3$ for electrons, and $\gamma^{ortho} \leq 2.6 \cdot 10^6$ for protons. θ_0 and ϕ_0 are the coordinates of those positions, where the fieldlines under consideration end on the surface of the neutron star,

But the situation turns out to be different for *parallel* rotators, where

$$\gamma^{para} \leq 2l_0^{-1/2}(r_N/r_T) \operatorname{ctg}\theta_0 \cdot \frac{\sqrt[4]{\cos^2\theta_0(3\cos^2\theta_0 + 1)}}{(1 + \cos^2\theta_0)}. \qquad (23)$$

In the polar region, $\theta_0 \cong 0$, of parallel rotators, this limit is $\gamma^{para} \leq 3.0 \cdot 10^3$ ctg θ_0 for electrons, and $\gamma^{para} \leq 5.5 \cdot 10^6$ ctg θ_0 for protons. Thus, radiation reaction permits *aligned* rotators, (unlike *orthogonal* rotators,) to develop very narrow *nozzles* around their axis, $\theta_0 = 0$, through which ultra high energy particles *can* escape from the surface.

For an estimate on the energy values which may actually be achieved by particles escaping through that nozzle one has to refer to the well-known fact that a *parallel* rotators, in the vacuum, establish an electric quadrupole of which the two polar cap regions are negatively charged, while the equatorial region is positively charged. *Anti-parallel* rotators establish a quadrupole of the alternative polarity. Electric and magnetic field lines coalesce along the axis of rotation and, in the absence of transverse acceleration, radiation losses and radiation reaction are insignificant. The resulting asymptotic 'typical' Lorentz factor is

---

[39] Remarkably, in this limit, the particle is found to move, *practically*, on a straight line at constant speed while *radiating!* In the MRS, Thomson force is compensating Coulomb force through friction in an upstream motion of that particle against photons [7]. K.O. Thielheim, in: Proc. 24th Int. Cosmic Ray Conference, Rome, Italy, **3**, 398 (1995); in: Proc. EPAC, London, **1**, 808 (1994); in: Conf. on Some Unsolved Problems in Astrophysics, Princeton, N.J., April 27 - 29, 1995, p.32.

$$\gamma_T = 6 \ (r_T/r_L)^2 \cdot (r_L/r_N) \qquad (24)$$

and thus turns out to be considerably higher than the corresponding Gunn-Ostriker Lorentz factor of particles from the wave zone[40].

Electric vacuum fields around rotating magnetized neutron stars, of course, are modified through the fields generated by global charge clouds forming under the influence of the force-free surface (FFS) inside magnetospheres [15]. We have developed a numerical iterative procedure to reproduce quasi-stable plasma configurations[41] [7] From there, we hope to derive information on the possible existence of polar gaps [16,17] and the possible *local* persistence of very strong electric fields permitting particle acceleration and beam formation[42].

REFERENCES

1. J. Linsley, L. Scarsi and B. Rossi, Phys. Rev. Lett., **133**, 572 (1961).
2. D.J. Bird, S.C. Corbató, H.Y. Dai, J.W. Elbert, K.D. Green, M.A. Huang, D.B. Kieda, S. Ko, C.G. Larsen, E.C. Loh, M.Z. Luo, M.H. Salomon, J.D. Smith, P. Sokolsky, P. Sommers, J.K.K. Tang, and S.B. Thomas, ApJ., **441**, 144 (1995).
3. J.W. Elbert and P. Sommers, ApJ., **441**, 151 (1995).
4. N. Hayashida, K. Honda, M. Honda, S. Imaizumi, N. Inoue, K. Kadota, F. Kakimoto, K. Kamata, S. Kawaguchi, N. Kawasumi, Y. Matsubara, K. Murakami, M. Nagano, H. Ohoka, M. Takeda, M. Teshima, I. Tsushima, S. Yoshida and H. Yoshii, Phys. Rev. Lett., **73**, 3491 (1994).
5. K.O. Thielheim, Fundamentals of Cosmic Physics, **13**, 357 (1989).
6. K.O. Thielheim, Nucl. Phys. (Proc. Suppl.), **22B**, 60 (1991).
7. K.O. Thielheim, Physica Scripta, **T52**, 123 (1994).
8. K.O. Thielheim, Proc. PAC, Dallas, Texas, 3370 (1995).
9. F. Pacini, Nature, **216**, 567 (1967); Nature, **219**, 145 (1968).
10. J.P. Ostriker and J.E. Gunn, ApJ., **157**, 1395 (1969); **165**, 523 (1971).
11. A. Hewish, S.J. Bell, P.F. Pilkington and R.A. Collins, Nature, **217**, 709 (1968).
12. T. Gold, Nature, **218**, 731 (1968).
13. P. Goldreich and W.H. Julian, ApJ. **157**, 869 (1969).
14. J. Ostriker, Nature, **217**, 1227 (1968).

---

[40] For the 'standard set of parameters', $\gamma_T = 3.9 \cdot 10^9$, for protons. Obviously, $\gamma_T \propto \omega \cdot \mu$, so that a millisecond pulsar with the 'standard' magnetic dipole moment $\mu = 10^{30} Gcm^3$ would accelerate protons up to $\gamma_T \ mc^2 = 3.7 \cdot 10^{20} eV$!

[41] K.O. Thielheim and H. Wolfsteller, ApJ. Suppl., **71**, 583 (1989); ApJ., **431**, 718 (1994); K.O. Thielheim, Proceedings of the 17th Texas Symposium on Relativistic Astrophysics, München, Germany, 12-17 December 1994, Annals of the New York Academy of Sciences, **795**, 257 (1965); Recently, plasma simulations are being performed with the same intention: [P. Biltzinger and K.O. Thielheim; H. Laue and K.O. Thielheim, to be published].

[42] In that case, one could speculate on possible contributions to dark matter from a halo population of aligned neutron stars, able to create UHECRP (if anti-parallel) or, alternatively, Gamma Ray bursts (if parallel.) Still, there would be the problem, how such an old population could have preserved enough angular momentum.

15. E.A. Jackson, ApJ., **222**, 675 (1978); **237**, 198; **238**, 1081 (1980); **251**, 665 (1981).

16. P.A. Sturrock, ApJ., **164**, 529 (1971).

17. M. Ruderman, Phys. Rev. Lett., **27**, 1306 (1971). M. Ruderman and P.G. Sutherland, ApJ., **196**, 51 (1975).

18. F.C. Michel, ApJ., **192**, 713 (1974); **227**, 579 (1979); Rev. Mod. Phys., **54**, 1 (1982); The Theory of Neutron Star Magnetospheres, The University of Chicago Press (1991).

Photodetectors for OWL

John W. Mitchell
for the OWL Study Collaboration[1]

Laboratory for High Energy Astrophysics
NASA Goddard Space Flight Center
Greenbelt, Maryland 20771

Abstract: At its focal plane, an Orbiting Wide-Angle Light Collector (OWL) will require a highly segmented large area photon detector, able to detect ultraviolet light at the single photoelectron (few photon) level. Depending on the OWL field-of-view and collecting optics, the focal plane detector must have 6 to 20 square meters of contiguous active area segmented into between 0.4 and 1.7 million individual pixels. Each pixel must have a time response of ten nanoseconds to one microsecond, depending on the readout technique, and be able to resolve and record single photoelectrons. The technical requirements of the OWL focal plane detector and possible approaches to its realization will be presented and discussed. A baseline photodetector system is identified.

INTRODUCTION

The Orbiting Wide-Field Light-Collector (OWL) instruments, will study the energy, arrival direction, and interaction characteristics of the highest energy ($> 10^{20}$ eV) individual particles of the cosmic radiation yet observed. The origin of particles at these energies is poorly understood and their measurement will yield important insights into the fundamental physics of ultra-high-energy astrophysical processes.

The flux of cosmic rays at these energies is on the order of one per square kilometer per century and a huge detection area is required to give good statistical significance in a limited amount of time. By observing from space the particle showers induced by highly energetic particles, OWL makes use of the Earth's atmosphere as a huge calorimeter. The showers are detected and measured using the UV fluorescence of atmospheric nitrogen caused by the passage of shower particles. OWL must detect and measure the shower tracks over a large area of atmosphere and in the presence of significant light contamination. In the baseline configuration, this is accomplished by an optical system with a collecting aperture of about five square meters viewing more than a million square kilometers of atmosphere.

1. Louis M. Barbier, Kevin Boyce, Eric R. Christian, John F. Krizmanic, John W. Mitchell, Jonathan F. Ormes (PI), Floyd W. Stecker, Donald F. Stilwell, and Robert E. Streitmatter: *NASA / Goddard Space Flight Center;* Hun-yu Dai, Eugene C. Loh, Pierre Sokolsky, and Paul Sommers: *University of Utah;* Russell A. Chipman, John Dimmock, Lloyd W. Hillman, David J. Lamb and Yoshiyuki Takahashi: *University of Alabama*; Mark J. Cristl and Thomas A. Parnell: *NASA / Marshall Space Flight Center.*

Particular challenges in the development of OWL are the spacecraft, the optical system, the trigger/electronics, and the focal plane detector. The OWL collaboration is studying the technical requirements of each of these systems and developing approaches for their realization. The general technical requirements of OWL are reviewed by R. Streitmatter [1] and the optical system is discussed by D. Lamb, et al. [2], elsewhere in this volume. In this paper we concentrate on the focal plane detector system.

DETECTOR PERFORMANCE REQUIREMENTS

As discussed by Streitmatter [1], each OWL "eye" must be capable of resolving both the temporal and spatial development of the particle showers. Thus, the focal plane detector system must contain from 4.3×10^5 (0.1 degree angular resolution) to 1.7×10^6 (0.05 degree resolution) pixels. The area of the focal plane will be between 6 and 20 m^2, depending on the f number of the optical system, so each pixel will have an area between 3.5 and 46 mm^2. The dead area between pixels or groups of pixels must be minimized both to maximize the detected signal and to insure that most showers will produce contiguous tracks. Each pixel, and its associated readout electronics, must have a time response of about 1 μs to track the shower as it crosses the field of view of the pixel (3 μs for a shower perpendicular to the viewing direction to cross 1 km).

The OWL detector must have a large live-time fraction. During the 1 μs viewing interval, each pixel will integrate both signal and background. Since the signal-to-noise ratio will ultimately limit the lowest signal (and hence the lowest particle energy) that can be measured by OWL, devices which integrate background over many pixels are unsuitable. The pixels must be independent and individually read out.

The principal limitations to the low end of the range of incident particle energies accessible to OWL are the photon collecting power and detection sensitivity of the optical and focal plane detector systems. For the focal plane detector, this means that the effective quantum efficiency for the N_2 fluorescence lines at 337, 357, and 391 nm must be as high as possible, and that the detector must be capable of resolving single photoelectrons. The signal dynamic range of the detector must extend from single photoelectrons to the shower maximum signal at the highest expected incident energy.

Single photoelectron detection requires the detector to have low intrinsic noise without extensive cooling. It also dictates that it should have a high intrinsic gain so that amplifier gain (with associated noise and power consumption) is minimized or eliminated. This is especially true if photon counting (see below) is employed since extremely fast amplifiers would be required.

These requirements can be summarized as follows:

- High quantum efficiency in UV (300-400 nm): require ≥ 17%, goal > 30%
- Single photoelectron sensitivity
- Total area: 6 to 20 m^2
- High degree of segmentation: 4.3×10^5 to 1.7×10^6 pixels
- Individual pixel readout
- Short response time: ~ 10 ns to ≤ 1 μs
- High gain without (or before) amplification: require 10^3, goal ≥ 10^6
- Minimal dead area: require ≤ 20%, goal ≤ 5%

In order to satisfy OWL mission requirements, the total power consumption must be minimized (< 0.25 mW/pixel for a 1000 W maximum power target), cooling requirements must be moderate (~300 K operation), and the device must be space qualifiable for the anticipated orbit. In addition, (based on preliminary background estimates and assuming that the detector is gated on only during the estimated 8% of the time when viewing conditions are good) the photon detector must be able to deliver about 1 C/cm^2 (at 10^5 gain) to the readout electronics over a 5-year OWL mission.

The detector should be modular in construction to reduce the production cost and to facilitate on-orbit deployment. An array of close-packed detector modules (probably hexagonal but possibly square) with high individual reliability offers lower cost through increased yield and can conform to variety of focal plane shapes.

READOUT SYSTEM

Two approaches to the readout system are under consideration. In one approach, the signal recorded by each pixel during a viewing interval (nominally 1 μs) is delivered to a time-sliced analog storage device (e.g. an analog shift register or linear CCD) as well as to the trigger electronics. The data in the analog storage for the entire OWL array must be shifted each viewing interval. The full depth of each (pixel) storage chain must be sufficient to accommodate the full detectable duration of the particle shower event.

In order to reduce the power required by the readout electronics, it is planned that the full storage depth will be broken into two parts, the first slightly greater than the time necessary to make a trigger decision and the second covering the time between the trigger decision and the full duration of the shower. In operation, the data from the full focal plane array will be shifted until a trigger decision is made. The second storage section is only enabled (for the full detector or for a selected portion) following a positive trigger decision. The trigger electronics will identify a small region of the focal plane where the shower must have occurred. After the full shower development time is reached, the stored signals from this region can be shifted (at a much slower rate) into digitizers and then transmitted to the ground. Although the trigger design is still under consideration it will certainly be possible to make a trigger decision no later than a few viewing intervals after shower maximum and probably much sooner.

In the second approach to the readout, digitization of the detected signals is accomplished by photon counting applied to each pixel during the viewing interval. Counting thresholds will be established (taking pixel gain variations into account and possibly using dynamic feedback from background integration) corresponding to one photoelectron and perhaps to the first few additional photoelectrons (with the number based on photon arrival statistics and the time response of the detector). The resulting digital data will be stored in digital shift registers and treated in much the same way as in the analog approach described above. The main effect of this approach on the focal plane detector would be that for single photoelectron counting its time response would have to be less than the instantaneous interval between arriving photons at the maximum of the shower. This would require a resolving time (for the detector and the electronics) of at most a few tens of nanoseconds. For multiple threshold counting the time response could be longer but would still have to be fast compared to the nominal 1 μs.

It should be noted that the OWL focal plane system will be self-triggering. The only signal on which a trigger decision can be based is the same N_2 fluorescence light that is detected by OWL itself. In principle, it would be possible to provide an independent trigger detector but signal-to-noise considerations limit the utility of such an approach since it would still have to be a very complex device with a large number of pixels.

BASIC DETECTOR TECHNOLOGIES

Detectors for UV photons generally fall into two basic categories: those that directly detect the electron-hole pairs liberated in a semiconductor by the photon and those that detect photoelectrons emitted from a photocathode (generally a solid but can be a liquid or gas).

Direct Semiconductor Detection

Direct-to-semiconductor devices (excluding cryogenic detectors such as visible light photon counters or VLPCs) include photodiodes (PD), avalanche photodiodes (APD), and charge-coupled devices (CCD) [3]. APDs differ from normal photodiodes in that they are designed to have bias voltages far in excess of the depletion voltage applied across the diode. The resulting internal electric fields are high enough to accelerate the liberated electrons and cause an avalanche within the diode. This can result in a gain of up to 10^3 (usual APD gain is much lower).

The principal advantages of direct-to-semiconductor devices are a "simple" monolithic structure, long life, and low power consumption (although this is increased when amplification is required). They can also have a quantum efficiency which is greater than that of conventional photocathodes. At longer wavelengths the quantum efficiency of semiconductor devices can be as high as 80%, but in the 300-400 nm range they offer little (if any) advantage.

The only direct-to-semiconductor technology that might be useful to OWL is the APD (although, as discussed below, semiconductor devices may be useful in a photocathode based detector). Neither PDs or CCDs are suitable for single photoelectron detection. In addition, CCDs integrate background unless used in a "snapshot" mode which results in increased dead time.

APDs can meet some of the OWL requirements. However, even at an APD gain of 10^3, amplification would be required. Partly as a result, resolving single photoelectrons would require cooling a silicon APD to below 260 K [4]. Power consumption would also increase and would prohibit the use of APDs in a photon counting mode even though the intrinsic time resolution of the APD is adequate.

Photocathode-Based Detectors

In photocathode-based detectors, photoelectrons liberated by incident photons are collected on an anode or array of anodes, usually after a multiplication stage. These devices, in the form of vacuum photomultipliers, are historically the most important detectors for small light signals.

Photocathodes can be made in solid, liquid, or gas (e.g. TMAE) form but for OWL, solid photocathodes made from a combination of metals are most suitable. "Conventional" bialkali or multi-alkali (e.g. S20-UV) photocathodes can give quantum efficiencies of between 17% and 30% (depending on the processing method) in the 300-400 nm band and are capable of satisfying all OWL requirements. There is also some possibility that materials (perhaps InN based) can be developed to give a "tailored" response which is insensitive above 400 nm and can reduce or eliminate the requirement of an optical filter to minimize background from longer wavelength sources.

A variety of electron multiplication methods can be employed and are generally independent of the type of photocathode that is used. The most common [5] are devices in which the photoelectrons are accelerated by an electric field and strike a metal surface such as the first dynode of a conventional vacuum photomultiplier tube (PMT). The resulting secondary electrons are again accelerated and impact another emitting surface. This process is continued until the desired gain is achieved. Conventional PMTs with twelve or more multiplication stages and gains of 10^7 are common. These can offer excellent single photoelectron resolution and time response of no more than a few nanoseconds.

Even with the relatively large (~3.5 to 46 mm^2) OWL pixels, tubes using individual chains of discrete dynodes for each pixel are impractical. However, multi-anode PMTs have been developed [6,7] using perforated metal foil, metal channel, and mesh dynode structures. All are capable of the image resolution necessary for OWL although the collection efficiency of perforated metal foil and metal channel PMTs are somewhat better than that of the mesh dynode PMT. A multi-anode PMT using foil or mesh dynodes is illustrated in Figure 1.

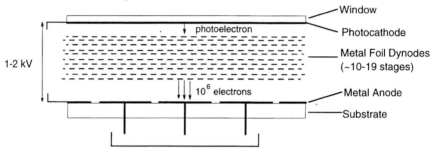

Figure 1: Schematic cross-section of proximity focused foil or mesh dynode photomultiplier.

Variations on the secondary emission multiplier include channel electron multipliers (including microchannel) and microsphere multipliers. In a channel multiplier the accelerating potential (after initial collection of electrons from the cathode) is developed along a thin tube or channel rather than between successive dynodes. The accelerated electrons strike the walls of the tube and gain is developed through a cascade of impacts and secondary emission. Channel multipliers can be manufactured from individual metal tubes or from insulating (usually glass) tubes coated on the inside with the emitting material. Microchannel plates (MCP) are made up of an array of thin tubes, typically 5-25 μm in diameter, formed in a glass plate and coated in a similar manner [8]. Because the electrons remain in the channels, MCPs preserve the pattern or image present at the entrance to the MCP. Although the output image may be degraded somewhat from the image at the photocathode by focusing effects and by photoelectrons striking between channels and "hopping", microchannel plates make excellent imaging detectors and are used in a wide variety of image intensifier applications ranging from night-vision goggles and high-speed oscilloscope screens to astronomical detectors. When an MCP is used with an array of anodes, the resulting position or image sensitive phototube is often referred to as multi-anode microchannel array or MAMA.

MCPs with areas of ~100 cm^2 are commercially available (an 80 cm^2 plate with 25 μm channels is used as an image intensifier in the Tektronix 2467B oscilloscope) and MAMA detectors have been used in a number of space-based experiments including SOHO and the STIS instrument for the HST.

Because ion feedback to the photocathode can be a major source of noise, individual channel multipliers are generally curved to reduce the chance of ions being transmitted back up the channel. Similarly, microchannel plates are usually made with the channels at an angle to the plate surface and are arranged in a "chevron" of two MCPs or a "Z" of three MCPs. This slightly degrades the image. However, the resulting image resolution of about 4 line-pairs/mm (equivalent to a position resolution of 125 μm) for a Z stack of 25 μm microchannel plates far exceeds OWL requirements.

Microsphere plates (MSP) are made up of small glass spheres, typically 20-100 μm in diameter, which are coated with a secondary emitting material and are sintered into a solid plate [9]. This plate is overcoated top and bottom with a conductor. A potential difference is established between the top and bottom. After initial collection electrons "percolate" through the plate, impacting many spheres and producing a cascade of secondary electrons. Large gains (up to 10^8) can thus be achieved. Ion feedback is strongly suppressed and vacuum requirements are reduced compared to MCPs. Because of the percolation process, the image resolution is lower than for an MCP and a high gain MSP with 50 μm spheres is limited to about 2 line-pairs/mm (equivalent to a resolution of 250 μm), still well in excess of OWL requirements. Although the MSP is new, it is a scalable and relatively robust technology and may be well suited to OWL.

A schematic of a representative PMT using MCPs or MSPs is shown in Figure 2.

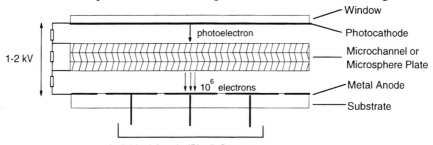

Figure 2: Schematic cross-section of a proximity focused microchannel/microsphere photomultiplier.

Regardless of the details of the multiplier structure, vacuum photomultipliers can be read out directly using an array of individual anodes. It should be noted that position sensitive photomultipliers are often built as derived-image devices in which a set of crossed wires are used to record the location of the incident photons. These are unsuitable for OWL since they would integrate background over the full length of the wire.

It is also possible to follow the multiplication section with a phosphor screen. The optical image on the screen can be viewed by a semiconductor array (CCD, PD, or APD) either directly or through an optical image reducer. This technique is often used in image intensified video systems. For OWL such an approach may be useful if the semiconductor pixel is the first element in the analog storage chain discussed above and if readout is by individual pixels.

Multiplication of photoelectrons can also be achieved using an ionization energy deposit multiplier [10], commonly known as a hybrid PMT (HPMT). In an HPMT (see schematic drawing), a photodiode or array of photodiodes is used as an anode plane. A large potential (10-15 keV) is established between the photocathode and the anode (diode) plane. Photoelectrons accelerated to high energy strike the diode and are stopped. The ionization energy loss of the accelerated photoelectrons in the photodiode results in gain of ~ 1 to 5×10^3, depending on the accelerating voltage and the window thickness of the diode. By replacing the PD plane with APDs, gain can be increased by an additional factor of up to 10^3.

As compared to the secondary emission multipliers discussed above, the HPMT offers greatly enhanced single photoelectron resolution. Although the structure of the HPMT can be very simple, as illustrated in Figure 3, designing the tube to sustain very high voltages increases the difficulty and may result in an increase in dead area. In addition, the cost per pixel can be much greater due to the PD (perhaps $10/cm^2) or APD array (currently about $1000/cm^2) although an effort is underway to reduce these costs. Since the pixel readout must be on an individual basis, semiconductor strip detectors are not useful as the detector plane in an HPMT for OWL.

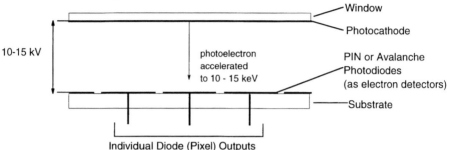

Figure 3: Schematic cross-section of a proximity focused hybrid photomultiplier.

In addition to the vacuum photodetectors described above, a number of gas avalanche multiplication devices are possible. Currently, gas electron multiplication (GEM) devices [11] which develop gain by gas avalanches at successive mesh electrodes are under development by a number of groups as a (relatively) cheap, large-area photodetector. However, for OWL, these offer no real advantages while adding expendables (gas) and complexity.

In general, it is expected that any photomultiplier developed for OWL will employ proximity focusing. In this technique the electric field lines are straight and the image is transferred in a 1-to-1 form with no reduction between sections of the PMT (e.g. between the photocathode and the multiplier or the multiplier and the anode array).

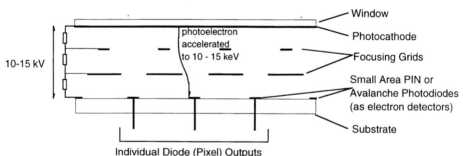

Figure 4: Schematic cross-section of an electrostatic focused hybrid photomultiplier.

However, in the case of the HPMT electrostatic focusing may make it possible to use smaller and more economical diodes or arrays of diodes. This is illustrated in Figure 4.

The principal advantage of photomultiplier devices is that there are a wide range of well developed technologies which can meet all of the OWL performance requirements. All of these offer clean intrinsic gain to > 10^6. However, all of these devices have relatively complex structures.

Vacuum photon detectors generally require an optical vacuum window. For OWL this represents a considerable difficulty due to the large area of the focal plane array and places a practical upper limit on the dimensions of each detector module. The structure necessary to support the vacuum windows represents the major source of dead area in the focal plane system

It is possible to operate a vacuum photomultiplier open to the vacuum of space and small MCP-based VUV detectors have been made for space applications using the photocathode material deposited directly on the MCP surface. The thermal emission from a bialkali photocathode due to joule heating of the multiplier would likely preclude a simple application of this technique to OWL. However it would be possible to deposit the photocathode on a thin leading MCP separate from the main multiplier and operated only at a high enough voltage to provide reliable secondary electron collection. The photocathode could also be deposited on a high-transparency pierced foil. It is unlikely that the photocathode could be generated on-orbit and so, regardless of the multiplier technology employed, a vacuum window would still be required. This could be either opaque or transparent (to facilitate ground tests) but would be removed on-orbit.

WINDOW CONSIDERATIONS

Due to the required field of view, light can strike an OWL pixel from a relatively large angle. Since the same pixel will receive light from all parts of the optical system, the window cannot be curved to compensate for the incident angle and hence refraction errors from the window can be significant, as illustrated in Figure 5.

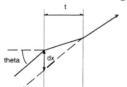

Figure 5: Error (dx) for n = 1.5 window

Theta	dx
10°	5.97×10^{-2} t
20°	9.28×10^{-2} t
30°	2.24×10^{-1} t

For "large" photon incident angles, the thickness of the window, t, must be much smaller than the pixel dimension. Unfortunately, a large plate thickness is needed to support the external pressure of the atmosphere while on the ground. For a 30 cm wide tube, for example, the window must be > 1.5 cm thick to support this load.

One solution to this problem is to provide an internal support structure, probably consisting of several small "posts" joining the anode plane to the window for mutual support. These could serve a dual role as supports for the multiplier.

Another solution to this problem is a glass honeycomb structure which combines thin window regions with a thicker glass web or honeycomb for structural strength. The window and honeycomb walls could be cast together as a unit or honeycomb could be fused to the window. For increased photon detection efficiency, the photocathode could be deposited on both the inner surface of the window regions and on the walls of the honeycomb.

BASELINE OWL PHOTODETECTOR

From the discussion above, the baseline OWL photodetector can be identified. This will be a vacuum photomultiplier with a transparent vacuum window to facilitate ground testing. The vacuum window will use glass honeycomb technology to reduce both optical thickness and weight and will be made of conventional UV-transparent borosilicate glass (Corning 9741 or equivalent) unless radiation damage considerations dictate using UV grade fused silica.

The photodetector modules will use a semitransparent bialkali or S20-UV photocathode unless "tailored" formulations currently under consideration prove superior. The photocathode will be deposited on the window before assembly and then transferred to the tube in vacuum. This will provide the highest photocathode uniformity and quantum efficiency.

The multiplication will be provided by a Z-stack of 25 μm channel MCPs or by a MSP approximately 1.4 mm thick and using 50-100 μm spheres. These will probably be distributed in a tiled arrangement to cover the area of the module, although scaling up the multiplier technology to cover a full module will be investigated. This is most likely to be practical with the MSP.

The anode plane will be made up of individual pixels deposited on a ceramic substrate. Connection to the electronics will be made through cast-in connections incorporated into the substrate.

The baseline detector is shown in cross-section in Figure 6 and in an exploded schematic in Figure 7.

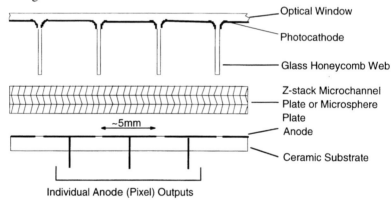

Figure 6: Schematic cross-section of the baseline OWL photodetector.

It is expected that the sidewalls of the detector will be stainless steel and that the internal structure (if required) will largely be ceramic or glass cast along with the anode substrate or the window.

Although the dimensions of each detector module are the subject of ongoing study, a 30 cm hexagon (apex-to-apex) provides a reasonable balance of yield and module number. It is estimated that using the techniques outlined above, each module could be made about 4 to 5 cm thick.

The column density could be as low as an average of about 2 g/cm^2 for the detector. The tube structure would add about 400 g for a 30 cm hexagonal tube. A 30 cm hexagonal tube with 25 mm^2 pixels (as a representative area) would total about 1.6 kg for 2.3 x 10^3 pixels. A full 4.3 x 10^5 pixel array would require ~200 (187 plus edge

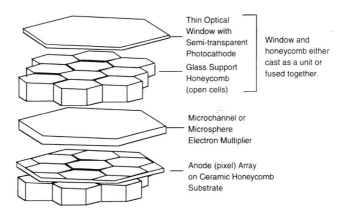

Figure 7: Exploded schematic of the baseline OWL photodetector.

losses) tubes weighing 320 kg. Thus, with a 25 % allowance for joining structure the focal plane detector would weigh about 400 kg.

The cost of the focal array can be estimated roughly from the cost of current MCPs. At present a single layer 25 μm MCP can be produced in volume for ~ $10/cm^2. Three MCPs are required for the Z-stack. When the cost of the window, photocathode, anode array, structure, and assembly are added the tube cost will be about $60/cm^2 or $35 k/tube. Without considering the cost of support or deployment structures this results in a base cost of $7 M/array.

To summarize the discussion above, the baseline photodetector for OWL is a close-packed array of about 200 large area (~585 cm^2) vacuum photomultiplier modules. Bialkali or S20-UV semitransparent photocathodes deposited on a UV transparent window using a transfer technique provide ~ 25 - 30 % quantum efficiency. The window optical thickness is reduced by using a glass honeycomb structure. A Z-stack of three microchannel plates with large channels (25 μm) and 40:1 length-to-diameter ratios results in a gain of >10^6 (microsphere plates are being evaluated as an alternative). Readout is by individual anodes with areas of 3.5 to 46 mm^2 coupled through a ceramic substrate.

CONCLUSION

Development of the OWL focal plane detector system will present a variety of challenges ranging from the technical details of the photodetector modules and readout system to the difficult engineering issues involved in launching and deploying such a large and relatively delicate detector. These issues will receive careful study by the OWL team in the coming years. We will also take special care to remain current regarding developments in the art of photon detection which may alter the technologies finally adopted for OWL. At present, however, we can conclude that a photodetector concept which is fully capable of meeting all of the OWL performance requirements has been identified and established as a baseline.

REFERENCES

1. R.E. Streitmatter, this volume (1998).
2. D.J. Lamb, *et al.*, this volume (1998).
3. See, reviews in e.g., G.F Knoll, Radiation Detection and Measurement 2nd. Edition, John Wiley and Sons, New York, 1989.
4. R. Farrell, *et al.*, *Nucl. Inst. and Meth.* **A353** 176 (1994).
5. See, e.g., R.W. Engstrom, Phomultiplier Handbook, Burle Industries Inc., Lancaster, PA, (1980); Photomultiplier Tubes Principals and Applications, Philips Photonics, Brive, France, (1994).
6. Philips Photonics, e.g. information on XP1700 perforated metal foil dynode multi-channel PMT.
7. Hamamatsu Photonics, e.g. information on R4549 fine mesh dynode multi-anode PMT, and H6564 metal channel dynode multi-anode PMT.
8. Galileo Electro-optics Corp., e.g. information on Long-life Microchannel Plates; Hamamatsu Photonics; Philips Photonics; Photek Limited; also ref. 4.
9. A.S. Tremsin, *et al.*, *Nucl. Inst. and Meth.* **A368** 719 (1996).
10. G. Anzivino, *et al.*, *Nucl. Inst. and Meth.* **A365** 76 (1995).
11. R. Bouclier, *et al.*, CERN-PPE/97-23 (1997).

Polymer Selection Criteria for the Orbiting Wide Angle Light-collector (OWL) Project Lens Material

Thomas M. Leslie*‡, Elizabeth Burleson*, John Dimmock†‡,
David J. Lamb†, Lloyd W. Hillman†‡, Yoshiyuki Takahashi†‡,
Michael D. Watson◊

*Department of Chemistry, †Department of Physics, and the ‡Center for Applied Optics
The University of Alabama in Huntsville, Huntsville, Alabama 35899
◊NASA Marshall Space Flight Center, Huntsville, Alabama 35812

Abstract. Optical quality polymers normally incorporate functional groups that provide physical integrity that are incompatible with deep UV radiation. Groups like the phenyl ring found in polystyrene (PS) increase the refractive index of the polymer, provide a bulky group that inhibits crystallization rendering the material amorphous, and adds impact strength to the polymer by coupling vibrations of the ring into fracture mechanisms. However these groups also absorb UV light at wavelengths starting at about 330 nm or longer repressing ultra-violet transparency. Our search for a polymer with properties similar to polystyrene or poly(methyl methacrylate) but with better ultraviolet transparency led us to study poly (methylpentene) as a potential lens material candidate.

INTRODUCTION

Polycarbonate, one of the most common optical polymers, is the polymer that impact resistant safety glasses are made of. It is a great material from a health and safety point of view. It has a very high impact strength with a cut-off of UV radiation to almost 400 nm. This provides the worker with both particle and UV protection. Poly(methylmethacrylate) (PMMA,) is bullet proof when thick enough and has been used as unbreakable spectrometer cuvettes in the laboratory. PMMA has better UV transparency than polycarbonate but still transmits only 82 percent of the available light at 320 nm. With cosmic ray air shower photons not being very plentiful, it is preferable if there are no losses to absorption by the lens.

Polystyrene and poly(methylmethacrylate) are both available as "optical grade" polymers for use in applications like visible spectroscopy, but are even less appropriate since the stabilizers mentioned above are left out of the formulation. The functional groups previously mentioned will absorb UV light and may cause photochemistry that will

CP433, *Workshop on Observing Giant Air Showers from Space*
edited by J.F. Krizmanic et al.

eventually lead to degradation of the polymer. The transmission characteristics of PMMA and PS disposable spectrometer cells are highlighted in Table 1.

TABLE 1. The Transmission Characteristics of Polymer Spectrophotometer Cells[a]

Wave-length (nm)	PS % T	PMMA % T	Wave-length (nm)	PS % T	PMMA % T
800	88	92	360	82	89
700	88	92	350	80	88
600	85	92	340	75	86
500	85	91	320	60	82
400	85	90	300	35	70
380	85	90	280	<5	47

[a]The Fisher Catalog, under the heading "Spectrophotometer Cells", page 1520, 1993/1994.

Fused silica on the other hand offers excellent transmission over the wavelength range of 190 to 2500 nm depending on the grade. It is brittle and can be broken. It can be machined and polished to shape.

The best choice of polymer for this application with regard to UV stability is to minimize any type of functional group. Therefore, one might expect polyethylene to be the best material. It has only carbon hydrogen and carbon carbon single bonds. Polyethylene (PE) however is a partially crystalline polymer in its native state and therefore scattering. It is comprised of two types of internal structure. The one being crystalline areas with one refractive index imbedded in an amorphous matrix of lower refractive index. Any crystallites will produce scattering centers which are unacceptable in a lens even if it need not be diffraction limited. Also, the amorphous phase in PE has a glass transition temperature far below room temperature making it impossible to make a lens whose properties would not change very much.

PMP is also comprised of only carbon-hydrogen and carbon carbon single bonds but it does not have two phases with different refractive indexes. Also, the glass transition temperature is above that of PMMA and polystyrene giving it similar mechanical-physical properties.

Properties of Poly(methyl pentene), PMP

PMP sample jars with approximately two mm thick walls were obtained for the determination of spectral characteristics. These jars were manufactured by an injection molding process and were studied as received. The jars were first broken with a hammer to fit into the spectrophotometer. All spectra were obtained on a HP 8452A diode array spectrophotometer over the range of 200 to 820 nm. A background (blank) spectrum for the spectrophotometer was obtained using a fused silica (quartz) absorption cell to account for surface reflections. The spectrum of the samples were then obtained. Figure 1 shows the PMP as obtained from the supplier. The arrows on the spectrum are at 344, 298, 280, 252, and 234 nm. The peak with an absorption maximum at 280 nm is indicative of a low molecular weight compound that has been added to the polymer. As previously stated, the polymer should not have any absorption until almost 200 nm.

To test this hypothesis, the sample was soaked in tetrahydrofuran (THF) for 3 days, removed from the solvent, soaked for an additional 2 days, removed then dried. The THF aliquots were combined and evaporated. A waxy residue was obtained. The spectrum of the sample jar was obtained once again and the absorption peak at 280 nm was gone. The absorption of the sample at all marked wavelengths was markedly decreased as seen in Figure 2. This indicates the major contribution to the absorption seen in the sample are due to the additive, not the polymer.

TABLE 2. Comparison of PMP With and Without the Small Molecule Additive.

Wavelength (nm)	Absorbance With Additive	Absorbance Without Additive
344	0.073	0.019
298	0.227	0.066
280	1.160	0.122
252	0.486	0.212
234	1.905	0.361

Careful inspection of the spectra shown in Figures 1 and 2 show a slight absorbance at about 660 nm. This is an artifact of the instrument not a true absorption. The detector is made of two different diode arrays and this "peak" is where they are joined. More importantly, the absorption of the sample with the small molecule compound removed at 298 nm is only 0.066 corresponding to a transmission of 86% and the transmission at 344 nm is 96% (an absorbance of 0.0187). At this time, it is not possible to state if the absorption is due to true absorptions of the material or due to scattering by the sample.

Further tests (infrared spectroscopy) on the extracted waxy small molecule show it to have a carbonyl functional group consistent with the UV absorption at 280 nm.

Conclusions

The sample of PMP obtained from the manufacturer contains an additive that has some UV absorption in the region of interest. This material can be extracted from the polymer and the transmission characteristics greatly improved. The transmission of PMP at the wavelengths of interest is almost 10% better than optical grade PMMA that can be obtained as disposable spectrophotometer cells. Since the samples jars obtained were formed using a precision injection molding process, it is reasonable to consider manufacturing lenses using this technique when precision optics are not necessary.

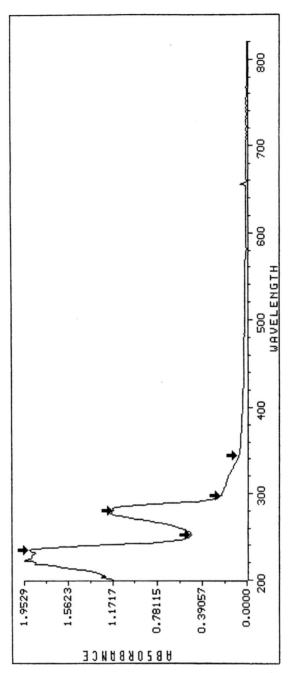

FIGURE 1. The absorption spectrum from 200 to 820 nm of poly(methypentene) as received from the manufacturer containing the small molecule additive.

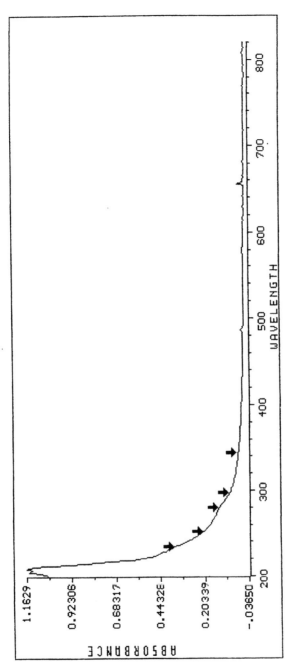

FIGURE 2. The absorption spectrum from 200 to 820 nm of poly(methypentene) after removing the small molecule additive from sample with tetrahydrofuran.

Workshop on *"Observing the Highest Energy Particles (>10^{20} eV) from Space"*

The Inn and Conference Center
University of Maryland University College

November 13-15, 1997

Thursday, November 13, 1997

8:00 AM   **Registration**

8:30 AM   **Opening Session**
Chair: Alan Bunner

> A. Bunner   *Opening Remarks*
> J. Linsley   *Early History and Challenge*

9:10 AM   **Session #1: Acceleration of >100 EeV Particles**
Chair: Floyd Stecker

> I. Axford   *Overview on Shock Acceleration*
> M. Livio   *Acceleration of CR by rotating AGN*

10:10 AM   **Morning Break** (provided in the concourse)

10:30 AM   **Session #2: Acceleration of >100 EeV Particles II**
Chair: Floyd Stecker

> P. Biermann   *Acceleration of CR in Halo of AGN Jets*
> F. Jones   *Acceleration of CR by Colliding Galaxies*
> L. Scarsi   *Gamma Bursts and EHCR*

11:50 AM   **Lunch** (provided in the Founders Room)

1:00 PM   **Session #3: Early Work**
Chair: Gene Loh

G. Tanahashi	*Early Air Fluorescence Work: Cornell and Japan*
P. Sokolsky	*Utah's Realization of the Air Fluorescence Detector: Fly's Eye*
M. Nagano	*Summary of AGASA Results*

2:15 PM   **Afternoon Break** (provided in the concourse)

2:30 PM   **Session #4   Potential Observations From Space**
Chair: Thomas Gaisser

C. DeMarzo	*MASS, Airwatch*
R. Streitmatter	*OWL*
B. Khrenov	*Russian Plan for the Detection of $CR > 10^{19}$ eV on ISS*
Y. Takahashi	*Planned Efforts in Japan*

3:50 PM   **Afternoon Break** (provided in the concourse)

4:05 PM   **Session #5: Phenomena at High Energy**
Chair: Robert Streitmatter

I. Sarcevic	*Interactions of 10^{20} eV Neutrinos*
S. Klein	*LPM Effect*
L. Gonzales-Mestres	*Physics Opportunities Above the GZK Cutoff*

6:00 PM   **Banquet** at the Inn and Conference Center
Founders Room

After Dinner Talk by Dr. George Clark

Friday, November 14, 1997

8:00 AM   **Registration**

8:30 AM   **Session #6: New Physics and Astrophysics**
Chair: Frank Jones

P. Bhattacharjee	*Strings, Monopoles, Necklaces and Other TDs*
P. Kronberg	*Extra-galactic Magnetic Fields*
F. Stecker	*Propagation of Heavy Nuclei Through Intergalactic Space*

10:00 AM   **Morning Break** (provided in the concourse)

10:20 AM   **Session #7: Astrophysics I**
Chair: Francis Halzen

S. Yoshida	*Propagation of CR and Neutrinos Through Space*
G. Farrar	*Can UHECR's Be Evidence for New Particle Physics?*
G. Sigl	*Angle-Time-Energy Images of Ultra-High Energy Cosmic Ray Sources*

11:50 AM   **Lunch** (provided in the Ft. McHenry Room)

| 1:00 PM | **Session #8: Astrophysics II** |
| | Chair: Motohiko Nagano |

	T. Weiler	*Physics Opportunities Above the GZK Cutoff*
	D. Cline	*Physics Potential of the Detection of UHE Neutrinos with OWL*
	F. Halzen	*Physics and Neutrinos at OWL Energies*

| 2:30 PM | **Afternoon Break** (provided in the concourse) |

| 2:50 PM | **Session #9: Astrophysics III** |
| | Chair: Motonhiko Nagano |

| | V. Berezinsky | *Ultra-high Energy Cosmic Rays from Decaying Particles* |
| | T. Gaisser | *Capabilities and Limitations of Air Shower Modeling* |

| 4:00 PM | **Poster Papers** (in the concourse) |
| | **Wine & Cheese** |

Saturday, November 15, 1997

8:00 AM   **Registration**

9:00 AM   **Session #10: Properties of the Atmosphere as a Detector Seen from Space**
Chair: Thomas Parnell

K. N. Liou	*Radiation Transport* (presented by D. Starr)
L. Hillman	*Wide-angle Optics, Principles and Development*
K. Driscoll	*Lightning*
D. Starr	*Clouds/Other Atmospheric Phenomena*

10:20 AM **Morning Break** (provided in the concourse)

10:35 AM **Session #11: Plans for Ground Based Observations**
Chair: Jonathan Ormes

G. Giannini	*Initial Results from the CHOOZ, Long Baseline Reactor Neutrino Oscillation Experiment*
M. Teshima	*Status of the Telescope Array/Japan-US Consortium*
C. Pryke	*Auger: What, Why and How*
J. Cronin	*Status of the Auger Project*

11:20 AM **Session #12: Closing Remarks**

D. Schramm	*Summary*
J. Ormes	*Closing Remarks / Adjournment*

Listing of Poster Papers

Airwatch Collaboration   #15
*Scenarios for a Space Observatory for Detection of Extreme
Energy Cosmic Rays and Atmospheric Phenomena*

Airwatch Collaboration   #16
The Optical System for the Exploratory Mission

Airwatch Collaboration   #16
The Fast Detector

Airwatch Collaboration   #17
Background Measurement with UVSTAR

Airwatch Collaboration   #17
Air Fluorescence Efficiency Measurements: Experimental Set-up

R. Antonov, et al.   #01
*Balloon Borne Detector for the Cosmic Ray Energy Spectrum
Measurements*

V.S. Berezinsky and P. Blasi   #02
Low Frequency Radio Radiation from Clusters of Galaxies

H.Y. Dai and P. Sokolsky   #03
OWL Aperture and Resolution

G. Domokos and S. Kovesi-Domokos
Observation of UHE Neutrino Interactions from Outer Space

E. Flyckt   #06
Focal Plane Dectectors

L. Gonzales-Mestres   #07
Observing Air Showers from Cosmic Superluminal Particles

ILC Dover   #07
Space Inflatable Structure

D.J. Lamb, et al.   #14
Wide Angle Refractive Optics for Astrophysics Applications

D.J. Lamb, et al.   #14
Computer Analysis of Optical Systems Containing Fresnel Lenses

D.J. Lamb, et al.   #15
Focal Plane Reduction of Large Aperture Optical Systems

E.C. Loh, et al.   #08
Concepts for OWL Inflatable Optics

J.W. Mitchell for the OWL Study Team $\#\pi$
Photodetectors for OWL

L. Popova and G. Kamberov   #11
*Connection Between the Statistical Parameters of Hadronization
and the QCD Coupling Constant*

O.A. Saprykin, et al.   #29
*Computer Aided Optimal Design of Space Reflectors and
Radiation Concentrators*

Y. Takahashi   #12
*Study of Correlations of EHE Cosmic Ray Events with Gamma
Ray Bursts*

Y. Takahashi   #13
*The Multi-OWL, an Assembly Concept by Using EVA's
on the ISSA*

K.O. Thielheim
Cosmic Particle Acceleration of the Highest Energies

D.J. Wagner and T.J. Weiler   #28
*Neutrino Oscillations from Cosmic Sources: A Possible
Nu Window toTiny Mass and to Cosmology*

Workshop Participants

Tareq Abuzayyad
University of Utah
Dept. of Physics
201 JFB
Salt Lake City, UT, 84112
tareq@franny.physics.utah.edu

Ivone F.M. Albuquerque
University of Chicago
Astronomy & Astrophysics
5640 S. Ellis Ave.
Chicago, IL 60637
ifreire@mafalda.uchicago.edu

Glenn E. Allen
NASA/GSFC
Code 662
Greenbelt, MD 20771
glenn.allen@gsfc.nasa.gov

John L. Anderson
NASA HQ
Advanced Concepts Exec.
Code SM
Washington, DC 20546
john.anderson@hq.nasa.gov

Rem A. Antonov
Skobeltsyn Inst. of Nuclear Physics
Moscow State University
Vorobjevy Gory
Moscow, RUSSIA
antr@dec1.npi.msu.su

Primo Attiná
Alenia Aerospazio
Space Division
C. so Marche. 41
Turin, 10146, ITALY
pattina@to.alespazio.it

W. Ian Axford
MPAe Lindau
Max-Planck-Str. 2
D-37191 Katlenburg-Lindau
GERMANY
axford@linmpi.mpg.de

Louis M. Barbier
NASA/GSFC
Code 661
Greenbelt, MD 20771
lmb@cosmicra.gsfc.nasa.gov

Matthew Baring
NASA/GSFC
Code 661
Greenbelt, MD 20771
baring@lheavx.gsfc.nasa.gov

Venya Berezinsky
INFN
Lab. Nazionali del Gran Sasso
SS 17/bis km 18+910
67010 Assergi (AQ)
ITALY
veniamin.berezinsky@lngs.infn.it

Dave Bertsch
NASA/GSFC
Code 661
Greenbelt, MD 20771
dlb@mozart.gsfc.nasa.gov

Pijush Bhattacharjee
NASA/GSFC
Code 661
Greenbelt, MD 20771
pijush@milkyway.gsfc.nasa.gov

Peter L. Biermann
Max-Plank-Inst.. fur Radioast.
Auf dem Hugel 69
D-Bonn 1, GERMANY
plbiermann@mpifr-bonn.mpg.de

Pasquale Blasi
University of Chicago
Dept. of Astronomy & Astrophysics
E. Fermi Institute
Chicago, IL 60637-1433
blasi@oddjob.uchicago.edu

Elihu Boldt
NASA/GSFC
Code 661
Greenbelt, MD 20771
Elihu.Boldt@gsfc.nasa.gov

Jerry Bonnell
NASA/GSFC/USRA
Code 660
Greenbelt, MD 20771
bonnell@grossc.gsfc.nasa.gov

Kevin Boyce
NASA/GSFC
Code 663
Greenbelt, MD 20771
Kevin.Boyce@gsfc.nasa.gov

A. Lyle Broadfoot
University of Arizona
Lunar & Planetary Lab.
901 Gould Simpson Bldg.
Tucson, AZ 85721
broadfoot@looney.lpl.arizona.edu

Alan Bunner
NASA HQ
Code S
Washington, DC 20546
Alan.Bunner@hq.nasa.gov

Francesco S. Cafagna
INFN, Bari
c/o Dip. di Fisica
v. Amendola 173
Bari, 70126 ITALY
francesco.cafagna@ba.infn.it

Osvaldo Catalano
CNR
Via U. La Malfa 153
Palermo, 90146 ITALY
catalano@ifcai.pa.cnr.it

Eric R. Christian
NASA/GSFC/USRA
Code 661
Greenbelt, MD 20771
erc@cosmicra.gsfc.nasa.gov

Marco Circella
INFN, Bari
Via Amendola 173
Bari, I-70126 ITALY
circella@ba.infn.it

George W. Clark
M. I. T.
Physics Dept.
MIT 37-611
Cambridge, MA 02139
gwc@space.mit.edu

David B. Cline
UCLA
Physics Dept.
405 Hilgard Ave.
Los Angeles, CA 90095-1547
dcline@physics.ucla.edu

Tom Cline
NASA/GSFC
Code 661
Greenbelt, MD 20771
cline@apache.gsfc.nasa.gov

Alberto Cordero
University of Puebla
A.P., 1152
Puebla, MEXICO

Jim Cronin
University of Chicago
Astronomy & Astrophysics
5640 S. Ellis Ave.
Chicago, IL 60637
jwc@uchepa.uchicago.edu

Hongyue Dai
University of Utah
Physics Dept.
201 JFB
Salt Lake City, UT 84112
dai@booboo.physics.utah.edu

Francesco De Paolis
Bartol Research Institute
University of Delaware
Newark, DE 19716
depaolis@bartol.udel.edu

Carlo De Marzo
INFN
Dipartimento di Fisica
via Amendola 173
70126 Bari, ITALY
demarzo@ba.infn.it

Kevin Driscoll
Global Hydrology & Climate Ctr.
977 Explorer Blvd.
Huntsville, AL 35806
driscoll@kchange.msfc.nasa.gov

David Ehrenstein
Science Magazine
AAAS
1200 New York Ave., NW
Washington, DC 20005
ehrenstein@nasw.org

Robert Ellsworth
George Mason University
Physics Dept.
Fairfax, VA 22030
ellswort@gmu.edu

Ralph R. Engel
Bartol Research Institute
University of Delaware
Sharp Lab.
Newark, DE 19716
eng@lepton.bartol.udel.edu

Joseph A. Esposito
NASA/GSFC/USRA
Code 661
Greenbelt, MD 20771
jae@egret.gsfc.nasa.gov

Glennys R. Farrar
Rutgers University
Physics Dept.
P.O. Box 849
Piscataway, NJ 08855-0849
farrar@farrar.rutgers.edu

Thomas K. Gaisser
Bartol Research Institute
University of Delaware
Newark, DE 19716
gaisser@bartol.udel.edu

Gianrossano Giannini
INFN
Via Valerio
2 Trieste 34127, ITALY
giannini@trieste.infn.it

Salvo Giarrusso
IFCAI/CNR
Via U. La Malfa 153
90146 Palermo, ITALY
jerry@ifcai.pa.cnr.it

Luis Gonzales-Mestres
Laboratoire de Physique Corpusculaire
College de France
11 pl. Marcellin-Berthelot
Paris, FRANCE
lgonzalz@vxcern.cern.ch

Jordan Goodman
University of Maryland
Dept. Of Physics
College Park, MD 20742
goodman@umdgrb.umd.edu

Mark Grahne
ILC Dover, Inc
One Moonwalker Rd.
Frederica, DE 19946
grahnm@ilcdover.usa.com

Anna Gregorio
C.A.R.S.O.
Via Davis, 116
Trieste 34100 ITALY
a.gregorio@ts.infn.it

Sunil K. Gupta
NASA/GSFC
Code 661
Greenbelt, MD 20771
gupta@cosmicra.gsfc.nasa.gov

Francis L. Halzen
University of Wisconsin, Madison
Physics Department
1150 University Ave.
Madison, WI 53706
halzen@pheno.physics.wisc.edu

Alice K. Harding
NASA/GSFC
Code 661
Greenbelt, MD 20771
harding@twinkie.gsfc.nasa.gov

Lloyd Hillman
University of Alabama, Huntsville
Dept. Of Physics
Huntsville, AL 35899
hillmanl@email.uah.edu

Carlos Hojvat
MS 105
Fermilab
P.O. Box 500
Batavia, IL 60510
hojvat@fnal.gov

Frank Jones
NASA/GSFC
Code 661
Greenbelt, MD 20771
Frank.Jones@gsfc.nasa.gov

W. Vernon Jones
NASA HQ
Code SR
Washington, DC 20546
wvjones@hq.nasa.gov

Demos Kazanas
NASA/GSFC
Code 661
Greenbelt, MD 20771
kazanas@lheavx.gsfc.nasa.gov

Boris A. Khrenov
Skobeltsyn Institute of Nuclear Physics
Moscow State University
Vorobjevy Gory
Moscow, RUSSIA
khrenov@eas.npi.msu.su

Spencer R. Klein
Lawrence Berkeley National Lab
70A-3307 LBNL
1 Cyclotron Road
Berkeley, CA 94720
srklein@lbl.gov

John F. Krizmanic
NASA/GSFC/USRA
Code 661
Greenbelt, MD 20771
jfk@cosmicra.gsfc.nasa.gov

Philipp Kronberg
University of Toronto
Dept. of Astronomy
60 St. George St.
Toronto, Ontario, M5S 1A7 CANADA
kronberg@astro.utoronto.ca

David J. Lamb
University of Alabama, Huntsville
Dept. Of Physics
Optics Building, Room 201C
Huntsville, AL 35899
lamb@linger.uah.edu

529

Anna Lenti
Laben S.P.A.
S.S. Padama Superiore 290
20090 Vimoprone
(Milan) ITALY
lenti.A@laben.it   .

John Linsley
University of New Mexico
Dept. Of Physics & Astronomy
Albuquerque, NM 87131
quargnali@apsicc.aps.edu

Mario Livio
Space Telescope Science Inst.
3700 San Martin Drive
Baltimore, MD 21218
mlivio@stsci.edu

Gene Loh
NASA/GSFC
Code 661
Greenbelt, MD 20771
loh@cosmic.physics.utah.edu

Rebeca Lopez
Preparatorio Emiliano Zapata
BUAP
4 Norte No. 6, Col Centro
7200 Puebla
Puebla, MEXICO
rlopez@fcfm.buap.mx

David L. Manion
NASA/GSFC/HSTX
Code 661
Greenbelt, MD 20771
dmanion@pop3.stx.com

Francesco Miniati
University of Minnesota
116 Church St., S.E.
Minneapolis, MN 55455
min@msi.umn.edu

John W. Mitchell
NASA/GSFC/USRA
Code 661
Greenbelt, MD 20771
mitchell@lheavx.gsfc.nasa.gov

Alexander A. Moiseev
NASA/GSFC
Code 661
Greenbelt, MD 20771
moiseev@cosmicra.gsfc.nasa.gov

Harm Moraal
South Africa

Motohiko Nagano
Institute for Cosmic Ray Research
University of Tokyo
3-2-1 Midoricho
Tanashi-shi, Tokyo 188, JAPAN
mnagano@icrr.u-tokyo.ac.jp

Jay Norris
NASA/GSFC
Code 661
Greenbelt, MD 20771
norris@groax0.gsfc.nasa.gov

Angela V. Olinto
University of Chicago
5640 S. Ellis Ave.
Chicago, IL 60605
olinto@oddjob.uchicago.edu

Jonathan Ormes
NASA/GSFC
Code 660
Greenbelt, MD 20771
ormes@lheapop.gsfc.nasa.gov

Thomas Parnell
NASA/MSFC
ES 84
Huntsville, AL 35812
parnell@ssl.msfc.nasa.gov

Lara Pasquali
University of Iowa
Physics & Astronomy Dept.
501 Van Allen Hall
Iowa City, IA 52242-1479
pasquali@hepsun1.physics.uiowa.edu

Lilia M. Popova
Institute of Nuclear Research &
Nuclear Energy
Tsarigradsko Shosse 72
Sofia, 1784 BULGARIA
kambrov@math.wustl.edu

Clement L. Pryke
University of Chicago
LASR
933 East 56th Street
Chicago, IL 60637
pryke@hep.uchicago.edu

Humberto Salazar
University of Puebla
A.P., 1152
Puebla, MEXICO
hsalazar@fcfm.buap.mx

Bruno Sacco
IFCAI/CNR
Via U. La Malfa 153
90146 Palermo, ITALY
sacco@ifcai.pa.cnr.it

Ina Sarcevic
University of Arizona
Physics Department
Tucson, AZ 85721-0065
ina@physics.arizona.edu

Livio Scarsi
IFCAI/CNR
Via U. La Malfa 153
90146 Palermo, ITALY
scarsi@ifcai.pa.cnr.it

David N. Schramm
University of Chicago
5640 S. Ellis Avenue
AAC 140
Chicago, IL 60637
dns@oddjob.uchicago.edu

M. M. Shapiro
205 Yoakum Pkwy. #1514
Alexandria, VA 22304

Guenter H.W. Sigl
University of Chicago
5640 S. Ellis Ave.
Chicago, IL 60637
sigl@oddjob.uchicago.edu

Pierre Sokolsky
University of Utah
Physics Dept.
Salt Lake City, UT 84112
ps@cosmic.physics.utah.edu

David Starr
NASA/GSFC
Code 913
Greenbelt, MD 20771
starr@climate.gsfc.nasa.gov

Floyd Stecker
NASA/GSFC
Code 661
Greenbelt, MD 20771
stecker@lheavx.gsfc.nasa.gov

Alfred Stephens
Space Physics Division
TATA Institute of
Fundamental Research
Homi Bhabha Road
Mumbai 400 005 INDIA
alfred@tifrvax.tifr.res.in

Don Stilwell
NASA/GSFC
Code 661
Greenbelt, MD 20771
Donald.Stilwell@gsfc.nasa.gov

Steven J Stochaj
New Mexico State University
Particle Astrophysics Lab
Las Cruces, NM 88001
stochaj@nmsu.edu

Robert Streitmatter
NASA/GSFC
Code 661
Greenbelt, MD 20771
streitmatter@lheavx.gsfc.nasa.gov

Yoshi Takahashi
University of Alabama, Huntsville
Dept. Of Physics
OB 212
Huntsville, AL 35899
takahashi@ssl.msfc.nasa.gov

Goro Tanahashi
The President
Miyako-Tanki-Daigaku College
1-5-1 Kanan, Miyako-shi
Iwate 027 JAPAN

Masahiro Teshima
Institute for Cosmic Ray Research
University of Tokyo
Midori-cho 3-2-1
Tanashi-shi, Tokyo, 188 JAPAN
mteshima@icrr.u-tokyo.ac.jp

K.O. Thielheim
Arbeitsgruppe Mathematische Physik
Universitdt Kiel
Otto-Hahn-Platz 3
Postfach 5151
24 063 Kiel, GERMANY
thielheim@email.uni-kiel.de

Thomas J. Weiler
Vanderbilt University
Box 1807B
Nashville, TN 37235
weilertj@ctral1.vanderbilt.edu

Shigeru Yoshida
University of Utah
ICRR
1155 East 300 South #17
Salt Lake City, UT 84102
syoshida@icrr.u-tokyo.ac.jp

Arnulfo Zepeda
Cinvestav
P.O. Box 14-740
07000 Mexico D.F.
MEXICO
zepeda@fis.cinvestav.mx

Author Index

A

Alippi, E., 353
Antonov, R. A., 367
Attiná, P., 353

B

Berezinsky, V. S., 279, 373
Bhattacharjee, P., 168
Biermann, P. L., 22
Biondo, B., 361
Blasi, P., 373
Bonanno, G., 353, 361
Bosisio, L., 353
Bunner, A., 0
Burleson, E., 511

C

Catalano, O., 353, 361
Celi, F., 361
Chernov, D. V., 367
Chipman, R. A., 304, 428, 434, 439
Clark, G. W., 159
Cline, D. B., 262
Cosentino, R., 361

D

Dai, H. Y., 382
DeMarzo, C. N., 87
Di Benedetto, R., 361
Dimmock, J. O., 304, 428, 434, 439, 511
Domokos, G., 390

F

Farrar, G. R., 226
Fazio, G., 361
Fedorov, A. N., 367
Flyckt, E., 394

G

Gaisser, T. K., 297
Garipov, G. K., 108, 403
Giannini, G., 353
Giarrusso, S., 353, 361
Gonzalez-Mestres, L., 148, 418
Gorshkov, L. A., 108
Gracco, V., 353
Gregorio, A., 353, 358, 361

H

Halzen, F., 265
Hillman, L. W., 304, 428, 434, 439, 511

J

Jones, F. C., 37

K

Kamberov, G., 446
Khrenov, B. A., 108, 403
Kinelev, V. G., 460
Klein, S. R., 132
Korosteleva, E. E., 367
Kovesi-Domokos, S., 390
Kronberg, P. P., 196

L

Lamb, D. J., 304, 428, 434, 439, 511
La Rosa, G., 361
Lenti, A., 353
Leslie, T. M., 511
Linsley, J., 1
Loh, E. C., 382

535

M

Mangano, A., 361
Mitchell, J. W., 500

N

Nagano, M., 76
Nikitsky, V. P., 403

P

Panasyuk, M. I., 108, 367, 403
Petrolini, A., 353
Petrova, E. A., 367
Piana, G., 353
Popova, L., 446
Pryke, C., 312

R

Richiusa, G., 361

S

Saprykin, O. A., 108, 403, 460
Scarsi, L., 42, 358
Schramm, D., 321
Scuderi, S., 361

Sholokhov, A. V., 403
Sigl, G., 237
Sokolsky, P., 65, 382
Spirochkin, Y. K., 460
Stalio, R., 353, 358
Stecker, F. W., 212
Streitmatter, R. E., 95
Sulimov, V. D., 460
Syromyatnikov, V. S., 108, 403

T

Takahashi, Y., 117, 304, 428, 434, 439, 469, 511
Tanahashi, G., 54
Thielheim, K. O., 483
Trampus, P., 353, 358

V

Vacchi, A., 353

W

Watson, M. D., 511
Weiler, T. J., 246

Y

Yoshida, S., 217